The Physiology of
Thirst and
Sodium Appetite

NATO ASI Series

Advanced Science Institutes Series

A series presenting the results of activities sponsored by the NATO Science Committee, which aims at the dissemination of advanced scientific and technological knowledge, with a view to strengthening links between scientific communities.

The series is published by an international board of publishers in conjunction with the NATO Scientific Affairs Division

A	**Life Sciences**	Plenum Publishing Corporation
B	**Physics**	New York and London
C	**Mathematical and Physical Sciences**	D. Reidel Publishing Company Dordrecht, Boston, and Lancaster
D	**Behavioral and Social Sciences**	Martinus Nijhoff Publishers
E	**Engineering and Materials Sciences**	The Hague, Boston, and Lancaster
F	**Computer and Systems Sciences**	Springer-Verlag
G	**Ecological Sciences**	Berlin, Heidelberg, New York, and Tokyo

Recent Volumes in this Series

Volume 98—Structure and Function of the Genetic Apparatus
edited by Claudio Nicolini and Paul O. P. Ts'o

Volume 99—Cellular and Molecular Control of Direct Cell Interactions
edited by H.-J. Marthy

Volume 100—Recent Advances in Nervous System Toxicology
edited by Corrado L. Galli, Luigi Manzo, and Peter S. Spencer

Volume 101—Chromosomal Proteins and Gene Expression
edited by Gerald R. Reeck, Graham A. Goodwin,
and Pedro Puigdomènech

Volume 102—Advances in Penicillium and Aspergillus Systematics
edited by Robert A. Samson and John I. Pitt

Volume 103—Evolutionary Biology of Primitive Fishes
edited by R. E. Foreman, A. Gorbman, J. M. Dodd,
and R. Olsson

Volume 104—Molecular and Cellular Aspects of Calcium in Plant Development
edited by A. J. Trewavas

Volume 105—The Physiology of Thirst and Sodium Appetite
edited by G. de Caro, A. N. Epstein, and M. Massi

Series A: Life Sciences

The Physiology of Thirst and Sodium Appetite

Edited by

G. de Caro

University of Camerino
Camerino, Italy

A. N. Epstein

University of Pennsylvania
Philadelphia, Pennsylvania

and

M. Massi

University of Camerino
Camerino, Italy

Plenum Press
New York and London
Published in cooperation with NATO Scientific Affairs Division

Proceedings of a NATO Advanced Research Workshop on
The Physiology of Thirst and Sodium Appetite,
held July 12–20, 1984,
in Camerino, Italy

Library of Congress Cataloging in Publication Data

NATO Advanced Research Workshop on the Physiology of Thirst and Sodium
Appetite (1984: Camerino, Italy)
    The physiology of thirst and sodium appetite.

    (NATO ASI series. Series A, Life Sciences; v. 105)
    "Proceedings of a NATO Advanced Research Workshop on the Physiology
of Thirst and Sodium Appetite, held July 12–20, 1984, in Camerino, Italy"—T.p.
verso.
    "Published in cooperation with NATO Scientific Affairs Division."
    Includes bibliographies and index.
    1. Thirst—Physiological aspects—Congresses. 2. Drinking (Physiology)—
Congresses. 3. Sodium—Metabolism—Congresses. 4. Appetite—Congresses. 5. Water—Metabolism—Congresses. 6. Osmoregulation—Congresses. I.
De Caro, G. II. Epstein, Alan N. III. Massi, M. IV. North Atlantic Treaty Organization. Scientific Affairs Division. V. Title. VI. Title: Sodium appetite. VII. Series.
[DNLM: 1. Drinking Behavior—physiology—congresses. 2. Sodium—physi-
ology—congresses. 3. Thirst—physiology—congresses. WI 102 N279p 1984]
QP139.N38   1984   599'.013   86-4882

ISBN 978-1-4757-0368-9   ISBN 978-1-4757-0366-5 (eBook)
DOI 10.1007/978-1-4757-0366-5

© 1986 Plenum Press, New York
Softcover reprint of the hardcover 1st edition 1986
A Division of Plenum Publishing Corporation
233 Spring Street, New York, N.Y. 10013

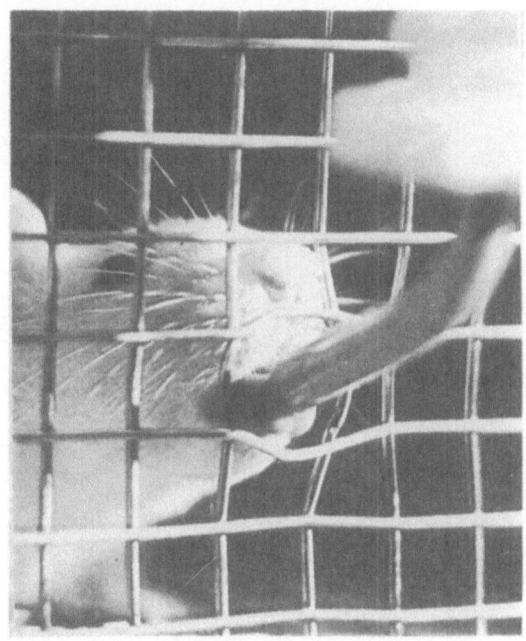

The behavior that made
this workshop possible.

PREFACE

The behavioral neuroscience of thirst and sodium appetite are research ventures that have expanded dramatically in recent years. Work done in the mid-1950s and early 1960s made it clear that drinking behavior could be affected by direct manipulations of the brain, especially by brain damage and by pharmacological treatments. Since that time experimental approaches have diversified and the research enterprise has attracted the interest of a broad international community of scientists.

Many aspects of both thirst and sodium appetite are being studied. The most prominent of these are:
1) phylogenetic and ontogenetic aspects of the phenomena of drinking behavior,
2) the mechanisms of a variety of dipsogenic and antidipsogenic treatments, both drugs and hormones,
3) the biological controls of drinking and their interaction with the regulation of blood volume and blood pressure,
4) the peripheral signals of drinking including the role of the baro- and volume-receptors,
5) the receptor systems within the brain and the neuroanatomical circuitry for thirst and sodium appetite, and
6) the possible roles of brain sodium and of the hormones of sodium conservation in the arousal of sodium appetite.

This acceleration of basic research activity has given insights into the clinical disorders of thirst and salt appetite and has produced pharmacological agents of potential therapeutic use. All of the above topics affect the circulatory physiology and the problem of hypertension. Consequently, there is now a rich and complex scientific literature in which the many aspects of the behavioral neuroscience of thirst are discussed, and a large international network of senior investigators and their students who are pursuing these problems.

The first international meeting on Thirst was held in 1963 at Florida State University (see:"Thirst", M.J. Wayner, Ed., Pergamon, 1964), and the last time that the problems of thirst and sodium appetite were discussed exclusively in an international setting was in 1973 at Lugano (see: "Control Mechanisms of Drinking", G.Peters & J.T. Fitzsimons, Eds., Springer-Verlag, 1975). Many of the topics mentioned above had only begun to be studied at that time and others had not yet been initiated. Another international meeting was therefore long overdue, and for this reason the NATO Advanced Research Workshop held in Camerino on July 12 to 20, 1984 was of great interest to neuroscientists of thirst and sodium appetite.

The investigation of the physiology of thirst and sodium appetite is an international scientific enterprise. There are major productive laboratories in the USA, the UK, Canada, France, and Italy, and at least one important research group in each of West Germany, Sweden, Australia, Brazil, Argentina and Poland. Except for the Swedes who declined our invitation, all of these laboratories were represented at Camerino. The meeting therefore had among its participants the world leaders in research on thirst and sodium appetite and it included all of the major themes of their recent research. Moreover, we sought out participants from other countries (e.g. Portugal, Spain, Denmark, and Turkey) in which the study of drinking behavior is not yet well developed and were fortunate to have them with us.

The meeting was organized to fulfill all of the objectives of a Nato Advanced Research Workshop. First, with only a few exceptions, all the participants attended the entire meeting. And, second, the program, which included review lectures, volunteered research reports, panel discussions and open discussions, provided ample opportunity for an exchange of views at the frontiers of knowledge and for presentation of the latest scientific results among the most experienced scientists working on the behavioral neuroscience of thirst. The daily review lectures, given at the beginning of each working day, set the theme of the day's work and provided critical assessments of the state of our scientific art. The panel discussions, with which each day was concluded, had as one of their specific tasks the formulation of recommendations for new research directions.

Lastly, the bringing together of the community of experts in the neuroscience of thirst for ten days of both scientific and social close encounters in the heuristic atmosphere of a NATO Workshop, lead to formulations of plans for international scientific collaborations. This NATO Workshop had one very special asset. The number of participants was far larger than the usual Workshops because the participants were not only senior scientists but also young and talented people who are just beginning to work in our field. The meeting, therefore, gave these young scientists the opportunity to profit from the experience of older colleagues and gave these more experienced workers the challenge of new enthusiasms and new ideas.

The following is the text of the review lectures and volunteered papers. It is a detailed summary of the state of our research enterprise.

Camerino, Italy
G. De Caro
M. Massi

Philadelphia, Pennsylvania
A.N. Epstein

AKNOWLEDGEMENTS

This Advanced Research Workshop was sponsored by NATO and co-
sponsored by the University of Camerino, Italy.
The Directors of the Workshop also aknowledge the financial
assistance of:

        Assessorato alla Sanità, Regione Marche, Ancona
        Consiglio Nazionale delle Ricerche, Rome
        Pharkos s.r.l., Cisterna di Latina
        Nutrition Foundation of Italy, Milan
        Ministero della Pubblica Istruzione, Rome
        Italfarmaco s.p.a., Milan
        Janssen Farmaceutici s.p.a., Rome
        Squibb s.p.a., Rome
        Davide Campari, Milan
        Comunità Montana "Zona I", Camerino
        Rotary Club, Camerino
        Rotary Club, Civitanova Marche

CONTENTS

PHYLOGENY AND ONTOGENY OF THIRST

The Phylogeny and Ontogeny of Thirst. 1
    D.A. Denton, M.J. McKinley, P. Osborne, E. Tarjan, and
    R.S. Weisinger

The Ontogeny of Angiotensin Dipsogenesis in the Rat 13
    A.N. Epstein

Perinatal Sodium Chloride Intake Modifies the Fluid
        Intake of Adult Rats. 15
    R.J. Contreras, and E. Bird

The Ontogeny of Thirst Mechanisms in Albino Rats. 19
    E.M. Blass

Behavioral Responsivity to Tastes in Developing Rats. 25
    P. Kehoe, and E.M. Blass

The Ontogeny of Salt Intake in Rats 31
    K. Moe

Effects of the Tachykinins Eledoisin and Physalaemin on
        Drinking Behaviour in Baby Rats 37
    F. Cantalamessa, G. de Caro, A.N. Epstein, and
    M. Perfumi

DRINKING AND WATER HOMEOSTASIS

Integrative Rostromedial Diencephalic Neurons are
        Comodulated by Vasopressin and Angiotensin. 43
    A.C. Jeulin, and S. Nicolaïdis

Thyroid Activation Following Dehydration 59
   L. Di Bella, and M.T. Rossi

Influence of Thyroid on Water and Food Intake 65
   M.T. Rossi, and L. Di Bella

Role of Antidiuretic Hormone in Volume Regulation 71
   Z. Arad, G.E. Rice, and E. Skadhauge

Posterior Hypothalamic Polydipsia: Differential Effects
     to Several Dipsogenic Treatments 77
   A. Morales, and A. Puerto

HYPEROSMOTIC AND HYPOVOLEMIC THIRST

Hyperosmotic and Hypovolemic Thirst 83
   D.J. Ramsay, and T.N. Thrasher

Osmotic Thirst Suppression after Central Administration
     of Vasopressin Antagonists 97
   E. Szczepańska-Sadowska, J. Sobocińska, and
   S. Kozłowski

Hyperosmotic Thirst: an Osmoreceptor Mechanism,
     a Sodium Receptor Mechanism or Both 103
   S.N. Thornton

Relationship between Right Atrial Stretch and Plasma
     Renin Activity 109
   S. Kaufman

A Single Experience with Hyperoncotic Colloid Dialysis
     Persistently Alters Water and Sodium Intake 115
   S.P. Frankmann, D.M. Dorsa, R.R. Sakai, and
   J.B. Simpson

CENTRAL RENIN-ANGIOTENSIN SYSTEM

The Brain-Renin-Angiotensin System: Update 123
   T. Unger, D. Ganten, G. Ludwig, and R.E. Lang

Angiotensin-Sensitive Sites in the Central Nervous System . . 135
    D. Felix, M.C. Gambino, Y. Yong, and P. Schelling

Central Metabolism of Angiotensins: Potential Functional
    Significance. 141
    J.W. Harding, R.H. Abhold, C.G. Camara, J.B. Erickson,
    and E.P. Petersen

Aspects of Cerebrospinal Fluid Pressure Control in Con-
    scious Rats During Central Infusions of Angiotensin
    and Vasopressin. 149
    W.B. Severs, H.J. Spaeth, J.N.D. Wurpel, R.L. Dundore,
    R.T. Henry, and L.C. Keil

Decreased Binding Capacity of Central Angiotensin II
    Receptors Following Long-Term Administration of
    Angiotensin II. 155
    E. Richards-Sumners, and M.I. Phillips

PERIPHERAL RENIN-ANGIOTENSIN SYSTEM

The Role of the Renal Renin-Angiotensin System
    in Thirst 161
    A.K. Johnson, M.M. Robinson, and J.F.E. Mann

Renin Dependence of Insulin-Induced Thirst. 181
    M. Costales, M. Vijande, B. Marìn, J.I. Brime and
    P. Lopez-Sela

Comparison of Angiotensin II and III Induced
    Dipsogenicity and Pressor Action. 187
    J.W. Wright, S.L. Morseth, M.J. Sullivan and
    J.W. Harding

Angiotensin II and Arterial Pressure in the Control
    of Thirst 193
    M.M. Robinson and M.D. Evered

Angiotensin Dependent Thirst Following Polyethyleneglycol
    Treatment in the Rat. 199
    J.F.E. Mann, S. Eisele, D. Ganten, A.K. Johnson,
    R. Rettig, E. Ritz, and T. Unger

A General Review of the Effects of Angiotensin-Peptides
        and 8-Substituted Analogs of Angiotensin II 205
    R.K. Türker

                BRAIN PEPTIDES AND OTHER ENDOGENOUS SUBSTANCES

Effects of Peptides of the "Gut-Brain-Skin Triangle" on
        Drinking Behaviour of Rats and Birds 213
    G. de Caro

Suppression of Water Intake by the E Prostaglandins 227
    N.J. Kenney

Benzodiazepine and Endorphinergic Mechanisms in Relation
        to Salt and Water Intake 239
    S.J. Cooper

Selective Antidipsogenic Effect of Kassinin in Wistar
        Rats 245
    G. de Caro, and L.G. Micossi

Sensitivity to Dipsogenic Peptides of Pigeons Bearing
        Lesions Directed to the Subfornical Organ (SFO) 251
    M. Massi, G. de Caro, A.N. Epstein, and L. Mazzarella

Inhibition of Ang II-Induced Drinking by Dermorphin Given
        into the SFO or into the Lateral Ventricle of Intact
        or of SFO Lesioned Rats 257
    M. Perfumi, G. de Caro, M. Massi, and F. Venturi

                PERIPHERAL MECHANISMS IN MAINTENANCE AND
                      TERMINATION OF DRINKING

Peripheral Mechanisms for the Maintenance and Termina-
        tion of Drinking in the Rat 265
    G.P. Smith

Disturbances in Water Balance Controls Following Lesions
        to the Area Postrema and Adjacent Solitary Nucleus . . . 279
    R.R. Miselis, T.M. Hyde, and R.E. Shapiro

Preabsorptive and Postabsorptive Factors in the
        Termination of Drinking in the Rhesus Monkey 287
    J. Gibbs, B.J. Rolls, and E.T. Rolls

Histamine Plays a Role in Drinking Elicited by Eating in
        the Rat 295
    F.S. Kraly

Satiety and the Effects of Water Intake on Vasopressin
        Secretion 301
    D.J.Ramsay, and T.N. Thrasher

            NATURE AND LOCALIZATION OF CENTRAL RECEPTOR SYSTEMS

The Nature and Localization of Central Receptor Systems . . . 309
    J.B. Simpson

Adipsia in Sheep Caused by Cerebral Lesions 321
    M.J. McKinley, D.A. Denton, M. Leventer, R.R. Miselis,
    R.G. Park, E. Tarjan, J.B. Simpson, and R.S. Weisinger

The Organum Vasculosum Laminae Terminalis and Water
        Balance in Dogs 327
    T.N. Thrasher, and D.J. Ramsay

In Vitro Down Regulation and Possible Internalization of
        Central Angiotensin Receptors 333
    J.W. Harding, J.B. Erickson, J.W. Wright, C.G. Camara,
    and R.H. Abhold

Effect of Intracerebroventricular Administered Vanadate
        on Salt and Water Intake and Excretion in the Rat . . . 339
    E. Chiaraviglio, and C. Lozada

                    NEUROANATOMICAL CIRCUITRY

The Visceral Neuraxis in Thirst and Renal Function 345
    R.R. Miselis

The Role of the Zona Incerta in Water Intake Regulation . . . 355
    S.P. Grossman

Neural Circuits that Contribute to Procurement of Water . . . 361
    G.J. Mogenson and B. Box

The Role of the Septal Area in the Regulation of
        Drinking Behavior and Plasma ADH Secretion 367
    M. Iovino, and L. Steardo

The Surface Morphology of the Cat Subfornical Organ 375
    D. Felix, and H. Felix

SODIUM APPETITE

Endogenous Angiotensin and Sodium Appetite 383
    J.T. Fitzsimons

Hormonal Synergy as the Cause of Salt Appetite 395
    A.N. Epstein

Sodium Appetite Induced by Sodium Depletion is Suppressed
        by Intracerebroventricular Captopril 405
    M.L. Weiss

Increased Sodium Appetite and Polydipsia in Goldblatt
        Hypertension 413
    M. Vijande, M. Costales, and J.T. Fitzsimons

Influence of Sodium Load on Angiotensin-Induced Sodium
        Appetite 419
    L.M. Fuller, and J.T. Fitzsimons

The Hormones of Renal Sodium Conservation Act Synergis-
        tically to Arouse a Sodium Appetite in the Rat 425
    R.R. Sakai

Sodium Appetite During Captopril Blockade of Endogenous
        Angiotensin Formation 431
    K. Moe, and A.N. Epstein

The Renin-Angiotensin-Aldosterone System and Sodium
        Appetite in Rats 441
    M.J. Fregly, and N.E. Rowland

Sodium Appetite after Adrenalectomy or Deoxycorticosterone-
    Treatment in Diabetes Insipidus (Brattleboro) Rats . . . 447
  L.M. Fuller

Dopaminergic Modulation of Choice in Salt Preference Tests . . 453
  S.J. Cooper, and D.B. Gilbert

Control Mechanisms of Salt Appetite 459
  S. Kaufman

The Nature of the Salt Appetite of Adrenalectomized Rats . . . 465
  C. Meuli, and G. Peters

The Effect of Local Change in CSF [Na] in the Anterior
    Third Ventricle on Salt Appetite 473
  E. Tarjan, P. Cox, D.A. Denton, M.J. McKinley,
  and R.S. Weisinger

Peripheral Gustatory Mechanisms of Salt Intake in the Rat . . 479
  R.J. Contreras, and T. Kosten

Cerebral Na Sensors and Na Appetite of Sheep 485
  R.S. Weisinger, D.A. Denton, M.J. McKinley,
  J.B. Simpson, and E. Tarjan

The Evolution and Expression of Salt Appetite 491
  J. Schulkin

Behavioral Dynamics in the Appetite for Salt in Rats 497
  J. Schulkin

Effect of Cerebroventricular Infusion of Hypertonic
    Sodium Solutions on Sodium Intake in Rats 503
  E. Chiaraviglio, and M.F. Pérez Guaita

PHARMACOLOGY, PATHOLOGY AND CLINICAL ASPECTS

Role for α_2-Adrenoceptors in Experimentally-Induced
    Drinking in Rats 509
  M.J. Fregly, and N.E. Rowland

Human Thirst: the Controls of Water Intake in
        Healthy Men 521
    B.J. Rolls, P.A. Phillips, J.G.G. Ledingham
    M.L. Forsling, J.J. Morton, and M.J. Crowe

Effects of Narcotic Analgesics on Water and Food
        Intake in Normal Rats 527
    Y. Arslan, R. Burckhardt, K. Jawaharlal,
    K. Ornstein, and G. Peters

Biochemical, Physiological and Pathological Properties
        of Endogenous Digitalis-Like Compounds 535
    J.F. Cloix, G. Deray, M.A. Devynck, M.G. Pernollet,
    M. Crabos, G. Henning, M. Rieux, I.W. Wainer, and
    P. Meyer

Disorderly Drinking During Development in Spontaneously
        Hypertensive Rats 541
    F.S. Kraly

Participants 547

Author Index 549

Subject Index 551

THE PHYLOGENY AND ONTOGENY OF THIRST

D.A. Denton, M.J. McKinley, P. Osborne, Eva Tarjan and
R.S. Weisinger

Howard Florey Institute of Experimental Physiology and
Medicine, University of Melbourne
Parkville, Victoria 3052, Australia

This subject is so wide that in the context of a symposium, the selection of a few aspects only is practical - rather than an encompassing review. One strategy of approach to analysis of phylogenetic emergence is to identify major mechanisms in the higher mammalian species, and then, where feasible in the light of data available, to trace the mechanisms down the phylogenetic tree. In so doing, attention can be focussed on characteristics of the econiche inhabited by a particular species, and thus the survival advantages which might accrue as a result of specific mechanisms of thirst.

In higher mammals, an initial theory of genesis of thirst derived from experiment was cellular dehydration[1]. The cephalic receptors were eventually hypothesized to be osmoreceptors in the light of Verney's (1947) elegant demonstration of control of anti-diuretic hormone[2]. Oropharyngeal receptors were thought to have only a contributory causal role in adding to the primarily central drive, in contrast to Cannon's original ideas. Progressively it was recognized that there were many systems involved in the control of water intake, and that a delicate interplay of neural and humoral stimuli occurred. The pioneering work of Andersson[3] showing profound polydipsia of goats following hypothalamic injections of hypertonic saline could have been interpreted as consistent with osmoreceptor theory, though Andersson[3,4] in discussing this cautioned against any conclusion on specificity of response. This included the grounds that hypertonic NaCl possibly had a general stimulatory effect on nerve cells. However, Blass and Epstein[5] showed that injections of hypertonic saline or sucrose[6] bilaterally in the medial preoptic area of rats, caused water drinking. Suggesting that an osmotically

determined response is one element of the central sensor system.
Andersson, Olsson, Rundgren, Leksell and colleagues have by a series
of skilful experiments involving dysjunction of the parallel movement
of osmotic pressure and [Na] of cerebral arterial plasma and cerebrospinal fluid (CSF) obtained evidence for the existence of specific
Na receptors in the juxtaventricular region which control thirst
and ADH secretion[7]. Included in a spectrum of manipulations, it
has been shown that slow intraventricular infusion of isotonic or
hypertonic saccharides (sucrose or fructose) which reduces CSF [Na],
attenuated or abolished the desire of the dehydrated goat to drink,
and they reported infusion of hypertonic sucrose alone into the
CSF did not cause water drinking - an argument against osmoreceptors.
The question of thirst as generated by either osmoreceptors or specific Na receptors has been placed in another light by data from
sheep. The results lead to a postulate that the existence of both
receptors is essential to encompass the experimental data. This
leads to an interesting phylogenetic question. But first the findings[8,9,10,11] which were : (a) Infusion or injection of hypertonic
sucrose into the CSF did cause water drinking provided it was prepared in an artificial CSF which prevented concomitant fall in CSF
[Na] or other ions[8,9,10]. However, equiosmotic hypertonic NaCl
in artificial CSF was nearly three times more effective than the
saccharides dissolved in artificial CSF in causing thirst (Fig.1).
This has been comprehensively confirmed in the cow (Fig.2) and the
data[12] in the rat suggest a similar conclusion. Thrasher et al[13]
however, find that intracerebroventricular (ICV) infusion of equiosmolar hypertonic sucrose or NaCl are equally effective in causing
water drinking in the dog. (b) Lowering CSF]Na] by 10-20 mmol/1

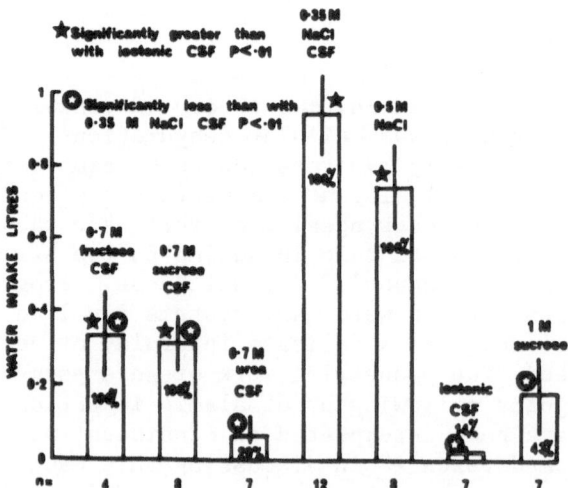

Fig. 1. Volume (mean ± SEM) of water drunk by sheep during infusion
         of various hypertonic solutions into the lateral ventricle
         at 0.05 ml/min for 20 min.

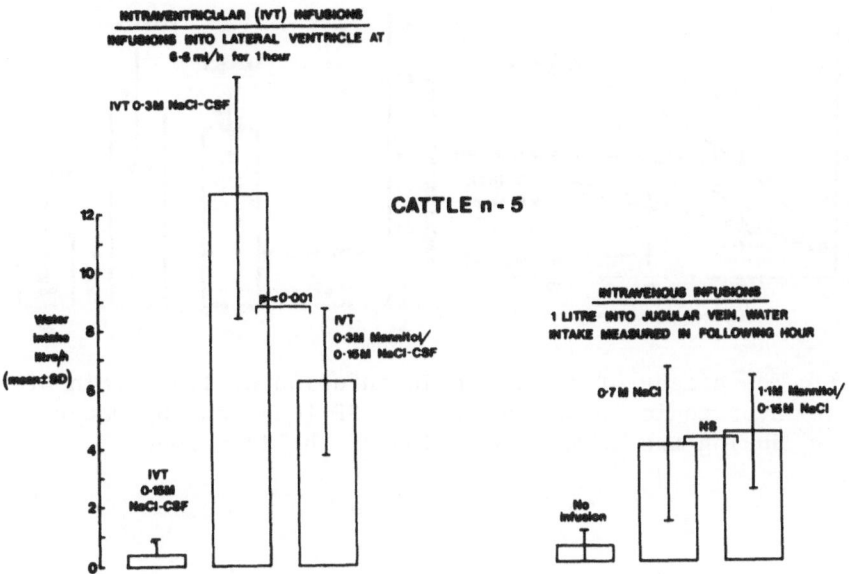

Fig.2.   Comparison of the effect of infusion of equiosmotic solut-
         ions of hypertonic mannitol or sodium salts into the
         lateral ventricle or into the jugular vein on drinking
         of water by cows during one hour.

by ICV infusion of hypertonic mannitol in artificial CSF extinguished
thirst in water deprived sheep despite audited sustained rise in
CSF osmotic pressure[11]. It also increased free water clearance.
(c) Intracarotid infusion of 1 M NaCl, 2 M sucrose, or 4.6 or 2.0
M urea caused progressive increase in CSF [Na] with 4.6 M urea
giving the largest rise, whereas 2 M glucose did not increase CSF
[Na][9] (Fig.3). Only 1 M NaCl and 2 M sucrose caused rapid large water
intake (Fig.3)[9]. Intracarotid infusion of 4.6 M urea caused the
largest rise in CSF [Na] (6 - 8 mmol/l) (Fig.3) and though its dipsogenic effect was less than that of 2 M sucrose where CSF [Na] rise
was 4 mmol/l, there was water drinking in 3 of 10 trials (Fig.3).

     Summarizing, these data showed unequivocally that immediate
water drinking occurred in response to raised CSF osmotic pressure.
Also there was a clear effect of raised CSF [Na] in causing thirst
in sheep, rats and cows which was much greater than that caused
by equal rise of CSF osmotic pressure produced by a saccharide.
Furthermore, in sheep a decrease of CSF [Na] counteracts an osmotic
stimulus whether it is inside or outside the blood brain barrier.

     Given two types of receptors as proposed[10] the fact that the
differences in water drinking with intracarotid infusion of solutes
are consistent with known cell membrane permeability of the infused
solutes, coupled with the common denominator of increased CSF [Na],
has led to the postulate that there are osmoreceptors outside the
blood brain barrier[10,13]. It should perhaps be noted here that
the action of intracarotid 4.6 M urea on drinking appears quantitat-

3

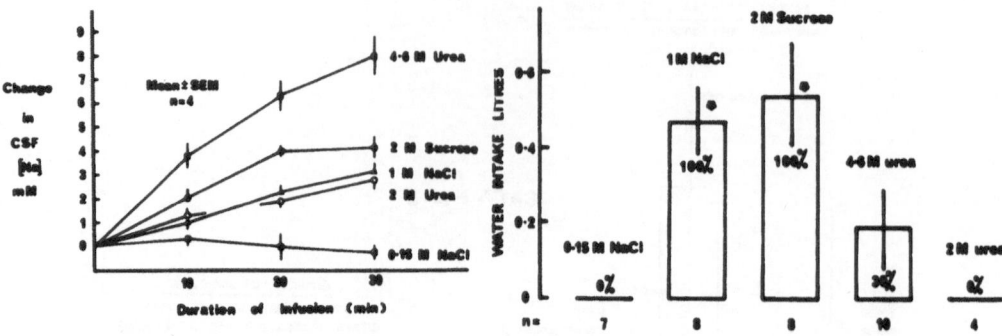

Fig. 3   The effect of intracarotid infusion at 1.6 ml/min of
hypertonic solutions on (a) CSF [Na] and (b) water intake
during the 30 min   duration of the infusion.

ively paradoxical since CSF [Na] was increased. Increased CSF [Na]
has been confirmed in the rat with hypertonic urea[14]. However,
it has to be noted that substantial water drinking did occur in
sheep but it was inconsistent. If there are osmoreceptors which
can be accessed by rise in CSF osmotic pressure, and most data
agree that delivery of such a stimulus into the CSF is an effective
dipsogen, then a 5-8 mmol rise of the [Na] is also an osmotic
stimulus, and should have been consistently dipsogenic unless,
perhaps, some change outside the barrier was affecting its action,
or the change is just approximating the dipsogenic threshold.
Thrasher et al[13] who also espoused the idea of an osmoreceptor outside the blood brain barrier in dogs attribute the dipsogenic action
of hypertonic urea to the high [Na] of CSF dehydrating osmoreceptors
close to the ventricle.

A purely schematic representation of the possible interactions
(Fig.4) shows osmoreceptors outside the blood brain barrier, presumably in a circumventricular region. Being close to the wall of
the third ventricle, a rise in osmotic pressure of the CSF produced
by the addition of a saccharide to the CSF draws water across the
blood brain barrier represented by tight-junction ependyma and then
stimulates them. If an equiosmotic change in CSF is made with Na
(Fig.4a), not only are osmoreceptors similarly stimulated but the
high sodium acts on a juxtaventricular sodium receptor freely accessible through the ependyma. The sodium receptor is stimulated
specifically by increased [Na] and the two stimuli are additive
causing a greater effect on the neural systems subserving control
of water drinking than that of increased osmotic pressure alone.
On the other hand there are also juxtaventricular receptor neurones
specifically responsive to decrease of CSF [Na], and they exert
an inhibiting effect on drinking. Thus decrease in CSF [Na] can
override and inhibit any excitatory effect of osmoreceptors.

4

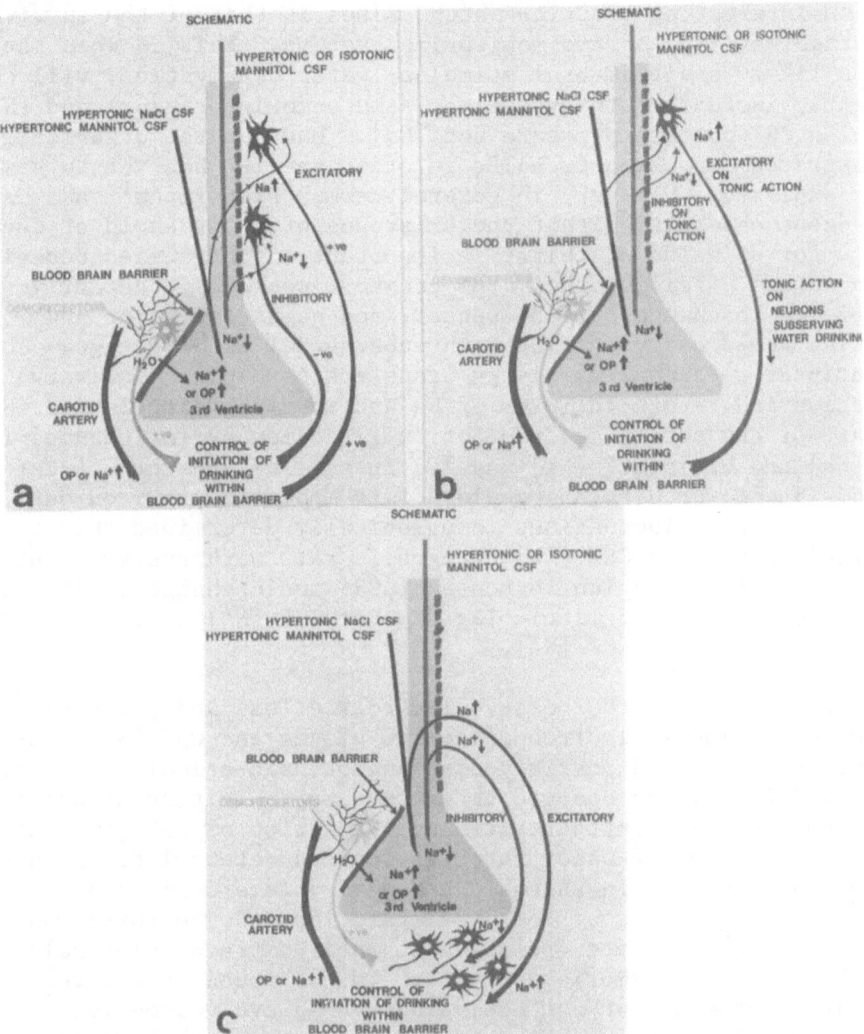

Fig.4   Schematic representation of a hypothetical functional
         organization of control of drinking : (a) excitatory and
         inhibitory Na sensors and osmoreceptors, (b) Na sensor
         with tonic influence on drinking control and osmoreceptor,
         (c) direct effect of [Na] of brain ECF on neurones control-
         ling drinking, and osmoreceptors.

Rather than an excitatory-inhibitory organization, an alternative
parsimonious schema could envisage the juxtaventricular Na sensors
having a tonic effect on the control of water drinking with
stimulatory increase of firing with increased [Na], and inhibitory
decrease of firing with decreased [Na] (Fig.4b). A further
possibility schematically shown, depicts the neurones subserving
drinking control being themselves directly responsive to change
in the environmental [Na] amplifying stimulatory inflow from
osmoreceptors and with decreased [Na] having an inhibitory effect
(Fig.4c).

An interesting question which arises is that of the survival advantage, if any, of two monitoring systems. This is when the common likely environmental stimulus, water deprivation, will concurrently increase both systemic plasma osmotic pressure and [Na], and also CSF osmotic pressure and [Na]. Whether the organization demonstrated in ruminants holds in other species has yet to be fully explored. However, in general terms, Fitzsimons[15] has made the cogent observation that the existence of a threshold of chemical change for stimulus of thirst is important for organized behaviour of an animal. Failing this, the creature would be constantly responsive to changes at the sensor, and quest for water and drinking would be overriding with substantial disadvantage. Thus, in ruminant species where large transfers of Na into salivary secretion and return influxes of Na and into the circulation may be part of the eating and rumination circadian cycle, changes in CSF [Na] and brain ECF [Na] may lag in time behind that of carotid blood. The lower CSF [Na] could act to moderate or "iron out" otherwise more rapid fluctuations in osmotically determined thirst drive. Epstein[13] measuring CSF [Na] changes in rats with systemic infusions remarked on the sluggish response of CSF ionic changes, although our own data[10] showed little lag time of CSF [Na] and, therefore, possibly also brain ECF [Na].

To consider another case, electrolyte loss and extracellular volume depletion will often cause low plasma and CSF [Na]. If atrial volume receptors and possibly also angiotensin stimulate thirst, the low CSF [Na] may operate to cause some inhibition of water intake. This would obviate further extensive reduction of CSF [Na] and the risk of the onset of water intoxication and cerebral disorientation which could ensue. In sheep made Na and volume deplete by parotid cannulation, analysis of total salivary loss showed there was loss of Na in excess of water relative to their extracellular relation, and therefore, no osmotic stimulus. It was found that water drinking increased but not equivalent to saliva volume lost[16] - a finding consistent with the hypothesis. Whereas the influence of low CSF [Na] has not been extensively explored in the rat, it is interesting that Stricker[17] has shown that hypovolaemic rats stop drinking water despite the continued presence of substantial volume deficits, with no more than 8 - 10% dilution of body fluids[18] There is a relation between the point at which inhibition occurs and the extent of plasma deficit, the greater deficit allows more water drunk before inhibition. If given 0.15 M NaCl to drink instead of water the rats drink continuously and at higher rates than with water, and restore plasma volume most effectively[19]. Such observations could be consistent with the postulate advanced above, if reduced CSF [Na] in the rat is shown to inhibit some types of thirst. Overall, it seems feasible that a survival advantage might accrue as a result of there being two receptor systems - at least in some species.

6

Considering phylogenetic comparisons of response to cellular dehydration, Fitzsimons details in his monograph[15], which is the most comprehensive treatise in this field, that drinking is an important factor in water balance in bony fishes, it is not present in amphibia, but reappears in reptiles. The Myxinoedea of the class Agnatha (jawless fish) emergent in Ordivician times, are represented today by the hagfish (exclusively marine) and the lamprey. Both drink. When living in fresh water, the lamprey does not drink, but on entering the sea they maintain relative hypertonicity by drinking sea water, excreting sodium and chloride extrarenally, and retaining water. The cartilagenous fish maintain tonicity relative to sea water by a physiological uraemia and do not drink. The migratory eels and teleosts (salmon) do not drink in fresh water and produce a copious urine, but on entering sea water drink and excrete the monovalent ions via the gills. It is interesting that in the eel which drinks immediately on being placed in sea water, the stimulus is not hypertonicity. There being no drinking when placed in hypertonic mannitol or sucrose, but a direct, probably, reflex response to the Cl^- ion[20].

The marine teleost which drinks continuously stops drinking when placed in fresh water. It can be made to drink by infusing hypertonic saline, suggesting mechanisms responsive to cellular dehydration exist. However, drinking in fish occurs after removal of prosencephalon and mesencephalon, so that it can be proposed that encephalization of drinking occurred with the migration to land and the problems of finding water (Fitzsimons 1979). Little is known with regard to any water seeking behaviour of amphibians.

Snakes drink immediately in response to water deprivation. Fitzsimons and Kaufman[21] have shown that iguanas drink as a result of cellular dehydration. The accuracy of response, suggests that the basic organization for initiation of water intake in the face of dessication is well developed in the reptile.

Birds emerged in the Jurassic period about 150 million years ago. The carnivorous, insectivorous and omnivorous species generally obtain enough water from food but granivorous species drink. Their diet is low in water and salt, and their ranging activities in hot areas may be restricted to close proximity to water. They will often drink salty water or seek out salt[22].

In relation to comparative study of organization of control systems, Fitzsimons et al[23] have shown that the pigeon drinks in response to IV administration of hypertonic NaCl or sucrose. ICV infusion of hypertonic NaCl caused drinking but equiosmolar sucrose did not cause drinking to continue unless contained in saline solution in order to avoid decrease in CSF [Na]. Their abstract appears to indicate, in contrast to the sheep, that ICV hypertonic sucrose with Na added was as effective a dipsogenic stimulus as equiosmotic NaCl. It was noted that IV hypertonic urea was not dipsogenic though CSF [Na] was increased. The interpretation by the authors is that CSF [Na] is permissive of drinking in response to cellular dehydration in the pigeon. However, their results could be consistent with a contributory causal as distinct from a permissive role. Further data are required to show whether these systems are the same in birds as in mammals.

Consideration of the role of angiotensin as a vector of thirst in the higher mammals has eventually focussed on the issue of the concentrations which will cause water drinking. The initial proposal concerned angiotensin II (AII) of renal origin as a systemically active dipsogen, but attention has now focussed on a proposed brain renin-angiotensin system. If AII were a neurotransmitter in systems subserving thirst, or if it were generated locally from a renin-substrate reaction in response to specific stimuli in order to activate a neural system subserving thirst, it may well be that very high concentrations do occur locally. In this case very high concentrations may need to be applied to activate the system, and many experiments reported in the literature fulfil this requirement.

However, on the primary question of systemic blood levels of AII in higher mammals and initiation of drinking, the evidence suggests considerable species variation. The data on dogs suggest they are the most sensitive species yet studied and blood concentrations required to initiate drinking are in the physiological range. Excessive water drinking in dogs with circulatory impairment is reduced by infusion of a competitive blocker of angiotensin[24]. Ledingham et al[25] found intravenous AII infusion to be dipsogenic in man but the levels required were pharmacological, in contrast to the clear-cut effect of small changes in plasma [Na]. IV infusions of AII which would contrive high physiological levels in goats and sheep do not consistently stimulate water intake, but this does not preclude an effective contribution when the action of other factors also operates to stimulate thirst[26].

At a comparative level, it can be said in general terms in the light of data of Japanese workers[27,28], that angiotensin is an effective dipsogen in those animals which drink water in nature, but is ineffective in creatures which do not. These include amphibians. The eel which responds to drinking to infusion of hypertonic saline or blood withdrawal, also drinks with mammalian or eel angiotensin. In sea water where it drinks continuously, systemic infusion of angiotensin increases the rate. However, these effects still occurred in decerebrate eels after the encephalon, diencephalon and part of the mesencephalon were removed[29]. The iguana drinks water in response to large doses of angiotensin. With reservations attaching to the interpretation of use of large intraperitoneal injections[28], most tetrapods respond to angiotensin. Exceptions, include hibernating reptiles, animals from arid areas such as the mongolian gerbil, the mouse (mus musculus), and the carnivorous birds. Data from our laboratories[30] show that the wild rabbit, shows no augmentation of water intake with ICV infusion of AII, though it drinks avidly to extracellular hypovolaemia.

Since there is a paucity of data regarding blood or CSF levels of AII in lower species, there is little basis for comparison with higher animals. On teleological grounds there are reasons for proposing that angiotensin has a physiological role in drinking in the vertebrates, but, among the many matters needing resolution, a principal one is whether a physiologically activated brain renin-angiotensin system exists and at what stage in the phylogenetic scale it becomes emergent.

With ontogeny of thirst, data is sparse. As Rolls and Rolls[31] point out, in mammals, life immediately after birth is characterized by ability to survive on mother's milk alone. It is not critical that mammals be born with thirst control mechanisms. Wirth and Epstein[32] found 6 day old suckling rats made thirsty by cellular or extracellular dehydration will readily swallow water placed in the mouth but will not seek it - perhaps analogous to drinking by fish. Three week old rats will seek water. Their experiments showed water drinking in response to cellular dehydration at 2 days, to hypovolaemia at 4 days, and to AII injected into the ventricles at 4 days. Thus neurophysiological controls for drinking were functional prior to their being essential for survival. Bruno[33] has shown that before 2 weeks of age, dehydration did not influence suckling, but, after 2 weeks, dehydration reduced suckling analogous to the fashion in which dehydration will reduce food intake in the adult.

Questions of ontogeny involve many other aspects of drinking little explored. Adolph[34] showed clear differences in young and adult rabbits in satiation behaviour - possibly involving learning of effects. A most interesting facet comes from work of Stricker and Sterrit[35] indicating the taste of water and regulation of intake are 'prewired' in the chicken but do not come into play until water is visually recognized in the environment and ingested. Young chickens, dehydrated and naive in experience of water, will run through water puddles without recognition. They have an innate tendency to peck at small irregular objects (grains). When this happened to occur at an irregularity in water, the association between visual water and "hard wired" water taste was made rapidly, and they immediately drank an amount commensurate with deficit.

Overall, it is clear that the behavioural response of water drinking in response to cellular dehydration emerged early in evolution and the threshold for effect - about 2-4% change in plasma

osmotic pressure - has not altered significantly. Thus, there is
an impressive qualitative similarity in animals and quantitative
response to dessication. Though these statements are by no means
universally applicable, the suggestion is strong that the neurological mechanisms subserving the behaviour emerged early in
phylogenesis, are of high survival value, and relative to other
systems have not changed radically with irradiation of life into
widely divergent environments. Drinking responses to hypovolaemic
stimuli also appear early in vertebrate evolution, as suggested
by drinking of eels in response to haemorrhage.

Much further work is required prior to assigning AII a
physiological role in the genesis of thirst in lower orders. In
mammalian, as in lower species, there are striking variations in
its effect. Substantial elucidation of its role must await resolution
of the question of whether there is a cerebral renin-angiotensin
system or an AII neurotransmitter in neural systems subserving thirst.
Such knowledge seems most likely to emerge decisively from molecular
biological techniques such as hybridization histochemistry[36] where
it can be shown whether or not specific genes of the renin-angiotensin cascade are transcribed in response to appropriate changes.
Notwithstanding, there is evidence that systemic AII is a physiological stimulus in some higher species - the action probably
always being concurrent and additive to other dipsogenic influences.

ACKNOWLEDGEMENTS

Experimental studies from the Howard Florey Institute mentioned in
this review were supported by the National Health and Medical
Research Council of Australia.

REFERENCES

1. A. Gilman, The relation between blood osmotic pressure, fluid
    distribution and voluntary water intake. Am. J. Physiol. 120:
    323 (1937).
2. E. B. Verney, The antidiuretic hormone and factors which
    determine its release, Proc. R. Soc. B 135: 25 (1947).
3. B. Andersson, The effect of injections of hypertonic NaCl
    solutions into different parts of the hypothalamus of goats,
    Acta Physiol. Scand. 28: 188 (1953).
4. B. Andersson, M. Jobin and K. Olsson, A study of thirst and
    other effects of an increased sodium concentration in the
    third brain ventricle, Acta Physiol. Scand. 69: 29 (1967).
5. E. M. Blass and A. N. Epstein, A lateral preoptic osmosensitive
    zone for thirst in rat, J. Comp. and Physiol. Psychol, 76:
    378 (1971).

6. E. M. Blass, Evidence for basal forebrain thirst osmoreceptors in rat, _Brain Res._ 83: 69 (1974).
7. B. Andersson, Regulation of water intake. _Physiol. Rev._ 58: 583 (1978).
8. M. J. McKinley, E. H. Blaine and D. A. Denton, Brain osmoreceptors, cerebrospinal fluid electrolyte composition and thirst, _Brain Res._ 70: 532 (1974).
9. M. J. McKinley, D. A. Denton and R. S. Weisinger, Sensors for antidiuresis and thirst - Osmoreceptors or CSF sodium detectors? _Brain Res._ 141: 89 (1978).
10. M. J. McKinley, D. A. Denton, L. G. Leksell, E. Tarjan and R. S. Weisinger, Evidence for cerebral sodium sensors involved in water drinking in sheep, _Physiol. Behav._ 25: 501 (1980).
11. L. G. Leksell, M. Congiu, D. A. Denton, D. T. W. Fei, M. J. McKinley, E. Tarjan and R. S. Weisinger, Influence of mannitol-induced reduction in CSF sodium on nervous and endocrine mechanisms involved in control of fluid balance, _Acta Physiol. Scand._ 112: 33 (1981).
12. J. Buggy, W. E. Hoffman, M. I. Phillips, A. E. Fisher and A. K. Johnson, Osmosensitivity of rat third ventricle and interactions with angiotensin, _Am. J. Physiol._ 236: R75 (1979).
13. T. N. Thrasher, R. G. Jones, L. G. Keil, C. J. Brown and D. J. Ramsay, Drinking and vasopressin release during ventricular infusions of hypertonic solutions. _Am. J. Physiol._ 238: R340 (1980).
14. A. N. Epstein, Consensus, controversies and curiosities. _Fed. Proc._ 37: 2711 (1978).
15. J. T. Fitzsimons, The physiology of thirst and sodium appetite. Cambridge University Press (1979).
16. S. F. Abraham, J. P. Coghlan, D. A. Denton, J. G. McDougall, D. R. Mouw and B. A. Scoggins, Increased water drinking induced by sodium depletion in sheep. _Quart. J. Exp. Physiol._ 61: 185 (1976).
17. E. M. Stricker, Thirst, sodium appetite and complementary physiological contributions to the regulation of intravascular fluid volume, _in_: "The neuropsychology of thirst - New findings and advances in concepts", A. N. Epstein, H. R. Kissileff and E. Stellar (eds), V. H. Winston and Sons, Washington, (1973).
18. E. M. Stricker, Osmoregulation and volume regulation in rats; inhibition of hypovolaemic thirst by water, _Am. J. Physiol._, 217: 98 (1969).
19. E. M. Stricker, J. E. Jalowiec, Restoration of intravascular fluid volume following acute hypovolaemia in rats, _Am. J. Physiol._, 218: 191 (1970).
20. T. Hirano, Some factors regulating water intake by the eel (Anguilla Japonica), _J. of Exp. Biol._, 61: 737 (1974).
21. J. T. Fitzsimons and S. Kaufman, Cellular and extracellular dehydration and angiotensin as stimuli to drinking in the common iguana (Iguana iguana), _J. Physiol._, 265: 443 (1977).

22. T. J. Cade, Water and salt balance in granivorous birds. in: "Thirst; Proc. of the First International Symposium on thirst in the regulation of body water", M. J. Wayner, ed., 237, Pergammon Press, Oxford, (1964).

23. J. T. Fitzsimons, M. Massi and S. M. Thornton, Permissive effect of cerebrospinal fluid sodium on drinking in response to cellular dehydration in the pigeon. Columba livia, J. Physiol, 315: 14P (1981).

24. D. J. Ramsay, B. Rolls and R. J. Wood, The relationship between elevated water intake and oedema associated with congestive cardiac failure in the dog. J. Physiol., 244: 303 (1975).

25. Ledingham, J. G., J. J. Morton, P. A. Phillips and E. J. Rolls, Effects of hypertonic saline and angiotensin (AII) on thirst in man. J. Physiol., 345: 114 (1983).

26. D. A. Denton, The Hunger for Salt, Springer Verlag, Heidelberg (1982).

27. T. Hirano, Y. Takei and H. Kobayashi, Effect of angiotensin on drinking in the eel and frog, in: "Volume and osmotic regulation, Alfred Benzon Symposium XI", C.B. Jorgensen and E. Skadhauge eds., p123, Acad. Press, New York/London (1978).

28. H. Kobayashi, H. Uemura, M. Wada and Y. Takei, Ecological adaptation of angiotensin-induced thirst mechanisms in tetrapods. Gen. Comp. Endocrinol., 38: 93 (1979).

29. Y. Takei, T. Hirano and H. Kobayashi, Angiotensin and water intake in the Japanese eel (Anguilla japonica), Gen. Comp. Endocrinol., 38: 466 (1979).

30. E. Tarjan, D. A. Denton, M. J. McKinley, J. F. Nelson and R. S. Weisinger, What makes wild rabbits drink? Symposium on Body Fluid Homeostasis, J. de Physiologie (Paris) Suppl., (In press).

31. B. J. Rolls and E. T. Rolls, Thirst, Cambridge University Press, (1982).

32. J. B. Wirth and A. N. Epstein, Ontogeny of thirst in the infant rat, Am. J. Physiol., 230: 188 (1976).

33. E. M. Blass, W. G. Hall and M. H. Teicher, The ontogeny of suckling and ingestive behaviour, in: "Progress in Psychobiology and Physiological Psychology, Vol.8, J. M. Sprauge and A. N. Epstein ed., pp 243, Acad. Press, New York (1979).

34. E. F. Adolph, Thirst and its inhibition in the stomach, Am. J. Physiol., 161: 374 (1950).

35. E. M. Stricker and G. M. Sterrit, Osmoregulation in the newly hatched domestic chick, Physiol. Behav. , 2: 117 (1967).

36. P. Hudson, J. P. Coghlan, J. Shine and H. D. Niall, Hybridization Histochemistry : Use of Recombinant DNA as a "Homing Probe" for tissue localization of specific mRNA population. Endocrinol. 108: 1 (1981).

THE ONTOGENY OF ANGIOTENSIN DIPSOGENESIS IN THE RAT

Alan N. Epstein

Department of Biology
University of Pennsylvania
Philadelphia, PA 19104

The invention of a benign method[2] for verified injection of chemical agents into the anterior cerebral ventricles of the unanesthetized rat pup of any age has allowed us to describe the ontogeny of the dipsogenic effect of intracerebroventricular (ICV) angiotensin.

Pups in groups of four are separated from their dam, satiated on milk by pulsatile infusion into the mouth while they are warmed inside an incubator, injected by pulse ICV with angiotensin (100 ng in 1 μl) in a vehicle of India-Ink, and offered water or milk for 20-30 min again by pulsatile oral infusion. Their weight gain measures their fluid intake. They are encouraged to urinate and defecate prior to the last infusion and are "diapered" (with rubber cement) during it to prevent weight loss by evacuation. Data is accepted only from animals in which India-Ink is found in the anterior ventricles when their brain is autopsied shortly after completion of the test of drinking.

As shown in Figures 1 and 2 drinking was activated by ICV angiotensin in rat pups as early as the 2nd day after birth. But their drinking is indiscriminate between water and milk (see 4-6 day olds in Figure 2). Pups that are less than one week old drink as much milk as water in response to ICV angiotensin (milk is largely water) until the 8th day when the characteristic adult response emerges. Pups of 8 days or older drink twice as much water as milk after ICV angiotensin. They are, in addition, as sensitive at this age to lower doses of the hormone as are adults.[2]

The brain's sensitivity to angiotensin dipsogenesis undergoes a staged ontogeny. Shortly after birth fluids are drunk indiscriminately. Roughly 1 week later, when the brain has achieved adult

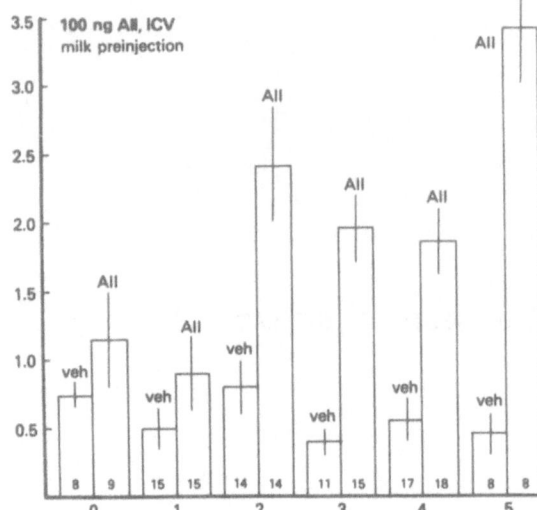

Figure 1. Development of Ang II-induced drinking. By 2 days of age milk-sated pups drink water in response to ICV Ang II. There is further increase in responsiveness at 5 days of age suggesting that a second system of intracranial sensitivity has matured. Numbers within bars are numbers of animals tested at each age, open bars denote pups given water after ICV injection. From: Ellis, et al. J. Neurosci. (1984).

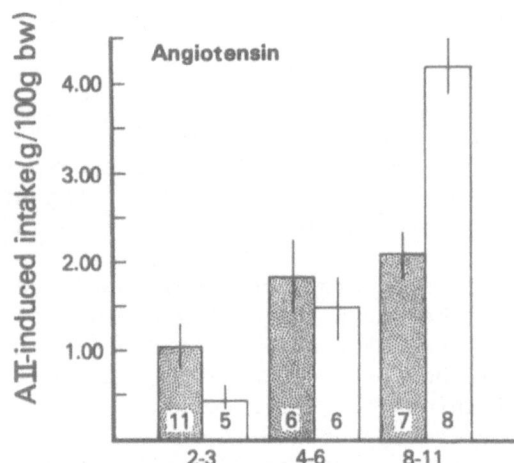

Figure 2. Development of the selective effect of Ang II on water intake which is the characteristic adult response. Note that until 8 days of age pups ingest equal volumes of water (open bars) and milk (stipled bars). Thereafter twice as much water as milk is drunk to ICV Ang II. From: Ellis et al. J. Neurosci. (1984).

sensitivity to the hormone, the true dipsogenic response is expressed. The animal will drink milk when it is the only fluid offered, but water is recognized as such and is drunk in greater volumes.

Like other neurochemical systems for control of ingestive behavior (e.g. the orexigenic effect of intracranial norepinephrine[1]) the system for angiotensin dipsogenesis matures well before it is required for adult drinking, and may, like norepinephrine orexigenesis, be a nascent control of adult ingestion rather than an operating control of suckling.

REFERENCES

1. Ellis, S., Axt, K. and Epstein, A.N. The arousal of ingestive behaviors by chemical injection into the brain of the suckling rat. J. Neurosci. (1984) 4: 945-955.

2. Misantone, L., Ellis, S.B. and Epstein, A.N. Development of angiotensin-induced drinking in the rat. Brain Res. (1980) 186: 195-202.

PERINATAL SODIUM CHLORIDE INTAKE MODIFIES THE FLUID INTAKE OF ADULT RATS

Robert J. Contreras and Edythe Bird

Yale University
Department of Psychology
Box 11A Yale Station
New Haven, CT   06520

Our interest is in determining whether differences in early salt intake produce long-term changes in intake in the rat. Perinatal salt consumption may influence long-term intake either through changes in taste sensitivity or through changes in the regulatory mechanisms mediating salt and water balance. Both taste and regulatory mechanisms only gradually acquire typical adult structural and functional characteristics over the course of development, providing a possible basis for plasticity in intake control mechanisms. With regard to taste, taste buds begin to appear during the last trimester of pregnancy and develop to their adult form by 14 days after birth[1]. Furthermore, electrophysiological studies of taste afferents suggest that salt taste sensitivity changes during development[2,3,4,5]. The regulatory mechanisms for maintaining salt and water balance are also in transition to their adult form during prenatal and early postnatal life[6,7].

To study the influence of early salt intake, adult female rats were maintained on diets containing either low (0.12%), mid (1.0%) or high (3.0%) amounts of sodium chloride. The mid salt diet contains a sodium level equivalent to that found in commercial diets for rodents; both the low and high salt concentrations are but modest deviations from normal levels, as these salt concentrations do not disrupt the reproductive success of the animals. The dams were maintained on their respective diets throughout gestation and lactation periods, and their offspring were continued on these salt diets until 30 days of age. Thereafter, all offspring were maintained on a mid salt diet through adulthood (180 days of age). We chose this time period for the experimental manipulation to maximize the influence of early experience, since we did not know the perinatal time window of greatest sensitivity.

Differences in dietary sodium levels did not affect energy, water or electrolyte balance of the dams during gestation. Energy balance was maintained as there were no differences between the low, mid and high salt dams in daily or total food intake and body weight. Sodium balance was also not influenced by dietary condition as the dams in the low salt condition had reduced (relative to the mid salt dams) urinary sodium output in conjunction with a lowered sodium intake; the dams in the high salt condition had elevated (relative to the mid salt dams) urinary sodium output in conjunction with an elevated sodium intake. Similarly, water balance was maintained, as differences in intake were compensated for by differences in urine volume. Surprisingly, however, fluid turnover levels of the low salt dams were higher than those of the mid salt dams but lower than those of the high salt dams. The low salt dams' elevated water intake and urine volume were apparent particularly during the last trimester of pregnancy.

Although the dams were able to maintain energy, water and electrolyte balance throughout gestation, the dietary salt conditions had some physiological consequences for the offspring. First, the sodium concentration of the amniotic fluid reflected differences in dietary sodium levels: the average sodium concentration was 121.0 ± 2.1 (mEq/l) for females in the low salt group, 124.4 ± 3.3 for females in the mid salt group, and 131.2 ± 4.8 for females in the high salt group. There were no differences in the potassium concentration of the amniotic fluid. Second, there was a significant trend for body weights of the offspring at birth to be positively related to dietary salt content; as the amount of salt increased so did the pups' body weights. The average body weight was 5.75 ± 0.15 g for the low salt pups, 6.16 ± 0.17 g for the mid salt pups, and 6.39 ± 0.14 g for the high salt pups. Third, the effects of the low salt diet were evident in the serum sodium levels of the rat pups at birth as they were significantly lower (120.6 mEq/l) than those of the mid salt (140.0 mEq/l) and high salt (139.2 mEq/l) pups. Although not measured in the present study, milk sodium levels are reduced in dams fed a low salt diet[8]. Thus, these four physiological measures reflect some of the effects of differences in maternal salt intake. The possible implication of these effects is the creation of a variant physiological environment responsible for producing permanent changes in the fluid consumption of the offspring.

The low, mid and high salt offspring were examined for differences in water intake and salt tast preference as adults. The animals were given a two-bottle preference test between water and either 0.3 M NaCl or 0.1 M glucose. Glucose was considered to be a control taste solution since maternal sodium intake, not

glucose intake, was varied. During this two-bottle test, the 24-h
water intakes of the male offspring were found to be inversely related
to the level of maternal salt intake; the low salt males consumed the
most and the high salt males the least. These differences in water
consumption are likely responsible for the lowered NaCl preferences
exhibited by the low salt animals and the elevated NaCl preferences
exhibited by the high salt animals relative to the intermediate NaCl
preferences of the mid (control) salt animals. These results are
identical to our previously reported findings where we used several
different molar concentrations of NaCl[9]. In the present study,
glucose intake was similar for the three groups of animals.

These results confirm our previously reported findings showing
that the offsprings' water intake was inversely related to maternal
salt intake. The present data are also consistent with the findings
of Mouw, Vander and Wagner[10], who showed that perinatal sodium deprivation (extreme low salt diet) was associated with an elevated water
intake (but an unaltered NaCl intake) of the offspring as adults.
It seems unlikely that changes in the physiological sensitivity of
the taste receptors are responsible for the differences in water
intake reported here and found previously. We suggest, instead,
that differences in maternal salt intake altered permanently the
regulatory mechanisms for maintaining fluid balance of the offspring.
This notion is corroborated further by our results showing that the
male rats in the low and high salt conditions had elevated urine
sodium levels. We propose that the alterations in intake and output
due to differences in maternal salt intake are indicative of permanent alterations in brain mechanisms of thirst.

REFERENCES

1. C. M. Mistretta, Topographical and histological study of the
        developing rat tongue, palate, and taste buds, in: "Oral
        Sensation and Perception. III. The Mouth of the Infant,"
        J. F. Bosma, ed., Charles C. Thomas, Springfield (1972).
2. D. L. Hill and C. R. Almli, Ontogeny of chorda tympani nerve
        responses to gustatory stimuli in the rat, Brain Res.
        197:27 (1980).
3. T. Yamada, Chorda tympani responses to gustatory stimuli in
        developing rats, Jpn. J. Physiol. 30:631 (1980).
4. M. F. Ferrell, C. M. Mistretta, and R. M. Bradley, Development
        of chorda tympani taste responses in rat, J. Comp. Neurol.
        198:37 (1981).
5. D. L. Hill, C. M. Mistretta, and R. M. Bradley, Developmental
        changes in taste response characteristics of rat single
        chorda tympani fibers, J. Neurosci. 2:782 (1982).
6. M. Friedman and B. A. Campbell, Ontogeny of thirst in the rat:
        effects of hypertonic saline, polyethylene glycol, and vena
        cava ligation, J. Comp. Physiol. Psychol. 87:37 (1974).

7. J. B. Wirth and A. N. Epstein, Ontogeny of thirst in the infant rat, <u>Am</u>. <u>J</u>. <u>Physiol</u>. 230:188 (1976).

8. M. C. Ganguli, J. D. Smith, and L. E. Hanson, Sodium metabolism and requirements in lactating rats, <u>J</u>. <u>Nutr</u>. 99:395 (1969).

9. R. J. Contreras and T. Kosten, Prenatal and early postnatal sodium chloride intake modifies the solution preferences of adult rats, J. Nutr. 113:1051 (1983).

10. D. R. Mouw, A. J. Vander, and J. Wagner, Effects of prenatal and early postnatal sodium deprivation on subsequent adult thirst and salt preference in rats, Am. J. Physiol. 234:F59 (1978).

THE ONTOGENY OF THIRST MECHANISMS IN ALBINO RATS

Elliott M. Blass

Department of Psychology
Johns Hopkins University
Baltimore, MD 21218

The ontogeny of ingestive behavior, especially in laboratory rats, has been the subject of recent intensive investigation. Considerable information is now available about the determinants of suckling, feeding and drinking mechanisms during development and a number of reviews concerning suckling and feeding have recently appeared.[1,2,3] The purpose of the present chapter is to review the development of hydrational controls over suckling and fluid intake.

DEHYDRATION AND SUCKLING

Suckling is the mammals' exclusive behavior for obtaining nutrients and fluid until the start of the weaning period. The question has arisen therefore, as to the hydrational influences over nipple attachment and the volume of milk that is taken from the nipple. Concerning dehydration, Drewett and Cordall,[4] Bruno, Craigmyle and Blass,[5] and Cramer, Pfister and Blass[6] have demonstrated that rats 5 and 10 days of age neither increase nor decrease milk intake at the nipple when dehydrated either cellularly or extracellularly. They failed to replicate Friedman's[7] finding of increased milk intake following extra-cellular hypovolemia. In rats 15 days of age and older both cellular and extracellular dehydration markedly reduces nipple attachment and milk intake at the nipple.[5,6] These findings suggest therefore, that sometime at about the beginning of the

weaning process, when food and water are both sampled directly
from the environment, suckling starts to take on certain
characteristics of feeding behavior. In this particular case,
it becomes vulnerable to "dehydration-induced anorexia". There
is reason to believe that the anorexogenic effect on suckling is
biological. Dose response functions were obtained with increasing dehydration and, more severely deprived rats were less
affected by dehydration, a finding that accords well with the
adult literature on hydrational determinants of food intake.[8]
It is not surprising that milk intake is not affected by dehydration prior to the onset of the weaning period. According to
McCance,[9,10] newborn mammals maintain normal fluid balance
during the first days of life when receiving mothers milk and
the occasional slight increase in blood urea that occurrs during
this period is transient. Clearly, inhibition of milk intake in
response to these slight increases in osmolality would be very
maladaptive, especially during the early postnatal weeks when
almost all calories obtained from the mother are dedicated to
growth.

THE DEVELOPMENT OF THIRST MECHANISMS

    Research in the development of drinking can be categorized
according to three classes of experimental analysis: (1) At
what age does the drinking system become available to the
developing animal? (2) At what age does the system normally
appear? and, (3) at what age does it take on the full complement
of adult properties? Let us start with the last questions. In
their classic study on multiple determinants of thirst, Adolph,
Barker and Hoy[11] demonstrated that by 35 days of age rats compensated for a 5.3% bw deficit caused by deprivation with the
same precision as adults. Moreover, they adjusted their intake
in adultlike fashion to a 5.3% intragastric preload delivered at
the initiation of drinking. Adolph et. al.,[11] also noted
that drinking in response to acute cellular dehydration was also
adultlike by Day 35 when rats were allowed to excrete the hypertonic load, a finding confirmed by Friedman and Campbell.[12] To
my knowledge the precision of the cellular thirst system has not
been fully assessed in isolation during development as has the
adult drinking system by causing anuria via nephrectomy or
ureteric ligation.[13]

    According to Friedman and Campbell[12] drinking caused by
hyperoncotic colloid administration, depleting the extracellular
space, is normal by 35 days of age, although that caused by
acute ligation of the inferior vena cava is not even at 6 weeks
postnatal. This may be due to the failure of collateral

20

circulation to adjust to the deficit, causing some incapacitation. In any event, it would appear that by the end of the natural weaning period of 35 days, i.e., under conditions of adequate natural resources and an uncrowded nest area, the rat can probably fend well for itself hydrationally, especially when considering its now fully developed renal system. The interesting question arises as to how rats fare when weaned by the mother following her delivery of a litter conceived during her postpartum estrous? These rats may be 23-27 days of age when forced out on their own. This common, premature weaning may pose special problems to the young, to which I will return shortly.

As to the question of when does drinking normally take place in the nest, the data from a number of laboratories suggest that it first occurs at 14-15 days of age in Norway rats.[14] This, of course, is the time that rats start to also take dry food from the environment directly.[15] The thirst system may be spontaneously driven by postprandial concerns. It is the case, however, that it can be expressed in rats that had never eaten. According to Almli,[16] Day 14 rats, if kept warm, drink to acute cellular dehydration, but this dipsogen appears to gain control over behavior under "normal" laboratory conditions by 19-20 days of age.[12,14] Almli's[16] study was prescient because it demonstrated that when the special needs of the developing animal are met (e.g., thermal liability) then unexpectedly rich behaviors can be obtained. The precocial development of drinking mechanisms is a case in point, to which I now turn.

Wirth and Epstein[17] were the first to show that rats during the suckling phase of development drank in response to various forms of dehydration. Pups, beginning at 2 days postnatal, were challanged with cellular dehydration, colloid dialysis, or isoproterenol, to cause beta-adrenergic activation, were hand-held to a soft tube, inserted 3mm past their lips, and allowed to drink from its continuous flow. Wirth and Epstein[17] infused water, milk or 3% NaCl into dehydrated and control pups mouths and assessed amount of weight gained as a measure of the behavioral expression of the various thirst mechanisms. They reported that drinking in response to cellular dehydration started by Day 2, that hypovolemia became dipsogenic by Day 4 and that isoproterenol caused enhanced fluid intake by Day 6. In a subsequent study, Misantone, Ellis and Epstein[18] demonstrated enhanced fluid intake to intracranial injections of angiotensin appeared by Day 2, with adult levels of sensitivity (in terms of threshold) being achieved by Day 5 when 10-12

Moles of Angiotensin delivered intraventricularly caused reliable drinking from the fount. Thus, according to Wirth, Epstein and colleagues,[17],[18] each hydrational system appears to have its own developmental timetable, at least in regards to a hydrational stimulus increasing the rate and/or efficiency of obligate swallowing.

Bruno[19] more extensively analyzed the behavioral characteristics of dehydrated rats by assessing "voluntary" intake of various solutions, water, milk and .1M NaCl after cellular and extracellular dehydration. Bruno[19] revealed a number of important motivational characteristics of the drinking system. He replicated and extended the major findings of Wirth and Epstein[17] by demonstratng that dehydrated pups would actively lick fluid off the surface of a saturated mat kept in a warm (32°C) ambient temperature. The system was not yet adultlike, however pups until 15 days of age did not approach a fluid source from a distance, when dehydrated. After Day 15, they did so. Moreover, unlike deprived pups that pressed a lever to obtain milk, dehydrated pups did not. Drinking behavior lacked the adult characteristic of selectivity. That is, unlike adults, infants drank water, milk or hypertonic saline. Selectivity was expressed starting at about 10 days of age.

The issue still remains, however, as to how infants forced to evacuate the nest maintained normal body fluid homeostasis despite a relatively immature drinking mechanism. Although they may leave the nest, they do not abandon it and appear to maintain contact with the dam at the nest's periphery. This maintenance is significant. According to Bruno, Blass and Amin[20] dehydrated rats will rarely drink if they can not maintain contact with the mothers. When such contact is permitted, even though suckling is prevented, then dehydration induces significant drinking.

In summary, as in the development of feeding[21],[22] drinking mechanisms are available at a very early point during ontogeny, i.e., at a time that the system is probably never called upon. Under normal nest conditions drinking, as well as feeding, first becomes expressed at about two weeks of age but does not become fully mature until about the end of natural weaning, circa 7 weeks in rats. But drinking, like feeding is a social behavior so that the early weaned animal, in addition to calling upon its well developed renal capacities, drinks freely in the presence of its mother, thereby ensuring body fluid balance.

REFERENCES

1. E. M. Blass, W. G. Hall, and M. H. Teicher, The ontogeny of suckling and ingestive behaviors. in: Progress in Psychobiology and Physiological Psychology (Vol. 8), J.M. Sprague and A.N. Epstein (Eds.), Academic Press, New York (1979).
2. E. M. Blass, and M. H. Teicher, Suckling. Science, 210: 15-22 (1980).
3. W. G. Hall, and C. L. Williams, Suckling isn't feeding, or is it? A search for developmental continuities, Advances in the Study of Behavior, Academic Press, 1983, 13:219-254 (1983).
4. R. F. Drewett, and K. M. Cordall, Control of feeding in suckling rats: Effects of glucose and osmotic stimuli. Physiol. and Behav., 16:711-717 (1976).
5. J. P. Bruno, S. S. Craigmyle, and E. M. Blass, Dehydration inhibits suckling behavior in weanling rats, J. Comp. Physiol. Psychol., 96:405-415 (1982).
6. C. P. Cramer, J. F. Pfister, and E. M. Blass, Transitions in the dehydration-induced inhibition of milk intake in suckling rats, Physiol. and Behav., (1984, In Press).
7. M. I. Friedman, Some determinants of milk ingestion in suckling rats. J. Comp. Physiol. Psychol., 89: 636-647 (1975).
8. K. Oatley, and F. M. Toates, Osmotic inhibition of eating as a subtractive process. J. Comp. Physiol. Psychol., 82:268-277, (1973).
9. R. A. McCance, Overnutrition and undernutrition, Lancet 2: 685-706 (1953).
10. R. A. McCance, and M. Otley, Course of the blood urea in newborn rats, pigs and kittens. J. Physiol., London 113:18-22 (1951).
11. E. F. Adolph, J. P. Barker, and P. A. Hoy, Multiple factors in thirst. Am. J. Physiol., 178:538-562 (1954).
12. M. I. Friedman, and B. A. Campbell, Ontogeny of thirst in the rat: Effects of hypertonic saline, polyethylene glycol, and vena cava ligation. J. Comp. Physiol. Psychol., 87:37-46 (1974).
13. J. T. Fitzsimons, The effects of slow infusions of hypertonic solutions on drinking and drinking thresholds in rats. J. Physiol. (London), 167:344-354 (1963).
14. J. Křeček, and J. Krečeková, The development of the regulation of water metabolism: III. The relation between water and milk intake in infant rats. Physiol. Bohem., 6:26-34 (1957).

15. A. Babicky, J. Ostadalova, J. Parizek, R. Kolar, and B. Bibr, Onset and duration of the physiological weaning period for infant rats reared in nests of different sizes. Physiol. Bohemo., 2:449-456 (1973).
16. C. R. Almli, The ontogeny of the onset of drinking and plasma osmotic pressure regulation. Devel. Psychobiol., 6:147-158 (1973).
17. J. B. Wirth, and A. M. Epstein, Ontogeny of thirst in the infant rat. Am. J. Physiol., 320:188-198 (1976).
18. L. T. Misantone, S. Ellis, and A. N. Epstein, Development of angiotensin-induced drinking in the rat. Br. Res., 186: 195-202 (1980).
19. J. P. Bruno, Development of drinking behavior in preweanling rats. J. Comp. Physiol. Psychol., 95:1016-1027 (1981).
20. J. P. Bruno, E. M. Blass, and F. Amin, Determinants of suckling versus drinking in weanling albino rats: Influence of hydrational state and maternal contact, Devel. Psychobiol., 16:177-184 (1983).
21. W. G. Hall, The ontogeny of feeding in rats: I. Ingestive and behavioral responses to oral infusions. J. Comp. Physiol. Psychol., 93:977-1000 (1979).
22. W. G. Hall, and T. E. Bryan, The ontogeny of feeding in rats: II. Independent ingestive behavior. J. Comp. Physiol. Psychol. 94:746-756 (1980).

BEHAVIORAL RESPONSIVITY TO TASTES IN DEVELOPING RATS

Priscilla Kehoe and Elliott M. Blass

The Department of Psychology
Johns Hopkins University
Baltimore, Maryland   21218

The development of taste responsivity has recently attracted considerable attention from neurophysiologists' who have studied taste ontogeny in the sheep,[2,3] an animal with a lengthy gestation period (circa 150 days) and rats,[4,5,6] whose gestation period is considerably shorter, (22 days). In general, it appears that, in sheep, taste development is a prenatal phenomenon such that at birth the system is mature, morphologically and electrophysiologically at least.

Rat gustatory ontogeny follows a markedly different pattern that is characterized by morphological and electrophysiological immaturity at birth followed by postnatal development and differentiation of taste receptors on the tongue and soft palate. The profile of responsivity also undergoes an important developmental progression. Initially, the electrophysiological response to Ammonion salts is much greater than to NaCl. In comparison with Ammonium, NaCl undergoes a marked change in reactivity with adult profiles only being attained by 40 days of age. Thus, on the basis of electrophysiological data obtained from integrated recordings of the entire chorda tympani nerve,[4] from single chorda tympani fibers,[7] from single second-order neurons in N. Tractus Solitarius,[8] and in the pontine taste area,[9] one would predict much greater taste reactivity to NH_4Cl than to NaCl and relatively little reactivity to QHCl.

These considerations, in conjunction with the development of a technique by Hall and Rosenblatt[10] that allows introduction of fluid into the posterior oropharynx of pups during suckling, allowed us[11] to systematically explore the emergent characteristics of behavioral responsivity to gustatory stimulation. Our strategy was to study responsivity in the context of suckling behavior because this represents the animals natural mode of obtaining fluid. We initially injected fluids through cannulae that, as in the Hall and Rosenblatt[10] study were seated 2-4mm caudal to the intermolar eminence. During preliminary experiments, two modifications were undertaken. In one we seated the cannulae in one of three positions, 2mm rostral to the intermolar eminence, 0-3 mm- and 3.5-6 mm caudal to it. Fluid injected into any of these positions in suckling rats elicited the hyperextension response that is normally caused by the delivery of milk from the nipple. The other modification consisted of manipulating certain characteristics of fluid delivery to 5-day-old rats. In one condition a fluid reservoir was suspended 6 cm above the cage. Sucking caused fluid delivery that was continuously sustained until suction terminated. Yet rats could remain attached to the nipple and not receive fluid. The second condition more closely approximated the discrete natural mode of delivery in that fluid from the reservoir was made available for 30 sec every 3 min. The third condition was identical to the second except that .05ml fluid was delivered via infusion pump for 8-10 sec. Solutions infused were tap water, 1.4×10^{-3}M QHCl, .5M NH$_4$Cl and .43M NaCl. Rats were studied at 5, 10, 15 and 20 days of age.

These findings can be most meaningfully discussed in terms of behavioral responsivity to various sapid solutions in Day 5 rats and the developmental changes in responsivity expressed in the suckling situation during the course of ontogeny. A number of general statements are in order. First, in comparison with the behavior of rats receiving injections of water through the tongue cannula, Day 5 rats receiving injections of QHCl, NaCl or NH$_4$Cl, in the concentrations specified above, were markedly reactive under certain situations. This was most dramatically expressed by rats receiving NH$_4$Cl continuously through the anterior tongue cannula when sucking. These rats became cyanotic upon receiving the ammonium solution. Second, there was a clearcut diminution in behavioral reactivity as the cannulae were more caudally seated in the tongue. Third, behavioral reactivity, as defined by the loss of the stretch response, that is the hallmark of milk intake from the nipple, and by the increased incidence of aversive responses became much more pronounced developmentally.

It is appropriate to discuss these findings in detail, first concerning Day 5 rats and then comparing them with older animals. Day 5 rats that received quinine in the continuous condition in any of the three positions on the tongue always took in significantly less fluid than control rats that received infusions of water. Moreover, they spent less time attached to a nipple than did their counterparts receiving water. They presented more aversive responses. These findings were pronounced in this sustained condition where the bitter fluid was continuously available and entered the mouth when the slightest suction was exerted. It is of considerable interest that rats receiving discrete infusions, or with fluid available every 3 min. were not affected by QHCl in any obvious way.

As expected, suckling in rats receiving NH_4Cl, was profoundly disrupted. For example, animals receiving fluid either continuously or discretely, only obtained less than !% of their body weight when the cannula was in the anterior position (rats receiving water took in 4% BW). As with quinine, the other measures revealed a marked sensitivity to the ammonium solution relative to water. An exception to this were rats that received their ammonium solution through the pump. Although they spent less time on the nipple than "water" rats, made fewer stretch responses and more aversive ones, their intakes did not differ significantly from rats receiving water.

The behavior of rats receiving NaCl was unexpected. These Day 5 rats were very much affected by all three modes of saline delivery. They, more than any of the other rats, showed a very distinctive rostral-caudal sensitivity, with injections into the posterior oropharynx causing behavioral patterns that were indistinguishable from those of animals receiving water, and anterior injections causing behaviors that were almost as severely disruptive as ammonium. Thus, regarding reactivity of Day 5 rats to hypertonic NaCl, it is clear that the relatively weak electrophysiological signals are gaining significant control over the animals behavior during suckling.

There were profound changes in behavioral reactivity during development, and the time course of these changes differed for each of the substances tested. Reactivity to NH_4Cl remained stable in all three cannula positions until about 20 days of age, at which time injections at all three placements caused near total rejection of the nipple. As to NaCl, the relative lack of sensitivity of the posterior cannula held until Day 20, at which time all three placements became very sensitive and only 1 of the 8 animals tested in each position emitted a stretch response. Quinine showed a different developmental pattern: the anterior placement reached its maximum level of sensitivity at Day 10, the middle placement at Day 15 and the posterior placement on Day 20.

In brief, these studies have demonstrated taste sensitivity in rats as young as 5 days of age during suckling. This is, to our knowledge, the first demonstration of the ability of rats of this age to respond to environmental contingencies while suckling. It also demonstrates that the behavioral response mechanisms, especially for hypertonic saline and quinine are more sensitive than the electrophysiological data have implied. Thus while peripheral nerves and initial central relays indicate immaturity, these signals either reach central systems that are sufficiently mature to gain control over nipple attachment and suckling, or the signals are amplified en route to central aversive systems. Despite this, the systems are not completely mature as evidenced by increased reactivity developmentally. Whether this represents a maturation of the gustatory mechanisms or a waning of the intensity of suckling behavior, remains an important issue.

The present data are compatable with data recently provided by Hall and Bryan[12] who demonstrated taste reactivity in 3 and 6 day old rats that received quinine or NaCl infusions through an anterior oral cannula that was seated in the lower jaw. Their animals were studied after either 1 or 24 hr. of privation and in a very warm ambient temperature (32-34°C.). Taken together these studies that were conducted under very different experimental circumstances, provide validity to the idea of behavioral sensitivity to quinine and sodium chloride in rats within a few days of birth.

Finally, to the extent that infant rats can detect and respond to the taste of various solutions while suckling, the question arises as to whether taste information can be utilized to form learned aversions. They can. Day 5 rats that received saccharine through anterior intraoral cannulae, while suckling, and then injected with the toxin lithium chloride, took in far less saccharine solution when retested 5, 10, or 15 days later.[13]

REFERENCES

1. R. M. Bradley, and C. M. Mistretta, Fetal sensory receptors. Physiol. Rev., 55, 352-382 (1975).
2. C. M. Mistretta, and R. M. Bradley, Neural basis of developing salt taste sensation: Response changes in fetal, postnatal and adult sheep. J. Comp. Neurol., 1983a, 215, 199-210.
3. C. M. Mistretta, and R. M. Bradley, Developmental changes in taste responses from glossopharyngeal nerve in sheep and comparisons with chorda tympani responses. Dev. Br. Res., 11, 107-117 (1983b).
4. M. F. Ferrell, C. M. Mistretta, and R. M. Bradley, Development of chorda tympani responses in rat. J. Comp. Neurol. 198, 37-44, (1981).
5. D. L. Hill, and R. C. Almli, Ontogeny of chorda tympani nerve responses to gustatory stimuli in the rat. Br. Res., 197, 27-38 (1980).
7. D. L. Hill, C. M. Mistretta, and R. M. Bradley. Developmental changes in taste response characteristics of rat single chorda tympani fibers. J. Neurosci. 2, 782-790 (1982).
8. D. L. Hill, R. M. Bradley, and C. M. Mistretta, Development of taste response in rat nucleus of solitary tract. J. Neurophys. 50, 879-895 (1983).
9. D. L. Hill, Development of pontine parabrachial nuclei taste responses in rat. S. Neurosci. Abst., 9, 378 (1983).
10. W. G. Hall, and J. S. Rosenblatt, Suckling behavior and intake control in the developing rat pup. J. Comp. Physiol. Psychol., 91, 1232-1247 (1977).
11. P. Kehoe, and E. M. Blass, Gustatory determinants of suckling in albino rats 5-24 days of age. Dev. Psychobiol (1984, in press).
12. W. G. Hall and T. E. Bryan, The ontogeny of feeding in rats: IV. Taste development as measured by intake and behavioral responses to oral infusions of sucrose and quinine. J. Comp. Physiol. Psychol., 95, 240-251 (1981).
13. P. Kehoe, and E. M. Blass, Evidence for learned preferences and aversions in 5-day-old suckling rats. Ms. submitted.

THE ONTOGENY OF SALT INTAKE IN RATS

Karen Moe[1]

Depatment of Biology and Institute of Neurological Sciences

University of Pennsylvania, Philadelphia, PA   19104

Many mammals eat salt irrespective of need. When depleted of sodium, they consume even more of it. In both conditions, the amount consumed depends on the concentration of salt offered, with weak concentrations being preferred and strong ones avoided.

The expression by neonatal animals of both the need-free preference for salt and the need-induced salt appetite has received little experimental attention even in rats, the species most often used in studies of salt appetite. Yet there are unique advantages associated with a developmental approach to physiological mechanisms of behavior. For example, the effect on adult behavior of early experience or intervention of some kind can be ascertained.

So little has been done using this approach with regard to salt intake that it is not even clear whether neonates are capable of demonstrating these two phenomena of salt intake. The following experiments address this question, using rat pups.

METHODS

Pups used in these experiments were offspring of Sprague-Dawley rats produced by a breeding colony maintained in the laboratory. The day after birth litters were culled to 10 pups.

The pups were removed from their dams on the day of testing. A catheter of PE-10 tubing was implanted in the anterior oral cavity of each pup, under the tongue, as described elsewhere[1]. Infusions made into this anterior location do not stimulate reflexive swallowing.

Pups were allowed to adapt to the presence of the catheter

---

[1] Present address: Department of Psychology, Washington State University, Pullman, WA 99164-4830.

for 30 min while in a warm (32–34° C), moist incubator. They were then labelled, sexed and weighed to the nearest 0.01 g. Change of body weight over the course of the infusion period was used as an accurate measure of fluid intake[2].

Following their return to the incubator, the catheter of each pup was attached to a length of PE-50 tubing. This tubing connected the catheter to a 10 cc syringe driven by a infusion pump.

Pups taken directly from their dams differ with respect to how recently they have been fed. Therefore, all pups except those treated with furosemide (see below) were sated with milk 3-4 hrs before the test solutions were offered. A room temperature, commercially available solution of half milk and half cream ("milk") was offered to the pups by pulsatile infusion. The infusion pump delivered a 15-sec pulse of milk once every 2 min, for 60 min. Pulse volume was chosen to offer pups somewhat more than they could swallow, and was therefore different for each age group tested. Upon completion of the infusion, the pups were disconnected, dried if necessary, weighed, and returned to the incubator.

Three-four hours later, the test solutions were infused, in exactly the same manner as above but for only 30 min. Just before the test, pups were gently stimulated to micturate and defecate. To prevent further evacuation and ensuing weight loss from occurring during the test infusion, a small amount of celloidin (12%) or rubber cement was painted on the anogenital region. Each pup was tested only once, with only one solution.

To facilitate comparisons among different age groups, all data are presented as grams of fluid ingested per 100 g body weight. A solution is defined as preferred if it is consumed in greater quantity than water; if less of a solution is consumed compared to water, that solution is said to be aversive.

Salt preference. The need-free preference for salt was assessed in 6, 12 and 18 day-old rat pups. During the test infusion, they were offered saline (0.3-18%), water or nothing. All concentrations of NaCl were not used with each group.

Salt appetite. Two classic techniques were used to arouse a need-induced salt appetite: adrenalectomy and furosemide treatment. For the adrenalectomy experiment, the adrenal glands were surgically removed from 8 day-old pups under hypothermia anesthesia. Half of each litter was adrenalectomized; the other half served as a sham surgical control. The pups were returned to their dam after surgery and remained with her until they were 12 days of age when they were tested as described above. The test solution offered to them was 6% NaCl.

In another experiment, salt appetite was aroused by furosemide-induced sodium depletion in 12 day-old pups. Treatment and testing occurred on the same day. These pups were not sated with milk prior to testing. Instead, a single subcutaneous injection of physiological saline or 1 mg/0.1 ml furosemide was administered in the morning, after the pups had been weighed. At 0.5, 1, 2, 3, 5, and 8 hours after the injection, micturation was stimulated and the pups re-weighed. At 8 hours, oral catheters were implanted.

To alleviate dehydration, the pups were then offered water by infusion for 30 min. This served to replace much of the water, but not the sodium, lost in the furosemide-induced diuresis. The test infusion occurred about 2 hours later, in the same manner as above, with 8% NaCl as the test solution.

RESULTS

Rat pups demonstrate a need-free preference for salt that is concentration-dependent, as do adult rats. However, their preference-aversion curves are shifted to the right along the concentration gradient in an age-related way (Figure 1), with the youngest age group tested showing the greatest shift and the oldest showing the least. Six day-old pups prefer saline at concentrations of 3 - 8%. Concentrations above 16% are aversive (less preferred to water). Twelve day-old pups prefer concentrations of 2% - 4%, and find concentrations above 7% aversive. Eighteen day-old pups show a preference-aversion curve that is very similar to the adult curve (shown by dotted line; data take from Young and Chaplin, 1956[3]). They prefer NaCl in the 0.9% - 2% range, and reject concentrations of 3% and above. Average water intake (g/100 g) at each age was: 1.0±0.1 (6 days), 0.4±0.1 (12 days) and 1.0±0.1 (18 days).

Rat pups also precociously demonstrate need-induced salt appetite. Twelve day-old pups that were adrenalectomized at 8 days of age drank 0.99±0.15 g/100 g of 6% NaCl, compared to 0.46±0.1 g/100 g consumed by the sham surgical controls (Figure 2). However, adrenalectomized pups drank the same amount of milk as control pups in the pre-test milk infusion (4.5±0.4 g/100 g versus 4.4±0.3 g/100g).

Furosemide treatment also elicited a salt appetite at 12 days (Figure 2). Furosemide-treated pups drank 0.75±0.15 g/100 g of 8% NaCl whereas vehicle-treated pups drank 0.40±0.1 g/100 g. Not surprisingly, they also drank more water in the pre-test water infusion (2.8±0.2 g/100 g for furosemide pups compared to 0.8±0.1 g/100 g for vehicle pups).

DISCUSSION

These experiments show that rat pups will precociously demonstrate both the need-free preference and the need-induced appetite for salt, in the same general manner as adults. However, there are some differences between the pups and adults.

First, though the preference-aversion curves for pups are of the same general shape as the curve for adults, they are shifted to the right along the concentration axis. Thus, the saline concentrations that are preferred by pups are generally much higher than those preferred by adults. Also, they reject saline relative to water only at very high concentrations.

However, the pup preference-aversion curves show a clear

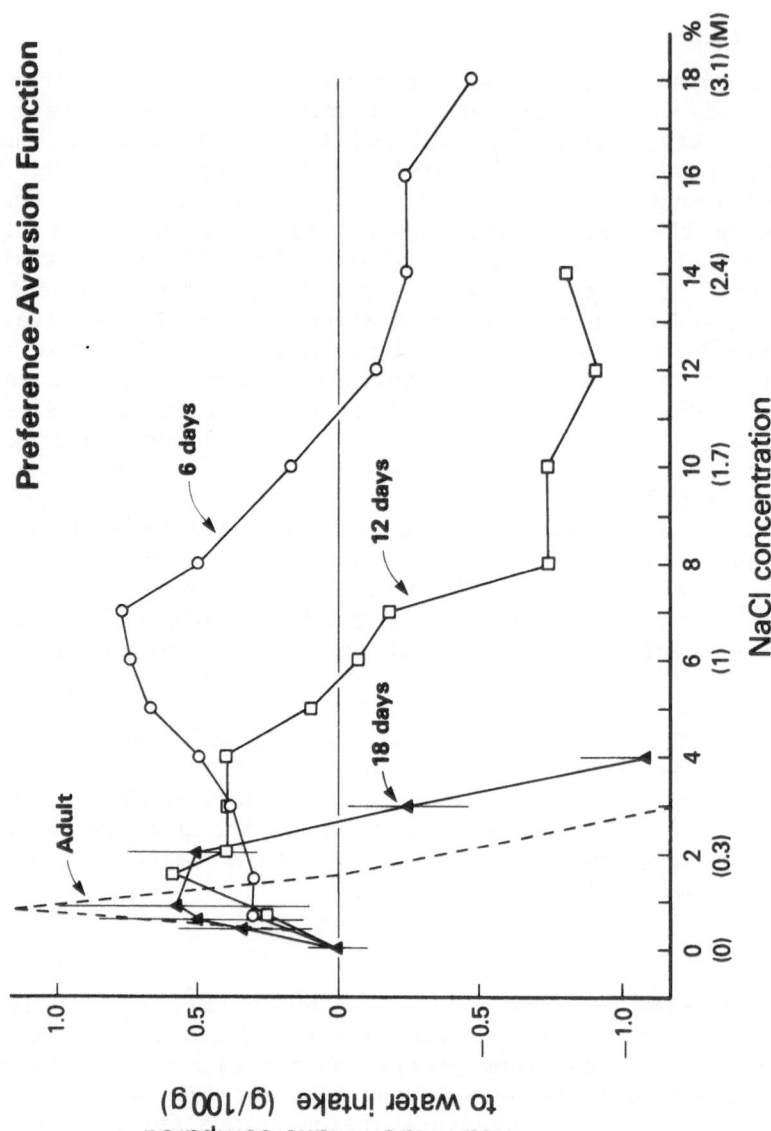

Fig. 1. NaCl intake (g/100 g body weight) of 6, 12 and 18 day old rat pups expressed relative water intake ("0" line). Each point represents the mean value for 12 pups (6 males, 6 females).

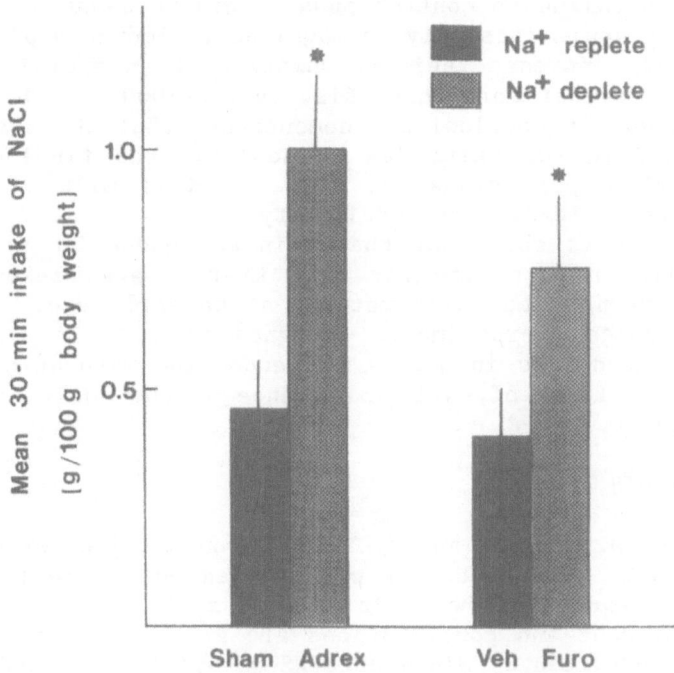

Fig. 2. NaCl intake (g/100 g body weight) of sodium-depleted 12
day old rat pups. <u>Left</u>: Consumption of 6% NaCl by adrenalectomized or sham-operated pups. <u>Right</u>: Consumption of 8%
NaCl by pups injected with furosemide or its vehicle.

developmental progression toward the adult curve, with the curve
for 18 day-olds being almost the same as the adult curve. This
progression parallels in time and is probably explained by the
development of the rat gustatory system. Though taste buds begin
to form on the 20th day of gestation, their adult structural
characteristics develop gradually during the first two postnatal
weeks[4]. In addition, the neurophysiological responses to NaCl on
the tongue show marked postnatal changes during the first five
weeks[5,6]; i.e., sodium chloride is relatively ineffective at
eliciting a neural response from the chorda tympani for the first
2 weeks. Thereafter it gradually becomes more effective.
These experiments also show that they increase their consumption of it under conditions of need. When depleted of sodium by
adrenalectomy or furosemide injection, 12 day-old pups drink about

twice as much saline as control pups. This response differs from that shown by adult rats only in magnitude. Sodium-depleted adult rats typically increase their consumption of an aversive saline solution by somewhat more than this, by 100-1000%[7]. Though these experiments do not conclusively demonstrate that the response of pups is specific for salts (as in adults), the finding that adrenalectomized pups drank the same amount of milk as control pups suggests at least some specificity.

These results show that the brain and endocrine mechanisms that underlie the appetite for salt in rats are present and functional, though not fully mature, at an early age. Furthermore, they suggest experiments designed to tie the emerging behavior to landmarks in neural and endocrine maturation and to evaluate the effects of early experience on the adult expression of the behavior.

ACKNOWLEDGMENTS

J.Butler, D. Steiner and M.L. Stevenson provided excellent technical assistance. This research was carried out while the author was a postdoctoral fellow in Dr. Alan Epstein's laboratory, and his encouragement and contributions are gratefully acknowledged. Support was provided by NIH grants NS-03469 (A.N.E.) and NS-07304 (K.E.M.).

REFERENCES

1. W.G. Hall, The ontogeny of feeding in rats: I. Ingestion and behavioral responses to oral infusions, J. Comp. Physiol. Psychol. 93:977 (1979).
2. K.A. Houpt and A.N. Epstein, Ontogeny of controls of food intake in the rat: GI fill and glucoprivation, Am. J. Physiol. 225:58 (1973).
3. P.A. Young and J.P. Chaplin, Preferences of adrenalectomized rats for salt solutions of different concentrations, Comp. Psychol. Monographs 19:45 (1949).
4. C.M. Mistretta, Topographical and histological study of the developing rat tongue, palate and taste buds, In: "Third symposium on Oral Sensation and Perception: The Mouth of the Infant," J.F. Bosma, ed., Charles C. Thomas, Springfield, Ill (1972).
5. M.F. Ferrell, C.M. Mistretta and R.M. Bradley, Chorda tympani taste responses during development in rat, J. Comp. Neurol. 198:37 (1981).
6. D.L. Hill and C.R. Almli, Ontogeny of chorda tympani nerve responses to gustatory stimuli in the rat, Brain Res. 197:27 (1980).
7. K.E. Moe, M.L. Weiss and A.N. Epstein, Sodium appetite during captopril blockade of endogenous angiotensin II formation, Am.J. Physiol. 247:R356 (1984).

EFFECTS OF THE TACHYKININS ELEDOISIN AND PHYSALAEMIN ON DRINKING BEHAVIOUR IN BABY RATS

F. Cantalamessa, G. de Caro, A.N. Epstein and M. Perfumi

Institute of Pharmacology, University of Camerino
Camerino-Italy, J. Leidy Labratory
University of Pennsylvania, Philadelphia

Substance P was until recently the only tachykinin known to occur in mammals, all the other members of the family having been found only in molluscan or amphibian tissues. Now we know that at least two other tachykinins, one physalaemin-like and the other kassinin-like, occur in mammalian tissues, including the brain (1, 2).

The tachykinins are potent antidipsogens in the rat, and we have suggested that those which occur in the rat brain play a physiological role in the control of water intake, acting as thirst inhibitors (3). An understanding of the ontogeny of this antidipsogenic effect may be important for the confirmation of our hypothesis. We therefore studied the antidipsogenic effect of eledoisin in baby rats. Eledoisin, a tachykinin of molluscan origin which has not yet been found in mammalian tissues, was chosen because of the wide spectrum of its effects and because of its reliability. The other tachykinins, as well as the antidipsogenic agents which belong to other peptide families, will be the object of future ontogenic studies. However, in some experiments reported here we employed physalaemin.

METHODS

We employed the technique described by Ellis et al. (4). In brief, pups aged 1 to 15 days were separated from their dams, presatieted on milk or water by pulsatile infusion into the mouth, injected into the brain ventricles (ICV) with eledoisin in a vehicle of India-ink, and eventually offered water or milk for 20 min, again by pulsatile oral infusion. The experiments were carried out in a glass-front incubator kept at 32 ± 0.5 °C. Relative humidity was 100%.

The location of the injected material in the pups' brains was verified as described by Epstein et al. (5).

Since animal's anogenital area had been previously covered with celloidin (12%, in ether-alcohol) to prevent loss of weight by urination and defecation, weight gain indicated pups' fluid intake.

Drinking was induced by ICV angiotensin II, 100 ng/rat, or by sc NaCl, 3 mOsmoles per 100 g bw. In other experiments, increased drinking of water or milk was produced by deprivation: pups were separated from the dam (and warmed in the incubator) for 3 hrs before the experiments began.

RESULTS

Pups previously satiated on milk received a single pulse injection of either angiotensin II, 100 ng per rat, or the same dose of angiotensin II plus eledoisin in doses ranging between 10 and 300 ng per rat. The rats ages were 3, 6, 9 and 12 days.

Drinking response to angiotensin II was not modified by 10 ng of eledoisin, but was significantly reduced by larger doses. The inhibition was easily reproducible, was related to the dose and was not influenced by age (fig. 1). Similar results were obtained

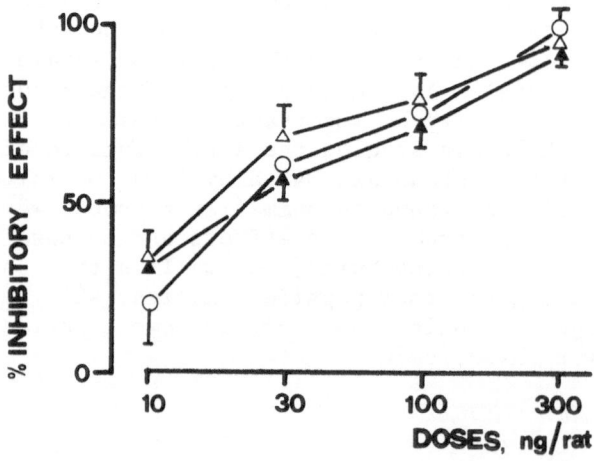

Fig. 1.   AII-induced drinking: inhibitory effect elicited by pICV-
        -eledoisin in rats of 3 (△-△), 6 (▲-▲) or 12 (○-○)
        days.
        Values are means of 8 to 10 data ± s.e.m.

with physalaemin. However, this peptide was about 30 times less effective than eledoisin and did not at all inhibit drinking of 3 day old rats, which normally responded to eledoisin.

Pups aged 1, 3, 6, 9 and 12 days were injected subcutaneously with NaCl, 3 mOsmoles per 100 g bw, and 15 min later received a pulse ICV injection of eledoisin, in doses ranging between 10 and 300 ng per rat. In 1 day old rats eledoisin was not effective at all, but it inhibited drinking in 3 to 12 day old pups. The effect was easily reproducible, depended on the dose, and lasted a considerable time (fig. 2). In the same experimental conditions phy-

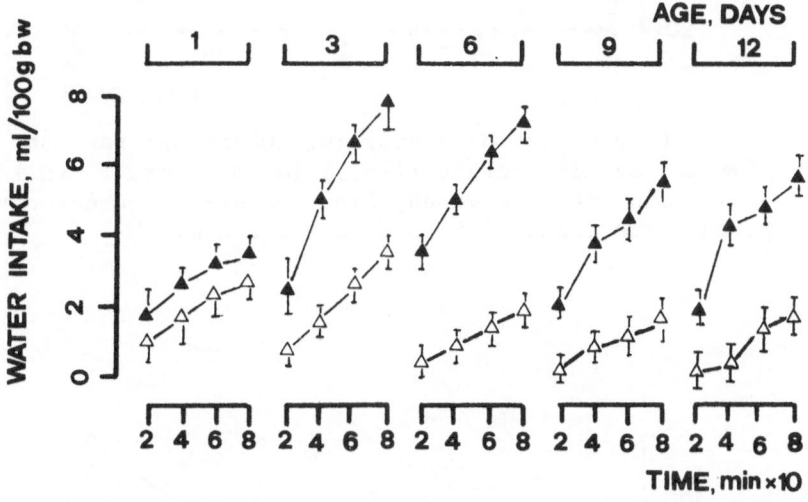

Fig. 2.   Cell dehydration-induced drinking: water intake following pICV injection of vehicle (▲-▲) or eledoisin, 300 ng per rat (△-△) to rats of 1 to 12 days.
Values are means of 6 to 10 data ± s.e.m.

salaemin, administered to rats of 6 or 12 days in doses of 100 and 400 ng per rat, did not elicit any evident inhibitory effect.

To have some information about the specificity of these effects, we checked whether eledoisin also modified fluid intake in baby rats which had been off dam for 3 hr and which were offered either milk or water, without receiving any fluid before the treatment. In these experiments, eledoisin inhibited both milk and water intake to the same extent in 3 day old rats. However, while water intake inhibition increased with age, milk intake inhibition decreased with age and completely disappeared at 12 days, when the inhibition was less than 20% and not statistically significant (fig. 3). These results suggested that in baby rats eledoisin may be both antidipsogenic and anorexigenic. We therefore checked

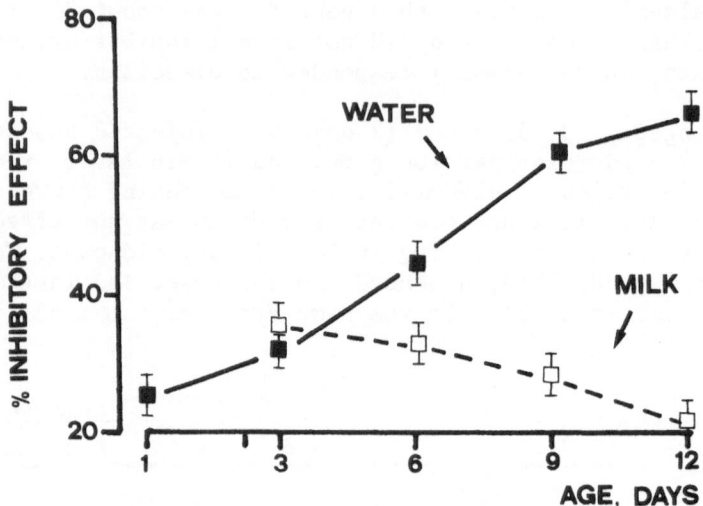

Fig. 3. Inhibitory effect of eledoisin, 100 ng per rat, on water
(■-■) or milk intake (□-□) in rats of 1 to 12 days
which did not receive any fluid before the treatment.
Values are means of 6 to 8 data ± s.e.m.

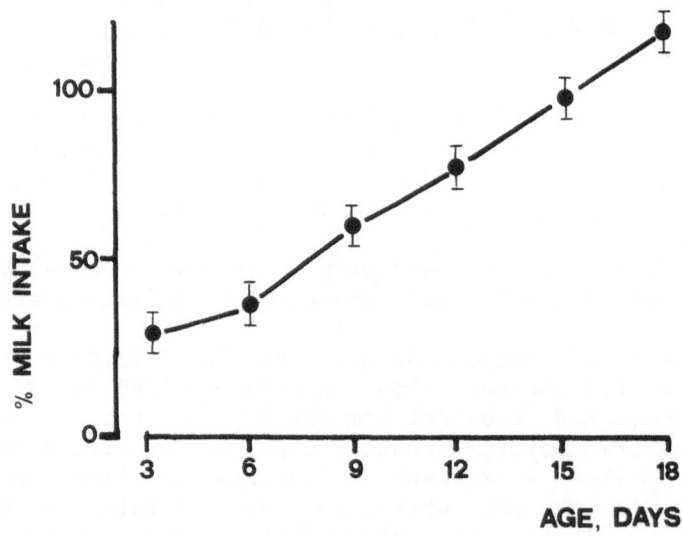

Fig. 4. Per cent milk intake in water satiated rats of 1 to 18
days following pICV of eledoisin, 100 ng per rat (milk
intake of controls: 100%).
Values are means of 6 to 8 data ± s.e.m.

whether the peptide inhibited milk intake in water-satiated rats. The animals were off dam for 3 hrs, then received water for 20 min and, after having been injected with eledoisin 100 ng per rat, were offered milk. In these experimental conditions eledoisin inhibited milk intake in the earliest stages of postnatal life, but its effect decreased with age and completely disappeared in 12-15 day old rats (fig. 4).

Similar results were obtained with physalaemin which produced a clear cut inhibition of milk intake in 6 day old rats, but did not affect it in 12 day olds.

DISCUSSION

The results of our experiments give us information about the ontogenesis of the antidipsogenic effect of eledoisin and physalaemin and offer suggestions about the biological significance of the effect that brain tachykinins elicit on water intake.

As far as the developmental calendar of the antidipsogenic effect is concerned, our data demonstrate that the effect of eledoisin is extremely precocious. In fact, the ingestion of water is inhibited as early as the third day of neonatal life, and in some instances a certain degree of thirst inhibition, even if not statistically significant, may be seen at the age of 1 day.

We do not have as much data with physalaemin, however the developmental calendar for this peptide seems to be similar to that of eledoisin.

In adult rats eledoisin and physalaemin show almost the same spectrum of effects on water intake, the main difference between the two peptides being the superior effectiveness of the former and the fact that the latter practically does not affect cell dehydration thirst. Our experiments confirm in baby rats both the different potency of the two peptides in inhibiting water intake and the different effectiveness they have on cell-dehydration thirst.

An important piece of information is that in the earliest stages of neonatal life eledoisin and physalaemin not only inhibit drinking (eledoisin suppresses cell-dehydration drinking, both suppress angiotensin-induced drinking) but inhibit also milk intake in water satiated rats. Ellis et al. (4) observed that in neonatal rats angiotensin II stimulates both water and milk intake, but that the animals drink more water than milk as early as 8 days. More or less the same thing happens for the inhibitory effect of eledoisin and physalaemin, which originally involves milk and water but in less than two weeks becomes selective for water. In other words, the anorexia produced by tachykinins somehow overlaps the orexia induced by angiotensin II, in that the two effects are both evident in the earliest stages of neonatal life, both progressively decrease with age and both completely disappear when rats are about 12 days old.

In our opinion, this confirms the specificity of the effect of tachykinins. In fact, these peptides inhibit fluid intake, without discriminating between milk and water, when rats fulfill their requirements for food and water by taking milk, but as the rat matures the peptides become selectively inhibitory for water and when the animal progresses from a fluid to a solid diet, they suppress only water intake.

In conclusion, our experiments demonstrate that in suckling rats the inhibitory effect of tachykinins is precocious, involves different kinds of thirst, according to the biological characteristic of the peptides, and is specific for water intake, since the inhibition of milk intake is limited to the earliest stages of neonatal life.

REFERENCES

1. L. H. Lazarus, R. Ilona Linnoila, O. Hernandez and R. P. Di-Augustine, A neuropeptide in mammalian tissues with physalaemin-like immunoreactivity, Nature 287:555 (1980).
2. A. J. Harmar, Three tachykinins in mammalian brain, Trends NeuroSci. 7:57 (1984).
3. G. de Caro, Effects of peptides of the "gut-brain-skin triangle" on drinking behaviour of rats and birds. This Symposium.
4. S. Ellis, K. Axt and A. N. Epstein, The arousal of ingestive behaviors by chemical injection into the brain of the sukling rat, J. Neurosci. 4:945 (1984).

Supported by NATO Research Grant N. 0502/82

INTEGRATIVE ROSTROMEDIAL DIENCEPHALIC NEURONS

ARE COMODULATED BY VASOPRESSIN AND ANGIOTENSIN

Anne Catherine Jeulin and Stylianos Nicolaïdis

Lab. de Neurobiologie des Régulations
CNRS, Collège de France
11 Place Marcelin Berthelot 75231 Paris C 05 France

INTRODUCTION

Although there is little doubt on the duality of the central system controlling drinking and diuresis there are old and new data indicating that both systems are intermingled and share common properties and common neuroendocrine factors. They are intermingled around the rostromedial osomreceptive structures located not only within the supraoptic nucleus (SON) and paraventricular nucleus (PVN)[1,2] but also in the surrounding region as shown by osmoactive microinjections[3,4,5,6] and electrophysiological techniques outside the above nuclei[7,8,9,10,11,12]. The involvement of rostral circumventricular organs (CVO) in triggering responses to hydromineral deficiencies was also recognised. Cells in the subfornical organ (SFO) were found to react to extracellular challenges[13,14] and to be very sensitive to the dipsogenic action of carbachol[15] and angiotensin II (AII)[16]. The responsiveness of neurons surrounding the organum vasculosum laminae terminalis (OVLT) to hypovolaemic challenges was first revealed in 1970 by extracellular unit recordings[17]. The same anteroventral periventricular area was shown to be highly sensitive to local application of AII, which triggered drinking more effectively than a similar application to the SFO[18,19]. The SFO seems to be more important for sensing blood-borne AII whereas the OVLT seems to be more concerned with detecting AII in the cerebrospinal fluid, but both structures initiate drinking, antidiuretic and blood pressure responses. There is considerable controversy at present on the respective roles of these two CVOs on drinking and other regulatory responses, and this is probably because both of them and a number of other intermediary structures form a functional unit, being widely interconnected by means of vascular as well as neural links[20,21]. Among the numerous rostral interconnected structures we find the classic PV and SO nuclei and also the nucleus medianus. The latter seems to play an important role in the hydromineral regulation since, in an auto-historadiographic study using labelled 2-DG to measure glucose uptake rate in dehydrated rats, it was found to increase its activity to a level comparable to that of the SFO and OVLT[22]. This finding has been substantiated recently by lesion work. Damage restricted to this nucleus was followed by total abolition of drinking responses to AII[23].

On the other hand one of the main properties of the rostro-median area was to be able to integrate information from various sensors concerning the hydromineral changes. The first integrational property to be described was that of specific gustatory messages of water or salt intake converging upon neurons which in addition are similarly responsive to NaCl hypertonic or hypotonic intracarotid pulse injection[7,8]. Neurons bearing integrative properties towards various messages of intracellular and/or extracellular dehydration were subsequently found to be more largely distributed along the rostro-medial diencephalic area[24] and to respond during spontaneous drinking when the unit recordings were made on unanesthetized animals[11,25,26]. We are now reporting that neurons of the same region sense vasopressin (AVP) and they do it in a way showing that they use this information concomitantly with AII for integrative purposes.

44

PEPTIDIC MODULATION OF NEURONAL ACTIVITY

Dealing with an area which is able to receive various messages
of hydromineral imbalances one could expect its neurons to be also
sensitive to hormones known to play an important role on the
corrective responses to the same hydromineral imbalances. Whatever
the various converging signals of the hydrational state of the body
are, they are accompanied by increases in secretion of hormones,
including ADH, AII, ACTH, aldosterone. Although they are
secreted in response to the prevailing hydrational imbalance, they,
in turn, modulate the neuronal responsiveness to the hydrational
imbalance. N.B. It is interesting to note that this is a positive
feedback mechanism. However, such a positive feedback action leads
finally to the achievement of homostasis by subsequently triggering
the negative feedback response of drinking and antidiuresis in the
case of dehydration.

This is particularly well documented in the case of AII.
Although AII increases in response to extracellular dehydration, it
also enhances the responsiveness of neurons concerned in drinking
at the level of the anterior third periventricular area (A3pV)
structures. As a result one would expect to find within this area
not only hormone-sensitive neurons but also neurons integrating
both hormonal and other information about the hydrational state
such as plasma osmolality. Nonetheless most studies have only been
devoted to the exploration of hormone-sensitive neurons.

A great number of investigations used iontophoretic applications of a single hormone in order to test neuronal responsiveness
[25,27,28,30,31,32,33]. As far as AVP is concerned, its effectiveness
on neurons was tested within the SON and PVN[27] and also outside
these specific structures[34]. In this study made with Thornton and
De Beaurepaire we found 18 neurons in the superior edge of the
OVLT and anterior portion of the medial preoptic nucleus, the
medial aspect of the diagonal band of Broca and the nucleus medianus

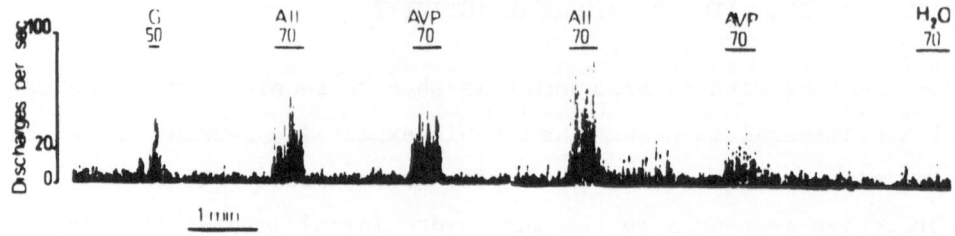

Fig. 1. Recording trace of cellular activity following iontophoretic
application (Io) of various solutions. The line underneath
each substance represents the length of lime the current
(in namp), indicated by the figure, was applied for. Whether
positive (+) or negative (-) current is also shown. G =
glutamate, AII = angiotensin II, AVP - vasopressin ,
H_2O = control.

which responded to AVP. Twelve of these neurons responded to AVP
and also to AII when it was applied separately (Fig. 1). The possibility that AVP and AII could comodulate neuronal activity has been
tested subsequently and preliminary data have been reported elsewhere[35].

PEPTIDIC COMODULATION OF NEURONAL ACTIVITY

An increasing number of morphological investigations has been
showing that many neurons are the site of colocalization of more
than one peptide or of one peptide and one monoamine[36,37]. Besides
the controversy on the role of these peptides, there is good
evidence supporting the idea that they both act on brain structures
in order to trigger drinking[18,38] although to our knowledge the
combined effect of AII and AVP on the brain has not been investigated.

In order to demonstrate that these two peptides can affect one
neuron's activity in an interdependent way, we used single unit
recordings and osmophoretic-iontophoretic (Io) ejections through
multibarrelled electrodes.

46

Methods

Male Wistar rats (250–300 g) were anaesthetized with urethane (1.3 g/kg body weight, i.p.) and the femoral artery and vein were exposed and catheterized. The rat was placed in the stereotaxic frame and the skull exposed. A tiple cannula system (3 x 30 gauge needles fixed together) was implanted in the right lateral cerebral ventricle (A = 3.5 mm ; L = 4.5 mm ; V cortex = 4.0 mm, head positioned according to Paxinos and Watson's atlas). After verifying that the position of the cannula tip was correct by otaining a rise in femoral artery pressure in response ICV injection of 800 pmol AII, the cannula was fixed in place with dental cement. An opening was made in both sides of the skull and dura over the region to be explored (around A = 7 to 8 mm ; L = 0, V cortex = 5 to 8.5 mm). The superior longitudinal sinus was gently displaced by a lever moved horizontally in a micromanipulator so that the flow of blood was not interrupted and cerebral oedema did not develop. The electrode was then introduced in the interhemispheric space or in a parasagittal vertical traject crossing the cortical layers at that level.

A seven barrel glass electrode was used for Io applications. Each barrel had a tip diameter of less than 1μm. The 7 pipettes were heated and twisted together before pulling. In order to record unit discharges an eighth glass pipette, bent at 45°, was sealed to the multibarrelled microelectrode so that its tip projected below the multibarrelled tips by 10 to 15 μm (as in Oomura et al.,[39]). The impedence of the recording electrode was 6 to 10 MΩ. The solutions filing the 7 barrels were : glutamate (1M Carlo Erba), AII (Asp-Arg-Val-Tyr-Ile-His-Pro-Phe) 1 mg/ml dissolved in distilled water and adjusted to pH 8 with NaOH (Biochem Corp), vehicle for AII (water adjusted to pH 8), AVP (Sigma 10 U/ml), AVP analogue (d(CH$_2$)$_5$ Tyr (Me) AVP 3 μg/ml-Manning) blocking specifically the V_1 effect of AVP, AVP analogue d(CH$_2$)$_5$ Leu VAVP (3μg/ml-Manning) blocking specifically the V_2 effect of AVP and finally the universal

vasoplegic agent sodium nitrite 1 M (Coop. Pharma. française).
These compounds were chosen not only for the present experiment but
also to another complementary investigation. The solutions were
either made up freshly on each experimental day or kept in the
freezer (- 30°C) in plastic microcontainers and unfrozen before
filing the micropipette on the day of the experiment. The impedance
of injecting electrodes was 90 to 180 MO.

For all substances, ejected currents from 40nA and up to 110nA
were used. No iontophoretic effect of a substance on neuron discharge was considered without testing the effect of the solvent (control)
at the same current. Other classical criteria of specificity of the
effect, as revealed by ejections or ICV injections of substances or
changes in blood pressure, were routinely applied[40].

Single unit discharges were recorded using a conventional
impedance change FET pre-amplifier followed by a digital integrator
(Hewlett Packard). Spike monitoring on an oscilloscope was always
used in order to avoid summation of activities of distant units,
particularly during blood pressure changes. Single units discharges
were selected by means of electronic gates and the digitally
integrated output was recorded on heat sensitive paper as shown on
the Figures.

Blood pressure was monitored through the femoral artery
catheter PE3 (Biotrol) using a calibrated transducer (Sanei).

No experimental data were used without anatomical identification of the recording tip, which was carried out by iontophoretic
ejections of pontamine made through the recording electrode
subsequent to recording. The brain was removed after perfusion of
20 ml of 10 % formalin, preceded by saline to wash the blood away,
through a ventriculo-aortic cannula. The 30μm serial slices were
cut from the frozen brain and stained with a 30 % glycerol solution.
This produced a sharp contrast between somato- and fibre-dominated
areas and left the sky blue marking spot produced by the pontamine
clearly apparent.

Results

Two hundred and eighteen neurons have been tested in 49 rats. They are located along a rostral area between 7.2 and 8.7 mm anterior to the interaural line and extended also to units from the cortex used as a control area as compared to deeper structures supposed to be more implicated in sensing AII and AVP. As shown previously[33], AII responsive neurons were found in various locations including the cortex ; only the proportions varied, confirming our previous report.

The main goal of this experiment was to investigate hormonal comodulations of neuronal activity. More specifically the question : Do AVP and AII coact on one neurons'activity and if they do so, do they coact in an additive, substractive or another way ?

From the total number of 218 neurons, 54 % responded either to AVP or to AII. Forty % neurons responded to both of these hormones when they were applied simultaneously. Hormonal coaction was of three different types : a) The most usually observed was additive (34 %)(Fig. 2).

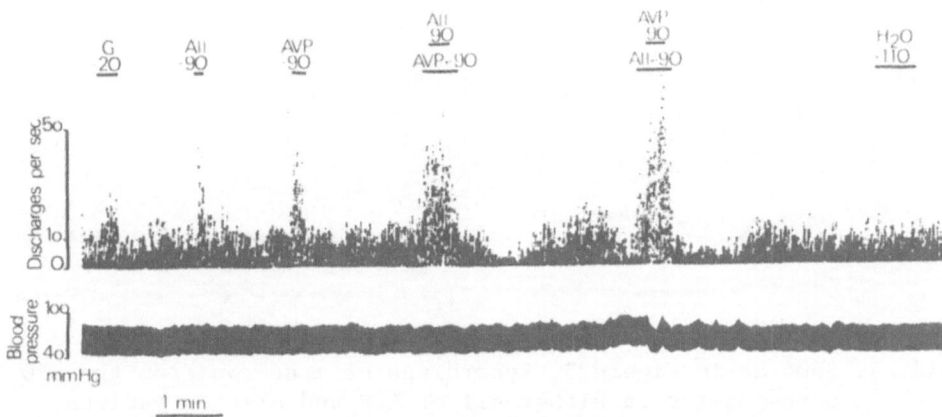

Fig. 2. One second integrated unit discharges from one neuron (glutamate (G) responsive) localized in the medial septal nucleus. This neuron responded to either or both AII and AVP. When the electrophoresis of one of these peptides was superimposed on the other the response was additive (double or less). Equal or stronger application of the vehicle (H_2O) was unresponsive.

Fig. 3. One second integrated unit discharges of one neuron
(glutamate (G) responsive) from the corticohypothalamic
tractus responding to either AII or AVP. When AVP was
applied during AII's action the effect was substractive.

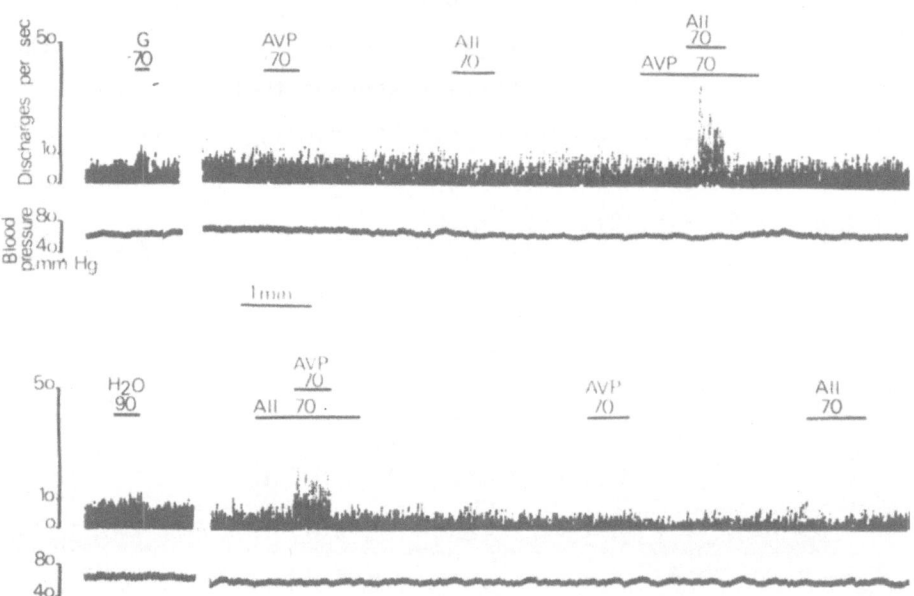

Fig. 4. Same as in Figure 3. Recordings of a neuron from the SFO
unresponsive to either AII or AVP and also to various
currents through the vehicle. Only when either AVP or AII
were applied during the other peptides'application was a
clearcut response observed. This response was decreased as
soon as the second molecules' application was discontinued.
Blood pressure was added in this case to show that accidental
fluctuations did not affect the neuronal activity. This is
not a barosensitive neuron.

The additivity was not always a simple addition but had either
more or less effect than doubling the individual effects. b) More
seldom the effect was substractive (3%)(Fig. 3). Finally seven units
(3%) responded to AVP and to AII only if they were applied simultaneously. Repeated application of only one of the coacting peptides
was ineffective (Fig. 4) as if one of these peptides was gating the
effect of the other.

The usual tests for specificity of the observed effects were
made. In particular the vehicle was always applied and the effect
had to be reproducible. Whenever the unit could be observed for a
long enough time the effect of increasing doses of current was also
tested.

The above coactive effects were observed along a sagittal area,
although their occurrence decreased when neurons were located within
areas more lateral than 0.5 mm. Their dorso-ventral localization
extended from the SFO to the lower part of the area within the suprachiasmatic nucleus. The anatomical structures in which these units
were recorded included : for additive coaction, medial preoptic area,
periventricular hypothalamic nucleus, medial corticohypothalamic
tractus and SFO ; the substractive effect : medial corticohypothalamic nucleus, paraventricular thalamic nucleus ; for gating effect :
SFO, periventricular hypothalamic nucleus.

Two peculiar observations need a special reference. a) the
effect of combined application of AVP and AII can be both delayed
and prolonged as shown in Fig. 5 from a neuron located within the
limit between the medial preoptic area and the bed nucleus laminae
terminalis. b) three of the neurons reported here were located
within the OVLT. This is the first time we have been able to record
from this structure as we can assess it on the basis of precise
histological localization of the recording tip shown in Fig. 6. In
addition to the marking point we show the next, more rostral
section where the OVLT appears in its ascendant part. In a previous
study we could assess we have been able to record from the OVLT

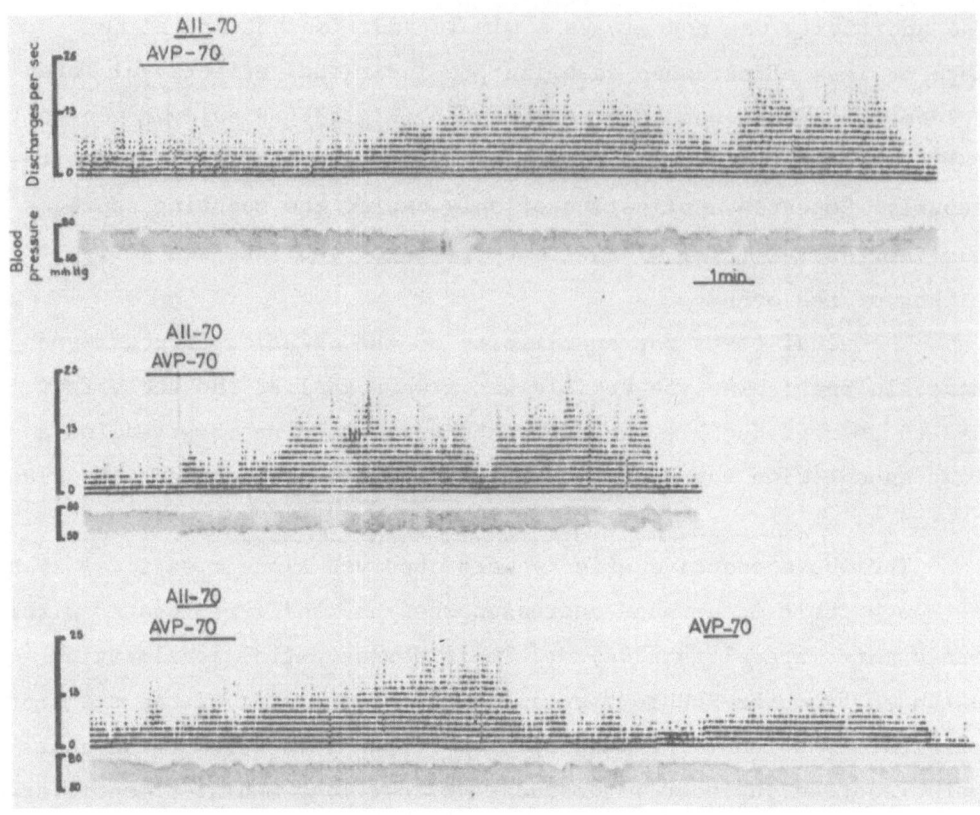

Fig. 5. Neuronal activity in response to repeated applications of
both AVP and AII followed by a delayed and prolonged
activation. Both the delay and the timecourse where
shortened when the Io applications have been reported. AVP
alone was less effective than in association with AII. This
neuron was located within the limit between the medial
preoptic area and the bed nucleus laminae terminalis.

itself only in two cases without precise histological demonstration
which also misses in other previous studies. All of the three
neurons recorded from the OVLT were responsive to AII but not AVP.
The concomitant application of these two hormones had no effect on
these three neurons.

52

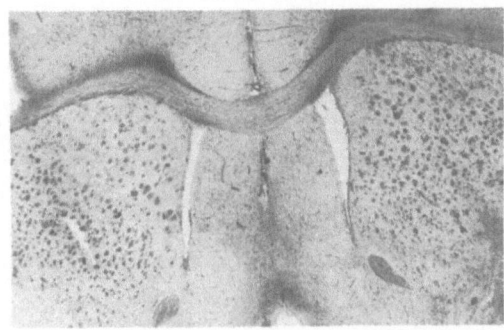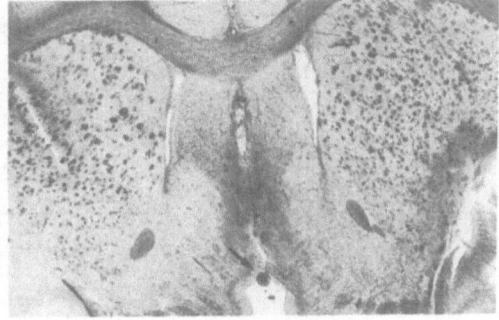

Fig. 6. The arrow shows the pontamine sky blue mark made electrophoretically after the conclusion of the recording
experiment. These neurons are located within the OVLT.

DISCUSSION

These data confirm an earlier report that in the sagittal
diencephalon close to the anterior wall of the third ventricle one
single neuron may have its activity comodulated by two peptides.
This microphenomenon at the neuronal level might be the expression
of a more general property of the CNS. This is the property which
allows convergence of various regulatory signals on single neurons,
or on a small number of interconnected neural closters, in order
to add, substract or combine differently these signals before
processing an integrated response towards the efferent circuit. In
other words, the type of regulatory response (drinking, antidiuresis, blood pressure adjustment, etc) may take into account each
deficiency (additive coaction of AVP and AII) and also the coeffi-
cient of its relative importance (which could be expressed by the
non additive comodulations of neuronal activity by AVP and AII).
The fact that the phenomenon of neuronal comodulation by these
hydromineral deficiencies bound hormones was restricted to an area
surrounding the rostral limit of the axial ventricular system favors

the idea that it plays an integrative role. In the same area , neurons were shown to receive osmotic messages from blood stream and the mouth[41] or messages on the osmotic, volaemic and blood pressure deficiencies[24]. With the additional integrative properties towards hormonal composition it is apparent that the rostral area around the embryonic laminae terminalis is one of the richest centers of information of hydroionic state including complementary thermal as well as photic (opto-hypothalamic) messages, all known to affect regulatory responses. This idea fully expressed in 1971 [42] is being increasingly substantiated not only from data showing convergence of afferent signals but also because various manipulations of the anterior wall of the third ventricle result in efferent meaningful responses. Lesions and chemical or electrical stimulations at various levels result in modification of drinking or sodium appetite or of diuresis or of haemodynamic parameters[43], [44,45]. Such a multi target regulator could hardly be indifferent to the main regulatory hormones comodulating its responsiveness as it is now shown.

ACKNOWLEDGEMENTS

This work was partially supported by La Fondation pour la Recherche Médicale, Evian Co. and Sandoz Co. The authors thank M.J. Meile for histological assistance and J. Asmanis for secretarial assistance.

REFERENCES

1.   B.A. Cross and J.D. Green, Activity of single neurons in the hypothalamus effect of osmotic and other stimuli. J. Physiol. (London), 148, 554-569 (1959).
2.   E.B. Verney, The antidiuretic hormone and the facotrs which determine its release. Proc. Roy. Soc., 135B, 26-106 (1947).
3.   B. Andersson, Polydipsia caused by intrahypothalamic injection of hypertonic NaCl solution. Experientia, 8, 157-158 (1952).
4.   B. Andersson and S.M. McCann, A further study of polydipsia evoked by hypothalamic stimulation in the goat, Acta Physiol. Scand., 333-346 (1955).
5.   E.M. Blass and A.N. Epstein, A lateral preoptic osmosensitive zone for thirst in the rat. J. Comp. Physiol. Psychol., 76, 378-394 (1971).
6.   J.W. Peck and D. Novin, Evidence that osmoreceptors mediating drinking in rabbits are in the lateral preoptic area. J. Comp. Physiol. Psychol., 74, 134-147 (1971).

54

7. S. Nicolaïdis, Réponses des unités osmosensibles hypothalamiques aux stimulations aqueuses et salines de la langue. C.R.Acad. Sci. Paris, 267, 2352-2355 (1968).

8. S. Nicolaïdis, Discriminatory responses of hypothalamic osmosensitive units to gustatory stimulation in cats. Olfaction and Taste, vol. 3, p. 569-573, Rockefeller Univ. Press, N.Y., (1969).

9. R.B. Malmo and W.J. Munol, Osmosensitive neurons in the rat's preoptic area : medial-lateral comparison. J. Comp. Physiol. Psychol., 58, 161-175 (1975).

10. G.I. Hatton, Nucleus circularis : is it an osmoreceptor in the brain ? Brain Res. Bull., 1, 123-131 (1976).

11. M. Sessler and M.D. Salhi, Interaction of hypertonic NaCl and neural stimuli on lateral preoptic neurons. Neurosci. Lett., 26, 319-324 (1981).

12. J.N. Hayward and J.D. Vincent, Osmosensitive single neurons in the hypothalamus of unanesthetized monkeys. J. Physiol. (London), 210, 947-992 (1970).

13. H. Legait, Etude caryométrique de la pars intermedia de l'hypophyse et de l'organe subfornical au cours des états d'hyperactivité de l'hypothalamus neurosécrétoire chez quelques mammifères, C.R.Soc.Biol. (Paris), 156, 1662 (1962).

14. M. Palkovits, L. Zaborszky and P. Magyar, Volume receptors in the diencephalon, Acta Morphol. Ac. Sci. Hung., 16, 391-401 (1968).

15. J.B. Simpson and A. Routtenberg, The subfornical organ and carbachol induced drinking, Brain Res., 45, 135-152 (1972).

16. J.B. Simpson and A. Routtenberg, Subfornical organ : site of drinking elicitation by angiotensin II. Science, 181, 1172-1175 (1973).

17. S. Nicolaïdis, Mise en évidence de neurones barosensibles hypothalamiques antérieur et médian chez le chat, J. Physiol. (Paris), 62, 199-200 (1970).

18. S. Nicolaïdis and J.T. Fitzsimons, La dépendance de la prise d'eau induite par l'angiotensine II envers la fonction vasomotrice cérébrale locale chez le rat. C.R.Acad.Sci., 281D, 1417-1420 (1975).

19. J. Buggy, A.E. Fisher, W.E. Hoffman, A.K. Johnson and M.I. Phillips, Ventricular obstruction : effect on drinking induced by intracranial injection of Angiotensin. Science, 190, 72-74 (1975).

20. H. Duvernoy and J.G. Koritke, Die Gefässversorgung der Lamina Terminalis bei einigen Vögeln. Verh. Anat. Ges., 112, Suppl. 391-404 (1963).

21. R.R. Miselis, The efferent projections of the subfornical organ of the rat. A circumventricular organ within a neural network subversing water balance. Brain Res., 230, 1-23 (1981).

22. S. Nicolaïdis, M. Le Poncin-Lafitte, J. Danguir, C. Grosdemouge and J.R. Rapin, Specific behaviors bound brain cartography of the glucose uptake rate. Eur. Neurol., 20, 180-182 (1981).
23. A.K. Johnson, Periventricular structures of the lamina terminalis: their role in angiotensin induced thirst. In: Abstracts Evian Symposium, Body Fluid Homeostasis, p. 23(Ed. S.Nicolaïdis, J.T. Fitzsimons), (1983).
24. S. Nicolaïdis, Réponses unitaires dans les aires antérieures et médianes de l'hypothalamus antérieur associées à des variations de pression artérielle et de volémie, C.R. Acad. Paris, 270, 839-842 (1970).
25. J.D. Vincent, E. Arnauld and B. Bioulac, Activity of osmosensitive single cells in the hypothalamus of the behaving monkey during drinking, Brain Res., 44, 371-384 (1972).
26. E.T. Rolls, Neurophysiology og feeding, Dalhem Workshop on Appetite and Food Intake, Berlin 1975, T.6, Silverstone Ed., Dalhem Univ. Press, 21-42 (1976).
27. R.A. Nicoll and J.L. Barker, Excitation of supraoptic neurosecretory by Angiotensin II, Nature New Biol., 233, 172-174 (1971).
28. M.J. Wayner, T. Ono and D. Nolley, Effects of angiotensin applied electrophoretically on lateral hypothalamic neurons, Pharmacol. Biochem. Behav, 1, 223-226 (1973).
29. D. Felix and K. Akert, The effect of angiotensin II on neurons of the cat subfornical organ. Brain Res., 76, 350-353 (1974).
30. M.I. Phillips and D. Felix, Specific angiotensin II receptive neurons in the cat subfornical organ, Brain Res., 109, 531-540 (1976).
31. P. Buranarugsa and J.I. Hubbard, The neuronal organization of the rat subfornical organ in vitro and a test of the osmo- and morphine-receptor hypotheses, J. Physiol. (London), 291, 101-116 (1979).
32. S. Nicolaïdis, S. Ishibashi, B. Gueguen, S.N. Thornton and R. de Beaurepaire, Iontophoretic investigation of identified SFO angiotensin responsive neurons firing in relation to blood pressure changes, Brain Res. Bull., 10(3), 357-363 (1983).
33. S. Ishibashi, Y. Oomura, B. Gueguen and S. Nicolaïdis, Neuronal Responses in Subfornical Organ and other regions to Angiotensin II applied by various routes, Brain Res. Bull., 14(4) (1985).
34. S.N. Thornton, A.C. Jeulin, R. de Beaurepaire and S. Nicolaïdis, Iontophoretic application of Angiotensin II, Vasopressin and Oxytocin in the region of the anterior hypothalamus in the rat, Brain Res. Bull., 14(3) (1985).
35. S. Nicolaïdis and A.C. Jeulin, Converging projections of hydromineral imbalances, J. Physiol. (Paris), vol. 79, n°6, 406-415 (1984).

36. T. Hökfelt, J.M. Lunberg, M. Schultzberg, O. Johansson, A. Ljundahl and J. Rehfeld, Coexistence of peptides and putative transmitters in neurons, In : Costa E., Trabucchi M., (eds), Neural peptides and neuronal communication, Raven Press, N.Y., p. 23 (1980).

37. V. Chan-Palay, S. Palay, Coexistence of neuroactive substances in neurons, 432, Wiley and Sons publishers (1984).

38. A.N. Epstein, J.T. Fitzsimons and B.J. Rolls, Drinking caused by intracranial injection of angiotensin into the rat, J. Physiol. (London), 200, 98-100 (1969).

39. Y. Oomura, A. Ooyama, M. Sugimari, K. Yoneda and A. Simpson, Constant current device for drug application studies in the CNS, Physiol. Behav., 16, 799-802 (1976).

40. D.R. Curtis and K. Koizumi, Chemical transmitter substances in the brain stem of the cat, J. Neurophysiol., 24, 80-90 (1961).

41. S. Nicolaïdis, Int. Conf. on "Neural Regulation of food and water intake", N.Y., 1967, Ann. N.Y. Acad. Sci., 157, 1176-1203 (1969).

42. S. Nicolaïdis, Réflexe poto-hidrotique et étude de son mécanisme neurophysiologique, J. Physiol. (Paris), 63, 359-361 (1971).

43. M.L. Mangiapane and J.B. Simpson, Subfornical organ : site of pressor and drinking effects of angiotensin II. Soc. Neurosci. Abstr., 3, 351 (1977).

44. S. Ishibashi and S. Nicolaïdis, Hypertension induced by electrical stimulation of the subfornical organ (SFO), Brain Res. Bull., 6(2), 135-139 (1981).

45. A.K. Johnson, Neurobiology of the periventricular tissue surrounding the anteroventral third ventricle AVEV and its role in behavior fluid balance and cardiovascular control. In : Circulation, Neurobiology and Behavior, Ed. Smith D.A., Galosy R.A., Weiss S.M. Elsevier Press, 277-295 (1982).

THYROID ACTIVATION FOLLOWING DEHYDRATION

L. Di Bella and M.T. Rossi

Istituto di Fisiologia Umana
Cattedra di Fisiologia Generale
Via G.Campi 287, 41100 Modena

INTRODUCTION

ß-adrenergic stimulation can stimulate drinking (1),partly by activating the renal renin-angiotensin system (2,3,4,5),partly by activating the vascular receptors just as the systemic blood pressure starts to fall (6).

Thyroid activity can be controlled by the sympathetic nervous system (7,8) directly,since in addition to adrenergic nerve terminals in thyroid blood vessels in mice,nerve endings have also been seen in close apposition to the basal side of the follicular cells of the thyroid (9,10). It is unlikely that sympathetic effects on thyroid activity are secondary to changes in thyroid blood flow; indeed aromatic monoamines can stimulate the synthesis of thyroid hormones by acting directly on α-adrenergic receptors in the follicle cells (11,12,13). Cellular dehydration and extracellular fluid depletion, both drinking stimuli,probably activate thyroid secretion. The experiments we have set up tend to give some support to this idea.

MATERIAL AND METHODS

Male adult Wistar rats were housed in separate steel cages,kept at the ambient temperature of 21.5 ± 0.5 °C and at the L:D schedule of 12:12 (L:8 a.m.),fed a complete,balanced and dry powder diet offered ad libitum in a special glass container. The rats,food and water intake were weighed (scale sensitivity to 0.1 g) twice a day (8 a.m. and 8 p.m.). The dehydration was total (no drinking water) for 3.5

days,and at the end the following data were sought: T4,TSH,FT4,TBK
with ELISA methods,osmolarity of blood serum,thyroid weights. The
thyroids were fixed by immersion in neutral formaldehyde and stained
with PAS-hematoxylin.

Three groups,each of 4 rats,were partially dehydrated,by offering at 8 a.m. and 8 p.m. a quantity of drinking water corresponding
to 20%,40%,80% the average amount they had drunk in the preceding
days,separately during light and dark time,as follows:

Table 1: Water amount (ml) corresponding to 20%,40%,80% the normal
mean

		1st Group:20%				2nd Group:40%				3rd Group:80%			
Rat		1	2	3	4	5	6	7	8	9	10	11	12	
Light		4.0	1.8	1.8	2.3	3.8	1.8	5.6	3.6	13.1	11.4	7.1	9.8	
Dark		5.7	6.0	4.4	5.0	11.1	6.0	10.5	12.7	26.1	26.3	25.0	19.4	

The partially dehydrated rats drank in a few minutes the measured amounts of water,and carefully licked the walls of the container
at the end; they however behaved apparently normally.

A group of 4 rats was injected daily i.p. with Methymazole, a
known thyroid inhibitor (ml 0.25/100 g b.w. of a solution of mg 100
Methymazole in 100 ml physiological saline).

A group of 4 rats was thyroidectomized,and a control group of
5 rats was sham-thyroidectomized.

RESULTS

a) Partial dehydration: At the end of the period of 17 days of
partial dehydration the average b.w. changes were:

Table 2: Δ b.w. in the course of 17 days of partial dehydration

Group		%water		during	before	
1		20%		-2.536 ± 2.552	6.454 ± 5.462	
2		40%		0.656 ± 3.134	6.637 ± 5.008	
3		80%		3.102 ± 3.352	6.130 ± 4.818	

The rats grew more in proportion to the available drinking
water. The reduced bodily growth cannot be ascribed to a smaller
food intake,which was on the contrary lower in the 3rd group that
received a higher proportion (80%) of the average normal water,than
groups 1 (20%) and 2 (40%).

Table 3: Daily food intake (g) by partially dehydrated rats

Group	%water	Daily food intake (g)	
		Partially dehydrated	No dehydration
1	20%	47.1 ± 14.1	46.0 ± 17.3
2	40%	50.1 ± 22.1	50.5 ± 15.4
3	80%	33.3 ± 10.0	43.5 ± 15.7

It is therefore likely that a partial dehydration implies a more severe self consumption ("autocannibalic",Di Bella et Al.,14) whose purpose resides in developing a higher intracellular production of water,through an exaggeration of cellular oxidations.

This metabolic increase that follows partial dehydration can be exploited in order to combat obesity in some people.

b) Total dehydration: This was accomplised by taking away the drinking water for 3.5 consecutive days. The experiments were performed: 1°) on rats whose thyroid was functionally depressed by i.p. injection of Methymazole (Meth) and in control rats,daily i.p. injected with the same volume of physiological saline (C); 2°) on Thyroidectomized (Th) and Sham-Thyroidectomized (S-Th) rats.

Table 4: Δ b.w. and daily food intake in completely dehydrated rats

Rats	Δ b.w.	g food intake		
Meth	-9.6 ± 7.76	19.1 ± 15.7	*	* $0.01 < P < 0.05$
C	-10.0 ± 7.43	8.1 ± 6.2	*	
Th	-8.6 ± 3.07	18.7 ± 11.1	§	§ $0.05 < P < 0.10$
S-Th	-10.1 ± 3.82	30.2 ± 18.5	§	

B.w. changes are more negative in totally than in partially dehydrated rats. As concerns the food intake,there is an apparent discrepancy between Meth and S-Th,in as much as Meth eat a significantly higher ($0.01 < P < 0.05$) food amount than C,whereas Th eat a less significant ($0.05 < P < 0.1$) amount of food than S-Th. The discrepancy may depend upon the significantly higher levels of blood serum TSH in Th than in Meth; however an injury to the sympathetic fibers that innervate the thyroid in S-Th cannot be ruled out.

c) Thyroid weight: The thyroid glands were weighed as soon as excised (scale sensitivity to 0.01 mg).

Table 5: Thyroid weights (mg: M ± SD):

	Meth	C	20%	40%	80%
weight	27.26± 1.9	22.99± 2.7	32.00±10.5	34.50± 9.9	25.98± 6.1

Meth thyroids weigh significantly more (0.01< P <0.05) than C thyroids,whereas the difference between the weights of thyroids of partially dehydrated rats are hardly significant. In two groups of 4 and 3 rats which had received by mouth 100 and 50 µg daily of T4 (Na-salt) for 17 days the thyroids weighed mg 16.9±14.5 and 27.6±7.5. The difference was not significant.

    d) <u>Thyroidal and thyreotropic hormones</u>: The dosages were accomplished with ELISA method on blood serum,after centrifugation of blood clotting,decanting and freezing.

Table 6: Blood serum hormones:

Rats	T4(pg/ml)	TSH(µIU/ml)	FT4%	TBK%
Meth	16.00±0.80	1.47±0.21	13.36±1.41	0.83±0.07
C	14.37±1.57	1.56±0.30	12.22±1.06	0.85±0.10
Th	15.56±2.05	6.15±3.30	13.85±1.55	0.89±0.06
S-Th	14.32±1.87	0.90±0.24	12.76±1.91	0.90±0.16
20%	10.37±2.40	0.50±0.08	9.76±2.30	0.94±0.08
40%	10.43±2.60	0.52±0.12	10.99±3.35	1.04±0.09
80%	6.92±2.50	0.72±0.22	9.33±3.40	1.48±0.20

    e) <u>Osmolarity</u>:

Table 7: Osmolarity of Blood serum (mOsm):

	Meth	C	20%	40%	80%
mOsm	263.0± 4.3	277.4±32.77	316.2± 9.7	312.7± 9.9	306.2± 3.4

Osmolarity shows a tendency to decrease,but not significantly, as soon as the available drinking water rises from 20 to 40,to 80%.
    Bleeding of partially dehydrated rats was difficult apparently owing to the high viscosity.
    The difference between osmolarity of Meth and C was not significant,but the difference between the haematocrits was highly significant: against a haematocrit of 49.2±2.17 of Meth,a haematocrit of 54.3±2.08 was found (0.001< P <0.01) of C.

CONCLUSIONS

Total dehydration is perhaps too serious a stress. It deranges many functions as well as those relative to thirst and water turnover. Partial dehydration on the contrary provides a conveniently long period of time during which many behavioural and analytical procedures can progressively be performed on apparently healthy animals.

Partially dehydrated rats behave from several point of view as if they suffered hyperthyroidism. This condition is probably unleashed by peripheral or central osmoreceptors working through the hypothalamic centers, transduced both to adenohypophyseal thyreotropes with resulting higher excretion of TSH, and to the sympathetic adrenergic system which is activated. The transient thyroid hyperfunction stimulates intracellular oxidation processes, and an overproduction of water ensues that repairs the intracellular dehydration.

Many questions must be solved along this line.

REFERENCES

1. D.Lehr,J.Mallow and M.Krukowski,Copious drinking and simultaneous inhibition of urine elicited by ß-adrenergic stimulation and contrary effect of α-adrenergic stimulation,J.Pharm.Exp. Ther.,158:150 (1967).
2. A.K.Johnson,J.F.E.Mann,W.Rascher,J.K.Johnson and D.Ganten,Plasma angiotensin II concentrations and experimentally induced thirst,Amer.J.Physiol.,240:R229 (1981).
3. K.A.Houpt and A.N.Epstein,The complete dependence of ß-adrenergic drinking on the renal dipsogen,Physiol.Behav.,7:897 (1971).
4. B.J.Rolls and D.J.Ramsay,The elevation of endogenous angiotensin and thirst in the dog,in: "Control mechanisms of drinking", G.Peters,J.T.Fitzsimons and L.Peters-Haefeli Eds,Springer V., Berlin (1975).
5. E.Chiaraviglio,Drinking behaviour in rats treated with isoprenaline, angiotensin II or angiotensin antagonists, J.Physiol., (London),296:193 (1979).
6. J.A.Hosutt,N.Rowland and E.M.Stricker,Hypotension and thirst in rats after isoproterenol treatment,Physiol.Behav.,21:593 (1978).
7. T.S.Harrison,Adrenal medullary and thyroid relationships,Physiol. Rev.,44:161 (1964).
8. G.G.Waldstein,Thyroid-catecholamine interrelations,Ann.Rev.Med., 17:123 (1966).
9. L.W.Tice and C.R.Creveling,Electron microscopic identification

of adrenergic nerve endings on thyroid epithelial cells,
Endocrinology,97:1123 (1975).

10. A.Melander,L.E.Ericson,F.Sundler and S.H.Ingbar,Sympathetic in
    nervation of the mouse thyroid and its significance in thyroid
    hormone secretion,Endocrinology,94:959 (1974).

11. A.Melander,Thyroid hormone secretion,Acta Phisiol.Scand.,Suppl.
    370:31 (1971).

12. A.Melander,F.Sundler and U.Westgren,Intrathyroidal amines and
    the synthesis of thyroid hormones,Endocrinology,83:193 (1973).

13. A.Melander,L.E.Ericson and F.Sundler,Sympathetic regulation of
    thyroid hormone secretion,Life Sci.,14:237 (1974).

14. L.Di Bella,M.T.Rossi and G.Scalera,A contribution to a correla
    tion between drinking and feeding behaviour,8th Internat.Conf.
    on the Physiology of Food and Fluid Intake,IUPS,Melbourne,1983.

INFLUENCE OF THYROID ON WATER AND FOOD INTAKE

M.T. Rossi and L. Di Bella

Istituto di Fisiologia Umana
Cattedra di Fisiologia Generale
Via G. Campi 287, 41100 Modena

INTRODUCTION

Thirst is a subjective sensation (1) which is aroused by peripheral (2,3,4) or central receptors (5-18),is afferented by several vegetative nervous fibers (19-22),integrated by several nervous and neuroendocrine systems (5,10,13),controlled and modulated by various physiological conditions (23-30). In such a prominent function as the preservation of a constant body water amount and turnover,thirst has a predominant but not exclusive role. The intensity and kind of catabolism can repair the noxious effects of a mild dehydration (31, 32). It is moreover possible that even the initiation or termination of drinking are associated with miniature metabolic changes. We therefore investigated what functional deviations the thyroid suffers when various degrees of dehydration are experimentally brought about in the rat.

MATERIAL AND METHODS

Male adult Wistar rats were housed in separate steel cages,kept at the ambient temperature of $21.5 \pm 0.5°C$,and at the L:D schedule of 12:12 (L:8 a.m.). The diet had a white powder consistency; was complete and balanced (Casein 47.88%; Maize starch 35.91%; Butter 4.79%; Pig fat 7.18%; Salt mixture 3.83%; Vitamins 0.41%; 1 g diet developped 21.1786 J); it was offered ad libitum in a special glass container. The rats and the amounts of deionized water and food intake were weighed at least 0.1 g twice a day (8 a.m. and 8 p.m.).

Methymazole (1-methyl-2-thyioimidazol; a well known thyroid inhibitor) was dissolved in physiological saline; mg 100/100 ml; the solution was held at 0°C and injected i.p. at 8 p.m. at the dose of ml 0.25/100 g b.w.. The control rats received the same volume of physiological saline i.p..

The thyroid was extirpated according to the technique of Bomskov (33); a high death ratio occurs owing to the irritation of the inferior laryngeal nerve,which produces laryngeal spasm and dyspnea. The Sham-Thyroidectomized rats underwent identical operation with the exception that the thyroid was not removed. The thyroids were fixed by immersion in neutral formol and stained with PAS-Haematoxylin.

T4 (Merck;Na-salt) was administered in the late afternoon (8 p. m.) by mouth,after incorporation into a tiny scrap of pig fat,sweetened with some crystal saccharose.

RESULTS AND DISCUSSION

a) Methymazole treatment: Table 1: Effect of Methymazole treatment on daily b.w. changes (g),food and water intake (g): M±SD

Table 1.

		body weight			food intake			water intake	
Rats	before	during		before	during		before	during
C	0.21±7.91	2.93±6.56		37.8±18.3	29.0±9.9		21.8±6.8	27.7±7.8
Meth	1.02±8.91	1.83±7.02		33.7±18.8	36.3±16.4		24.6±6.8	23.9±7.3

Methymazole treatment (Meth) does not change significantly either b. w.,or food and water intake of the same rats before treatment; on the contrary,the control rats (C) grow significantly more,eat significantly less and drink significantly higher water amounts after treatment with saline. This peculiar behaviour pattern of C may depend upon the NaCl injected (mg 2.25/100 g b.w.). However since equal amounts of NaCl were i.p. injected also to Meth which suffered the same stress,it can be inferred that a depression of thyroid function stiffens the metabolic functions of the cells,so that the reactions to stress or NaCl injection no longer occur and a tendency appears to kept both food and water intake at a constant level. The thyroid gland of Meth weighed mg 27.2±1.9,against 22.9±2.7 of C. The difference is signicant (0.01<P<0.05),and the higher weight of Meth thyroids can be assigned to expansion of follicle colloid.

b) Influence of thyroidectomy: Table 2: Effect of thyroidectomy on daily b.w. changes (g),food and water intake: M±SD

Table 2.

| | body weight | | | food intake | | | water intake | | |
Rats	before	during		before	during		before	during	
S-Th	3.17+6.56	1.40+5.40		35.4+10.6	34.9+13.0		30.1+7.6	25.3+6.9	
Th	2.73+5.9	1.5 +6.6		31.1+11.0	28.8+15.9		26.9+7.6	28.2+7.6	

The measures pertinent to both thyroidectomized (Th) and Sham-thyroidectomized (S-Th) were taken during 25 days preceding surgical intervention and from the 8th day after the operation,for 17 days. In
the same manner as Meth,Th,too,showed small tendency to change their
Δ b.w.,as well as their food and water intake amount after thyroid
removal; S-Th on the contrary underwent a significant reduction both
of Δ b.w. (0.01<P<0.05) and of water intake (P<0.001) after intervention. Food intake is significantly higher in S-Th than in Th,both
before (0.01>P>0.001) and after (0.01>P>0.001) intervention,whereas
water intake is significantly higher in S-Th than in Th before (0.01
>P>0.001),and significantly lower in S-Th than in Th (0.05>P>0.01)
after intervention. The relative discrepancy in food and fluid intake behavior in Meth as compared to Th is probably conditioned by
several factors,by the significantly higher levels of serum TSH in
Th than in Meth,as we shall show in a late article.

c) <u>Effects of T4 administration</u>: 50 and 100 µg/day T4 (Na-salt)
were administred by mouth to two groups of 4 and 3 rats during 17
consecutive days. The results were compared with the corresponding
measurements which were acquired during 15 days preceding the start
of T4 administration. Table 3: Daily Δ b.w.,food and water intake
(g) before (15 days) and after (17 days) the start of T4 administration.

Table 3.

| | body weight | | | food intake | | | water intake | | |
Rats	before	during		before	during		before	during	
T 50	6.8+4.1	2.4+2.6		53.4+14.0	45.0+11.6		34.5+4.8	33.1+ 4.6	
T100	6.8+3.0	1.7+3.1		49.3+16.9	43.1+18.4		39.6+9.3	41.5+17.9	

The relatively young rats showed a significant decrease of daily Δ
b.w.,independently of whether the daily T4 dose was 50 or 100 µg.
Food intake decreases significantly almost identically (P<0.001) during treatment with daily doses both of 50 and 100 µg T4; but no significant difference,either before or during the treatment,exists between the two groups of rats. The rats which received 100 µg T4
drank significantly higher amounts of water than rats which received

only 50 μg/day; but the difference between the two groups was as significant before as during the administration of T4.
The conclusion can be drawn that the administration of T4 does not significantly change △ b.w.,or food or water intake.

CONCLUSIONS

Thyroid seems to partake of the many mechanisms that control fluid and food intake. Its role is however different from that of the various factors which take part in such a complex and integrated function,in as much as it attends to the production of water,inside the cells,through regulation of the catabolism,and therefore influences the intracellular water content. If sufficient water was produced so as to satisfy water losses,then probably no thirst would arise. If combustion water is less than lost water,then dehydration occurs,and thirst will arise,thus replenishing the declining water reserve. Thyroid function can be controlled both by hypothalamic--hypophyseal axis,and by sympathetic adrenergic system (34-38). Thyroid activity can therefore validly support water metabolism through the catabolic intracellular production of water. The stimulation of thyroid secretion may arise from a dehydration state,by way both of the hypothalamic-hypophyseal axis,and the medial forebrain bundle and the sympathetic adrenergic system. When the tyroid mechanism adapts itself perfectly to the manifold peripheral and central drives,and water production covers water output,then thirst will not appear,and life may proceed for a long time without drinking.
Both hypothalamic-hypophyseal thyroid axis,and the sympathetic adrenergic-thyroid system probably play a relevant role in the intracellular water supply and therefore indirectly take an important place among the mechanisms of drinking.

REFERENCES

1. B.J.Rolls,"Thirst",Cambridge Univ.Press (1982).
2. J.T.Fitzsimons and M.J.Moore,Short-latency,graded drinking in response to reduction in venous return in the dog,J.Physiol., 295:76 (1979).
3. J.T.Fitzsimons and M.J.Moore,Pulmo-atrial junctional receptors and the inhibition of drinking,J.Physiol.,307:74 (1980).
4. J.T.Fitzsimons and M.J.Moore,Drinking and antidiuresis in response to reduction in venous return in the dog: neural and endocrine mechanisms,J.Physiol.,308:403 (1980).
5. B.Andersson,Regulation of water intake,Physiol.Rev.,58:582 (1978).

6. E.M.Blass,Evidence for basal forebrain thirst osmoreceptors in rat,Brain Res.,82:69 (1974).

7. B.Brown and S.P.Grossman,Evidence that nerve cell bodies in the zona incerta influence ingestive behavior,Brain Res.Bull., 5:593 (1980).

8. P.Buramarugsa and J.I.Hubbard,The neural organization of the rat subfornical organ in vitro and a test of the osmo- and morphino- -receptor hypothesis,J.Physiol.,291:101 (1979).

9. P.C.Coburn and E.M.Stricker.Osmoregulatory thirst in rats after lateral preoptic lesions,J.Comp.Physiol.Psychol.,92:350 (1978).

10. J.T.Fitzsimons,Thirst,Physiol.Rev.,52:468 (1972).

11. S.P.Grossman,D.Dacey,A.E.Halaris,A.E.Collier,T.and A.Routtenberg, Aphagia and adipsia after preferential destruction of nerve cell bodies in hypothalamus,Science,202:537 (1978).

12. F.J.Haberich,Osmoreception in the portal system,Fed.Proc., 27:1137 (1968).

13. J.N.Hayward,Functional and morphologic aspects of hypothalamic neurons,Physiol.Rev.,57:574 (1977).

14. A.K.Johnson and J.Buggy,Periventricular preoptic-hypothalamus is vital for thirst and normal water economy,Amer.J.Physiol., 234:122 (1978).

15. S.Kozlowski and K.Drzewiecki,The role of osmoreception in portal circulation in control of water intake in dogs,Acta Physiol. Polonica,24:325 (1973).

16. R.B.Malmo and W.J.Mundl,Osmosensitive neurons in the rat's pre optic area: medial-lateral comparison,J.Comp.Physiol.Psychol., 88:161 (1975).

17. Y.Oomura,T.Ono,H.Ooyama and M.J.Wayner,Glucose and osmosensitive neurones of the rat hypothalamus,Nature,222:282 (1969).

18. J.W.Peck and D.Novin,Evidence that osmoreceptors mediating drink- ing in rabbits are in the lateral preoptic area,J.Comp. Physiol.Psychol.,74:134 (1971).

19. A.Adachi,A.Niijima and H.L.Jacobs,An hepatic osmoreceptor mech- anism in the rat: electrophysiological and behavioral studies, Amer.J.Physiol.,231:1043 (1976).

20. F.S.Kraly,Abdominal vagotomy inhibits osmotically induced drink- ing in the rat,J.Comp.Physiol.Psychol.,92:999 (1978).

21. F.S.Kraly,J.Gibbs and G.P.Smith,Disordered drinking after abdo- minal vagotomy in rats,Nature,258:226 (1975).

22. J.Sobocinska,Abolition of effect of hypovolemia on the thirst threshold after cervical vagosympathectomy in dogs,Bull.Acad. Polonaise d.Sciences,17:341 (1969).

23. A.V.Wolf;"Thirst",Charles C.Thomas,Springfield (1958).

24. M.J.Wayner,ed.,"Thirst",Pergamon Press,Oxford (1964).

25. E.M.Blass,R.Jobaris and W.G.Hall,Oropharyngeal control of drink
ing in rats,J.Comp.Physiol.Psychol.,90:909 (1976).

26. A.N.Epstein,Oropharyngeal factors in feeding and drinking,
Handbook Physiol.,Vol.1°,Sect.6,Ch.15,197,Wash.D.C.,Amer.
Physiol.Soc. (1967).

27. N.E.Miller,R.J.Sampliner and P.Woodrow,Thirst reducing effects
of water by stomach fistula versus water by mouth,measured by
both a consummatory and instrumental response,J.Comp.Physiol.
Psychol.,50:1 (1957).

28. D.G.Mook,Oral and postingestional determinants of the intake of
variuos solutions in rats with oesophageal fistulas,J.Comp.
Physiol.Psychol.,56:645 (1963).

29. S.Nicolaïdis,Early systemic responses to orogastric stimulation
in the regulation of food and water balance: functional and
electrophysiological data,Ann.N.Y.Acad.Sci.,151:1176 (1969).

30. E.J.Towbin,Gastric distension as a factor in the satiation of
thirst in esophagostomized dogs,Amer.J.Physiol.,8:71 (1949).

31. L.Di Bella,G.Scalera,G.Tarozzi and M.T.Rossi,Correlation between
the food and the fluid intake,7th Internat.Conf.on the Physiol
ogy of Food and Fluid Intake,IUPS,Warsaw,1980.

32. L.Di Bella,M.T.Rossi and G.Scalera,A contribution to a correla
tion between drinking and feeding behaviour,8th Internat.Conf.
on the Physiology of Food and Fluid Intake,IUPS,Melbourne,1983.

33. C.Bomskov,Methodik der Hormonforschung,I B,153,G.THieme,Leipzig
(1937).

34. A.Melander,L.E.Ericson and F.Sundler,Sympathetic regulation of
thyroid hormone secretion,Life Sci.,14:237 (1974).

35. A.Melander,F.Sundler and U.Westgren,Intratyroidal amines and the
synthesis of thyroid hormones,Endocrinology,83:193 (1973).

36. A.Melander,R.E.Ericson,F.Sundler and S.H.Ingbar,Sympathetic in
nervation of the mouse thyroid and its significance in thyroid
hormone secretion,Endocrinology,94:959 (1974).

37. A.Melander,Thyroid hormone secretion,Acta Physiol.Scand.,Suppl.
370:31 (1971).

38. L.W.Tice and C.R.Creveling,Electron microscopic identification
of adrenergic nerve endings on the thyroid epithelial cells,
Endocrinology,97:1123 (1975).

ROLE OF ANTIDIURETIC HORMONE IN VOLUME REGULATION

Z. Arad[1], G.E. Rice[2] and E. Skadhauge

Department of Veterinary Physiology and Biochemistry
The Royal Veterinary and Agricultural University
Bülowsvej 13, 1870 Copenhagen V, Denmark

INTRODUCTION

Recent theories on integration between osmotic and volume regulation places the antidiuretic hormone (ADH) in a central position. In this paper we outline first the concepts of osmotic and volume regulation in higher vertebrates, second we describe our own experiments which test a crucial consequence of the hypothesis stated most clearly by Gauer[5].

Interaction of osmotic and volume control

Higher vertebrates regulate both tonicity and Na concentration of the extracellular volume (ECV) by <u>concentration receptors</u>. The wellknown control loops involving release of ADH from the neurohypophysis and aldosterone from the adrenals are outlined in figure 1 A. Both hormones act on the kidney in order to maintain homeostasis, that is constant extracellular tonicity and sodium concentration in response to lack or surplus of water and Na(Cl).

It is quite clear that these feed back loops cannot directly control the magnitude of ECV. To solve this problem the presence of atrial (and other) stretch receptors has been proposed which, via n. vagus, will modify the release of ADH in such a way that constant volume is maintained. This hypothesis and its supporting

Present address: 1) Dept. of Biology, College of Arts and Sciences, New Mexico State University, Box 3AF/Las Cruces, New Mexico 88003, USA, 2) Dept. of Physiology, Monash University, Clayton, Victoria, Australia 3169.

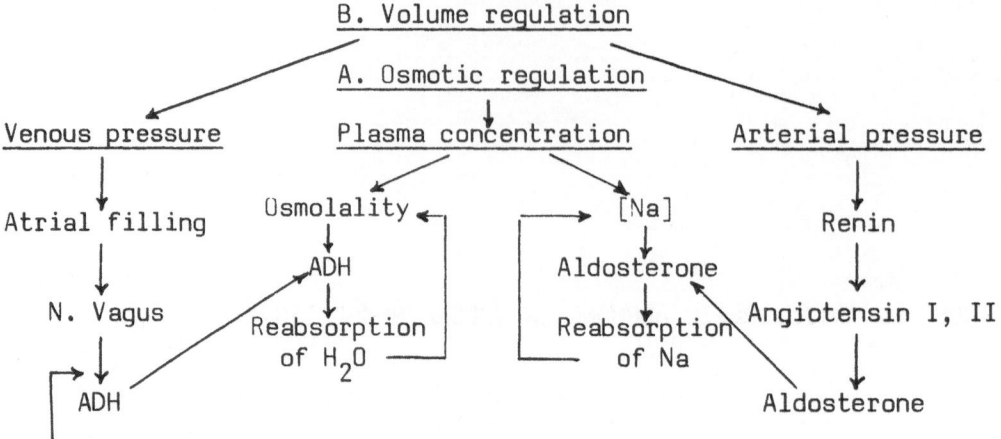

B. Volume regulation

A. Osmotic regulation

Venous pressure   Plasma concentration   Arterial pressure

Atrial filling   Osmolality   [Na]   Renin

                          ADH

N. Vagus   Aldosterone   Angiotensin I, II

          Reabsorption   Reabsorption
          of H$_2$O   of Na

ADH   Aldosterone

C. Modifies the hypothalamic sensitivity to plasma osmolality.

Figure 1.

evidence has been excellently presented by the late Otto Gauer[5].
The central concept is that the release of ADH in response to changes in plasma osmolality is modified by a vagal reflex so that less
ADH is released during volume expansion at the same increment in
plasma osmolality, more during contraction. When at the same time
the concentration of aldosterone is reduced if the NaCl intake is
augmented, and vice versa, volume homeostasis can be maintained.
The suggested control loop will, in conjunction with the modification of aldosterone secretion induced by the renin-angiotensin system, maintain constant EVC (figure 1 B).

     The main supporting evidence for the change in ADH sensitivity
with volume stimuli has come from experiments with inflated balloons
in the atria such as performed by Gauer and his colleagues[5] or Johnson et al.[8]. At an early stage acute volume changes were found in
the rat to have the expected effects on increment in plasma concentration of ADH for a given change in plasma osmolality[4]; subsequent
studies in man revealed change in zero set point for the osmoreceptor but no change in sensitivity when blood volume was changed moderately[9].

     It would therefore seem important to establish whether the postu-
lated volume controlling ADH-loop functions in day to day regulation
of ECV in response to varying NaCl content of the diet.

     For these experiments the domestic fowl was chosen as experimen-
tal animal. It is known that the renal effects of arginine vasotocin
(AVT), the antidiuretic hormone of birds, are grossly similar to
those of ADH in mammals[1] and the general pattern of release of AVT,
plasma concentration, etc. identical to the conditions in mammals[11].
On the other hand the osmotic regulation of birds seems to be slight-

ly "coarser" than in mammals[11]. An increase in plasma osmolality, and concentrations of Na and Cl, in response to a higher constant intake of NaCl in the diet has been observed in the domestic fowl[3,12]. These factors would make the domestic fowl suitable for experiments in which a relationship between plasma osmolality (and concentrations of Na and Cl) in relation to NaCl content of the diet is sought. The further aim of these experiments would be to induce increments in plasma osmolality and relate these to increments in plasma concentration of AVT following various osmotic stimuli in birds receiving different NaCl intake.

INVESTIGATIONS ON THE DOMESTIC FOWL

Prolonged dehydration

In our first study[3] hens were exposed to three levels of NaCl in the diet. Analysis of plasma showed that osmolality (and concentration of Na and Cl) as well as the concentration of AVT is augmented at higher NaCl content of the diet. The concentration of prolactin went up almost proportional to AVT, whereas both middle and high NaCl intakes suppressed plasma concentration of aldosterone (figure 2). These experiments confirm the suitability of hens for

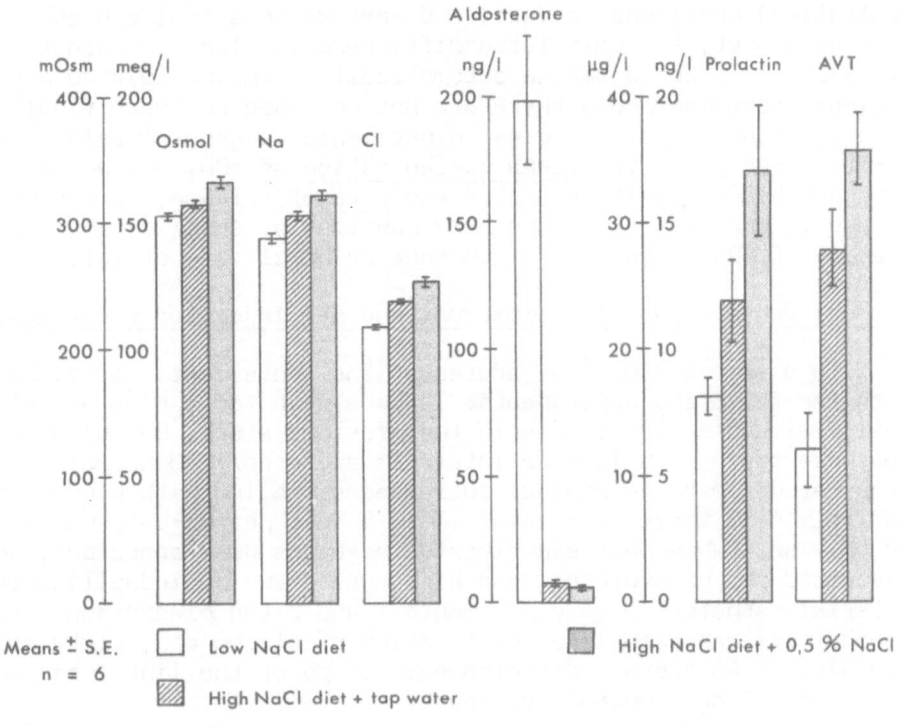

Fig. 2. Partly based on data from 3.

the intended study. Other investigators[7] observed that similar change from high to low NaCl diet induced an approximately 10 per cent fall in ECV.

The hens were deprived of water for several days during which blood samples were taken frequently and food intake controlled. The birds were rehydrated before food consumption declined. The results of these experiments were very clear: after two days the plasma concentration of AVT (and prolactin) reached a "plateau" which remained constant and at the same level for the three dietary groups throughout the period of dehydration. The plasma osmolality (and concentrations of Na and Cl) continued to rise almost linearly with time. These values were significantly lower on birds receiving the low NaCl diet than on the high. The concentration of AVT was calculated as function of plasma osmolality (and concentrations of Na and Cl). A higher slope (larger increment in AVT for a given increase in plasma osmolality) was observed on the low as compared to the high NaCl diet. The values were 0.50 ng/l/mOsmol and 0.27 ng/l/mOsmol, respectively. These sensitivities were of the same magnitude as observed by Gray and Simon in the duck[6]. Although our results apparently demonstrate the postulated change in sensitivity of ADH release, two objections to this conclusion are straightforward. First, since a "ceiling" of plasma concentration of AVT was reached already after two days, but plasma osmolality (and NaCl concentration) continued to rise, and were systematically higher on high NaCl diet, the calculated difference in slope between the regression lines would emerge automatically. The postulated physiological mechanism could therefore not be based on these findings. Second, although physiological experiments in general relate renal water reabsorption to plasma concentration of ADH, and do not distinguish between this parameter and rate of release, it may be questioned to what extent an observed constant or equal plasma concentration of ADH represents a constant or equal rate of release.

Gradual dehydration, salt loading, and AVT injection experiments

In a second study[2] we addressed the above mentioned problems with the following experiments: 1. Hens were fed a high or a low NaCl diet and dehydrated as in the previous study, but blood samples were taken with 12 hour intervals up to only 48 hours. 2. Hens were exposed to gradual intravenous loading with hyperosmotic NaCl given in three steps of 1 ml 18 % NaCl per kg body weight. In this and subsequent experiments the birds were conscious, undisturbed and prehydrated, and blood was drawn by indwelling intra-arterial catheters. In experiments 1 and 2 the plasma concentration of AVT was calculated as function of plasma osmolality at the two diets. An average difference in slope of the linear regression lines of 0.2 ng/l/mOsmol was found.

In experiment 3 a higher increment of plasma concentration of

AVT for a given increase in plasma osmolality was also observed on the low as compared to the high NaCl diet. In this experiment one hyperosmotic NaCl injection of 4 ml per kg was given. The plasma concentrations of AVT and osmolality were followed up to 4 hours. At 90 minutes the AVT concentration in the low NaCl diet birds was 16.1 ng/l <u>higher</u> whereas the plasma osmolality was 6 mOsmol <u>lower</u> than in the high NaCl diet birds. This experiment amply confirms the difference in release of ADH between the two dietary groups.

Finally, in experiment 4 the rate of inactivation of AVT in plasma was measured by frequent sampling of blood after a single injection of AVT to hens on the two diets. No difference in rate constant was found and a common half life in plasma of 6.3 minutes calculated. The results of this experiment makes it permissible to consider plasma concentration as equal to rate of release.

The frequent blood sampling in experiments 2-3 with simultaneous determination of AVT and prolactin showed that the concentration of prolactin rises several minutes <u>after</u> the increase in concentration of AVT. Furthermore, the injection of AVT did not result in any increase in plasma concentration of prolactin. These experiments seem therefore to exclude the possibility that prolactin release is stimulated directly by AVT. It seems to be caused by osmolality, Na concentration or some other "hyperosmotic" parameter.

CONCLUSION

These experiments involved a physiological stimulus which required no interference with the animal; the sole inducer of a change in ECV, etc. was a different NaCl content of the diet. It has been demonstrated beyond reasonable doubt that a lower/higher intake of NaCl leads to a larger/smaller release of ADH in response to an increment in plasma osmolality. The experiments support therefore the hypothesis that volume regulation is achieved partly as a result of changing sensitivity in the release of ADH (see figure 1 C). Although neither venous pressure was measured in these experiments nor the vagal nerve interfered with,it is reasonable to suggest that the observed response is mediated via n. vagus. Simon-Oppermann et al.[10] have in the duck demonstrated both an important effect of intact vagal function on plasma concentration of AVT (and urine flow and osmolality) and a lower slope of the linear regressions between serum AVT and plasma osmolality in volume loaded birds.

ACKNOWLEDGEMENTS

Major support came from The Danish Medical Research Council, NOVO's fond, a NATO Research Grant, and Thorvald Madsens legat. The collaboration of A. Chadwick and S.S. Árnason is gratefully acknowledged.

REFERENCES

1. E. Ames, K. Steven and E. Skadhauge, Effects of arginine vasotocin on renal excretion of Na^+, K^+, Cl^-, and urea in the hydrated chicken, Am. J. Physiol. 221: 1223 (1971).

2. Z. Arad, A. Chadwick, G. E. Rice and E. Skadhauge, Osmotic stimuli and NaCl-intake in the fowl; release of arginine vasotocin and prolactin, J. comp. Physiol. (1985). In press.

3. S.S. Árnason, G. E. Rice, A. Chadwick and E. Skadhauge, Plasma levels of arginine vasotocin, prolactin, aldosterone and corticosterone during prolonged dehydration in the domestic fowl, J. comp. Physiol. (1985). In press.

4. F. L. Dunn, T. J. Brennan, A. E. Nelson and G.L. Robertson, The role of blood osmolality and volume in regulating vasopressin secretion in the rat, J. clin. Invest. 52: 3212 (1973).

5. O. H. Gauer, Role of cardiac mechanoreceptors in the control of plasma volume, "Osmotic and volume regulation", C. Barker Jørgensen and E. Skadhauge, eds., Munksgaard, Copenhagen (1978).

6. D. A. Gray and E. Simon, Mammalian and avian antidiuretic hormone: Studies related to possible species variation in osmoregulatory systems, J. comp. Physiol. 151: 241 (1983).

7. K. Harris and T. I. Koike, The effects of dietary sodium restriction on fluid and electrolyte metabolism in the chicken (Gallus domesticus), Comp. Biochem. Physiol. 59A: 311 (1977).

8. J. A. Johnson, W. W. Moore and W. E. Segar, Small changes in the left atrial pressure and plasma antidiuretic hormone titers in dogs, Am. J. Physiol. 217: 210 (1969).

9. G. L. Robertson and S. Athar, The interaction of blood osmolality and blood volume in regulating plasma vasopressin in man, J. Clin. Endocrinol. Metab. 42: 613 (1976).

10. C. Simon-Oppermann, E. Simon, H. Deutsch, J. Möhring and J. Schoun, Serum arginine-vasotocin (AVT) and afferent and central control of osmoregulation in conscious pekin ducks, Pflügers Arch. 387: 99 (1980).

11. E. Skadhauge, Osmoregulation in birds, Springer-Verlag, Berlin, Heidelberg, New York (1981).

12. E. Skadhauge, D. H. Thomas, A. Chadwick and M. Jallageas, Time course of adaptation to low and high NaCl diets in the domestic fowl. Effects on electrolyte excretion and on plasma hormone levels, Pflügers Arch. 396: 301 (1983).

POSTERIOR HYPOTHALAMIC POLYDIPSIA : DIFFERENTIAL EFFECTS TO SEVERAL DIPSOGENIC TREATMENTS

Alberto Morales and Amadeo Puerto

Laboratory of Psychobiology
Department of Psychology
University of Granada   Cartuja, Spain

Water intake is a physiological process (primary or secondary) which can be triggered by several factors [1,2,3] At the same time, the cerebral mechanisms controlling drinking have been related to several brain structures, i.e., the septal area[4], circularis nucleus [5], lateral hypothalamus[6], subfornical organ[7], etc.

Several studies by the Santacana[8] and Grossman[9] groups have shown that lesions restricted to the caudal part of the hypothalamus produce an increase in the amount of water intake. The first of these reports suggested that this effect might be a primary polydipsia. Animals deprived of water for 48 hours are able to control urine excretion. On the other hand, Grossman and associates suggest that this hyperdipsia appears to be related to a disruption in the secretion of the anti-diuretic hormone (ADH) but at the same time other mechanisms may also be involved.

The present report describes a series of experiments designed to characterize this phenomenon. The first of these studies aimed to analyze the constancy and rhythmicity in water intake according to the following procedure :

1) The total amount of water consumed under the libitum conditions. The records were carried out every 12 hours (light/dark phases).

2) The procedure was identical to the previous one except that the animals had food available only during the dark phase.

3) Food was available only during the light phase.

The surgical procedure used in this study for the experimental and control groups (electrolitic and sham lesions) was the same as that reported by Santacana and associates[8].

Figure 1 shows the total amount and the percentage of water intake in the experimental (EG) and control (CG) group during daily dark/light phases.

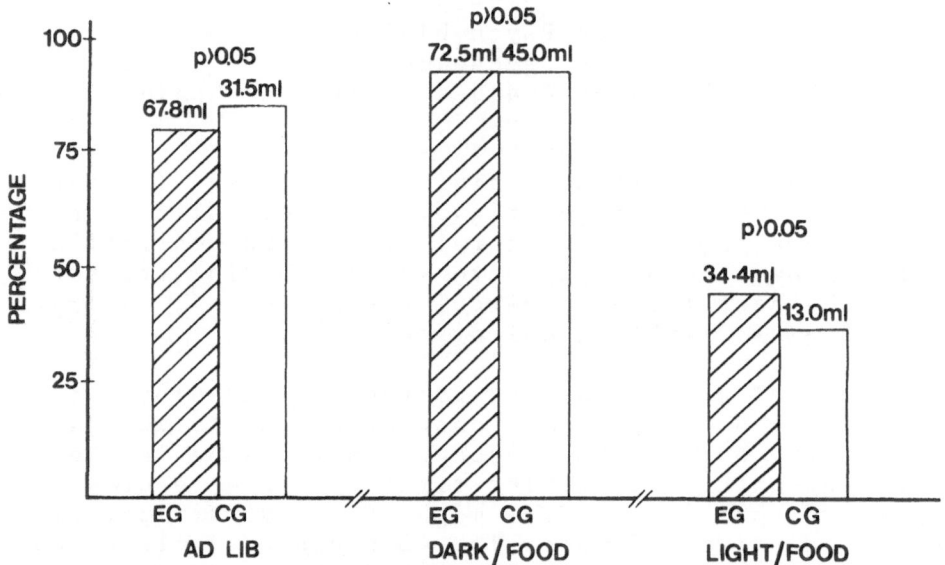

Fig. 1. Total amount and percentage of water intake of the experimental (EG) and control group (CG) during daily dark/light phases.

After this experiment and after bilateral ligation of the ureters, the amount of water consumed by both the experimental and control animals was recorded for a period of 24 hours (table 1).

In the next experiment we examined the relationship between the observed polydipsia and several dipsogenic treatments known to increase water intake. In this respect two groups of animals (experimental and control) were used following the same surgical procedure as in the first experiment.

TABLE 1. Amount of water consumed by the experimental and
control group for a period of 24 hours in three
different experimental conditions.

	Baseline	Pre-ligation **	Post-ligation *
EG	33.0 cc.	90.0 cc.	31.7 cc.
CG	31.0 cc.	32.5 cc.	19.9 cc.

$* p < 0.05$ $** p < 0.005$

The Dipsogenic treatments were :

1) 2 cc. of NaCl (2M, i.p.)
2) 2 cc. of sucrose (1.7M, i.p.)
3) 2 cc. of polyethylene glycol-20000 (PG, 40% i.p.)
4) sham injection i.p.

The total amount of water intake was recorded for
two hours after the administration of these treatments.
The amount of water consumed during the sham injection
procedure was considered to be an index of the polydipsia
itself, so it was subtracted from the water consumed in
response to each of the other treatments.

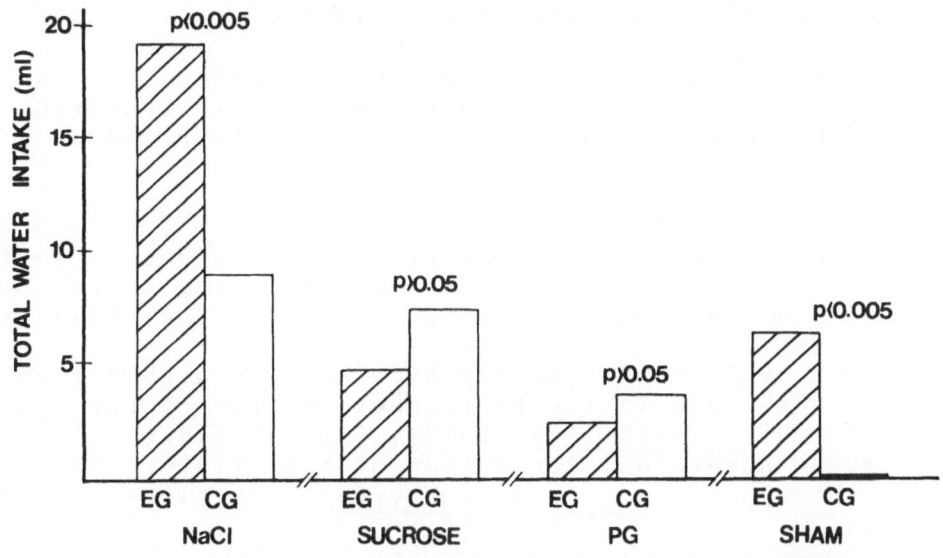

Fig. 2. Total amount of water intake of the experimental (EG) and control
group (CG) in response to three different dipsogenic treatments.

The present results indicate that this caudal hypothalamic polydipsia is caused by a non-permanent factor, since the experimental animals are hyperdipsic under all these three conditions, but the percentage of water intake during the light and dark phases is rhythmic and analogous to that of the control group (Fig. 1). Thus, this hypothalamic polydipsia seems to be different from those others characterized by an arrhythmic pattern. On the other hand although this polydipsia seems to be increased by food consumption, it does not appear to be food-related (prandial), since, for instance in the absence of food, a significant polydipsia still remains[8].

In addition to this, Grossman and associates[9] have shown that in hyperdipsic animals treated with Pitressin, the amount of water consumed is significantly reduced, but the experimental group still overdrinks the control animals. The bilateral ligation of the ureters also shows that even though the experimental animals return to their control baseline, the amount of water consumed is significantly higher than that of the post-ligature control group.

Following the procedure of Blass[10] and Blass and Hanson[4] the third experiment shows a differential effect to these dipsogenic treatments. That is, only NaCl, but not sucrose for instance, significantly increased the amount of water consumed by the experimental group. This result may suggest that, in some way, sodium could be related to this hypothalamic polydipsia.

Although this research may help to understand the physiological bases of this phenomenon, more research is needed in order to achieve a coherent integration of all these results.

REFERENCE

1. A.N. Epstein, D. Spector, A. Samman and C. Goldblum, Exaggerated prandial drinking in the rat without salivary glands. Nature. 201, 1342-1343 (1964)
2. H. Valtin, Hereditary hypothalamic diabetes insipidus in the Brattleboro strain of rat. Am. J. of Pathology. 83(3), 633-636 (1976)
3. B. Andersson, Regulation of body fluids and salt appetite. Ann. Rev. Physiol., 39, 185-200 (1977)

4. E.M. Blass and D.G. Hanson, Primary hyperdipsia in rat following septal lesions. J. Comp. Physiol. Psychol., 70(1), 87-93 (1970)
5. G.I. Hatton, Nucleus Circularis : is it an osmoreceptor in the brain?. Brain Res. Bull. 1(1), 123-131 (1976)
6. J. Kucharczyk and G.J. Mogenson, Separate lateral hypothalamic pathways for extracellular and intracellular thirst. Am. J. Physiol. 228(1) 295-301 (1975)
7. J.B. Simpson, The circumventricular organs and the central actions of angiotensin. Neuro-endocrinol. 32(4), 248-256 (1981)
8. M. P. Tejedor del Real, M.P. Santacana-Altimiras and R. Alvarez-Pelaez, Alteración de la toma de agua después de la lesión de los Cuerpos Mamilares. Rev. Esp. Fisiol. 28, 95-102 (1972)
9. J.W. Hennessy, S.P. Grossman and M. Kanner, A study of the etiology of the hyperdipsia produced by coronal knife cuts in the posterior hypothalamus. Physiol. Behav. 18, 73-80 (1977)
10. E.M. Blass, Separation of cellular from extra-cellular controls of drinking in rats by frontal brain damage. Science, 162, 1501-1503 (1968)

HYPEROSMOTIC AND HYPOVOLEMIC THIRST

David J. Ramsay and Terry N. Thrasher

Department of Physiology
University of California, San Francisco
San Francisco, CA 94143

The volume and composition of the extracellular fluid is controlled within very narrow limits in terrestrial mammals. This process depends upon the ability of the animal to vary urinary volume and concentration over a wide range and on the ability to regulate water intake homeostatically.[1] Additionally, it is becoming clear that sodium excretion and possibly sodium intake may also be regulated in order to maintain a constancy of plasma osmolality.[2] For example, if a dog is deprived of water for 24 hours while allowed access to food, its plasma osmolality increases by approximately 10 mosmol/Kg. This is accomplished by renal conservation of water together with a marked natriuresis. Dogs will rapidly make up their water deficits when allowed access to water.[3] The major emphasis in this review will be the mechanisms which regulate water intake.

HYPEROSMOTIC THIRST

In 1937, Gilman[4] demonstrated that intravenous infusions of hyperosmotic NaCl elicited drinking in dogs. In sharp contrast, equally hyperosmotic solutions of urea caused little or no drinking. Although both solutes raise the plasma osmolality to similar extents, only the NaCl solution caused a significant fall in plasma protein concentration. Gilman concluded that as urea could penetrate cell membranes, this solute could not result in osmotic withdrawal of water from cells. However, because NaCl is maintained in the extracellular fluid, osmotic withdrawal of water from cells occurs thus causing cells to shrink. This fundamental observation forms the basis of hyperosmotic thirst. Solutes introduced in the extracellular fluid compartment which do not penetrate cell membranes cause withdrawal of water from the

83

intracellular compartment and cellular dehydration. The signal of cellular dehydration has to be present for hyperosmotic thirst and associated drinking behavior to occur.

The importance of extracellular fluid osmolality in the control of drinking behavior was emphasized by Holmes and Gregerson in 1950.[5,6] These workers showed that hyperosmotic solutions of sucrose and sorbital were as effective as NaCl in stimulating drinking when given intravenously. This important observation has been confirmed by many groups of workers in different species over the last 35 years.[7] However, it is well worthwhile emphasizing that in normal animals, sodium and its attendant anions constitute well over 90% of the total osmolality of the extracellular fluid. Thus fluctuations in plasma osmolality are usually synonymous with changes in plasma sodium concentration.

The likely central location of osmoreceptors first came from work on the control of vasopressin secretion by Verney (1947).[8] Verney showed that intracarotid infusions of hypertonic NaCl into conscious dogs undergoing water diuresis resulted in an inhibition of that diuresis. He deduced that the inhibition of water diuresis resulted from secretion of vasopressin from the neurohypophysis. Later work has demonstrated that this deduction was correct.[9] Moreover, Verney demonstrated that the receptors necessary for this osmotic response were located in the anterior hypothalamus. Again, it was necessary to infuse solutes which could cause osmotic withdrawal of water from cells to elicit the antidiuretic response. Thus, both the antidiuretic and drinking responses to raised extracellular fluid osmolality depended on cellular dehydration.[10]

Since 1953 when Andersson[11] showed that injections of hypertonic NaCl into the hypothalamus of goats caused copius drinking, there has been general agreement that the osmoreceptors responsible for causing drinking are similar to those controlling vasopressin secretion. Indeed, many have assumed that the receptors are the same although this has never been demonstrated unequivocally. Although the existence of cerebral osmoreceptors is not in doubt, the participation of systemic osmoreceptors, particularly in the hepatic portal bed, have been claimed by some investigators[12,13] The essential importance of these proposed systemic osmoreceptors in the control of drinking may be questioned. Local injection of hypertonic NaCl or sucrose into the lateral preoptic area stimulates drinking in the rabbit[14] and the rat.[15] Moreover lesions in this area abolishes drinking following intraperitoneal injections of hypertonic NaCl. Another approach to this problem has been used in our laboratory. Dogs were prepared with bilateral exteriorized carotid loops. This preparation allows injections of various solutes to be made into the cerebral circulation, together with measurements of water intake, before significant changes in systemic plasma composition occur. Figure 1 shows the results of some of these experiments.[16] Jugular venous osmolality provides a good index of changes in cerebral plasma osmolality to which the osmoreceptors in the brain are exposed. There is a linear relationship between cerebral plasma osmolality and water intake. Sucrose is as effective as NaCl in causing drinking, whereas similarly hyperosmotic solutions of urea have no effect. These results clearly demonstrate that increases in cerebral plasma osmolality, unaccompanied by changes in systemic plasma osmolality, cause drinking.

In a second series of experiments the effect of infusing more strongly hypertonic NaCl intravenously was tested. Dogs were infused with 1.04 M NaCl intravenously which raised peripheral osmolality by 11 mosmol/Kg. This procedure resulted in the intake of 6.8 ± 2 ml/H20/Kg. This experiment was repeated during intracarotid infusion of water which prevented the rise in cerebral plasma osmolality, although the increases in systemic plasma osmolality were identical. During these experiments, the amount of water consumed was negligible (0.4 ± 0.2 ml H20/Kg). Although all tissues apart from those in the area of perfusion of the common carotid arteries were exposed to similar increases in osmolality, no drinking occurred. Thus, although these experiments do not rule out the presence of osmoreceptors in the periphery, it is clear that increases in systemic plasma osmolality alone do not cause drinking.

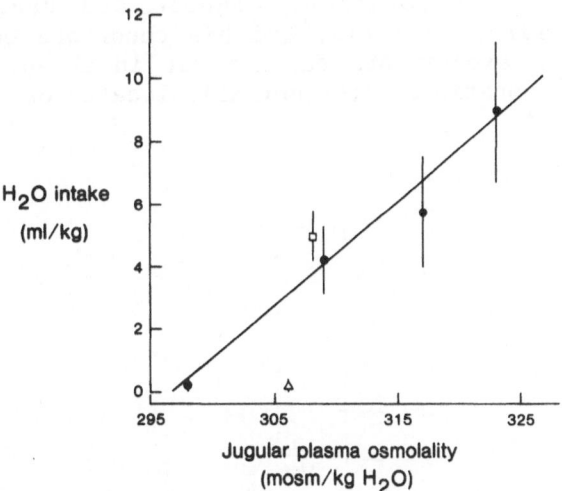

Fig 1. The solid circles show the relationship between cerebral plasma osmolality and water intake in dogs receiving intracarotid infusions of hypertonic NaCl. The open square represents hypertonic sucrose and the open triangle, hypertonic urea.

The nature of the osmoreceptors has been the subject of recent debate. Andersson[17] and his colleagues have proposed that centrally located osmoreceptors depend on specific changes in sodium concentration rather than osmolality. Moreover, they have proposed that the receptors are sensitive to the sodium concentration of CSF rather than the plasma. Thus, when extracellular fluid osmolality is increased by infusions of hyperosmotic solutes, water is withdrawn from CSF thereby raising sodium concentration and stimulating receptors. Experimental evidence in favor of this hypothesis is based on infusions of monosaccharides directly into the CSF which modified drinking in response to raised extracellular fluid osmolality.

Experiments in our laboratory challenge this hypothesis. In our experiments[18] equally hyperosmotic solutions of sodium, sucrose, urea and glucose were infused intravenously in dogs. In confirmation of previous work, only the infusions of NaCl and sucrose stimulated drinking and vasopressin secretion whereas glucose and urea did not. These latter solutes do not cause cellular dehydration. However, reference to Figure 2 shows that all four solutes cause similar changes in CSF osmolality and sodium concentration. Infusions of NaCl and of sucrose stimulated drinking following 14 and 16 minutes respectively. Although the infusions of urea and of glucose were continued for 45 minutes there was no effect on water intake. However, brain cells as judged by increases in CSF sodium concentration, were dehydrated to similar extents. If the osmoreceptors behave as sodium receptors sensitive to CSF composition, glucose and urea should have stimulated drinking. McKinley[19] and his coworkers came to similar conclusions from experiments carried out in sheep. Our results, therefore, are compatible with centrally located osmoreceptors

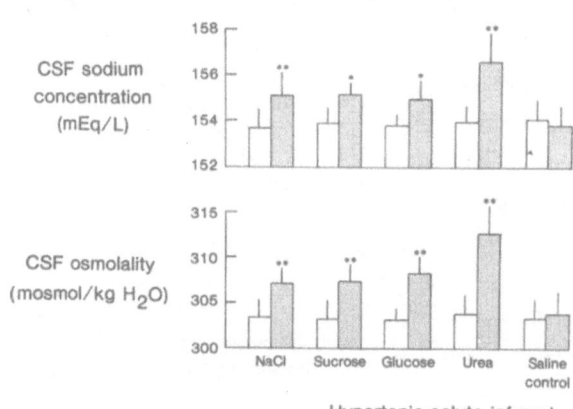

Hypertonic solute infused

Fig 2.   Changes in CSF sodium and osmolality following
         intravenous infusions of equally hyperosmotic
         solutions of NaCl, sucrose, glucose and urea
         From Thrasher et al (18).

86

sensitive to changes in plasma composition rather than CSF. These and similar experiments led to the postulation of osmoreceptors located in a circumventricular organ. Following the results of Buggy and Johnson,[20] whose work demonstrated that large lesions of the anterior ventral third ventricle (AV3V) caused profound disturbances in water balance, the likely location of these osmoreceptors is the organum vasculosum laminae terminalis OVLT. Details of experiments examining this possibility will be found in the chapter by Thrasher and Ramsay.

HYPOVOLEMIC THIRST

The general relationship between depletion of body fluids in clinical situations and drinking has been recognized for many years. However, many of these situations such as diarrhea or vomiting are complicated by changes in extracellular fluid composition. A seemingly straightforward method of examining the relationship between extracellular fluid volume depletion and water intake is the use of hemorrhage. However, whereas the relationship[21] between hemorrhage and thirst has been established in the rat attempts to demonstrate the relationship in other species such as man[22] and dog[23] were not successful. The likely explanation is that the quantity of blood which has to be removed in order to stimulate drinking causes sufficient non-specific weakening of the animal to obliterate the behavior. Consequently, other experimental models of hypovolemic thirst have been used.[24] These usually involve introduction of hyper oncotic colloid into the peritoneal cavity or subcutaneously in order to cause sequestering of extracellular fluids outside the vascular system. Such techniques reliably stimulate drinking and, if observations are made over a 24 hour period also stimulate sodium intake.

In 1969, Fitzsimons[25] demonstrated another method of simulating extracellular fluid thirst, ligation of the abdominal vena cava above the kidneys in the rat. This technique sharply reduces venous return and leads to stimulation of water intake until development of collateral circulation. Fitzsimons went on to show that the drinking response is inhibited by bilateral nephrectomy and restored by intravenous infusions of angiotensin II. Indeed these experiments led to the postulation of angiotensin II as an important dipsogen, a subject which is still debated.

Problems involved in establishing the mechanism of extra-cellular fluid thirst lies in the multiplicity of effects which accompany extracellular fluid volume depletion. Reduction in extracellular fluid volume may be sensed by low pressure cardiopulmonary receptors and also by arterial baroreceptors. Inputs from these receptors are known to project to nuclei in the hypothalamus.[26] Also, reductions in extracellular fluid volume may increase renin release by acting directly on the kidneys via the renal baroreceptor mechanism and also reflexly. Thus reductions in extracellular fluid volume might affect drinking by neural pathways and via the renin angiotensin system.

One approach to this problem has been to block the dipsogenic effects of angiotensin on the brain during reductions in extra-cellular fluid volume. In such experiments, blockers of the renin angiontensin system should not be given systemically. During hypovolemia, renin release is stimulated and the vasoconstrictor effects of the angiotensin II help to maintain arterial blood pressure. Thus, systemic administration of receptor blockers such as saralasin, or converting enzyme inhibitors such as captopril, cause a reduction of blood pressure. Such reductions in blood pressure may either be so severe as to cause non-specific behavioral depression of the animal or, if less severe, may actually add to the stimulus responsible for causing drinking.

One approach to this problem is to administer agents which block the action of angiotensin on the brain intracerebro-ventricularly.[27] In these experiments we first established the dose of saralasin infused directly into the lateral cerebral ventricals of rats which blocked the dipsogenic action of angiotensin II given either systemically or centrally. This method of giving saralasin had no effect on drinking elicited by systemic hypertonic NaCl. Moreover, the ability of a cavally ligated rat to maintain blood pressure was not compromised. Drinking following caval ligation was not blocked by simultaneous intracerebroventricular administration of saralasin. We, therefore, concluded that drinking following this stimulus to extracellular fluid thirst was not affected by proven blockade of the dipsogenic actions of angiotensin on the brain. Thus angiotensin must not be essential. Presumably in the absence of an angiotensin stimulus, cardiovascular reflexes alone are sufficient to sustain the drinking response. Although somewhat different results have been obtained using captopril, an agent which has other effects than blockade of the renin angiotensin system,[28] it is difficult to escape the conclusion that cardiovascular reflexes in the absence of the renin angiotensin system can still cause drinking.

Similar conclusions can be drawn in studies involving the dipsogenic effect of the B-adrenergic agonist isoproterenol. This agent causes profuse drinking when injected systemically. As the dipsogenic effects of isoproterenol in rats were blocked by bilateral nephrectomy but not by ureteric ligation[29] the drinking was thought to be an action dependent on the renin angiotensin system. However, when isopreterenol was given in dogs, the effects of nephrectomy were equivocal. The answer to this controversy lies in the dose of isoproterenol administered.[30] Small doses of isoproterenol cause stimulation of the renin angiotensin system with minimal cardiovascular effects. When central blockade of the

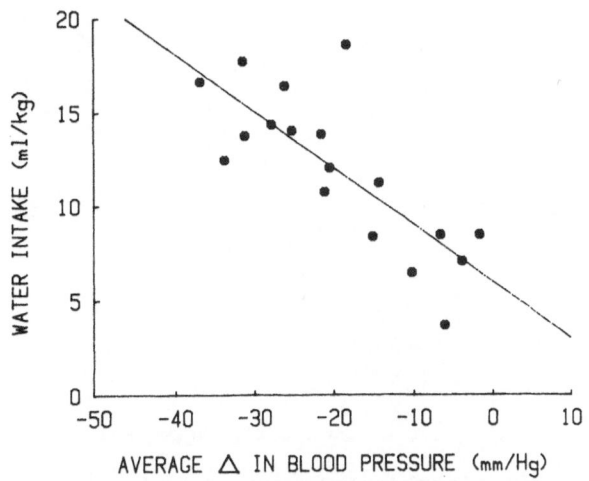

Fig 3. Relationship between water intake and blood
pressure in dogs following hemorrhage.

renin angiotensin system is carried out with these low doses
complete inhibition of drinking is achieved. Larger doses of
isoproterenol, however, cause similar stimulation of the renin
angiotensin system but marked decreases in arterial blood pressure.
Blockade of the renin angiotensin system under these circumstances
does not prevent drinking. Similar results have been reported from
studies in rats. [31] Presumably, just as with caval ligation, in the
intact animal both cardiovascular reflexes and stimulation of the
renin angiotensin are involved in drinking.

Recent experiments in our laboratory has shown that contrary
to earlier experience, graded hemorrhage does stimulate drinking in
dogs. Experiments were carried out on a group of dogs well used
to laboratory procedures. Blood was removed at 1.5 ml/Kg/min and
volumes ranging from 10 to 35 ml/Kg were withdrawn. Results of
these experiments are shown in Fig. 3. The dogs began drinking
soon after access to water was allowed (Latency 9 ± 3 min).
Drinking increased in a dose related manner (P 0.001). The
drinking was well correlated with reductions in mean arterial
pressure but showed no correlation with plasma renin activity. As
drinking occurred at levels of hemorrhage which did not stimulate
renin secretion, we have tentatively drawn the conclusion that
cardiovascular reflexes alone are responsible for the drinking.
Another approach employed in dogs has been to cause reduction in [33]
venous return by inflation of balloons[32] or constriction of cuffs
around the thoracic portion of the inferior vena cava. This method
of reducing venous return leads to stimulation of drinking as well
as vasopressin secretion. Results in our laboratory (Fig. 4) show
a good correlation between drinking and reduction in mean arterial
blood pressure. Systemic administration of saralasin in these

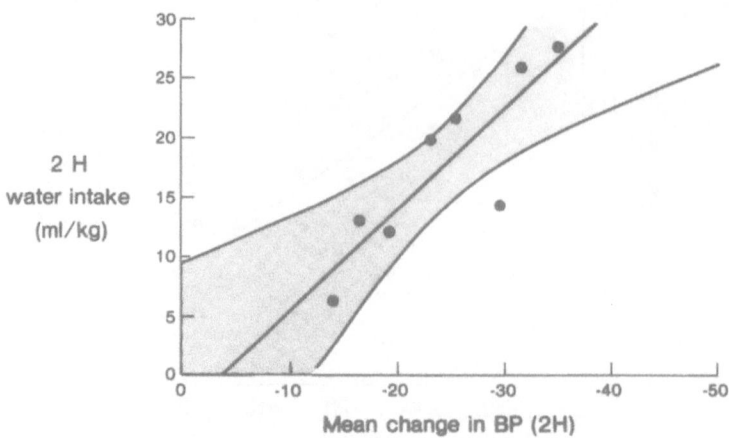

2 H
water intake
(ml/kg)

Mean change in BP (2H)

Fig 4. Relationship between water intake and blood
pressure in dogs following acute caval constriction.

experiments caused a reduction in the quantity of water consumed at
any given blood pressure.   This[32] was also the case with the
experiments reported by Fitzsimons. The conclusion might be drawn
that angiotensin II is involved in this dipsogenic response.
However, as it has been pointed out in experiments on the rat,
removal of the peripheral vasoconstrictor actions of angioteensin
II compromises interpretation of these experiments.   At any given
mean arterial pressure a lesser degree of reduction of venous
return will be required to maintain that pressure in the presence
saralasin.   In order to avoid these difficulties, saralasin was
given intracerebroventricularly during caval constriction in dogs.
The dose of saralasin necessary to block dipsogenic effects of
angiotensin II was first established.   Using this regime of
angiotensin blockade, drinking following acute reduction in venous
return was not affected.[33] Thus, again, participation of the renin
angiotensin system is not essential for hypovolemic thirst.

RESPONSES TO DEHYDRATION

     It is relatively rare under normal physiological circumstances
for hyperosmotic or hypovolemic stimuli to occur separately.   In
fact, the commonest perturbation of body fluid volume and
composition is offered by dehydration.   In the laboratory this may
be explored by examining the effects of depriving animals of water,
but not food, for 24 hours.   Under these circumstances
extracellular fluid osmolality will rise providing a cellular
dehydration stimulus and extracellular fluid volume will fall.   In
1977, Ramsay, Rolls and Wood[3] investigated the relative roles of
volume and osmolality in the drinking which follows 24-hour water
deprivation in dogs.   In these experiments, the volume stimulus was
removed in dehydrated dogs by infusing the appropriate quantity of
isosmotic NaCl solution before allowing dogs access to water.   Such
a maneuver reduced the water intake by 27%.   On another occasion,

90

the hyperosmotic stimulus to thirst was removed in dehydrated dogs
by infusing water via carotid loops to lower cerebral osmolality
to normal. Such a maneuver reduce subsequent water
intake by 72%. A combination of both procedures completely
eliminated the water intake of dehydrated dogs. In this respect,
it is of interest to note that similar contributions of extracellular fluid osmolality and extracellular fluid volume are
involved in the control of vasopressin secretion in dehydrated
dogs.[34] Although the contributions of osmolality and volume in
the control of drinking in dehydrated animals may vary from species
to species,[35,36] qualitatively similar results have been obtained in
all animals so far tested.

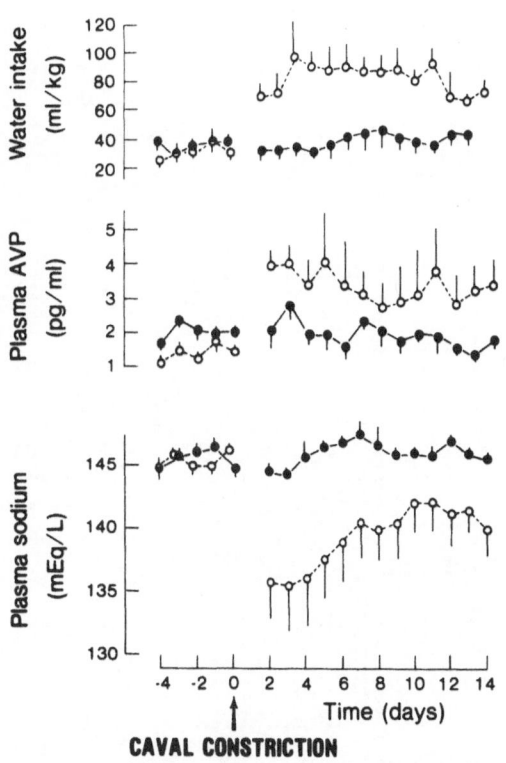

Fig 5. Effect of chronic caval constriction of water
intake, plasma vasopressin and plasma sodium in
dogs. Open circles - caval constriction. Closed
circles- controls.

J.O. Davis and his colleagues[37] have shown that constriction of the thoracic inferior vena cava in the dog causes increases in plasma renin activity (PRA) and aldosterone concentrations, sodium retention and the formation of large quantities of ascitic fluid. In this situation, which is often used as a model for the fluid balance changes which take place in low output congestive cardiac failure, the maintenance of arterial blood pressure is partly dependent upon angiotensin II. Apart from effects on sodium balance, both acute and chronic caval constriction are associated with large increases of water intake.[33,38] We have recently studied the effects of chronic caval constriction in a population of our dogs to establish whether the increased water intake is primary or is dependent on sodium retention.[38] These results are summarized in Fig. 5. Immediately following caval constriction, water intake day period of study. There was also a sustained increase in the

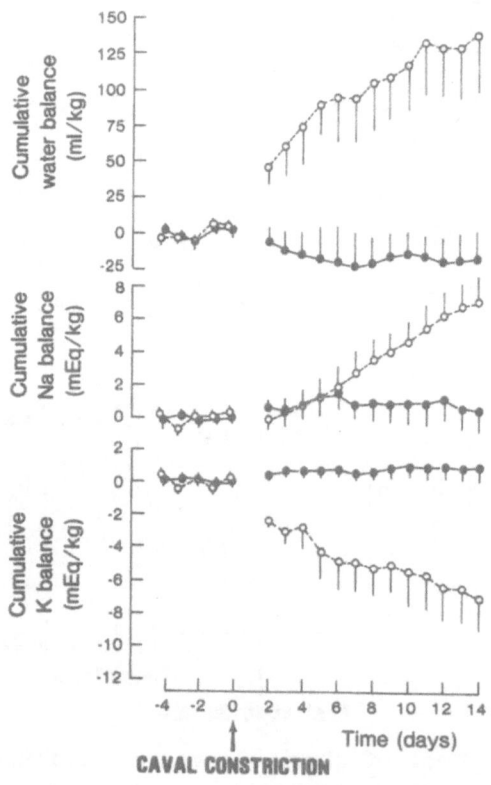

Fig 6. Effect of chronic caval constriction on water and electrolyte balance following caval constriction, Symbols - as Fig. 5.

plasma vasopressin concentration. These changes were assocated with the development of dilutional hyponatremia. Indeed at no time during the period of study was plasma osmolality or sodium elevated above normal. Thus the enhanced drinking and elevated plasma vasopressin concentration are inappropriate for the plasma osmolality. These data demonstrate that the effects of chronic caval constriction on water balance are primary and are not dependent upon changes in the plasma sodium. The importance of these findings is emphasized in Fig. 6. Over the period of 14 days there is a progressive increase in cumulative water intake which preceeds the development of positive sodium balance. The stimulus for the increased water intake and vasopressin secretion is still under review but our tentative explanation is the high PRA which was maintained at values between 40 and 50 ng/ml/3h.

We have therefore argued that the development and maintenance of the expanded extracellular fluid volume and peritoneal ascites associated with chronic caval constriction depends upon increased water intake and vasopressin secretion, as well as on sodium retention. Further studies in our laboratory have confirmed this view. When the fluid intake of dogs with fully developed peritoneal ascites was restricted to approximately precaval constriction levels, a marked diuresis and natriuresis occurred. This may be accounted for by a reduction in plasma aldosterone concentration in spite of the maintained high PRA. The explanation for this phenomenon is that when water intake is restricted, the ensuing increase in plasma sodium concentration inhibits aldosterone secretion at the level of the adrenal. There have been a number of reports in the literature during the last two years which emphasize the importance of plasma sodium concentration in the control of aldosterone secretion.[2,39,40] Thus during chronic thoracic caval constriction, not only is there a primary increase in water intake, but also prevention of excessive fluid intake reduces peritoneal ascites and edema. Thus the observations that patients with congestive cardiac failure exhibit thirst, and thus presumably increased water intake, may be important in our understanding of the pathogenecies of this condition.

REFERENCES

1. D.J. Ramsay, T.W. Thrasher and L.C. Keil, Stimulation and inhibition of drinking and vasopressin secretion in dogs, in Antidiuretic Hormone, S. Yoshida, L. Share and K. Yagi eds., University Park Press, Baltimore, 97, 1979
2. T.N. Thrasher, C.E. Wade, L.C. Keil and D.J. Ramsay, Sodium balance and aldosterone during dehydration and rehydration in the dog, Am. J. Physiol 247 (Regulatory, Integrative Comp. Physiol 16) R 76, 1984.

3. D.J. Ramsay, B.J. Rolls and R.J. Wood, Thirst following water deprivation in dogs. Am. J. Physiol 232 (Regulatory, Integrative, Comp Physiol 1) R 93, 1977.

4. A. Gilman, The relation between blood osmotic pressure, fluid distribution and voluntary water intake, Am. J. Physiol 120, 323, 1937.

5. J.H. Holmes and M.I. Gregersen, Role of sodium and chloride in thirst, Am. J. Physiol 162, 1950.

6. J.H. Holmes and M.I. Gregersen, Observations on drinking induced by hypertonic solutions, Am. J. Physiol. 162, 1950.

7. T.N. Thrasher, Osmoreceptor mediation of thirst and vasopressin secretion in the dog, Federation Proc. 41, 2528, 1982.

8. E.B. Verney, The antidiuretic hormone and the factors which govern its release, Proc. Royal Soc. London Series B. 135, 1947.

9. C.E. Wade, P. Bie, L.C. Keil and D.J. Ramsay, Effect of hypertonic intracarotid infusions on plasma vasopressin concentration, Am. J. Physiol 243 (Endocrinol Metab 6) E522, 1982.

10. A.V. Wolf, Osmometric analysis of thirst in man and dog, Am. J. Physiol 161, 75, 1950.

11. B. Andersson, The effect of injections of hypertonic NaCl-solutions into different parts of the hypothalamus of goats, Acta. Physiol. Scand 18, 1953.

12. S. Kozlowski and K. Drzewieski, The role of osmoreception in portal circulation in control of water intake in dogs. Acta. Physiol. Pol. 24, 1973.

13. P. Bie, Osmoreceptors, vasopressin and control of renal water excretion, Physiol Rev. 60, 1980.

14. J.W. Peck, and D. Novin, Evidence that osmoreceptors mediating drinking in rabbits are in the lateral preoptic area, J. Comp. Physiol. Psychol. 74, 1971.

15. E.M. Blass and A.N. Epstein, A lateral preoptic osmosensitive zone for thirst in the rat, J. Comp. Physiol. Psychol, 76, 1971.

16. R.J. Wood, B.J. Rolls and D.J. Ramsay, Drinking following intracarotid infusions of hypertonic solutions in dogs, Am. J. Physiol 232, (Regulatory, Integrative Comp Physiol 1) R 88, 1977.

17. B. Andersson, Regulation of water intake, Physiol. Rev. 58, 1978.

18. T.N. Thrasher, C.J. Brown, L.C. Keil and D.J. Ramsay, Thirst and vasopressin release in the dog: an osmoreceptor or sodium receptor mechanism? Am. J. Physiol, 238 (Regulatory, Integrative Comp. Physiol 7) R 333, 1980.

19. M.J. McKinley, D.A. Denton, and R.S. Weisinger, Sensors for antidiuresis and thirst - osmoreceptors or CSF sodium receptors? Brain, Res, 141, 1978.

20. J. Buggy and A.K. Johnson, Preoptic - hypothalamic periventricular lesions: thirst deficits and hypernatremia, Am. J. Physiol 233 (Regulatory, Integrative Comp. Physiol 2) R 44, 1977.

21. J.T. Fitzsimons, Drinking by rats depleted of body fluid without increase in osmotic pressure, J. Physol (Lond), 159, 1961.

22. J.H. Holmes and A.V. Montgomery, Thirst as a sympton, Am. J. Med. Sci. 225, 1953.

23. S. Kozlowski and J. Sobocinska, Thirst in regulation of blood volume in dogs, In Fourth Int. Cont. Regulation Food and Water Intake, Cambridge, 1971.

24. E.M. Stricker, Extracellular fluid volume and thirst, Am. J. Physiol 211, 1966.

25. J.T. Fitzsimons, The role of renal thirst factor in drinking induced by extracellular stimuli, J. Physiol (Lond) 201, 1969.

26. K. Yagi, H. Kannan and Y. Sawaki, Electrophynology of andiuretic hormone secreting neurons: afferent neural pathways from cardiovascular receptors, In Antidiuretic hormone, S. Yoshida, L. Share and K. Yagi eds., University Park Press, Baltimore, 81, 1979.

27. M.C. Lee, T.N. Thrasher and D.J. Ramsay, Is angiotensin essential in drinking induced by water deprivation and caval ligation? Am. J. Physiol 240 (Regulatory, Integrative Comp Physiol 9) R 75, 1981.

28. J.T. Fitzsimons and R.W. Elfont, Angiotensin does contribute to drinking induced by caval ligation in the rat, Am. J. Physiol 243 (Regulatory, Integrative Comp Physiol 12) R 558, 1982.

29. K.A. Houpt and A.N. Epstein, The complete dependence of beta-adrenergic drinking on the renal dipsogen, Physiol Behav. 7, 1971.

30. D.J. Ramsay, Beta-Adrenergic thirst and its relations to the renin angiotensin system, Federation Proceed. 37, 1978.

31. R. Rettig, D. Ganten and A.K. Johnson, Isoproterenol-induced thirst: renal and extrarenal mechanisms, Am. J. Physiol 241, (Regulatory, Integrative Comp. Physiol 10) 152, 1981.

32. J.T. Fitzsimons and M.S. Moore-Gillon, Drinking and anti-diuresis in response to reductions in venous return in the dog: neural and endocrine mechanisms, J. Physiol (Lond) 308, 1980.

33. T.N. Thrasher, L.C. Keil and D.J. Ramsay, Hemodynamic, hormonal and drinking responses to reducd venous return in the dog, Am. J. Physiol 243 (Regulatory, Integrative Comp. Physiol 12) R 354, 1982

34. C.E. Wade, L.C. Keil and D.J. Ramsay, Role of volume and osmolality in the control of plasma vasopressin in dehydrated dogs, Neuroendocrinology 37: 349, 1983.

35. R.J. Wood, E.T. Rolls and B.J. Rolls, Physiological mechanisms
     for thirst in the non-human primate, Am.J. Physiol 242
     (Regulatory, Integrative Comp. Physiol 11) R423, 1982.
36. D.J. Ramsay, B.J. Rolls and R.J. Wood, Body fluid changes which
     influence drinking in the water deprived rat, J. Physiol
     (Lond) 266:453, 1977.
37. J.O. Davis, The physiology of congestive heart failure in
     Handbook of Physiology, Circulation, Washington, D.C.,
     Am. Physiol. Soc., sect 2, vol III, 2071, 1965.
38. T.N. Thrasher, M. Moore-Gillon, C.E. Wade, L.C. Keil and
     D.J. Ramsay, Inappropriate drinking and secretion of
     vasopressin after caval constriction in dogs, Am. J.
     Physiol 244 (Regulatory, Integrative Comp Physiol 13)
     R 850, 1983.
39. J.R. Blair-West, A.H. Brook, A. Gibson, M. Morris and P.T.
     Pullan, Renin, antidiuretic hormone and the kidney in
     water restriction and rehydration, J. Physiol (Lond)
     294, 181, 1979.
40. E.G. Schneider, P.G. Davis and R.E. Taylor, Sodium modulation
     of aldosterone secretion by the isolated canine adrenal
     gland, Federation Proceedings 42, 1985.

OSMOTIC THIRST SUPPRESSION AFTER CENTRAL

ADMINISTRATION OF VASOPRESSIN ANTAGONISTS

Ewa Szczepańska-Sadowska, Jadwiga Sobocińska
and Stanisław Kozłowski

Microcirculation Research Unit and
Department of Applied Physiology
Institute of Physiological Sciences, School
of Medicine, Warsaw

INTRODUCTION

We have previously reported that administration
of vasopressin (AVP) to the third ventricle (3rdV)
stimulates intake of water in normally hydrated dogs[1].
It has been also found that concentration of AVP in
the cerebrospinal fluid (CSF) changes in parallel to
body fluid osmolality[2,3]. The present study was aimed
at elucidating whether centrally released endogenous
AVP may be of physiological significance for the control of the osmotic thirst. To this end the osmotic
thirst threshold and the postthreshold intake of water as well as restitution of plasma osmolality were
determined under control conditions and after administration of competitive antagonists of AVP.

METHODS

The experiments were performed on mongrel dogs
chronically implanted with a guide tube directed towards the 3rdV according to the technique described
previously[1]. The guide tube was placed vertically
in the sagittal plane 25-30 mm anteriorly to the
interaural plane, usually 1 mm behind bregma. The experiments were started 14 days after the surgery.

The osmotic thirst threshold was determined during intraventricular infusion of AVP antagonists
and compared to that found under control conditions.

The osmotic thirst threshold was determined as described previously[4,5] taking into account: 1/ the threshold osmotic load (L_{osmt}) and 2/ the threshold increment in plasma osmolality (ΔP_{osmt}) necessary to elicit drinking response. Water intake (WI) was stimulated by an intravenous (i.v.) infusion of 5.0% NaCl at a rate of 5.0 ml.min^{-1}. The postloading restitution of plasma osmolality (P_{osm}), the ratio of cumulative amount of water actually ingested during the experiment to the amount theoretically required for restoration of body fluid osmolality (WI_a/WI_r) and the osmotic load remaining in the body (L_{osm}) were determined at 30 and 60 min after the threshold was reached.

AVP antagonists were infused into the 3rdV during 30 min before and during the whole time of i.v. infusion of the hypertonic saline. They were dissolved in the artificial CSF. In control experiments the artificial CSF was infused without AVP antagonists during the same period of time. The rate of intraventricular infusion in control and proper experiments was 20 μl.min^{-1}. The following AVP antagonists were used: 1/d(CH_2)$_5$AVP (0.2 and 2.0 μg.min^{-1}), 2/ d(CH_2)$_5$Tyr (Me)AVP (1.2 μg.min^{-1}) and 3/ d(CH_2)$_5$D-Val-VAVP (2.0 μg.min^{-1}). The compounds were kindly supplied by Professor M. Manning (Medical College, Ohio, USA).

RESULTS

Intraventricular administration of d(CH_2)$_5$AVP at a rate of 2.0 μg.min^{-1} caused significant elevation of the osmotic thirst threshold (Table 1). The osmotic load remaining at 30 and 60 min after the threshold was significantly higher while the ratio WI_a/WI_r was significantly lower than under control conditions, indicating suppression of the postloading drinking. As a result the postloading restitution of plasma osmolality was delayed in comparison to control experiments (Table 1). Infusion of d(CH_2)$_5$AVP at 10 times lower rate (0.2 μg.min^{-1}) also elicited a significant elevation of P_{osmt} by 2.4 ± 0.7 mosm ($2P < 0.05$). The threshold osmotic load was elevated by 23.7 ± 10.9 mosm ($1P < 0.05 < 2P$). The postloading L_{osm} was not significantly elevated and WI_a/WI_r

Table 1. Effect of intraventricular infusion of d(CH_2)$_5$AVP on osmotic thirst. C-control, E - infusion of d(CH_2)$_5$AVP at a rate of 2.0 µg.min^{-1}. ΔP_{osm} - increment in plasma osmolality, L_{osm} - osmotic load, WI_a/WI_r - ratio of cumulative amount of water actually ingested during experiments to the amount theoretically required to restore body fluid isotonicity at thirst threshold and 30 and 60 min after the threshold. Means $\pm$ standard errors are presented. *2P<0.05, **2P<0.02, ***2P<0.005.

| Variable | | Threshold | after threshold | |
			30 min	60 min
ΔP_{osm}	C	3.1±1.6	-0.6±1.4	-2.2±0.9
mmol.kg^{-1}	E	7.4±1.7*	5.4±2.4**	4.2±1.8*
L_{osm}	C	63.9±9.0	46.3±14.8	26.7±14.5
mmol	E	111.5±14.9***	96.2±19.4***	79.5±18.8***
WI	C	79.4±17.8	121.0±24.7	211.3±16.2
ml	E	72.8± 8.3	154.9±26.9	202.8±30.2
WI_a/WI_r	C	0.54±0.14	0.89±0.19	1.38±0.21
	E	0.27±0.04*	0.68±0.09	0.87±0.08*

as well as the postloading restitution of P_{osm} were only slightly suppressed at the end of the experiment (1P<0.05<2P).

Intraventricular infusion of d(CH_2)$_5$Tyr(Me)AVP at a rate of 1.2 µg.min^{-1} influenced neither the osmotic thirst threshold nor the postloading restitution of P_{osm}.

Intraventricular infusion of d(CH_2)$_5$D-Val-VAVP at a rate of 2.0 µg.min^{-1} caused slight elevation of P_{osmt} and L_{osmt} and significant suppression of the postloading drinking and of the postloading restitution of P_{osm} (Table 2).

Administration of the above competitive antagonists of AVP into the 3rdV did not influence the rate of excretion of urine during the period preceding measurement of the osmotic thirst threshold.

Table 2. Effect of intraventricular infusion of $d(CH_2)_5$D-Val-VAVP on osmotic thirst. C - control, E - infusion of $d(CH_2)_5$D-Val-VAVP at a rate of 2.0 μg. min^{-1}. [x] 1P<0.05<2P, [*] 2P< 0.05, [**] 2P< 0.02, [***] 2P< 0.001. For other explanation see Table 1.

Variable		Threshold	after threshold	
			30 min	60 min
P_{osm} $mmol.kg^{-1}$	C	4.3±2.0	-2.0±1.8	-3.0±1.2
	E	10.0±2.9[x]	1.8±0.8[**]	-0.6±0.9[***]
L_{osm} mmol	C	56.8±12.9	48.6±12.6	32.6±12.9
	E	94.5±24.2[x]	78.0±28.4	64.1±29.7
WI ml	C	60.0±7.1	118.0±9.6	193.8±29.7
	E	68.0±11.8	132.0±26.8	181.0±33.4
WI_a/WI_r	C	0.51±0.14	1.01±0.25	1.87±0.46
	E	0.31±0.05[x]	0.56±0.11[*]	0.85±0.16[*]

DISCUSSION

The results of this study reveal that some of the competitive antagonists of AVP exert antidipsogenic effect. This action is manifested by an increase of the osmotic thirst threshold as well as by suppression of the postloading drinking. The antidipsogenic effect of vasopressin antagonists is strong enough to cause a significant delay in the postloading restitution of plasma osmolality. Thus, the results give support to the idea that endogenous AVP modulates sensitivity of the thirst system to osmotic stimuli[.]

Although both V_1 and V_2 antagonists were used in the present study, it cannot be concluded at present which kind of receptors mediates the dipsogenic effect of AVP. According to the effects observed in the experiments carried out on rats the properties and respective potencies (A_2) of the antagonists used in the present study were characterized as follows: $d(CH_2)_5$AVP-antivasopressor antagonist (A_2=8.35), $d(CH_2)_5$Tyr(Me) AVP - antivasopressor antagonists (A_2=8.6), $d(CH_2)_5$ D-Val-VAVP antiantidiuretic (A_2=7.48) and antivasopressor ($A_2 = 6.41$) antagonist[6,7,8]. Taking into account differences in potencies the rates of infusion of antagonists were so adjusted in the present study that the infusion of d(CH_2)$_5$ AVP at a rate of 2.0 μg.min^{-1} corresponded to infusion of d(CH_2)$_5$ Tyr(Me)AVP at a rate of 1.2 μg.min^{-1}, whereas infusion of d(CH_2)$_5$ AVP at a rate of 0.2 μg.min^{-1} corresponded to the infusion of d(CH_2)$_5$ D-Val-VAVP at a rate of 2.0 μg.min^{-1}. Thus, d(CH_2)$_5$ Tyr(Me)AVP seems to not exert any action on osmotic thirst whereas the corresponding dose of d(CH_2)$_5$AVP is strongly antidipsogenic. The latter peptide is also more potent in elevating the osmotic thirst threshold than d(CH_2)$_5$D-Val-VAVP. On the other hand d(CH_2)$_5$ D-Val-VAVP appears to be more effective in suppressing the postloading drinking. However, it should be emphasized that the above comparisons may be very inaccurate since they do not take into account the possible differences in the turnover rate of particular antagonists in the brain.

Some recent reports indicate existence of interspecies differences in the antagonistic and agonistic properties of various AVP analogues towards V_1 and V_2 receptors [9,10]. As far as it concerns the antagonists used in the present study there is as yet no information concerning the potency and selectivity of their anti V_1 and anti V_2 activities in the dog. The interspecies differences could perhaps account for the discrepancy between the effects exerted by d(CH_2)$_5$AVP and d(CH_2)$_5$ Tyr(Me) AVP with regard to the thirst system. Both these peptides were previously found to be potent antivasopressor antagonists in the rat [6,9].

Finally, the possibility that the central AVP receptor may have different properties than V_1 and V_2 receptors should be also considered.

REFERENCES

1. E. Szczepańska-Sadowska, J. Sobocińska, and B. Sadowski, Central dipsogenic effect of vasopressin. Am. J. Physiol. 242:R372 (1982).
2. E. Szczepańska-Sadowska, D.Gray, Ch. Simon-Opperman, Vasopressin in blood and third ventricle CSF during dehydration, thirst and hemorrhage. Am. J. Physiol. 245:R549, (1983)
3. E. Szczepańska-Sadowska, Ch. Simon-Opperman, D. Gray and E. Simon, Plasma and cerebrospinal fluid vasopressin and osmolality

in relation to thirst. Pflügers Arch. 400: 294 (1984).

4. A.V. Wolf, Osmometric analysis of thirst in man and dog. Am. J. Physiol. 161:75 (1950).

5. E. Szczepańska-Sadowska, W. Niewiadomski, J. Sobocińska, and S. Kozłowski, Thirst and solute excretion: their effectiveness in osmostatic control of body fluid. Am. J. Physiol. 244:R23 (1983).

6. M. Kruszyński, B. Lammek, M. Manning, J. Seto, J. Haldar, and W.H. Sawyer. 1-(β-Mercapto-β,β-cyclopentamethylenepropionic acid),2-(O-methyl-thyrosine-arginine-vasopressin and 1-(β-Mercapto-β,β-cyclopentamethylenepropionic acid)arginine vasopressin, two highly potent antagonists of the vasopressor response to arginine-vasopressin, J. Med. Chem. 23:364 (1980).

7. M. Manning, W.A. Klis, A. Olma, J. Seto, and W.H. Sawyer, Design of more potent and selective antagonists of the antidiuretic responses to arginine-vasopressin devoid of antidiuretic agonism. J. Med. Chem. 25:414 (1982).

8. M. Manning and W.H. Sawyer, Design of potent and selective in vivo antagonists of the neurohypophysial peptides. In: The Neurohypophysis: Structure, Function and Control, Progress in Brain Research, vol. 60, B.A. Cross and G. Leng ed., Elsevier, Amsterdam-New York (1983) p. 367.

9. J.F. Liard, O. Deriaz, P. Schelling, and M. Thibonnier, Cardiac output distribution during vasopressin infusion or dehydration in conscious dogs. Am. J. Physiol. 243:663 (1982).

10. F.L. Stassen, W. Bryan, M. Gross, B. Karanagh, D. Shue, L. Sulat, V.D. Wiebelhaus, N. Yim, and L.B. Kinter, Critical differences between species in the in vivo and in vitro renal responses to antidiuretic hormone antagonists. In: The Neurohypophysis: Structure, Function and Control. Progress in Brain Research, vol. 60. B.A. Cross and G. Leng, ed., Elsevier, Amsterdam-New York (1983) p. 395.

HYPEROSMOTIC THIRST : AN OSMORECEPTOR MECHANISM, A SODIUM RECEPTOR MECHANISM OR BOTH

Simon N. Thornton

Institute of Animal Physiology
Babraham
Cambridge CB2 4AT
England

Thirst has been divided into two categories, the first, primary, that which is involved in the regulation of body fluid homeostasis and the second that which is not involved in direct regulation of body fluids. Primary thirst has been further divided into having an extracellular origin and an intracellular origin. Extracellular thirst is involved during decreases in blood volume, resulting for example from haemorrhage, and appears to be mediated in part by angiotensin II (AII). Intracellular thirst is initiated by an increase in the 'effective' osmotic pressure of the blood i.e. when there is an increase in the concentration of solutes that do not readily penetrate cell membranes.

Verney[1] from observation of antidiuretic hormone (ADH) release in dogs proposed the existence of centrally located specific receptor cells, or osmoreceptors, responsible for ADH release. Wolf[2] suggested that these osmoreceptors, or cells very similar to them, may also be responsible for the initiation of drinking which followed carotid infusions of hypertonic NaCl or sucrose in conscious animals. These two solutions would have increased the effective osmotic pressure of the blood. Glucose or urea in hypertonic solution were found to be ineffective at initiating drinking or releasing ADH.

Andersson and his colleagues[3] have criticized the osmoreceptor theory and have proposed instead a sodium receptor hypothesis. The sodium receptor is thought to be located in the brain, possibly in the wall of the cerebralventricular system. Decreasing the concentration of sodium in the cerebrospinal fluid (csf) of goats with I.C.V. infusions of non-electrolyte solutions inhibited drinking in response to peripherally applied osmotic stimuli[4]. Andersson and his co-workers suggested that since a decrease in csf Na inhibited osmotically induced drinking then the increase in the concentration of sodium in the csf was the important stimulus to drinking follow-ing peripheral osmotic stimuli   However in dogs[5] and sheep[6] csf Na$^+$ concentration was increased following peripheral infusions of hyper-tonic NaCl, sucrose or urea, but only NaCl or sucrose caused drinking. Therefore there appears to be a controversy as regards the role of a csf Na receptor in osmotically induced drinking.   As yet there is no published evidence for the goat of changes in csf Na following peripheral infusions of hypertonic solutions of NaCl, sucrose or urea.

Osmotically induced drinking was investigated in the pigeon as they have large cerebroventricular spaces and are rapid drinkers. Under Equithesin anaesthesia a catheter was implanted in the jugular vein and a cannula implanted in the third cerebral ventricle. Both the catheter and the cannula were externalized on the head. Follow-ing 6-12 days of recovery, by which time the birds had regained their pre-operative weight, drinking in response to various hypertonic solu-tions infused intravenously (I.V.), intracerebroventricularly (I.C.V.) or simultaneously I.V. and I.C.V. was studied.

I.V. infusion of hypertonic solutions of NaCl, sucrose or man-nitol resulted in significant volumes of water being drunk. Infu-sions of hypertonic urea or glucose were much less effective in causing drinking.   Drinking in response to NaCl, sucrose or mannitol started after an amount of solute had been infused that was calcula-ted to have increased plasma osmolality by 2-3%.   The volume of water was calculated to have been sufficient (after sucrose) or al-most sufficient (after NaCl or mannitol) to dilute the osmotic load remaining in the body to isotonicity with plasma[7].

Drinking in response to I.V. infusion of increasing concentra-tions of NaCl, sucrose or mannitol was proportional to the amount of solute infused.   For all three solutions, as the concentration in-creased the latency to drinking decreased but the amount infused at the initiation of drinking was approximately the same and was calcu-lated to have increased plasma osmolality by 2-3%.   Sucrose caused a greater intake of water than either NaCl or mannitol and this may have been in response to the large loss of extracellular fluid due to the greater osmotic diuretic effect sucrose has on birds compared to NaCl (or mannitol)[7].

In 4 anaesthetized pigeons csf Na$^+$ was measured following I.V. infusions of hypertonic NaCl or urea.   The concentration of Na$^+$ was increased following the 2 infusions (NaCl + 22 mmol/1, urea + 11 mmol/1), however only NaCl produced drinking in conscious birds.

Therefore, pigeons appear to be able to respond to osmotic stimuli in a manner similar to mammals in that they start drinking in response to NaCl, sucrose or mannitol after a calculated increase in plasma osmolality of 2-3% (this value is similar to that obtained in rats[9], dogs and humans[2]). The drinking appeared to be initiated by stimulation of an osmoreceptor type mechanism since it started after a threshold increase in plasma osmolality had been reached and, as was found in dogs[5], csf Na^+ increased in response to hypertonic NaCl or urea yet urea did not cause drinking. The osmoreceptor therefore probably lies outside the blood brain barrier.

By contrast central infusions produced a different picture in that only NaCl was effective centrally and drinking was proportional to the amount of solute infused[8]. As the concentration increased the latency to drinking decreased. This could indicate that the pigeon has a central sodium receptor mechanism that may be responsible for the drinking. However I.C.V. infusion of a mixture of sucrose and NaCl in hypertonic solution produced the results shown in Fig. 1. Hypertonic sucrose caused drinking providing there was a certain amount, a background level, of Na in csf. Therefore it appears as though hypertonic solutions infused centrally may act through activation of an osmoreceptor mechanism but continued action of this osmoreceptor requires a certain level of Na in the csf.

The next step was to look at the effect on drinking induced by peripheral osmotic stimuli of infusions of various solutions into the csf that may have changed the csf Na^+ concentration. Solutions were infused simultaneously I.V. (at 0.34 ml/min) and intracerebroventricularly (I.C.V.) (at 2 µl/min) and drinking observed during the 15 min infusions and also for 60 min immediately afterwards. In Fig. 2 drinking in response to an I.V. stimulus of 0.5 M NaCl plus various I.C.V. infusions is shown. Hypertonic NaCl infused centrally increased drinking during the infusion and during the whole experiment. The three non-electrolyte solutions infused I.C.V. (0.3 M sucrose, 0.9 M sucrose or water) inhibited drinking during the infusion. Soon after the end of the combined infusion birds that had been infused I.C.V. with isotonic (0.3 M) sucrose or water began drinking. By the end of the experiment the intakes were not significantly different from those of the controls. Hypertonic sucrose I.C.V. attenuated drinking in response to the I.V. stimulus throughout the experiment. The inhibition of drinking during the combined infusion may have been caused by a decrease in csf Na concentration due to the I.C.V. infusion. These results are similar to those in the goat[4].

Drinking in response to I.V. 0.5 M NaCl and hypertonic NaCl infused simultaneously I.C.V. was not increased as was expected to a value approaching additivity. Further analysis of this response shows that drinking in response to the I.V. stimulus increased steadily over the whole experiment (curve a Fig. 1); hypertonic NaCl infused I.C.V. alone produced curve b, rapid drinking during and just after the infusion followed by very little intake for the remainder of the experiment; simple arithmetical addition of the responses a

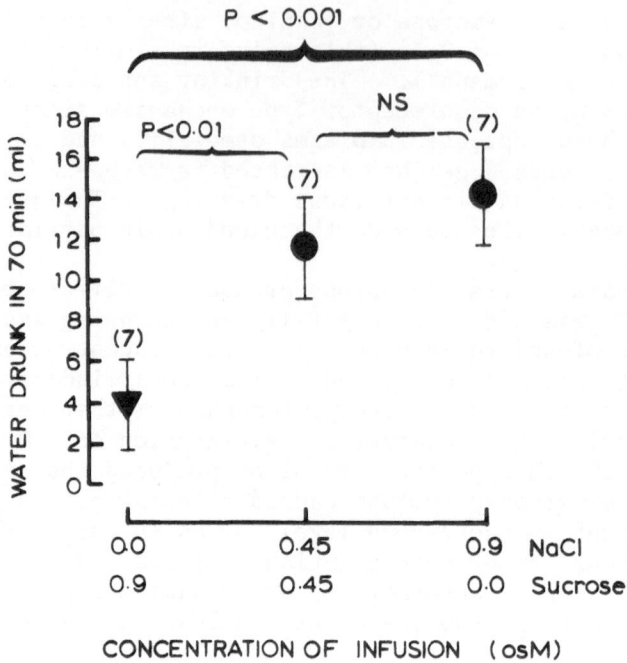

Fig. 1   Cumulative volume of water drunk in 70 min in response to
         I.C.V. infusions (2 μl/min for 10 min followed by 60 min)
         of hypertonic solutions of sucrose and NaCl, alone and
         together in equal concentrations.   Significance by paired
         t-test.

and b produced curve d;   whereas the final intake is shown by line
c.   The I.C.V. infusion of hypertonic NaCl produced a short duration
response (see line b, drinking was completed soon after the end of
the infusion) and when this was combined with the intravenous infusion the volume of water drunk by the end of the combined responses
appears to have been sufficient to switch off any further osmotically
induced intake.   The animal apparently 'knows' how much it has drunk
through oral metering and gastric distension[10] and this is enough to
dilute the only remaining stimulus to drinking, the infused osmotic
load, to isotonicity with the plasma.

    Simultaneous infusion of 1.0 M sucrose I.V. and solutions I.C.V.
produced much the same result as with I.V. 0.5 M NaCl.   Similar
I.C.V. infusions however had little influence on drinking in response

106

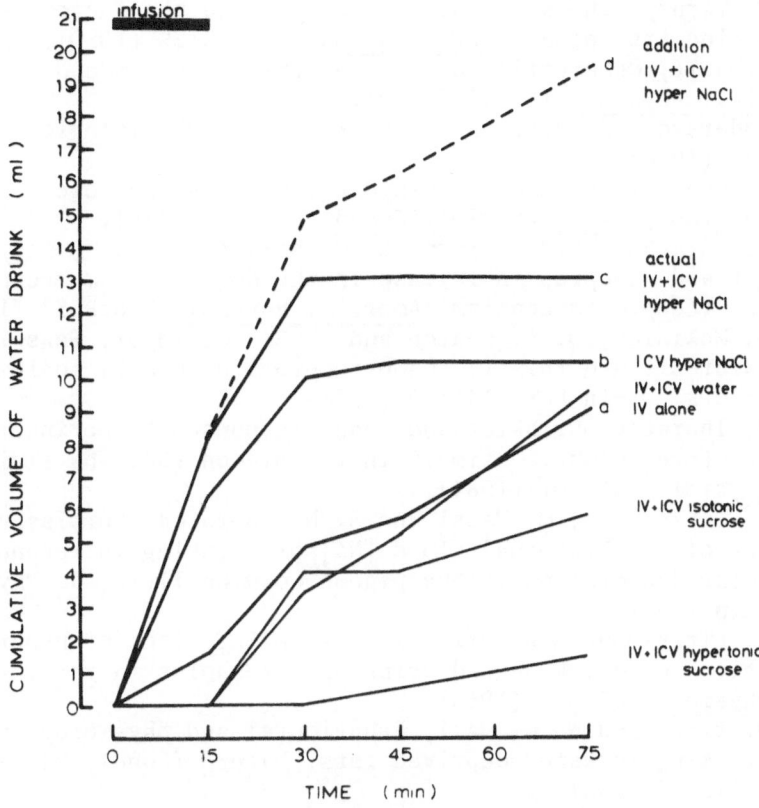

Fig. 2   Cumulative volume of water drunk in response to I.V. infusion
of 0.5 M NaCl with various solutions infused simultaneously
I.C.V.   Drinking was followed during the 15 min infusion and
for 60 min afterwards.   Error bars have been omitted to aid
clarity of the diagram.   I.V. alone is used as control
I.C.V. NaCl 0.3 M, I.C.V. isotonic sucrose 0.3 M, hypertonic
sucrose 0.9 M.

to two more concentrated solutions infused I.V. 1.0 M NaCl or 1.5 M
sucrose[11].   Thus the influence of changes in csf Na appears to be
dependent on the concentration of the solution infused I.V.   The
csf Na concentration may have been increased to such an extent by the
1.0 M NaCl and 1.5 M sucrose solutions that the I.C.V. infusions do
not significantly influence it.

In conclusion, drinking in the pigeon in response to osmotic
stimuli appears to be initiated by an osmoreceptor type mechanism but
it also appears to require a certain concentration of Na$^+$ in the csf
for it to operate.   Exactly how this mechanism operates remains to
be determined but it appears to be concerned with inhibiting water
intake following a decrease in Na$^+$ concentration rather than stimulating drinking in response to an increase in Na$^+$ concentration.

REFERENCES

1. E. B. Verney, The antidiuretic hormone and the factors which determine its release, Proc. R. Soc. B 135:25 (1950).
2. A. V. Wolf, Osmometric analysis of thirst in man and dog. Amer. J. Physiol. 161:75 (1950).
3. B. Andersson, Regulation of water intake, Physiological reviews 58:582 (1978).
4. K. Olsson, Attenuation of dehydrative thirst by lowering the C.S.F. (Na$^+$), Acta Physiol. Scand. 94:536 (1975).
5. T. N. Thrasher, C. J. Brown, L. C. Keil and D. J. Ramsay, Thirst and vasopressin release in the dog: an osmoreceptor or sodium receptor mechanism, Amer. J. Physiol. 238:R333 (1980).
6. M. J. McKinley, D. A. Denton and R. S. Weisinger, Sensors for antidiuresis and thirst: osmoreceptors or C.S.F. sodium detectors, Brain Res. 141:89 (1978).
7. S. N. Thornton, Drinking and renal responses to peripherally administered osmotic stimuli in the pigeon (Columba livia), J. Physiol. 351:501 (1984).
8. J. T. Fitzsimons, M. Massi and S. N. Thornton, Permissive effect of cerebrospinal fluid [Na] or drinking in response to cellular dehydration in the pigeon Columba livia, J. Physiol. 315:14p (1981).
9. J. T. Fitzsimons, The effects of slow infusions of hypertonic solutions on drinking and drinking thresholds in rats. J. Physiol. 167:344 (1963).
10. E. M. Blass and W. G. Hall, Behavioural and physiological bases of drinking in water deprived rats, Nature (Lond). 249:485 (1974).
11. Unpublished results.

RELATIONSHIP BETWEEN RIGHT ATRIAL STRETCH AND PLASMA
RENIN ACTIVITY

Susan Kaufman

Department of Medicine
University of Alberta
Edmonton, Alberta, Canada, T6G 2G3

INTRODUCTION

Stimulation of the right atrial receptors in rats attenuates
both water and salt intake[1,2]. It would be reasonable to propose
that this might be mediated by the renin-angiotensin system since
there is evidence that, in dogs, increased left atrial stretch
inhibits renin secretion [3,4,5]. However, the effects of right
atrial stretch have been less convincing. Although Brennan[6] et al
found that right atrial stimulation did suppress PRA, Lee et al[5]
found that it was the left atrium which primarily influenced renin
release. These experiments were carried out in dogs under very
different conditions from those used in the studies on thirst and
salt appetite. Therefore, it was decided to measure the changes
in PRA using the same model as had been used to study ingestive
behavior ie. inflation of a small balloon at the superior vena
caval/right atrial junction in normovolaemic and hypovolaemic rats
and after injection of the β -adrenergic agonist isoprenaline.

METHODS

Animals and experimental conditions

Male Wistar rats weighing between 269g and 423g, were housed
in temperature and humidity controlled rooms with the lights on
from 0700h. to 1900h. They had free access to food (Purina Rat
Chow) and water except where noted otherwise. At lease one week
after delivery they were prepared with indwelling venous cannulae,
atrial balloons and, in one group, peritoneal dialysis sacs. Six
days later the rats were placed in metabolism cages to familiarize

them with the experimental conditions. The next day, blood was taken for measurement of basal plasma renin activity (PRA). The following day the experiments were carried out as described below. The lines to the venous and balloon cannulae were led outside the metabolism cages so that blood could be withdrawn and the balloon inflated without disturbing the animals.

SURGERY

Surgery was performed at least one week after delivery. Under pentobarbital anaesthesia the right jugular vein was exposed and a 3 mm incision was made. Through this, a small inflatable balloon was advanced down the right superior vena cava so that its tip lay at the vein/atrial junction. This position was stabilized by ligating the cannula to the right clavicle (See Kaufman[1] for details). This balloon does not interfere with venous return from the head since, in the rat, there is a left superior vena cava which joins the inferior vena cava.

An indwelling cannula was also implanted non-occlusively into the inferior vena cava according to the method of Kaufman.[7]

Sacs were fashioned from a 12 cm. length of Spectrapor II dialysis tubing tied around a fenestrated cannula so as to allow material to be injected and withdrawn. These dialysis sacs were implanted into the peritoneal cavity at the time of venous cannulation. The three cannulae from the balloon, the inferior vena cava and the dialysis sac were led subcutaneously to the nape of the neck where they were attached to stainless-steel tubing which protruded through the skin. These tubes could be capped to prevent fluid escape.

Blood sampling

Blood was withdrawn into a clean syringe and rapidly transferred to Microtainer serum separator tubes (Becton & Dickinson, N.J., USA) to which had been added EDTA (0.5 mg). The samples were then centrifuged at 13,700g for 6 min. in a previously cooled rotor. The plasma was stored at -43°C.

Assay for PRA

Plasma renin activity was assayed using the RIANEN radioimmunoassay kit (New England Nuclear).

DRUGS

Isoprenaline hydrochloride (Isuprel, Winthrope) was freshly diluted with sterile isotonic saline to 20µg/ml immediately before injection. Polyethylene glycol (mol. wt. 20,000. Sigma) was dissolved with stirring in warm sterile isotonic saline (50% w/w/) about 1 hr before use.

STATISTICS

Results are expressed as mean ± S.E. of mean. In graphs, the S.E. of mean is delimited by the vertical bars. The significance of the difference between means was determined by Students' t-test.

RESULTS

Normovolaemic animals

Six rats bearing atrial balloons and venous cannulae were placed in metabolism cages with access to drinking water. At least 1 hr. later blood (0.5 ml) was withdrawn and the balloon was inflated. Thirty minutes later blood was again taken and the balloon deflated. One hour later a final blood sample was taken.

As can be seen from Fig. 1, there was no significant difference between the control and experimental groups after 30 min. of balloon inflation.

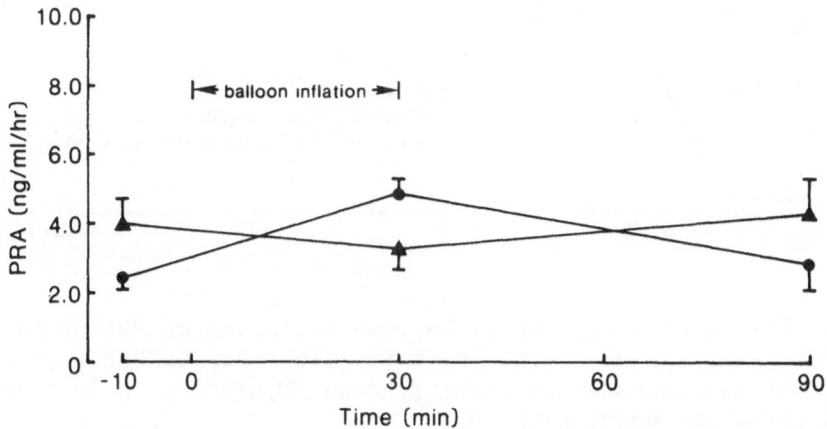

Fig 1. Effect of right atrial balloon inflation on PRA of normovolaemic rats. Control group (●,n=4), experimental group (▲,n=20)

Hypovolaemic animals

On the day of the experiment, a solution (25% w/w/) of polyethylene glycol was injected into the dialysis sacs (15ml/kg body
wt.) and the rats were returned to the metabolism cage without
access to water. Seven hours later blood was taken from all the
animals. The balloons were then inflated in the experimental
group. Fifteen and ninety minutes later blood was again sampled.
The balloons were then deflated and the sacs drained.

The mean volume deficit achieved in this manner was 16.3 $\pm$
0.5 ml (n=32) which corresponds to a 14% deficit in extracellular
fluid volume. As would be predicted, PRA increased as a result of
the depletion (Fig. 2). However, 15 min. after the balloons had
been inflated there was no significant difference between the
experimental (inflated) and control (non-inflated) groups. At 90
min., there was still no significant difference between the mean
values of the two groups although PRA had risen significantly more
in the control group (+3.5$\pm$0.7ng/ml/hr,n=14) than in the experimental group (+0.6$\pm$0.6ng/ml/hr,n=19,p$<$0.005).

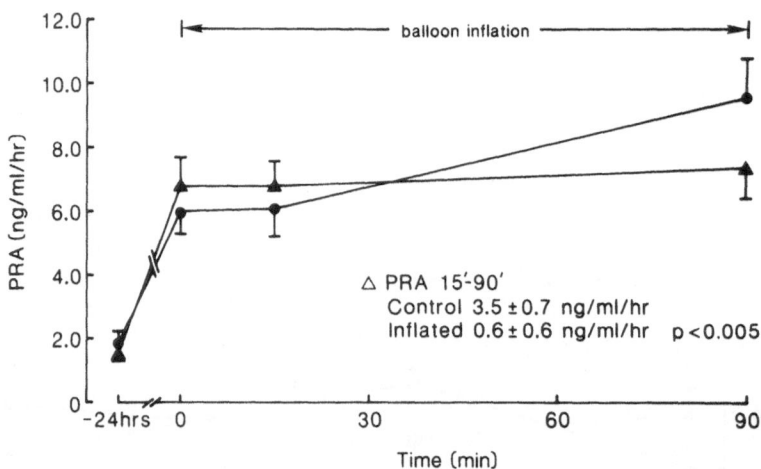

Fig 2. Effect of right atrial balloon inflation on PRA of rats
suffering extracellular fluid deficit. Increase in PRA from 15
min. to 90 min. greater in control group (●,n=14) than in experimental group (▲, n=19, p$<$0.005).

Isoprenaline

After blood was withdrawn for measurement of basal PRA,
isoprenaline was injected S.C. (10 µg/kg body wt) into 13 rats.

112

In seven of the animals the right atrial balloons were inflated. Twenty minutes later blood was taken and the balloons deflated. One hour later a final blood sample was taken.

Plasma renin activity increased as a result of the isoprenaline but there was no significant difference between the experimental (inflated) and control (non-inflated) groups (Fig. 3)

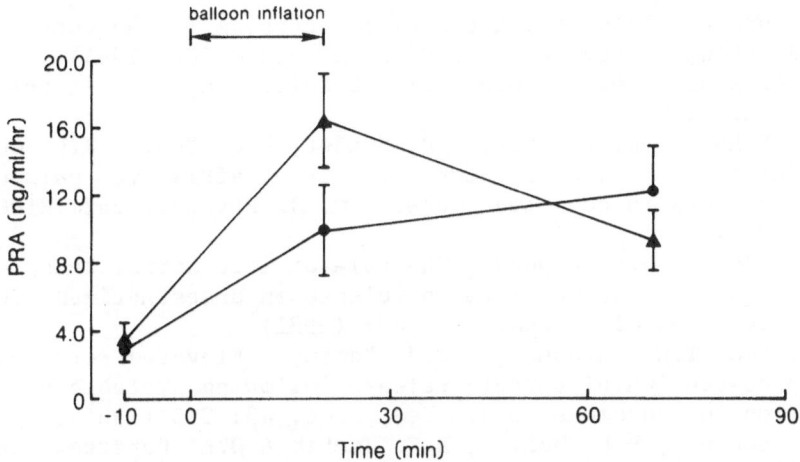

Fig 3. Effect of right atrial balloon inflation on PRA of rats injected at time 0 with isoprenaline (10 ug/kg body wt.). Control group (●,n=6), experimental group (▲,n=7).

DISCUSSION

Stimulation of the right atrial receptors appeared to have very little influence on plasma renin activity. This is not surprising in the normovolaemic animals where PRA is already low probably because their food (Purina rat chow) provides many times their minimum daily requirement of salt. Nor is it perhaps surprising that isoprenaline-induced release of renin should be refractory to reflex suppression since presumably a large component of the renin release resulted from a direct effect on the β-adrenergic receptors of the granular cells.[8] However, one might have expected to see an effect in the hypovolaemic animals. Inflation of the balloon did prevent the further increase in PRA which was observed in the control group and which probably occurred as a result of removing blood samples from an already depleted rat. However this effect was minimal and was only observed at the 90 min. sample. There was no inhibition after 15 min. of balloon inflation. This is important in light of the previous results

that the drinking response of similar hypovolaemic rats is completely inhibited at 15 min.[1] One must then conclude that the
attenuation of drinking to both hypovolaemia and isoprenaline that
results from stimulation of the right atrial receptors is not
exclusively mediated by reflex suppression of the renin angiotensin system but that other pathways, neural and/or hormonal, must
exist.

REFERENCES

1. S. Kaufman. Role of right atrial receptors in the control of
    drinking in the rat. J. Physiol. 349: 389 (1984).
2. S. Kaufman. Control mechanisms of salt appetite. In this
    publication.
3. H.D. Schultz, D.C. Fater, W.D. Sundet, P.G. Geer & K.L. Goetz.
    Reflexes elicited by acute stretch of atrial vs. pulmonary
    receptors in conscious dogs. Am. J. Physiol. 242: H1065
    (1982).
4. H. Holdaas & G.F. Dibona. The role of left atrial receptors
    in the regulation of renin release in anaesthetized dogs.
    Acta. Physiol. Scand. 111: 497 (1981)
5. M.E. Lee, T.N. Thrasher, & D.J. Ramsay. Elevated left heart
    pressure inhibits renin release following systemic hypoten-
    sion in conscious dogs. Fed. Proc. 43: 995 (1984).
6. L.A. Brennan, R.L. Malvin, K.E. Jochim & D.E. Roberts. The
    influence of right and left atrial receptors on plasma
    concentrations of ADH and renin. Am. J. Physiol. 221: 273
    (1971).
7. S. Kaufman. Chronic, nonocclusive and maintenance – free
    central venous cannula in the rat. Am. J. Physiol. 239:
    R123 (1980).
8. J.O. Davis & R.H. Freeman. Mechanisms regulating renin release.
    Physiol. Rev. 56: 2 (1976).

A SINGLE EXPERIENCE WITH HYPERONCOTIC COLLOID DIALYSIS PERSISTENTLY ALTERS WATER AND SODIUM INTAKE

Sandra P. Frankmann, Daniel M. Dorsa*, Randall R. Sakai, and John B. Simpson

Department of Psychology, and *GRECC, VA Medical Center and Department of Pharmacology; University of Washington Seattle, WA 98195

INTRODUCTION

Subcutaneous administration of high molecular weight colloids, such as polyethylene glycol (PEG), in hyperoncotic concentrations produces acute increases in water and saline intake, and is used extensively as a model system for hypovolemic thirst (1). Homeostasis in this state is defended both by renal conservation of water and sodium and by the ingestion of water and sodium. In the course of analyzing this effect, we noted that rats receiving an injection of PEG showed a persistently elevated preference for 0.3 M NaCl. Similar persistent increases of saline intake have been noted by other investigators (2,3,4). The experiments described here were designed to investigate several possible mechanisms underlying the prolonged increase in saline intake which followed a single experience with PEG.

GENERAL METHODS

Male Long-Evans rats, weighing 250-300 g at the beginning of an experiment, were individually housed in a temperature controlled room with a 12:12 hr D/L cycle. Purina chow, water and 0.3 M NaCl were available ad lib, except where otherwise noted. Daily fluid intakes and weekly body weights were recorded.

Injection of PEG (MW=20,000; 20% w/v; 16.7 ml/kg, s.c.) or the isotonic saline vehicle was made between the scapulae while the rat was under light ether anesthesia. Hourly intakes of water and 0.3 M NaCl, to the nearest ml, were recorded for the next 8 hours and then again at 24 hours following the injection.

EXPERIMENT I

    Baseline measures of water and 0.3 M NaCl on ad lib access
were obtained for five days. Rats were assigned to either the PEG
(n=5) or the vehicle (n=3) injection group such that mean saline
intake was the same in the two groups. Following the acute depletion, mean daily preference for saline (0.3 M NaCl intake/[water
intake + 0.3 M NaCl intake]) was significantly increased (Fig.1).

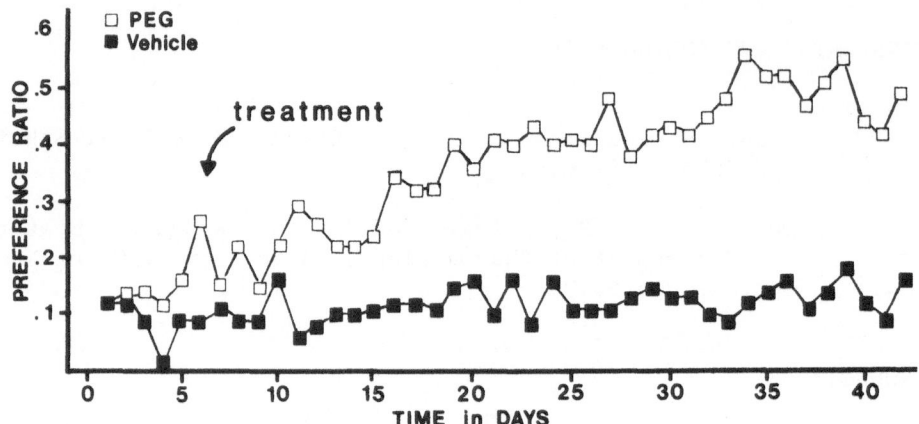

Fig. 1 Daily preference ratio for 0.3 M NaCl.

This was primarily accounted for by the intakes of 0.3 M NaCl,which
increased from a baseline of 5 ml/day to 30 ml/day over the 2-3
weeks following PEG treatment. Mean daily water intake was decreased
by 7 ml/day immediately following the acute depletion.

    These effects persisted for at least 3 months, during which
time the body weight gains of both groups were equivalent. Plasma
sodium concentration (138.15 vs 138.34 mEq/1), osmolality (297.7
vs 297.0 mOsmol/kg) and hematocrit (46.3 vs 46.7%) were not significantly altered in the vehicle vs. the PEG-treated rats at 5 weeks
following treatment.

    Thus, a single experience with an acute ECF depletion induced
by subcutaneous injection of PEG is followed by altered fluid intake
for extensive periods of time. Because the plasma sodium, plasma
osmolality, and hematocrit are not different, the increased saline
intake does not arise from major, chronic perturbations of key parameters of body fluid homeostasis, nor does it cause such perturbations.

The second experiment investigated the possibilitiy that the alterations in sodium intake were due to altered renal handling of sodium. The third experiment was a preliminary investigation into the possibility that some form of a learning mechanism underlies the altered daily fluid intake.

EXPERIMENT 2

At 6 weeks following PEG or vehicle injection, saline intake had stabilized at a new, higher level than control. At this time, the effects on fluid balance of restriction of food, water and saline for 24 hours or sodium deprivation for five days were measured.

In a first set of rats (n=5/group; PEG or vehicle treatments), all food, water and saline was removed from the home cage. Urine samples were collected at 30 and 60 minutes, hourly for the next 7 hours, and again at 24 hours after deprivation onset. There were no differences between groups in the urine volume or urine sodium loss at any measurement time over the 24 hour period (Fig.2). At 24 hours, body weight was recorded and saline and water returned. Intakes of both fluids were recorded at 30, 60 and 120 minutes. Body weight loss (9%) was the same in the two groups over the deprivation period. The saline intake did not differ significantly between groups at any time during the 120 minutes. At 30 minutes, the PEG-treated rats had consumed significantly more water than the vehicle-treated rats, although thereafter no differences between groups in water intake were present.

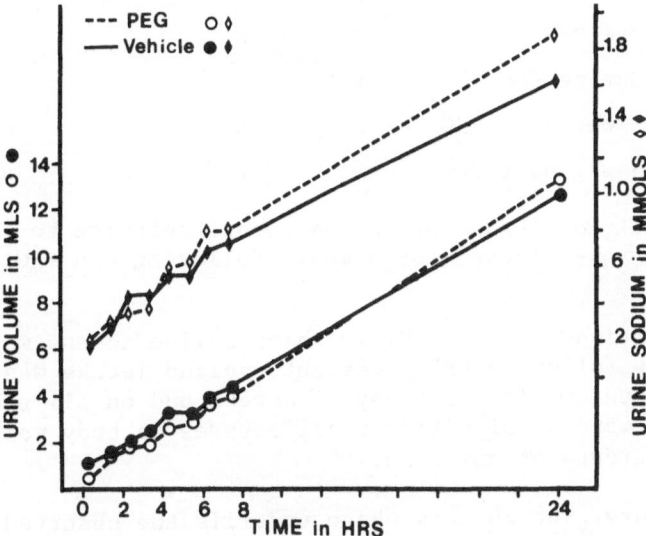

Fig. 2. Cumulative urine volume and sodium excretion over 24-hours of food and fluid deprivation.

In a second group of rats, Purina chow was replaced with a sodium deficient diet (ICN #902904) for 5 days, with water as the only available fluid. Urine volume was recorded every 24 hours and the sodium concentration was analyzed. Body weights and water intakes were recorded daily.

The water intake of the PEG-treated rats was significantly lower than vehicle-treated rats before, during and after the 5 days of dietary sodium restriction. Urinary volumes were also smaller. On the first night of sodium restriction, the PEG-treated rats had a significantly greater urinary sodium concentration than the vehicle treated rats, but this difference was not maintained on subsequent nights. Excessive sodium loss, then, does not occur in the PEG-treated rats during periods of dietary sodium restriction.

EXPERIMENT 3

The pairing of saline intake with a subsequent reduction of a sodium need could result in a learned preference for saline. In this experiment, access to saline was varied in time relative to the need for sodium induced by PEG treatment, as shown in Table 1.

Table 1.   Times During Which 0.3 M NaCl was Available, Relative to PEG Treatment

CONDITION	I	II	III	IV
TIME				
Pre-PEG	x	x	x	
During PEG	x		x	
3 wk. post-PEG	x	x		
5 wk. post-PEG	x	x	x	x

Under all conditions of saline access relative to PEG injection, saline intake was elevated at 5 weeks following the PEG treatment (Fig.3).

In conditions III and IV, in which saline access was delayed until 3 weeks following PEG treatment, saline intake was significantly elevated on the first day of access and on all subsequent days. There were no significant differences in body weight gain between the groups at any time.

In summary, the changes which underlie the observed behavior of increased saline intake are independent of the experience of saline intake before, during and/or immediately after acute depletion of the ECF by injection of PEG. These changes do not appear to be secondary to exaggerated renal sodium loss.

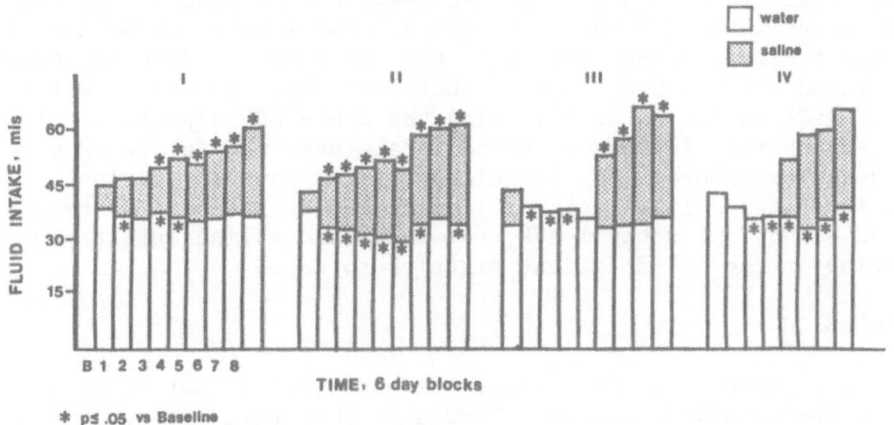

Fig. 3. Intake of water and 0.3 M NaCl, in six day blocks. First
block represents baseline intake; 24-hour intake following
PEG treatment is not included.

Changes of certain body fluid parameters, such as plasma osmolality
or sodium concentration, did not occur.

Water intake is decreased immediately following PEG treatment,
an effect which persists even in the absence of sodium intake.
This behavioral change may reflect an alteration in water metabolism.
One compensatory mechanism induced by an acute hypovolemia is the
secretion of the antidiuretic hormone, vasopressin (AVP). The
final experiments examine the relationship between AVP and the
altered fluid intakes.

EXPERIMENTS 4A and 4B

Trunk blood was collected from unanesthetized adult male Long-
Evans rats at 3 and at 5 weeks following an injection of PEG or
saline vehicle. Plasma AVP, measured by radioimmunoassay, was elevated in the PEG-treated rats (1.98 µU/ml) relative to the vehicle
treated rats (0.38µU/ml). Thus, the increased saline intake and
decreased water intake is accompanied by an increased plasma AVP
concentration.

In order to evaluate the necessity of the elevated AVP levels
in the development of the increased saline intake, a group of AVP-
deficient rats (Brattleboro strain) and their Long-Evans controls

were tested in the same manner as described previously and their subsequent intake of saline observed. The intakes of saline prior to PEG treatment were similar in the two groups. Both the Long-Evans and the Brattleboro rats increased their intake of water and 0.3 M NaCl in the 24 hr following the acute ECF depletion. The Long-Evans rats increased saline intake over the following 4 weeks, as described above. The Brattleboro rats, however, did not. Thus, the increased saline intake following PEG treatment is paralleled by an elevation in basal AVP; the increased saline intake does not develop in the AVP-deficient Brattleboro rat.

DISCUSSION

The present results extend the observation that acute body fluid perturbations can be followed by long-term alterations in fluid intake (2,3,4). We observed that a single depletion of the ECF by hyperoncotic colloid dialysis is followed by persistent elevations in the intake of 0.3 M NaCl as well as decrease in subsequent water intake.

These behavioral changes apparently are not secondary to a renal perturbation in sodium conservation. The elevated saline intake occurs in depletion, and persists in the absence of continuing experience with saline intake. These data suggest that the increase in saline intake is not the result of learning the restorative properties of saline during hypovolemia, or to some form of a learned preference for saline.

The observation that water intake is decreased while saline intake is increased following the PEG treatment is consistent with an enhanced ability to retain water. Similarly, Wright and Schulz (5) noted an enhanced ability to maintain plasma volume following repeated experiences with starvation-induced hypovolemia. Consistent with the above notion is the finding that plasma AVP concentrations are elevated at 3 and 5 weeks after PEG treatment and that homozygous Brattleboro rats, which lack endogenous AVP and acutely increase saline intake to PEG treatment, do not manifest a persistently increased saline intake. The Long-Evans rat, which does not lack AVP and which does develop an increased saline intake following an injection of PEG, may be making a behavioral adjustment to an enhanced ability to retain water. Thus, the enhanced saline intake may function by acting as a diuretic in the maintenance of water balance and not necessarily in the maintenance of sodium balance. Whether the elevated AVP levels are dependent on the increased saline intake and whether there is indeed an enhanced water retention remains to be determined.

An important implication of this research regards the common assumption that a body fluid depletion produces only acute and reversible effects. These data indeed question the use of multiple tests employing hyperoncotic colloid dialysis. They also question whether any experimental manipulation producing an acute body fluid depletion can occur without also producing persistent behavioral and physiological alterations.

REFERENCES

1. E. M. Stricker, Thirst and sodium appetite after colloid treatment in rats, J. Comp. Physiol. Psych. 95: 1 (1981).
2. R. W. Bryant, A. N. Epstein, J. T. Fitzsimons and S. J. Fluharty, Arousal of a specific and persistent sodium appetite in the rat with continuous intracerebro-ventricular infusion of angiotensin II, J. Physiol. (London). 301:365 (1980).
3. J. L. Falk and J. M. Lipton, Temporal factors in the genesis of NaCl appetite by intraperitoneal dialysis, J. Comp. and Physiol. Psychol. 63:247 (1967).
4. A.L.R. Findlay and A. N. Epstein, Increased sodium intake is somehow induced in rats by intravenous angiotensin II, Hormones Behav. 14:86 (1980).
5. J. W. Wright and E. M. Schulz, Influence of repeated deprivation upon starvation-induced hypovolemia and plasma aldosterone concentration in rats, Pharmacol. Behav. 16:697 (1982).

THE BRAIN-RENIN-ANGIOTENSIN SYSTEM: UPDATE

Thomas Unger, Detlev Ganten, Gerald Ludwig,
and Rudolf E. Lang

German Inst. for High Blood Pressure Research and Dept.
of Pharmacology, Univ. of Heidelberg, Im Neuenheimer
Feld 366, D-6900 Heidelberg, F.R.G.

Evidence for the presence of components of the renin angiotensin system (RAS) in the brain of various species has recently been reviewed (1-4). We discuss here some recent advances concerning the characterisation of renin, angiotensinogen, converting enzyme and angiotensin in the brain and their localization in areas which are relevant to volume homeostasis and cardiovascular control.

Renin (E.C. 3.4.23)

The primary structure of renin has been determined by classical amino acid sequence analysis (5) and was derived from the nucleotide sequences of cloned cDNAs complementary to their mRNA (6,7). The mature renin molecule consists of two chains. The heavy chain contains 288, and the light chain 48 amino acid residues. Data from different laboratories are in good agreement.
Mouse submaxillary gland renin exhibits a 43% sequence homology with porcine pepsin, 34% identity with bovine chymosin and 22% identity with penicillopepsin (5). The overall dimension and shape of the renin molecule appears to be similar to other acid proteases. Renins isolated from various sources including hog, rat and human kidney renin, bovine pituitary renin, and mouse submaxillary renin are all similar in such general molecular properties as amino acid composition, chain length, molecular weight, isolelectric points (see 3,5-10).

The fact that a number of acid proteases, including cathepsin D, can generate ANG I from angiotensinogen, has lead to a controversy whether the brain contained "true" renin. This controversy was terminated when it was demonstrated that brain renin is active at neutral pH, can be separated from cathepsin D and other acid proteases, is inhibited by specific renin antibodies and peptide inhibitors, and is active in vivo (3,8-10).

The ultimate proof for local synthesis of renin in the central nervous system will stem from recombinant cDNA techniques as has been shown for brain angiotensinogen. We have used a renin cDNA probe for hybridization studies with mRNA isolated from mouse tissue. The probe, plasmid pMSR 49 (7), contains a 700 bp insert of mouse submaxillary cDNA cloned into pBR 322. This approach was used for hybridization studies with RNA from brain and other organs of male mice (11).
Briefly, Messenger RNA (mRNA) was isolated from total RNA by chromatography on oligo-(dT) cellulose, denatured and then fractionated electrophoretically by size on formaldehyde-agarose gels. The mRNA was then blotted from the gel onto nitrocellulose. Hybridization on these filters was performed with nicktranslated heat denatured (alpha ^{32}p)-labeled plasmid pMSR 49 containing submaxillary gland renin cDNA (Northern blot). After washing hybrid bands were visualized by autoradiography using Kodak X-ray screen. Preliminary results show that the hybridization bands in brain corresponded to those found in submaxillary gland and in kidney.
Relative amounts of renin mRNA in mouse organs were determined by dotting analysis. For this purpose mRNA was spotted and fixed on nitrocellulose filters. Hybridization was done under the same conditions as for Northern blotting and the dried filters were counted directly in liquid scintillation fluid or exposed to X-ray film. Direct counting and densitometric analysis of the dot blots revealed the following rank order of hybridization: submaxillary gland > kidney > brain (11).
These data are in harmony with the idea that the renin gene is expressed and that the protein is synthetized locally in brain. However, the final proof will require one or more of various approaches including cell free translation of brain mRNA and identification of the newly synthetized protein as renin, further characterization of the specificity of hybridization by digestion with nuclease S1 ("S1 mapping"), or sequencing of cDNA transcripts synthetized by reverse transcriptase. Such

work is currently underway in several laboratories.

Angiotensinogen

The complete sequence of angiotensinogen was recently
determined by recombinant cDNA techniques from a clone
selected from a rat liver cDNA bank and subjected to
nucleotide sequence analysis (12). The deduced amino
acid sequence indicated that the precursor molecule
consists of a mature angiotensinogen and a putative
signal peptide of 24 amino acids. The predicted molecular weight and amino acid composition of angiotensinogen agreed well with those obtained by amino acid analysis of the purified protein (13). The ANG I moiety is
located at the amino terminal part of the molecule,
followed by a large carboxy-terminal sequence. This
carboxy-terminal sequence contains two small internally
homologous sequences and three potential glycosylation
sites. The possibility that the carboxyl-terminal region of angiotensinogen has some biological role after
the release of ANG I, still awaits investigation.

Using the technique of cell free translation of mRNA,
Campbell et al. (14) recently provided evidence that
the same angiotensinogen molecule is synthetized in the
liver and locally in the brain. In their studies, ^{35}S-
Methionine labeled angiotensinogen precursors were
synthetized by cell free translation of either rat
brain or rat liver mRNA and compared by immunoprecipi-
tation, sodium dodecyl sulfate polyacrylamide gel electrophoresis, and autoradiography. Rat liver mRNA
synthetized two angiotensinogen precursors: a major
precursor of molecular weight 52.5k and a minor precursor of MW 55.7k. Identical and similarly abundant precursor forms to those observed for liver, were synthetized by cell free translation of rat brain mRNA. Both
brain angiotensinogen precursors were cleaved by renin,
resulting in a single cleavage product with a molecular
weight of 47.5k, identical to that observed for liver.
Bilateral nephrectomy and dexamethasone administration
produced less than a two-fold increase in translatable
levels of brain angiotensinogen mRNA, in contrast to
the several-fold increase observed for liver. These results show that although rat brain and liver angiotensinogen mRNAs appear to be products of the same ge-
ne(s), the regulation of their transciption is tissue
specific (14).

Converting Enzyme (CE) (EC 3.4.15.1)

CE is widely distributed throughout the brain. The localizations obtained in microdissection studies measuring CE catalytic activity and those described using immunohistochemical techniques are in reasonably good agreement. Recent data were obtained by autoradiographic visualization of CE with ^{3}H-captopril.

The highest concentrations occur in the choroid plexus, subfornical organ, caudate-putamen, zona reticulata, substantia nigra, globus pallidus and median eminence. In certain areas (e.g. entopeduncular nucleus, medial habenula, median preoptic area), however, there is disagreement between the autoradiographic and biochemical or immunological data (16). Brain blood vessels contain CE, but the enzyme is clearly also located neuronally as evidenced in cell culture and by ultracentrifugation studies. The latter techniques produced evidence that CE is present in synaptosomes (17,19).

The striato-nigral localization of CE is particular interest.
Ibotenic acid, selectively destroys neuronal cell bodies intrinsic to the site of injection without damaging glial elements, extrinsic nerve terminals and axons of passage (15). Injections of ibotenic acid into the caudate-putamen produced a decrease of CE at the site of injection and, later on, a depletion in the substantia nigra. On the other hand, the same injections into the substantia nigra were without effect on CE activity (15,16). The findings show that CE has a neuronal localization within the corpus striatum and that the CE producing neurons (cell bodies) project to the ipsilateral substantia nigra. Glia seems to be devoid of CE. The decrease of CE in the candate-putamen was associated with an increase in renin activity which could represent a compensatory effect (15). The finding that the typical destruction of the corpus striatum in Huntington's disease is paralleled by a depletion of CE activity in the substantia nigra is noteworthy in this respect (20). Surprisingly it has not been possible to demonstrate ANG II-receptors and ANG II immunoreactivity in the striato-nigral structures. CE may nevertheless play an important role in these struture since the enzyme has a broad specifity and may hydrolyse other peptides as well.

Brain areas where ANG II has been shown to occur with no CE include parts of the spinal cord, the bed nucleus of the stria terminalis and the central nucleus of the amygdala. The significance of this remains to be investigated. The different ratio of ANG I/ANG II in various brain areas (21) would be consistent with the interpretation, that CE determines in certain areas the activity of the RAS in the brain.

The presence of CE in specific brain regions has become of particular interest since inhibitors of the enzyme have been introduced as antihypertensive drugs (22). These agents provide a new tool to study brain peptide metabolism and there is increasing evidence that they act on brain CE and thereby effect volume homeostasis and cardiovascular control even if given orally (22-25).

Angiotensin and its projections

Angiotensin I (ANG I) and Angiotensin II (ANG II) have been extracted from brain of neprectomized rats, rabbits and primates (21,26). Peptides were characterized with high performance liquid chromatography (HPLC) capable of separating all angiotensins and their fragments including (Val^5)-ANG II from (Ile^5)-ANG II. The peptides extracted from brain corresponded to synthetic (Ile^5)-ANG II and (Ile^5)-ANG I, with small amounts (approximately 10%) of (Ile^5)-ANG(2-8) (ANG III) being present in the brain. Brain angiotensin thus appears to have the same amino acid sequence as plasma angiotensin. The identical peptide was cleaved from brain and plasma angiotensinogen upon incubation with renin in vitro and in vivo (21).

The distribution of ANG II in the central nervous system has been investigated biochemically (21) as well as by immunohistochemical techniques (27-32). The main locations of ANG II are the hypothalamus, the limbic system, the medulla oblongata and the spinal cord. High density of ANG II-positive nerve terminals exists within the median eminence, in the nucleus paraventricularis, the supraoptic nucleus and the subfornical organ (SFO). Further, ANG II-positive brain areas are the substantia gelatinosa of the spinal cord, nucleus tractus spinalis nervi trigemini, nucleus amygdaloideus centralis, sympathetic lateral column, nucleus dorsomedialis hypothalami, locus coeruleus.

The presence of ANG II in the paraventricular nucleus of the hypothalamus (PVN) is of particular interest in view of the capability of ANG II to release adrenocorticotrophic hormone (ACTH) and in view of the projections from the PVN through the external layer of the median eminence to the portal blood circuit.

Lind et al. (32) recently studied the PVN/ANG II system in detail. It was confirmed that antisera to ANG II stain neurosecretory neurons that synthetize vasopressin in magnocellular parts of the PVN but it was also shown, that ANG II-immunoreactive neurons were scattered throughout the parvocellular division. A subpopulation of ANG II-immunoreactive parvocellular neurons in the PVN projects to the neurohemal zone via the external layer of the median eminence. These ANG II-stained projections were eliminated after bilateral destruction of the PVN. In contrast, the ANG II-stained magnocellular neurons in the PVN appear to project through the internal lamina of the median eminence to the posterior pituitary. Bilateral lesions of the PVN reduce, but do not eliminate ANG II staining in the internal lamina. The remaining fibers probably arising from ANG II-stained magnocellular neurons in the supraoptic nucleus.

These studies suggest that ANG II-stained fibers in the external and internal laminae of the median eminence arise from separate groups of neurons, a conclusion also supported by observations that these areas can be regulated independently. For instance, adrenalectomy leads to a selective enhancement of ANG II-stained fibers in the neurohemal zone as has also been described for CRF, oxytocin, and vasopressin. In contrast, the selective increase of ANG II staining in the internal lamina of the median eminence following water deprivation. supports a functional relationship of this ANG II pathway to the posterior pituitary in vasopressin and oxytocin release.

Another area of the brain, which is of particular interest for the regulation of fluid balance is the subfornical organ (SFO). The SFO is a small glomus-like convexity of the midline third ventricular ependyma near the interventricular foramen. It is densely vascularized, has relatively porous blood brain barrier and is strategically located to monitor plasma, cerebrospinal fluid (CSF) and neuronal inputs. ANG II-receptors have been demonstrated on the SFO (33,34) and stimulation of the receptors in water satiated animals results in copious drinking.

Recently, Lind et al. (31) examined the ANG II pathways to and from the SFO by immunohistochemical methods. ANG II immunoreactive cell bodies and fibers were clearly identified in the SFO of the rat. Cells were distributed in an annulus around the periphery of the SFO. Fibers were observed in a plexus, located centrally within the ring of cells. Knife-cuts through the ventral stalk of the SFO diminished but did not eliminate fiber staining in the SFO. Ventral to the cut, and to a lesser degree, also dorsal to the cut, bright varicose ANG II immunoreactive fibers were described. Combination of immunohistochemistry with retrograde transport, identified the perifornical zone of the lateral hypothalamus, the rostral zona incerta and the nucleus reuniens of the thalamus as the source of ANG II-stained inputs to the SFO, and the region of the median preoptic nucleus as a recipient of ANG II-immunoreactive SFO efferents. It was concluded that ANG II-stained pathways from the lateral hypothalamus and adjacent regions project to the SFO, and that, in turn, ANG II-stained neurons within the SFO project to the preoptic region of the hypothalamus. Interestingly, the perifornical region of the lateral hypothalamic area, and rostral parts of the zona incerta that project to the SFO are known to be involved in the regulation of thirst. Furthermore, the ANG II-stained projection from the SFO to the preoptic region are also thought to play a critical role in the initiation of drinking behavior (34).

The staining was neither altered by water deprivation nor by nephrectomy. Since the former procedure drastically increases and the latter drastically decreases circulating levels of ANG II, tissue bound peptide from the blood does not appear to have been responsible for the observed immunoreactivity. Nevertheless, circulating ANG II may also affect ANG II-receptors in the SFO through the relatively permeable blood brain barrier. Thus, the SFO could function as an integrating center where hormonal and neuronal angiotensin interact to control fluid blance.

Together with previously described data on the localisation of ANG II in the brain a picture begins to emerge which gives further evidence to a role for angiotensin pathways in blood pressure and volume control.

Acknowledgements

This work was supported by the Deutsche Forschungsge-meinschaft (DFG) within the Sonderforschungsbereich "Cardiovasculäres System" (SFB90). The secretarial help of I. Büchler is gratefully acknowledged.

REFERENCES

1   Rettig R, Lang RE, Rascher W, Unger Th, Ganten D (1982) Brain peptides and blood pressure regulation. Clin Sci 63:269s-283s.

2   Reid IA, Morris BJ, Ganong WF (1978) The renin-angiotensin system. Ann Rev Physiol 40:377-410.

3   Ganten D, Printz M, Phillips MI, Schölkens BA, eds (1982) The renin angiotensin system in the brain. Experimental brain research (Suppl 4). Berlin, Heidelberg, New York, Springer.

4   Printz M, Ganten D, Unger Th, Phillips MI (1982) Minireview: The brain renin angiotensin system 1982. In: Exp Brain Res (Suppl 4). The renin angiotensin system in the brain, eds Ganten D, Printz M, Phillips MI, Schölkens BA, Berlin, Heidelberg, New York, Springer, pp 2-52.

5   Misono KS, Chang J-J, Inagami T (1983) Structure of mouse submaxillary gland renin. Clin Exp Hyper A5(7/8):941-959.

6   Rougeon F, Chambraud B, Foote S, Pathier J-J, Nageotte R, Corvol P (1981) Molecular cloning of a mouse submaxillary gland renin cDNA fragment. Proc Natl Acad Sci USA 78:6367-6371.

7   Imai T, Miyazaki H, Hirose S, Murakami K (1983) Cell-free translation of human renin mRNA. Clin Exp Hyper A5(7/8):961-967.

8   Hirose S, Ohsawa T, Inagami T, Murakami K (1982) Brain renin from bovine anterior pituitary: isolation and properties. J Biol Chem 257:6316-6321.

9   Hirose S, Yokosawa H, Inagami T, Wokrman KJ (1980) Renin and prorenin in hog brain: ubiquitous distribution and high concentration in pituitary and pineal. Brain res 191:489-499.

10   Ganten D, Speck G (1978) The brain renin angioten-
     sin system: a model for the synthesis of peptides
     in the brain. Biochem Pharmacol 27:2378-2389.

11   Ludwig G, Lehmann E, Murakami K, Lang RE, Unger Th,
     Ganten D (1984) Demonstration of specific renin
     messenger RNA in mouse brain. Submitted for publi-
     cation.

12   Nakanishi S, Ohkubo H, Nawa H, Kitamura N, Kageyama
     R, Ujihara M (1983) Angiotensinogen and kininogen:
     cloning and sequence analysis of the cDNAS. Clin
     Exp Hyper A5(7/8):997-1003.

13   Hilgenfeldt U (1983) Structural aspects of rat an-
     giotensinogen. Clin Exp Hyper A5(7/8):1021-1035.

14   Campbell DJ, Bouhnik J, Menard J, Corval P (1984)
     Identity of angiotensinogen precursors of rat brain
     and liver. Nature 308(5955):206-208.

15   Fuxe K, Ganten D, Köhler Chr, Schüll B, Speck G
     (1980) Evidence for differential localization of
     angiotensin I-converting enzyme and renin in the
     corpus striatum of rat. Acta Physiol Scand
     110:321-323.

16   Strittmatter SM, Lo MMS, Javitch JA, Snyder SH
     (1984) Autoradiographic visualization of angioten-
     sin converting enzyme in rat brain with (^{3}-
     H)captopril: localization to a striatonigral path-
     way. Proc Natl Acad Sci USA: in press.

17   Yang HYT, Neff NH (1972) Distribution and proper-
     ties of angiotensin converting enzyme of rat brain.
     J Neurochem 19:2443-2450.

18   Rix E, Ganten D, Schüll B, Unger Th, Taugner R
     (1981) Converting enzyme in the chorioid plexus,
     brain and kidney: immunocytochemical and biochemi-
     cal studies in rats. Neurosci Lett 22:125-130.

19   Paul M, Hermann K, Printz M, Lang RE, Unger Th,
     Ganten D (1983) The brain angiotensin system: sub-
     cellular localization and interferences with con-
     verting enzyme inhibitors. J Hypertension 1 (suppl
     1):9-15.

20   Arregui A, Bennet JP, Bird EO, Aymamura HJ, Iversen LL, Snyder SH (1977) Huntington's chorea: selective depletion of activity of angiotensin converting enzyme in the corpus striatum. Ann Neurol 2:294-298.

21   Ganten D, Hermann K, Bayer C, Unger Th, Lang RE (1983) Angiotensin synthesis in the brain and increased turnover in hypertensive rats. Science 221,4613:869-871.

22   Unger Th, Ganten D, Lang RE (1983) Pharmacology of converting enzyme inhibitors: new aspects. Clin Exp Hypertension A5(7/8):1333-1354.

23   Unger Th, Kaufmann-Bühler I, Schölkens B, Ganten D (1981) Brain converting enzyme inhibition: a possible mechanism for the antihypertensive action of captopril in spontaneously hypertensive rats. Eur J Pharmacol 70:467-478.

24   Evered MD, Robinson MM, Richardson MA (1980) Captopril given intracerebroventricularly, subcutaneously or by gavage inhibits angiotensin converting enzyme activity in the rat brain. Eur J Pharmacol 68:443-449.

25   Unger Th, Ganten D, Lang RE, Schölkens BA (1984) Is tissue converting enzyme inhibition a determinant of the antihypertensive efficacy of converting enzyme inhibitors? Studies with the two different compounds HOE498 and MK421 in spontaneously hypertensive rats. J Cardiovasc Pharmacol 6: 872-880.

26   Balz W, Herrmann K, Unger Th, Lang RE, Ganten D (1984) Angiotensin in monkey tissue, in preparation.

27   Ganten D, Fuxe K, Phillips MI, Mann JFE, Ganten U (1978) The brain renin angiotensin system: biochemistry, localization and possible role in drinking and blood pressure regulation. In: Frontiers in neuroendocrinology, eds. Ganong WF, Marin LD, New York, Raven Press, pp 61-99.

28   Brownfield MS, Reid IA, Ganten D, Ganong WF (1982) Differential distribution of immunoreactive angiotensin and angiotensin-converting enzyme in rat brain. Neuroscience 7(7):1759-1769.

29   Zimmerman EA, Krupp L, Hoffmann DL, Matthew E, Nilaver G (1980) Exploration of peptidergic pathways in brain by immunocytochemistry: a ten year perspective. Peptides 1(Suppl1):3-10.

30   Fuxe K, Ganten D, Hökfelt T, Bolme P (1976) Immunohistochemical evidence for the existence of angiotensin II containing nerve terminals in the brain and spinal cord of the rat. Neurosci Lett 2:229-234.

31   Lind RW, Swanson LW, Ganten D (1984) Angiotensin II immunoreactivity in the neural afferents and efferents of the subfornical organ of the rat. Brain Res: in press.

32   Lind RW, Swanson LW, Bruhn TO, Ganten D (1984) The distribution of angiotensin II immunoreactive cells and fibers in the paraventriculo-hypophysial system of the rat. Brain Res: in press.

33   Felix D, Schelling P, Haas HL (1982) Angiotensin in single neurons. In: Exp Brain Res (Suppl 4) The renin angiotensin system in the brain, eds. Ganten D, Printz M, Phillips MI, Schölkens BA, Berlin, Heidelberg, New York, Springer, pp 255-269.

34   Lind RW, Johnson AK (1982) Central and peripheral mechanisms mediating angiotensin induced thirst. In: Experimental brain research (Suppl 4). The renin angiotensin system in the brain, eds. Ganten D, Printz M, Phillips MI, Schölkens B, Berlin, Heidelberg, New York, Springer, pp 353-364.

ANGIOTENSIN-SENSITIVE SITES IN THE CENTRAL NERVOUS SYSTEM

D. Felix[1], M. C. Gambino[2], Y. Yong[1] and P. Schelling[3]

1: Division of Animal Physiology, Zoology Institute
   University of Berne, Berne, Switzerland
2: Istituto di Ricerche Farmacologiche "Mario Negri"
   Milan, Italy
3: Exp. Medical Research, E. Merck, Darmstadt, Federal
   Republic of Germany

INTRODUCTION

The classical view that peptide hormones have singular functions has had to be modified because it has become clear that
peptides not only have multiple effects but are also widely
distributed throughout the body. Since Bickerton and Buckley (1)
demonstrated that angiotensin can act directly on the brain we
have begun to realize how much more needs to be known about the
interaction between the brain and the periphery. Angiotensin
injected directly into the brain can induce an increase in blood
pressure, vasopressin release and a remarkable drinking behaviour.
Multiple physiological effects have been reported to be mediated
by hypothalamic and circumventricular structures (2). The evidence
so far implies that angiotensin is formed intracellularly and is
concentrated in nerve terminals (3). The presence of angiotensin
receptors in the brain implies that the peptide produces a physiological response. By defining these receptors we should eventually
be able to study such diverse processes as thirst motivation and
hypertension.

RESULTS

In order to locate possible sites of action for angiotensin
II (ANG II) we applied this peptide to active cells that are considered to be receptor sites, such as the <u>subfornical organ</u> (Fig.
1A) in the third ventricle. Evidence for this organ being a receptor
site for thirst is based on the following findings: 1) a very low

135

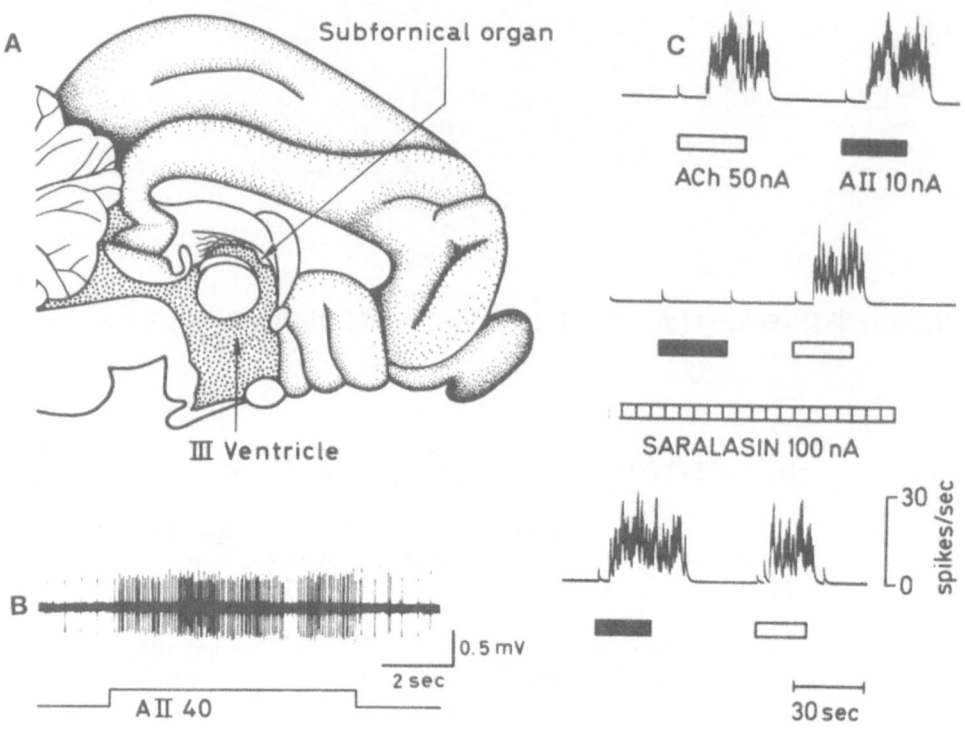

Fig. 1 A: Diagram of cat subfornical organ. B: Response of single SFO neurones to iontophoretically applied angiotensin II (A II, 40 µA). C: The effect of saralasin on acetylcholine (ACh) and angiotensin (AII) evoked responses of SFO neurones.

dose of ANG can elicit thirst, the threshold dose was 0.1 pg; 2) lesioning of the SFO apparently reduced thirst induced by ANG II; 3) injection of an ANG II antagonist, saralasin, into the SFO reduced intravenously induced thirst. One major feature of this organ is its lack of blood-brain barrier (for review see 4). Cell discharge was activated in the majority of neurones and a dose-response relationship was observed. In addition, intravenous and intraventricular application of ANG II excited neurones in this area. Excitation of neurones similar to that observed after application of ANG II was also seen with acetylcholine (ACh). We were able to differentiate between those cells which are specifically responsive to ANG II and those that responded to both ANG II and ACh. Similar results were obtained when ANG II and substance P were applied together (Fig. 2). ANG II-induced activation was specifically blocked by the antagonist saralasin (Fig. 1). In contrast, ACh-stimulated neurones were not affected by this antagonist, but blocked by atropine. The experimental system in the SFO was further used for structure activity studies with angioten-

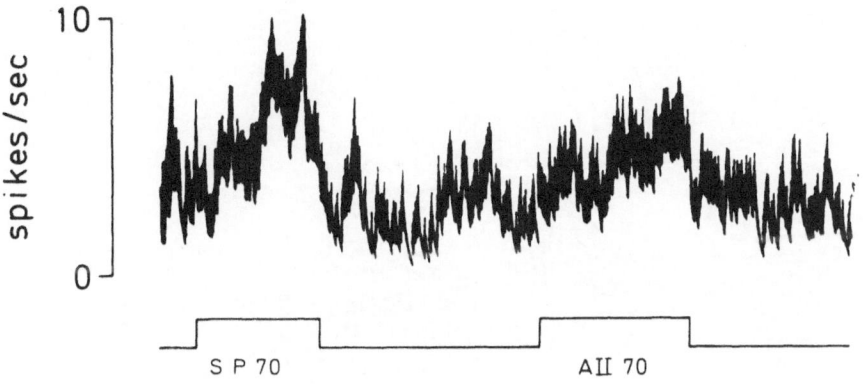

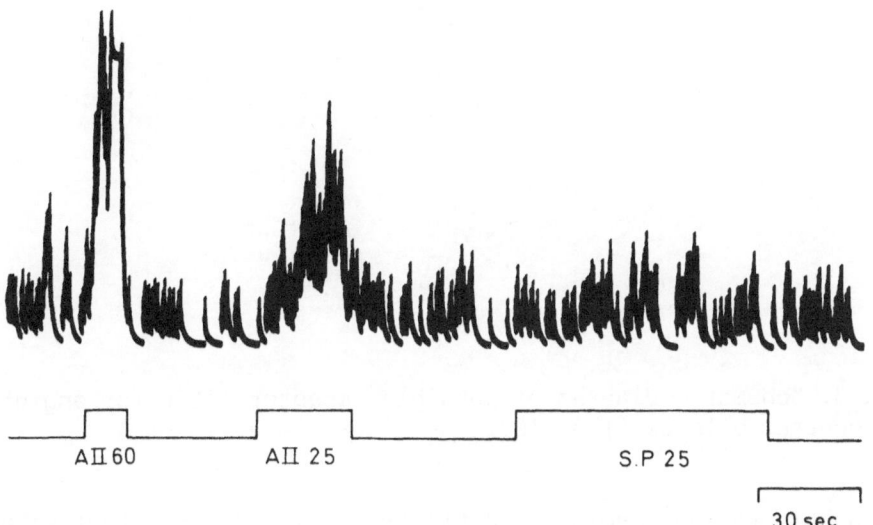

Fig. 2: The effect of angiotensin II (A II) and substance P on two SFO neurones.

sin fragments. We found that ANG [2-8] heptapeptide was more potent than ANG II, while the ANG [5-8] tetrapeptide was also active but was weaker than either the heptapeptide or the octapeptide. Saralasin blocked the action of both the heptapeptide and the tetrapeptide. In contrast, the ANG [6-8] tripeptide failed to enhance the firing rate. These results indicate that the tetrapeptide C-terminal sequence of ANG II is primarily responsible for receptor recognition at central neurones (5). The anterior third ventricle region has been identified as an important receptor site for the dipsogenic action of the peptide. The high proportion of neurones in the

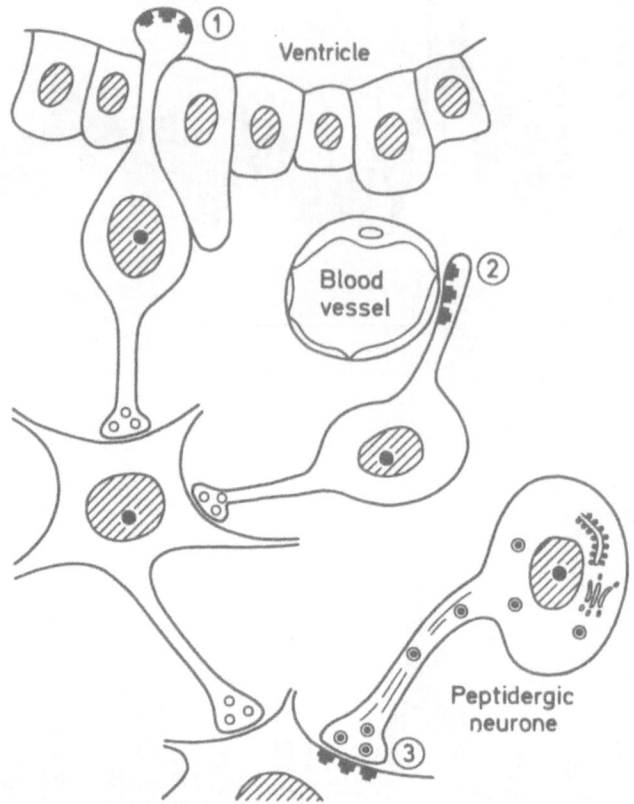

Fig. 3: Schematic diagram of possible receptor sites for angiotensin central effects (from 12).

supraoptic (6) and medial preoptic areas (7) is consistent with this evidence and suggests a direct mechanism of action on neuronal elements. We have recently observed that intravenous injections of ANG II stimulate preoptic neurones (Gambino and Felix, unpublished observations). The latency of neuronal activation, however, differs from the onset of the accompanying increase in blood pressure. Although we have concentrated on a ventricular site for the location of angiotensin receptors there is increasing evidence for the existence of ANG-receptive sites in regions lying within the blood-brain barrier. The fact that ANG-sensitive neurones have been located in regions with an intact blood-brain barrier is important in offering further evidence for the existence of an independent cerebral angiotensin system. We have recently investigated the septal areas of the rat, which contain cells having significant

ANG binding activity and which are one of the most sensitive sites for the dipsogenic response to ANG II. Our results indicate that the septum also contains cells having a specific chemosensitivity to ANG II (8).

In a recent study on hippocampal pyramidal cells we were able to show for the first time with intracellular recording techniques that ANG has a direct depolarizing effect on the neuronal membrane (9), an action which is antagonized by saralasin. In view of these results we are turning towards the question of the possible role of ANG II as a transmitter or a modulator. However, from the present study it is still too early to suggest an answer.

So far all our studies have been related to the dipsogenic action of ANG. Ganten (10) has shown that blocking central ANG II receptors or inhibiting cerebral ANG II synthesis lowers blood pressure in spontaneously hypertensive rats. Thus there is evidence for a functional role of the cerebral renin-angiotensin system in central blood pressure control. Furthermore, the specific binding activity of ANG II has been reported to be higher in spontaneously hypertensive rats. Using a microiontophoretic approach we intended to answer the question as to whether the receptor properties were changed in hypertensive animals. Neuronal firing evoked by the application of ANG II occurred at much lower thresholds in hypertensive rats and showed clearly extended postactivity as compared to age-matched normotensive Wistar-Kyoto rats. In contrast, activity induced by acetylcholine remained unchanged in both strains (11). In conclusion, the higher sensitivity of septal neurones in hypertensive rats contributes to the view that an activated cerebral renin-angiotensin system is involved in the genesis of hypertension in these rats.

Although no single piece of evidence alone proves or disproves the presence of an endogenous angiotensin-producing system in the brain, the electrophysiological experiments described here offer further evidence in favour of such a system. Angiotensin-sensitive neurones have been found in many areas in the brain and this specific chemosensitive action is confined to regions where the highest rates of angiotensin-binding activity were observed. So far, there is evidence that ANG reaches the receptor sites along different paths. Circumventricular organs may be target organs for ANG arriving either from the periphery or from the ventricular site. On the other hand, our results support the idea of the presence of peptidergic neurones within the central nervous system (Fig. 3, from 12). Although ANG II has been shown to be involved in water balance and blood pressure, it should be pointed out that ANG may have additional, more subtle effects which have so far remained undetected.

This work was supported by grant 3.627-0.84 from the Swiss National Science Foundation. Dr. Gambino was supported by a fellowship from the CEE. Y. Yong is a scholar from the University of Beijing.

REFERENCES

1. R. K. Bickerton and J. P. Buckley, Evidence for a central mechanism in angiotensin-induced hypertension, Proc.Soc.exp. Biol.(N.Y.) 106: 834 (1961).
2. M. I. Phillips, D. Felix, W. E. Hoffman and D. Ganten, Angiotensin-sensitive sites in the brain ventricular system, in: "Society for Neuroscience Symposia", W. M. Cowan and J. A. Ferrendelli, eds., (1977).
3. D. Ganten, M. Printz, M. I. Phillips and B. A. Schölkens, The renin angiotensin system in the brain, Exp.Brain Res. Suppl. 4, Springer (1982).
4. H. D. Dellmann and J. B. Simpson, The subfornical organ, Int. Rev.Cytol. 58: 333 (1979).
5. D. Felix and W. Schlegel, Angiotensin receptive neurones in the subfornical organ. Structure-activity relations, Brain Res. 149: 107 (1978).
6. R. A. Nicoll and J. L. Barker, Excitation of supraoptic neuro-secretory cells by angiotensin II, Nature new Biol. 223: 172 (1971).
7. R. J. Gronan and D. H. York, Effect of angiotensin II and acetylcholine on neurons in the preoptic area, Brain Res. 154: 172 (1978).
8. T. Huwyler and D. Felix, Angiotensin II-sensitive neurons in septal areas of the rat, Brain Res. 195: 187 (1980).
9. H. L. Haas, D. Felix and M. D. Davis, Angiotensin excites hippocampal pyramidal cells by two mechanisms, Cell.Molec. Neurobiol. 2: 21 (1982).
10. D. Ganten, J. S. Hutchinson and P. Schelling, The intrinsic brain iso-renin-angiotensin system in the rat: its possible role in central mechanisms of blood pressure regulation, Clin.Sci.Mol.Med. 48: 265s (1975).
11. D. Felix and P. Schelling, Increased sensitivity of neurons to angiotensin II in SHR as compared to WKY rats, Brain Res. 252: 63 (1982).
12. D. Felix, P. Schelling and H. L. Haas, Angiotensin and single neurons, Exp.Brain Res. Suppl.4: 255 (1982).

CENTRAL METABOLISM OF ANGIOTENSINS: POTENTIAL FUNCTIONAL

SIGNIFICANCE

J.W. Harding, R.H. Abhold, C.G. Camara, J.B. Erickson, and E.P. Petersen

Department of VCAPP
Washington State University
Pullman, WA 99164-6520

INTRODUCTION

Many species [2,7] have both pressor and dipsogenic responses to intracerebroventricularly (ICV) applied angiotensin II (AII) and III (AIII). Implicit in this central nervous system (CNS) responsiveness to AII and AIII is the existence of specific membrane-bound receptors. Furthermore these receptors would be expected to be present in those areas of the CNS which are known to be angiotensin sensitive, namely the circumventricular organs [4]. Surprisingly, this logical inference with regard to the distribution of receptors as determined by radioligand binding methods has not been consistently validated when ^{125}I-AII is used as the labeled ligand in the presence of chelating agents [1,3,6]. However, when ^{125}I-AIII is used as the radioligand, "apparent" binding is seen in many brain regions in every species examined. The term "apparent" is used because high performance liquid chromatographic (HPLC) analysis of bound label derived from ^{125}I-AIII incubated membranes clearly indicated that degradation products, especially ^{125}I-tyrosine (Tyr), make up the large majority of bound label. This finding, may, in fact, indicate that "apparent" AIII binding is artifactual in the sense that it has nothing to do with angiotensin binding sites but may represent nonspecific degradation of angiotensins and subsequent uptake of labeled products. On the other hand, this degradation and tyrosine transloction may be a concerted event that accompanies angiotensin binding, suggesting that specific peptidases may be a part of the receptor complex. This study which examines these possibilities strongly supports this notion.

MATERIALS AND METHODS

Adult female rodents 3 to 9 months old were used for most of the studies described in this report. Female rabbits 6 months to one year old and female African green monkeys 5 to 9 years old were also used. Most animals were killed by decapitation, the brains removed and dissected on ice. Crude membrane preparations used for the distribution studies were prepared as described elsewhere for AII binding[3] and AIII binding[7]. Purified synaptosomal membranes were prepared by first polytron homogenizing in 25 volumes of hypotonic buffer containing 50mM Tris, 5mM Na_2EDTA, and 5mM DTT, pH 7.4. Na_2EDTA was omitted where indicated in specific studies. The homogenate was spun at 1000 g for 15 minutes. This pellet was rehomogenized twice more in 25 volumes of homogenization buffer and recentrifuged. The final pellet was rehomogenized in isotonic homogenization buffer containing 150mM NaCl.

Binding assays were carried out at room temperature for 20 to 30 minutes, depending upon the ligand used. The details of the centrifugation assay have been described previously[3]. The filtration assay was identical except that free and bound ligand were separated by filtration through 934-AH glass fiber filters soaked in 0.3% polyethylenimine. Filters were washed 4 times with 5 ml of ice-cold isotonic homogenization buffer.

HPLC analyses were carried out on C-18 columns using a mobile phase of 40mM H_3PO_4 titrated to pH 3.0 with triethylamine. Depending on the actual analysis, gradients of acetonitrile (0-20%) or isocratic systems with 14-17% acetontrile were used.

All iodinated ligands were prepared using an immobilized glucose oxidase-lactoperoxidase system. ^{125}I-ligands were purified to greater than 99.5% purity by HPLC. Peptides and inhibitors were from U.S. Biochemical, Vega Biochemical and Sigma.

RESULTS

The distribution of ^{125}I-AII and ^{125}I-AIII binding has been examined in a number of species and is displayed in Table 1. When ^{125}I-AII is used as the radioligand, a varied pattern of binding is seen for the different species. The rat, which appears unique, exhibits binding throughout the CNS while all other species examined exhibit little or no binding.

Using ^{125}I-AIII as the radioligand, the distribution of apparent binding is very similar in all 5 species examined. Based upon a determination of Kd values, brain regions in the gerbil segregate themselves into two groups: those containing circumventricular organs (CVOs) and those without. [CVOs: Kd = 256 ± 27 nM (12); other brains: Kd = 813 ± 206 nM (12)]

Table 1. Brain region distribution of ^{125}I-AII and ^{125}I-AIII specific binding. Results are reported as means (fmole/mg protein) (n = 2-6). ND--not detectable.

Brain Reg	Rat		Gerbil		Rabbit		Monkey	
	AII	AIII	AII	AIII	AII	AIII	AII	AIII
Olfactory bulb	1.10	3.00	.36	1.46	ND	3.55	ND	2.90
Cerebellum	.32	.53	ND	.47	.16	.79	.24	.98
Striatum	.16	2.63	ND	3.07	ND	2.84	ND	4.90
Sup. colliculi	2.35	1.43	ND	1.10	ND	1.07	ND	1.12
Thalamus	2.24	2.67	ND	1.85	ND	1.36	ND	2.67
Area postrema	1.59	.61	ND	.85	.18	.59	ND	.70
Septum (+SFO*)	1.99	2.22	ND	2.74	ND	1.40	ND	5.00
Hypothalamus	.57	3.07	ND	3.36	ND	2.45	ND	2.04
AV3V**	.44	3.21	ND	4.12	ND	3.61	ND	5.00

*Subfornical organ
**Anterior ventral third ventricular region

HPLC analysis of bound ^{125}I-labeled material can be seen in Table 2. When tissues are incubated with ^{125}I-AIII, the ^{125}I-labeled material recovered from the membranes is predominantly ^{125}I-Tyr. This appears to be true no matter which species is examined or what technique is utilized to separate bound and free label.

Table 2. HPLC analysis of specifically bound* ^{125}I after incubation of the ^{125}I-angiotensins. Data is mean ± S.D. (n = 3-4).

Species	Tissue	^{125}I-Ligand	Separation Technique	%Tyr	%Angiotensin
Rat	Thalamus	AII	Centrifugation	88 ± 4	7 ± 3
	HSAT**	AII	Filtration	ND	91 ± 1
	Thalamus	AIII	Centrifugation	74 ± 6	24 ± 6
	HSAT	AIII	Filtration	90 ± 10	ND
	HSAT	Sar^1,Ile^8 AII	Filtration	7 ± 1	83 ± 1
Gerbil	Thalamus	AIII	Centrifugation	83 ± 5	9 ± 3
	Thalamus	AIII	Filtration	97 ± 5	ND
Monkey	Caudate	AIII	Centrifugation	81 ± 4	0 ± 5
	Adrenals	AIII	Centrifugation	81 ± 8	0 ± 4
	Cerebellum	AII	Centrifugation	73 ± 6	25 ± 3
	Adrenals	AII	Centrifugation	ND	104 ± 10
Rabbit	Striatum	AIII	Centrifugation	83 ± 12	7 ± 0

*Specifically bound is defined as total binding - nonspecific binding.
**HSAT = hypothalamus + septum + AV3V + thalamus.

A different picture emerges if one examines the composition of the labeled bound material when ^{125}I-AII is used as the radio-ligand. When filtration is used for separation of bound and free ligand, over 90% of the bound label is ^{125}I-AII. When centrifugation is used over 70% of the bound label is ^{125}I-Tyr. The amount of specific binding determined using centrifugation is generally 100-150% more than that measured by filtration. In spite of the dramatic differences in total specific binding and composition of bound material, the Kds of binding determined by equilibrium methods are nearly identical for the two separation techniques. [Centrifugation: 0.285 ± .029nM (15); Filtration: 0.222 ± .058nM (3)]

The majority of bound label following incubation with degradation resistant ^{125}I-Sar^{1},Ile^{8}-AII is the peptide itself when the filtration method is used. When centrifugation was used, significant ^{125}I-Tyr was detected (data not shown).

The effects of a number of peptidase inhibitors on the binding of ^{125}I-AII, -AIII, and -Sar^{1},Ile^{8}-AII to brain membranes was examined (Table 3). Although many inhibitors were tested, only four (puromycin, amastatin, bestatin, and bacitracin) had significant effects on angiotensin binding. The ineffective peptidase inhibitors included phenylmethylsulfonyl fluoride, aprotinin, chloroquine, and leupeptin. All were used at 6 times their published Kis. Amastatin, bestatin, and bacitracin were all effective at inhibiting both ^{125}I-AII and -AIII binding in the absence of EDTA. However, bacitracin was less effective in the presence of 5mM EDTA. Interestingly, gerbils which exhibit little AII binding in the presence of EDTA show substantial binding in its absence. The binding of ^{125}I-Sar^{1},Ile^{8}-AII was also inhibited, but to a lesser degree.

In addition to examining the ability of peptidase inhibitors to block the binding of ^{125}I-angiotensins, their effectiveness as inhibitors of angiotensin degradation (Table 4), ^{125}I-Tyr production (Table 4), and ^{125}I-Tyr transport into synaptosomes was also assessed. None of the peptidase inhibitors were effective at blocking ^{125}I-AII or -AIII degradation or ^{125}I-Tyr production. Also, none of the inhibitors altered the uptake of ^{125}I-Tyr into synaptosomes prepared from gerbil thalamus at 10 times the concentration used in the binding inhibition studies (data not shown).

DISCUSSION

Data from several laboratories, including our own, indicates that many AII-sensitive mammals exhibit little or no ^{125}I- AII binding to brain membranes [1,3,6]. This paradox suggested the possibility that AII is normally degraded to AIII before binding occurs.

With this notion in mind, we examined the binding of ^{125}I-AIII to brain membranes from several species [5,7]. Unlike the AII binding,

Table 3. Inhibition of specific angiotension binding* to rat and gerbil brain membranes** by peptidase inhibitors. Data expressed as % control (mean of 2-3 triplicate determinations).

Species	Inhibitor	AIII		AII		Sar^1, Ile^8-AII	
		+EDTA†	−EDTA	+EDTA	−EDTA	+EDTA	−EDTA
Gerbil	Amastatin (5.9µM)	77.1	91.7	--	100	--	41.0
	Bestatin (5.4µM)	84.3	100	--	100	--	46.2
	Am + Be	89.5	100	--	100	--	50.0
	Bacitracin (100µM)	26.0	95.8	--	100	--	47.0
Rat	Amastatin (5.9µM)	74.3	83.7	65.7	66.3	45.5	42.4
	Bestatin	72.8	81.5	41.2	51.9	15.9	18.1
	Am + Be	82.9	91.5	67.1	87.0	--	58.2
	Bacitracin (100µM)	24.9	81.5	25.2	59.0	24.2	23.2

*Binding was determined using purified synaptosomal membranes and filtration separations.
**Membranes prepared from septum, AV3V, hypothalamus, and thalamus.
†5mM Na_2EDTA

Table 4. Effect of aminopeptidase inhibitors on ^{125}I-angiotensin degredation after incubation with membranes from gerbil thalamus.

Condition	^{125}I-Tyrosine	%^{125}I-Angiotensin Remaining
^{125}I-AII	55	9
^{125}I-AII + 5.9µM Amastatin	40	15
^{125}I-AII + 5.4µM Bestatin	42	11
^{125}I-AIII	77	3
^{125}I-AIII + 5.4µM Bestatin	58	9

these results showed a consistent pattern of binding among all species. However, the significance of the results could be questioned on three accounts. (1) The bound material was shown to be ^{125}I-Tyr by HPLC analysis. (2) Brain regions not normally considered sensitive to AII or AIII showed considerable binding (e.g., striatum). And (3) the "apparent" Kds were much too high to be physiologically relevant. At first glance, these three considerations suggested that "apparent" AIII binding was a result of nonspecific AIII degradation and uptake of labeled degradation products. However, three additional observations weighed against this conclusion. (1) Overall, the ^{125}I-AIII distribution reflected the brain angiotensin sensitivity of various brain regions. (2) The circumventricular organs had unique binding characteristics that were different from other angiotensin insensitive brain regions. And (3) even though the composition of specifically bound label following ^{125}I-AII incubation with rat brain membranes was either all ^{125}I-AII or a mixture of ^{125}I-AII and ^{125}I-Tyr, the ^{125}I-Tyr accumulation and ^{125}I-AII binding had identical Kds. These results could be interpreted to mean that after AII (or AIII) binds, it is degraded at the recognition site, degradation occurs, and the products including tyrosine are taken up without ever dissociating from the receptor complex. The differential binding characteristics of CVOs compared to other brain regions indicates that nonspecific degradation of ^{125}I-AIII and random ^{125}I-Tyr uptake is occurring. This suggests that the AIII binding distribution would then also reflect the tyrosine uptake ability of various brain regions. It appears that the high Kds are actually composite Kds that include binding, nonspecific degradation, and tyrosine uptake.

The notion that peptidases may be associated with the angiotensin receptor complex was further investigated by examining the effect of specific peptidase inhibitors on the binding of angiotensins. The aminopeptidase inhibitors amastatin and bestatin blocked not only the binding of metabolizable angiotensins (AII and AIII) but also a degradation-resistant form (Sar^1,Ile^8-AII). This indicated that blockade of nonspecific angiotensin degradation and ^{125}I-Tyr uptake did not represent "apparent" binding. Furthermore, the receptor blocking ability of the peptidase inhibitors was not correlated with either inhibition of angiotensin degredaion or ^{125}I-Tyr uptake in general. These results support the idea that the degradation of angiotensins and the uptake of degredation products are concerted events that occur at the angiotensin receptor complex.

ACKNOWLEDGEMENTS

This study was supported by a research grant from the American Heart Ass'n. and the American Heart Ass'n. of Washington.

REFERENCES

1. J.P. Bennet Jr. and S.H. Snyder, Angiotensin II binding to mammalian brain membranes, J. Biol. Chem. 251:7423-7430 (1976).
2. J.T. Fitzsimons, Thirst, Physiol. Rev. 52:468-561 (1972).

3. J.W. Harding, L.P. Stone, and J.W. Wright, The distribution of angiotensin II binding sites in rodent brain, Brain Res. 205: 265-274 (1981).

4. A.K. Johnson, Neurobiology of the periventricular tissue surrounding the anteroventral third ventricle (AV3V) and its role in behavior, in "Circulation, Neurobiology, and Behavior," O.A. Smith, R.A. Galosy, and S.M. Weiss. eds. Elsevier, Amsterdam (1982).

5. E.P. Petersen, C.G. Camara, R.H. Abhold, J.W. Wright, Characterization of angiotensin binding to gerbil brain membranes using ^{125}I-angiotensin III as the radioligand, Brain Res. In press.

6. R.C. Speth, M.B. Vallotton, C. Chernicky, M.C. Khosla, and C.M. Ferrario, Angiotensin II receptors in dog brain, Fed. Proc. 42:494 (1983).

7. J.W. Wright, S. Morseth, M.J. Mana, E. LaCrosse, E.P. Petersen, and J.W. Harding, Central angiotensin III-induced dipsogenicity in rats and gerbils. Brain Res. 295:121-126 (1984).

ASPECTS OF CEREBROSPINAL FLUID PRESSURE CONTROL IN CONSCIOUS RATS

DURING CENTRAL INFUSIONS OF ANGIOTENSIN AND VASOPRESSIN*

W.B. Severs, H.J. Spaeth, J.N.D. Wurpel, R.L. Dundore,
R.T. Henry and L.C. Keil

Pharmacology Department, The Milton S. Hershey Medical
Ctr. of the Pennsylvania State Univ., Hershey, PA 17033
and NASA, Ames Res. Ctr., Moffett Field, CA 94035

INTRODUCTION

The cerebrospinal fluid (CSF) system has been described as a moderately distensible compartment, wherein pressure is maintained within relatively narrow limits.[1]   The principal components regulating CSF dynamics are synthesis rate, compliance, and drainage (mainly at arachnoid villa).   Angiotensin[2] and vasopressin[3] have been associated with choroid tissue, although a role for the peptides has not emerged; both peptides are endogenous to CSF.[4] Vasopressin, in anesthetized rabbits, reduced CSF pressure most likely by facilitating CSF transport to venous blood.[5]   This peptide may also alter intracranial water movement.[6]   Angiotensin has been reported to cause cerebral vasoconstriction by a central mechanism.[7] Thus, the peptides may affect intracranial pressure.

Previously, we reported that a 30 min intracerebroventricular (IVT) infusion of angiotensin (Ang) raised CSF pressure in conscious rats, whereas arginine vasopressin (AVP) was without effect.[8] However, AVP blocked the angiotensin-induced increase. Other investigators also concluded that IVT angiotensin elevates CSF pressure in rats and rabbits.[9]   Here, we report the effects of a 5 hr IVT infusion (2 µl/min) of CSF, Ang, AVP or sar[1]ile[8]-angiotensin II into conscious rats. Angiotensin promptly elevated CSF pressure which remained increased for 5 hr.   CSF infusions gradually raised CSF pressure, which was significant ($p < 0.05$) at 2 hr, but lower than Ang-treated rats. During the last 3 hr, CSF-infused rats developed CSF pressures equivalent to Ang-treated

---

*Supported by NASA-Ames Cooperative Agreement NCC-2-127.

animals.   IVT infusions of AVP, or the angiotensin receptor
antagonist, markedly inhibited (p < 0.01) the late-developing CSF
pressure increase evoked by CSF infusions.

MATERIALS AND METHODS

Adult male Sprague-Dawley rats (325-450 g) were anesthetized
with pentobarbital sodium and ketamine (45 mg/kg ip, 20 mg/kg im).
A guide cannula was inserted into a hole in the flat skull 1.5 mm
caudal and left lateral to the bregma to a depth about 1 mm superior
to the lateral ventricle.   The cannula and a small stainless steel
screw were cemented to the skull.   The left carotid artery was
catheterized and exteriorized at the scalp incision.   The ventricle
was penetrated with an "L" shaped length of 26 gauge hypodermic
tubing.   This tubing, an infusion pump, and an ultralow volume
transducer were connected to a small manifold and artificial CSF[10]
was infused (2 μl/min) at the time of ventricular puncture.   A sharp
fall in pressure signalled penetration of the ventricle.   The
infusion continued for 15 min while the needle, along with the
arterial catheter, were cemented to the guide cannula.   The
exteriorized cannulae were then sealed and rats were given a
minimum of 2 days for recovery.

On the test day, rats were placed in metabolism cages with
water available.   The arterial and ventricular cannulae were
connected to a polygraph for pressure recording.   Artificial CSF (2
μl/min) was infused IVT during a 30 min stabilization period.   The
infusion was then switched (same rate) for the next 5 hr to CSF
alone, or with one of the following peptide additions: ile-5 angiotensin II (0.6 μg/hr), arginine vasopressin (50 ng/hr), or
sar^{1}ile^8-angiotensin II (10 μg/hr).   Water intake during this time
was recorded.   Urine was collected in tubes containing 1 ml 1N HCl
for [Na$^+$] and [K$^+$] assay by flame photometry.   IVT dye was given at
the end of all experiments; dye distribution in the lateral, third
and fourth cerebroventricles was required for using data.   Blood
and CSF pressures were read at "0" time, .25, .50, 1, 2, 3, 4 and 5
hours.   Data were analyzed by 2-way ANOVA (infusate x time), using
repeated measures across time.   Significant F ratios (p < 0.05) were
evaluated with the Newman-Keul's range statistic.   Probabilities of
5% or less were considered significant.

RESULTS

The effects of 5 hr IVT infusions of Ang and CSF on CSF
pressure are shown in Figure 1. The CSF and Ang groups were similar
at 0, 3, 4 and 5 hr (p > 0.05) and different between 15 min and 2 hr
(p < 0.05). Within the Ang group, "0" differed from all times; 15
and 30 min differed from 2 and 3 hr, respectively (all p's < 0.05).
Within the CSF group, 2 hr was the 1st time to differ from "0"; 1 hr
differed from hrs 3, 4, and 5 (all p's < 0.05).   The CSF and AVP

groups are compared in Fig. 2. CSF pressure in AVP-treated animals was lower than CSF-infused rats at hrs 3, 4, and 5 ($p < 0.05$). Within the AVP group, the 5 hr pressure was different than times "0" through 1 hr ($p < 0.05$). Fig. 3 shows a comparison of CSF and sar[1] ile[8]-angiotensin II infused rats. The CSF pressures of these groups differed ($p < 0.05$) at hours 3, 4 and 5. Within the inhibitor group, times "0" through 1 hr, and 4 hr, differed from 5 hr ($p < 0.05$).

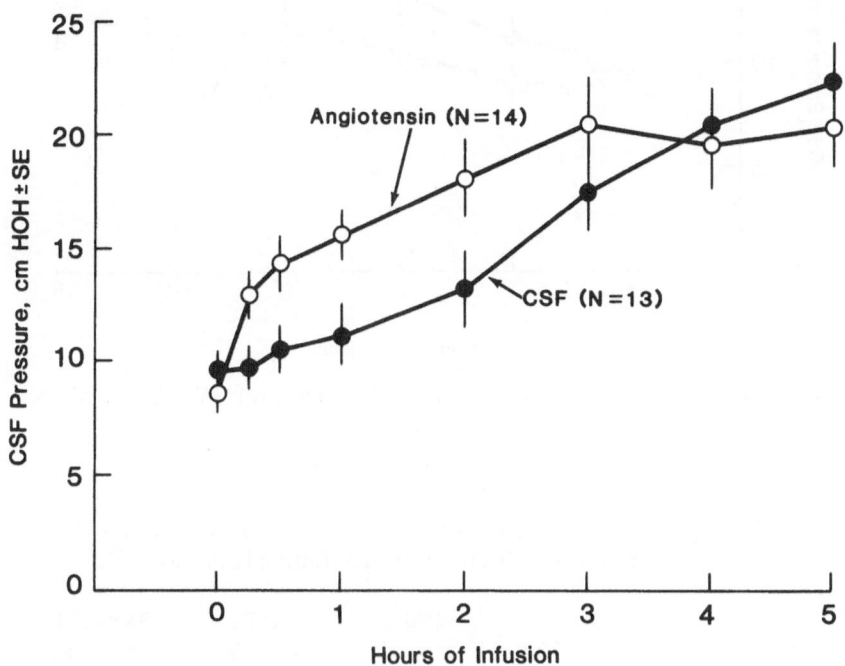

Figure 1: CSF Pressure During IVT CSF and Angiotensin Infusions

Blood pressures obtained during these studies appear in Table 1. The basal blood pressures of the AVP-infused rats was higher than the other groups ($p < 0.05$). Only Ang-infused rats revealed a within group effect: times "0" and 15 min differed from all other times ($p < 0.05$). Drinking volumes (ml/100 g ± SE) for the CSF, Ang, AVP and sar[1] ile[8]-Ang groups were, respectively: 1.02 ± .22, 4.84 ± .43, 1.20 ± 0.3 and 2.40 ± 0.5. Mann-Whitney U tests indicated that the Ang and sar[1] ile[8]-Ang groups differed from all others ($p < 0.05$). Urine volumes (ml/100 g ± SE), in the same order, were: 2.2 ± 0.3, 2.8 ± 0.4, 2.0 ± 0.4 and 2.2 ± 0.3. There were no significant differences ($p > 0.05$). Absolute Na[+] and K[+] excretion (μEq/100 g ± SE) were: CSF, 144 ± 26 & 93 ± 15; Ang, 249 ± 34 & 170 ± 12; AVP, 206 ± 46 & 126 ± 20; sar[1] ile[8]-Ang, 239 ± 27 & 180 ± 20. Group comparisons using the pooled error term were: CSF differed from Ang and sar[1] ile[8]-Ang treatments ($p < 0.05$).

The effect of sar[1] ile[8]-Ang on drinking was unlike Ang. There

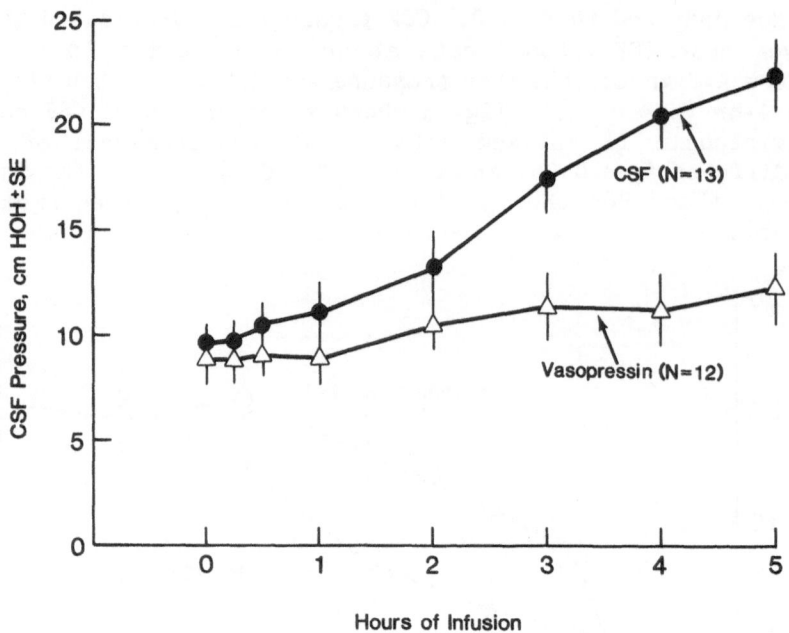

Figure 2: CSF Pressure During IVT CSF and AVP Infusions.

Table 1. Blood Pressures During 5-Hour Intraventricular Infusions in Conscious Rats[a]

Time (Hours)	CSF n = 11	ANG n = 12	AVP n = 9	SAR-ILE n = 11
0	111 ± 6	118 ± 4	128 ± 7	111 ± 5
.25	112 ± 7	144 ± 8	128 ± 7	109 ± 4
.50	111 ± 7	135 ± 6	129 ± 7	107 ± 4
1	112 ± 6	134 ± 6	127 ± 7	108 ± 4
2	110 ± 7	136 ± 5	126 ± 6	109 ± 4
3	107 ± 7	136 ± 7	125 ± 6	108 ± 4
4	107 ± 8	136 ± 5	123 ± 6	110 ± 4
5	108 ± 8	136 ± 5	123 ± 6	109 ± 5

[a]Data are Means ± SE (Units, mmHg). AVP group
(time 0): $p < 0.05$ vs CSF and SAR-ILE. Within
group analysis in ANG group showed 0 and 15 min
differed from all other data cells ($p < 0.05$).

was no consistent drinking at the beginning of experiments. The
increased electrolyte excretion after sar[1] ile[8] -Ang was unexpected;
the reason remains undefined. The drug may have some agonist
properties and may leak to the periphery.

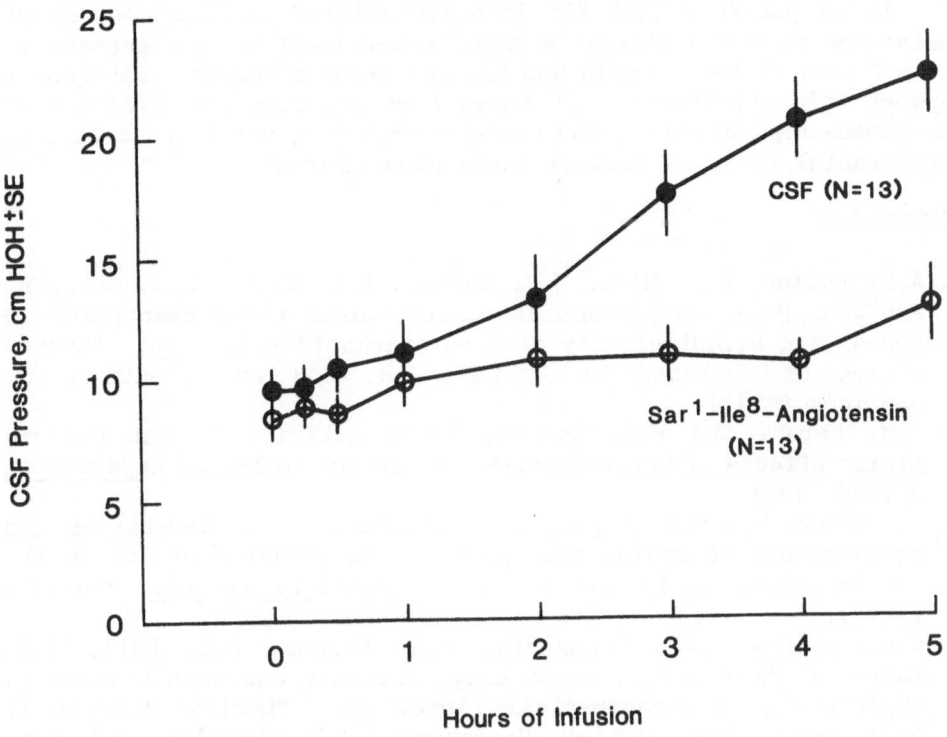

Figure 3: CSF Pressure During CSF and Sar[1]Ile[8]-Ang Infusions

DISCUSSION

IVT angiotensin promptly increased CSF pressure, confirming our earlier study.[8] The peptide effect was evident at 2 hr, compared to CSF-infused rats. Surprisingly, CSF infusion itself gradually elevated CSF pressure within 2 hr, and there was no difference vs Ang-treated rats at later times. However, CSF-treated rats had stable blood pressures with time, and drinking behavior was appropriate for the daytime experiment. Terminal plasma [Na^+] and [AVP] were also normal (data not shown), as was gross behavior. The mechanism of the CSF-induced rise in CSF pressure is not known. Lack of drinking behavior and normal p[AVP] indicates that the infusate was not hypertonic relative to the local brain environment. A volume effect by a 2 μl/min infusion seems unlikely. Pressure changes, such as produced by our infusion conditions, are evoked by bolus volume injections of 30 - 40 μl;[11] this represents the volume we infused over 15 to 20 min. Also, AVP and sar[1]ile[8]-Ang were given at the same rate and were dissolved in artificial CSF. These peptides markedly blunted the spontaneous rise in CSF pressure evoked by CSF.

It is possible that CSF infusion altered peptide control of resistance to CSF drainage, a major determinant of CSF pressure. The efficacy of vasopressin and the angiotensin receptor antagonist support this hypothesis. If these neuropeptides are involved as endogenous regulators of intracranial pressure, novel approaches to intracranial pressure control could be developed.

REFERENCES

1. A.B. Butler, J.D. Mann, C.J. Maffeo, R.G. Dacey, R.N. Johnson, and N.H. Bass, Mechanisms of cerebrospinal fluid absorption in normal and pathologically altered arachnoid villi, in: "Neuro-biology of Cerebrospinal Fluid 2", J.H. Wood, ed., Plenum Press, New York (1983).
2. J.M. Kapsha and W.B. Severs, Renin activity in rat choroid plexi: effects of water deprivation and hypovolemia, Experientia 39:429 (1983).
3. H. Davson and M.B. Segal, The effects of some inhibitors and accelerators of sodium transport in the turnover of Na^+ in the cerebrospinal fluid and brain, J. Physiol. (Lond). 209:131 (1970).
4. W.B. Severs, D.G. Changaris, J.M. Kapsha, L.C. Keil, D.J. Petro, I. Reid and J. Summy-Long, Presence and significance of angiotensin in cerebrospinal fluid, in: "Central Actions of Angiotensin and Related Hormones", J.P. Buckley and C.M. Ferrario, eds., Pergamon Press, New York (1977).
5. T. Noto, T. Nakajima, Y. Saji, and Y. Nagawa, Effect of vasopressin on intracranial pressure of rabbit, Endocrinol. Japon. 25:591 (1978).
6. M.E. Raichle and R.L. Grubb, Regulation of brain water permeability by centrally released vasopressin, Brain Res. 143:191 (1978).
7. A. Reynier-Rebuffel, E. Pinard, P. Aubineau, P. Meric and J. Seylaz, Generalized cerebral vasoconstriction induced by intracarotid infusion of angiotensin II in the rabbit, Brain Res. 269:91 (1983).
8. Y.R. Barbella, L.C. Keil, J.N.D. Wurpel and W.B. Severs, Cerebrospinal fluid pressure during cerebroventricular infusion of angiotensin and vasopressin, Exp. Neurol. 82:325 (1983).
9. L.C. Senay and D.L. Tolbert, Effect of arginine vasopressin, acetazolamide, and angiotensin II on CSF pressure at simulated altitude, Aviation, Space and Environ. Med. 55:370 (1984).
10. J.K. Merlis, The effect of changes in the calcium content of the cerebrospinal fluid on spinal reflex activity in the dog, Am. J. Physiol. 131:67 (1940).
11. J.E. Melton and E.E. Nattie, Intracranial volume adjustments and cerebrospinal fluid pressure in the osmotically swollen rat brain, Am. J. Physiol. 246:R533 (1984).

DECREASED BINDING CAPACITY OF CENTRAL ANGIOTENSIN II RECEPTORS

FOLLOWING LONG-TERM ADMINISTRATION OF ANGIOTENSIN II

Elaine Richards-Sumners and M. Ian Phillips

Department of Physiology, College of Medicine
University of Florida, Box J-274 JHMHC
Gainesville, FL 32610

INTRODUCTION

A substantial body of evidence supports the existence of a brain renin-angiotensin system in the rat. All the components of the renin-angiotensin system have been demonstrated in the brain.[1] Although a primary physiological role for the brain renin-angiotensin system (BRS) has not been clearly defined, the evidence strongly suggests that it plays a part in controlling water balance in the rat. So, it has been suggested that a dysfunctioning BRS could contribute to hypertension.[2] Therefore, an understanding of all aspects of the BRS is important. The aspect considered here was the regulation of central Ang II receptors after chronic infusion of Ang II into the brain. As chronic infusion of Ang II into the lateral ventricles alters water intake,[3] salt intake,[4] natriuresis,[5] and blood pressure when given in high enough doses or in combination with a high sodium diet.[6,7]

MATERIALS AND METHODS

Male Sprague-Dawley rats weighing 250 - 350 g were used in this study. The rats were housed individually in metabolism cages with free access to food and water from weighed drinking bottles. Rats began receiving either intracerebroventricular (i.v.t.) infusions of Ang II or 0.9% saline, the Ang II vehicle, 7 days before Ang II binding studies were performed. Rats were anesthetized under chloral hydrate anesthesia, an L-shaped cannula was placed in the lateral ventricle, and fixed to the skull. An Alzet minipump filled either with 1 mg/ml Ang II or 0.9% saline was attached to the cannula. Placed subcutaneously, this pump delivered either

1 μl of saline or 1 μg/μl of Ang II per hour, for 7 days.

Control and experimental rats were set up in parallel. Daily
intakes of tap water were measured from weighed water bottles.
Daily urine output was recorded by collection of urine. Urinary
sodium and potassium concentrations were determined by flame
photometry, and thus daily urinary sodium and potassium output was
measured.

On the seventh day of pump implantation, the rats were killed
by decapitation. The brains were quickly removed and the hypo-
thalamus-thalamus-septum-midbrain (HTSM) dissected as described
by Sirett et al.[8] Also taken were portions of the brainstem
(BS), cerebellum and cortex. These were homogenized in 20 volumes
of ice cold saline. The homogenates were then centrifuged at 600
g at 4°C for 10 minutes, the supernatants decanted and centrifuged
at 50,000 g and 4°C for 30 minutes. The supernatant was discarded
and the pellet resuspended in 10 - 20 volumes of buffer containing
150 mM NaCl, 5 mM ethylenediaminetetracetic acid (EDTA) and 50 mM
Tris-HCl, pH 7.2. Portions of this particulate fraction (100-350
ug protein) were incubated for 30 minutes at 22°C in 200 ul assay
buffer containing 150 mM NaCl, 5 mM EDTA, 5 mM dithiothreitol, 50
mM Tris-HCl, pH 7.2, 0.2% bovine serum albumin (BSA) and 0.15 nM
^{125}I-Ang II (1500 uCi/ug). After incubation the tubes were
placed on ice and diluted with 1 ml cold buffer. Separation of
bound and free radioactive Ang II was performed by centrifugation.
Samples were then counted in a Beckman 5500 gamma radiation
counter. Specific binding was considered as that binding
displaced by an excess of unlabelled Ang II (final concentration
150 nM) and all incubations were determined in triplicate.
Protein contents of the brain particulate fraction were estimated
by a standard method,[9] and the results expressed as fmol Ang II
bound/mg protein. Angiotensin II retains full activity throughout
the pumping period.[10]

RESULTS

The infusion of Ang II into the lateral ventricle of rats at
a rate of 1 μg/hr resulted in a significant increase in water
intake compared to saline infused controls, $F = 83.7$ with 1, 10
degrees of freedom (df), $p < 0.01$ (see Fig 1). Urine volume
closely followed water intake and thus was significantly increased compared to saline treated controls throughout the infusion
period, $F = 68.6$, with 1, 10 df, $p < 0.01$. The drinking and urine
output were maximal on day 4 and thereafter showed a tendency to
decline, remaining significantly greater than control levels
throughout the seven days, however (see Fig 1). Ang II infusion
also caused a significant increase in both sodium and potassium
excretion compared to vehicle infusions. Sodium excretion $F =
38.3$ with 1, 10 df, $p < 0.01$, potassium excretion $F = 17.8$ with 1,

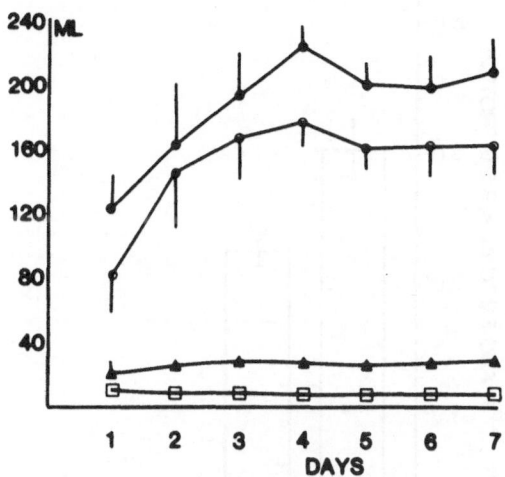

Fig. 1. Water intakes and urine output of rats receiving Ang II
infusion for seven days at a rate of 1 μg/μl/hr into the
lateral ventricle compared to saline infused controls.
● is the 24 hour water consumption of Ang II infused
rats, ▲ that of controls. ○ is the 24 hour urine
output of Ang II infused rats, and □ that of controls.
Mean ± standard error of the mean S.E.M., n = 6 rats at
each point.

10 df, p < 0.01. These data show that Ang II was delivered to the
lateral ventricle.[3,10,11]

The binding of angiotensin II to brain membranes was characterized. The binding conditions used thereafter were based on the
results of these initial experiments and were 300 μg protein/tube
with binding allowed to proceed for 30 minutes at 22°C. The effects of intracerebroventricular administration of Ang II for 7
days on Ang II binding to brain membranes was examined in 4 brain
areas. The HTSM region exhibited a significant decrease compared
to controls in Ang II binding following chronic Ang II administration t = 3.54 with 4 df p = 0.025 by Student's paired t test (see
Fig 2). Ang II infusion did not significantly alter central Ang
II binding in any of the other brain regions examined, eg.,
cortex, cerebellum and brain stem.

The HTSM area was further examined after chronic Ang II
treatment by investigating the effect of increasing Ang II concentration on binding. Thus, binding was performed at concentrations
of free Ang II from 0.04 to 1.0 nM. Binding was seen to increase

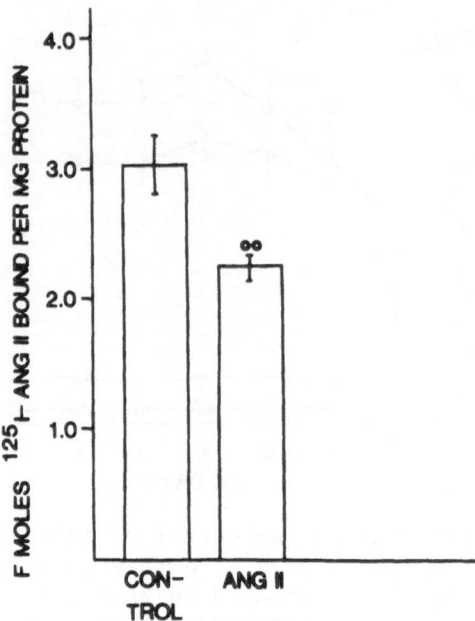

Fig. 2. Binding of ^{125}I-Ang II to membranes prepared from the HTSM of Ang II (1 µg/µl/hr for 7 days) treated rats compared to saline treated controls.   ●●= significantly different from control at p = 0.025 by Student's paired t test. Mean ± S.E.M. with n = 5 for each treatment.

to saturation in both control and treated membranes, but at low concentrations (0.04 to 0.2 nM) the membranes from Ang II treated rats exhibited significantly lower binding. Scatchard analysis of these data revealed that Ang II treatment reduced the affinity of the binding without significantly affecting the maximum number of binding sites. The Kd was altered from 0.174 nM for the control to 0.337 nM for the Ang II treated membranes. The B_{max} was 3.0 fmol/mg protein for the controls and 3.4 fmol/mg protein for the Ang II treated rats. Repeat Scatchard analysis showed the same decrease in affinity with little change in binding sites with Ang II treatment.

DISCUSSION

    These data suggest that administration of 1 µg/hr Ang II for 7 days into the lateral cerebroventricle caused a decrease in Ang II binding in the HTSM region. No changes were apparent in any of the other brain areas studied which included the brainstem, cerebellum and cortex. Further chronic Ang II infusion caused increased water intake and output, as well as increased sodium and potassium loss in the urine. Drinking and urine output exhibited maximum change on the fourth day after infusion, with a decreased but constant level after that time, which was still significantly greater than control drinking levels.

Sirett et al.[8] showed that the HTSM contains a high concentration of specific binding sites for Ang II. It is postulated to contain the areas of the brain responsive to Ang II which mediate the drinking,[12] sodium appetite[4] and the natriuresis[13] induced by central administration of Ang II. Similarly, the brain sites of the pressor responses to Ang II reside there.[12,14] Thus, while drinking and natriuresis showed no significant tendency to decrease, as would be expected if the receptors mediating these responses declined, it is possible that if the other parameters had been measured, for example sodium appetite and blood pressure, they would have shown a decrease over the infusion period.

Mann et al.[15] have shown that sodium depletion by a low sodium diet decreases Ang II binding in the rat brain by changing the affinity and maximum number of binding sites of the receptors. In the present study, significant urinary loss of sodium was seen, as previously reported by Halperin et al.[5] for this dose of Ang II infusion. Furthermore, rats chronically infused with Ang II have decreased food intakes.[3] Thus, it is entirely possible for sodium depletion to occur to these infusions. Also, decreased plasma sodium levels have been reported in chronically Ang II infused rats.[10,11,16] Thus, this study does not differentiate between a direct effect of Ang II on its own receptors, or an indirect effect of Ang II on its receptors via depletion of sodium. Further studies are necessary to differentiate between the two possibilities.

In summary, the present results show that Ang II administered chronically affects central Ang II receptor regulation. This is important because these receptors seem to play a role in regulating fluid balance and cardiovascular function. Possibly, pathological changes such as are seen in spontaneously hypertensive rats reflect the status of central angiotensin receptors.

This work was supported by NSF grant BSF 8025969 awarded to Dr. M.I. Phillips.

REFERENCES

1. D. Ganten, M. Printz, M. I. Phillips and B. A. Scholkens Eds., "The Renin Angiotensin System in the Brain," Springer-Verlag, Berlin (1982).

2. M. I. Phillips, The contribution of central angiotensin II to the hypertension of spontaneously hypertensive rats, In: "Nervous System and Hypertension," H. Schmitt and P. Meyer Eds., John Wiley and Son, New York (1978).

3. R. J. Gronan and D. H. York, Effects of chronic intraventricular administration of angiotensin II on drinking behavior and blood pressure, Pharmacol., Biochem. and Behav. 10:121 (1979).

4. R. W. Bryant, A. N. Epstein, J. T. Fitzsimons and S. J. Fluharty, Arousal of a specific and persistent sodium appetite in the rat with continuous intracerebroventricular infusion of angiotensin II, J. Physiol. 301:365 (1980).

5. E. S. Halperin, J. Y. Summy-Long, L. C. Keil and W. B. Severs, Aspects of salt/water balance after cerebroventricular infusion of angiotensin II, Brain Res. 205:219 (1981).

6. C. A. Bruner, J. M. Weaver and G. D. Fink, Will chronic intracerebroventricular saralasin infusion produce selective blockade of brain angiotnesin II receptors in the rat?, J. Pharmacol. Exp. Ther., 225:13 (1983).

7. G. D. Fink, W. J. Bryan and D. J. Mokler, Effects of chronic intracerebroventricular infusion of angiotensin II on arterial pressure and fluid homeostasis, Hypertension 4:312 (1982).

8. N. E. Sirett, A. S. McLean, J. J. Bray and J. I. Hubbard, Distribution of angiotensin II receptors in rat brain, Brain Res. 122:299 (1977).

9. O. H. Lowry, N. J. Rosebrough, A. L. Farr and R. J. Randall, Protein measurement with the Folin phenol reagent, J. Biol. Chem. 193:265 (1951).

10. R. DiNicolantonio, F. A. O. Mendelsohn, J. S. Hutchinson, Y. Takata and A. E. Doyle, Dissociation of dipsogenic and pressor responses to chronic central angiotensin II in rats, Am. J. Physiol. 242:R498 (1982).

11. G. H. Sterling, O. Chee, R. V. Riggs and L. C. Keil, Effect of chronic intraventricular angiotensin II infusion on vasopressin release in rats, Neuroendocrinology 31:182 (1980).

12. W. E. Hoffman and M. I. Phillips, The effect of subfornical organ lesions and ventricular blockade in drinking induced by Ang II, Brain Res. 108:59 (1976).

13. W. B. Severs, A. Daniels-Severs, J. Summy-Long and G. J. Radio, Effects of centrally administered angiotensin II on salt and water excretion, Pharmacology 6:242 (1971).

14. M. L. Mangiapane and J. B. Simpson, Subfornical organ: forebrain site of pressor and drinking actions of angiotensin II, Am. J. Physiol. 239:R382 (1980).

15. J. F. E. Mann, E. L. Schiffrin, P. W. Schiller, W. Rascher, R. Boucher and J. Genest, Central actions and brain receptor binding of angiotensin II: influence of sodium intake, Hypertension 2:437 (1980).

16. G. J. Radio, J. Summy-Long, A. Daniels-Severs and W. B. Severs, Hydration changes produced by central infusion of angiotensin II, Am. J. Physiol. 223:1221 (1972).

THE ROLE OF THE RENAL RENIN-ANGIOTENSIN SYSTEM IN THIRST

Alan Kim Johnson, Marilyn M. Robinson and Johannes F.
E. Mann

Departments of Psychology and Pharmacology and the
Cardiovascular Center, University of Iowa, Iowa City
IA, 52242, U.S.A.; Department of Physiology, University
of Western Ontario, London, Ontario, Canada, N6A 5C1
and Department of Medicine, University of Heidelberg
6900, FRG

INTRODUCTION AND BACKGROUND

Drinking is a homeostatic behavior that can correct body
fluid deficits. Water deprivation produces dehydration of both
the cellular and extracellular fluid compartments of the body.
Cellular dehydration is caused by osmosis whenever the osmolality
of the extracellular fluid is increased by solutes to which the
cell membrane is not freely permeable. It is well established
that this cellular dehydration is a potent stimulus to thirst and
antidiuresis[1]. Likewise, Fitzsimons[2] and others[3,4] have shown
that isotonic depletion of the extracellular fluid compartment by
the subcutaneous or intraperitoneal injection of a hypertonic
colloid solution such as polyethylene glycol (PEG) also causes
drinking.

Although depletion of either compartment alone will stimulate
drinking behavior, in recent years it has been recognized that
natural thirst arises as a result of multiple internal stimuli and
involves the integrative action of the central nervous system.
Cellular dehydration is detected by specific brain receptors which
are sensitive either to changes in their own cell volume[5,6,7] or
to the concentration of sodium in the extracellular fluid[8].
Extracellular fluid depletion is signaled by both neural and
hormonal mechanisms. Stretch receptors located on the low-
pressure side of the circulation monitor blood volume[9] and via
vagal afferents tonically inhibit central neural mechanisms
involved in the conservation and restoration of the extracellular

fluid[10,11]. The removal of this inhibition during underfilling of the capacitance vessels causes drinking by a direct activation of neural thirst systems in the brain[12,13,14] and by reflex stimulation of the renal sympathetic nerves which increases renin release and thereby increases the plasma concentration of the hormone angiotensin (ANG) II[15]. The arterial baroreceptors also play a role as receptors in the hypovolemic stimulation of thirst through reflex stimulation of renin release[16] and may also have a facilitory action on central thirst pathways through direct afferent activation.

Although work by several investigators during the 1950's and 1960's hinted that rats given renal extracts increased their water intake[17,18,19], it was the work of Fitzsimons[12,13] that unequivocally implicated the renal renin-angiotensin system (RRAS) in the drinking response to extracellular fluid depletion. Later, Epstein & Hsaio[20,21] showed that drinking could be induced in the rat with intravenous infusions of ANG II which probably produced plasma concentrations of ANG II within the physiological range[22]. Despite the demonstration of ANG II as a reliable dipsogen, its physiological role in the drinking response to extracellular fluid depletion remains controversial.

POSING THE QUESTION

The implication of ANG II as a dipsogenic substance in the mid 1960's was extremely provocative for several reasons. First, it identified a specific effector agent of an endogenous endocrine system in the control of a complex motivated behavior, drinking. Second, ANG was a peptide and peptides had received very little attention with regard to their action on the central nervous system and in terms of their relationship to behavioral phenomena. Third, the RRAS was a major focus of investigation in the field of hypertension research. Thus, there was already an established group with an interest in the biological actions of the RRAS and ANG. Finally, the dipsogenic action of ANG II was remarkably consistent in terms of the other known biological responses of the peptide. Water intake, like the ANG-induced pressor response and release of aldosterone and vasopressin contributes to the maintenance and restoration of extracellular fluid volume and blood pressure.

Despite the compelling evidence that ANG II was a potent dipsogen and the attractiveness of the hypothesis that it was a mediator of thirst, several investigators questioned the importance of its role under physiological conditions. Abraham and colleagues[23] were among the first to investigate whether drinking could be elicited by infusions of ANG II which produce plasma levels comparable to those seen under physiological or pathophysiological conditions (e.g., sodium depletion). These investigators reported that the levels of ANG II needed to induce drinking by intravenous, intracarotid or intracerebroventricular infusion or injection exceeded the levels which could be directly measured in arterial blood or cerebrospinal fluid after a variety of challenges. They argued on the basis of their data in sheep and data available in other species that the drinking response to ANG could only be produced by pharmacological doses of the peptide.

Subsequently, drinking was produced in both the rat[21] and the dog[24,25] with doses of ANG that maintain plasma concentrations within the physiological range[26]. Nevertheless, Stricker[27,28] has tended to minimize the significance of these observations. He estimated that these ANG-induced intakes would dilute body fluids by less than 1 or 2% and suggested that although statistically significant, such intakes would not have much physiological significance.

Furthermore, as an alternative to a dipsogenic role for the RRAS, it has been suggested that ANG may only play a permissive role in the genesis of drinking[27,28,29]. Specifically, Stricker and colleagues[28,31] have maintained that nephrectomized animals which failed to drink to treatments such as isoproterenol were unable to do so because of shock-related debility produced by the inability of the animal to defend blood pressure in the absence of the endogenous RRAS. Such observations were supported by demonstrating that systemic administration of pressor agents (e.g., epinephrine) to elevate pressure to 70 to 85 mmHg in nephrectomized animals given isoproterenol, was accompanied by increased water intake. Atkinson et al.[29], again using the nephrectomized rat preparation, have also concluded that the role of ANG in the drinking response to isoproterenol is only a permissive one.

Twenty years of research studying the role of the RRAS in drinking behavior has made it clear that ANG II probably never functions as the sole mediator of drinking under physiological conditions. Nevertheless, it is important to investigate the conditions under which the RRAS does make a contribution to drinking behavior and the way in which it participates and interacts with the many intrinsic stimuli acting through various receptor systems involved in thirst mechanisms. The remainder of this paper will be devoted (a) to describing and critically evaluating several different experimental approaches which have been applied to investigating this question and (b) to offering possible explanations for why different methods of stimulating or disabling the RRAS have led to different conclusions regarding the role of ANG in drinking behavior.

METHODS USED TO EVALUATE THE ROLE OF THE RENAL RENIN-ANGIOTENSIN
SYSTEM IN EXPERIMENTAL THIRST

The role of the RRAS in thirst has been evaluated previously
by three methods: 1) administration of exogenous components of
the RRAS (e.g., renin; ANG II); 2) measurment of components of the
RRAS in body fluids; and 3) disabling the RRAS. As is true for
virtually any single experimental method that we employ in the
behavioral and physiological sciences, no technique is totally
adequate to answer the question at hand. Each procedure has
associated with it certain experimental limitations which to the
unwary can constitute major pitfalls to providing valid data. It
is the intention of the discussion presented in this section to
examine both the strengths and weaknesses of these techniques as
they have been employed to address the question of the role of ANG
in drinking.

Drinking to Exogenous Renin and Angiotensin

As pointed out above, one of the first lines of evidence to
question a physiological role for the RRAS was advanced by
investigators who were unable to produce significant water intake
from intravascularly applied ANG II unless pharmacological amounts
were used. Abraham et al.[23] infused sheep with ANG II at doses
which elevated plasma ANG II concentrations to values produced by
various renin-dependent thirst stimuli, but found no drinking.
Further, drinking to intracarotid or intraventricular injection of
ANG II could only be elicited by supraphysiological dosages.
Similarly, Stricker[27,28,30] compared the water intake and plasma
renin activity in rats following the intraperitoneal injection of
hog renin with that seen following various thirst stimuli. He
found that renin injections at doses which produced plasma renin
activities comparable to those produced by hypovolemia or
hypotension, caused very little drinking; much less than that seen
following the thirst-inducing stimuli themselves. However, the
assumption made in the experiments of both groups of investigators
was that exogenously administered renin or ANG II has the same
effect on thirst mechanisms as does ANG II synthesized
endogenously in response to the physiological need following caval
ligation or treatment with PEG or isoproterenol. As suggested by
Robinson & Evered (this volume), however, the two conditions are
not comparable. Injections or infusions of renin or ANG II
increase arterial pressure above normal resting levels whereas
under physiological conditions similar concentrations of ANG II
would occur when blood pressure was at or below normal. The
pressor response to exogenous ANG II inhibits the drinking
response. When blood pressure is maintained at normal levels by
an intravenous infusion of a vasodilator, the drinking response to
exogenous ANG II is increased 2-4 fold (Robinson & Evered, this
volume).

Atkinson et al.[29] have suggested that the role of the RRAS in the drinking response to hypotensive agents such as isoproterenol is only a "permissive" one. They found that injections of renin at doses too low to cause nephrectomized rats to drink when given alone, did cause drinking when given with isoproterenol. They interpreted this to mean that normally the drinking response to isoproterenol depends on the presence of a small amount of renin or ANG II in the blood, but that the ANG II itself is not the stimulus to drink. The injections of renin given by Atkinson et al.[29], increased arterial pressure and the subsequent injections of isoproterenol reduced it. Therefore, in light of the finding by Robinson & Evered (this volume) that drinking to ANG II is inhibited by the pressor response, a more likely interpretation of the Atkinson data is that the effect of isoproterenol to offset the hypertensive effect of the renin injection "permitted" the maximal drinking response to ANG II.

Measurement of Components of the Renin-Angiotensin System after Exogenous Administration of Renin or Angiotensin

One way to estimate the likely contribution of the RRAS to thirst in physiological states is to directly measure circulating levels of components of the RRAS following administration of renin or ANG II at doses that produce drinking. Renin activity and renin concentration are two of the most widely used indices of RRAS activation. Both measures are enzymatic assays in which plasma is incubated and the amount of ANG I generated per unit time determined. Plasma incubated in the presence of excess angiotensinogen produces the measure of renin concentration and plasma incubated without the addition of any additional substrate provides an estimate of renin activity.

Stricker and his colleagues[30] administered 0.5 to 6 units of hog renin per 100 grams body weight to nephrectomized rats (i.p.) and measured water intake over the subsequent 3 hour period. At the conclusion of the drinking test, plasma was collected and a determination of plasma renin activity made. A regression analysis indicated that a plasma renin activity of 100 ng/ml/90 min was required for each ml intake of water consumed in a 1 hour period. Based on these results it would seem that the RRAS is not very effective in producing water intake.

Although plasma renin activity or plasma renin concentration has proved to be a valuable clinical tool as a general index of activity of the RRAS, there are potential limitations to the method which may be especially important when grossly unphysiological manipulations are made such as the administration of exceedingly high doses of renin. For example, the ultimate formation of the effector peptide ANG II depends upon the availability of appropriate amounts of endogenous angiotensinogen to form ANG I and in turn sufficient converting enzyme to transform ANG I to the effector peptide, ANG II. Thus, a better determination of the efficacy of the RRAS would be a measurement of the active component itself.

Table 1. Plasma ANG II concentrations after intravenous infusions of ANG II amide into nephrectomized rats

	Doses Infused, $pmol.kg^{-1}.min^{-1}$				
	0	1	25	50	100
ANG II, fmol/ml plasma at 15 min	34.6 ±3.2	104.9 ±7.0	184.9 ±10.5	363.2 ±24.6	766.2 ±104.2
at 60 min	32.3 ±0.4	116.2 ±12.8	201.7 ±10.1	377.2 ±41.0	805.9 ±83.1

Values are given as means ± S.E.; n, 5-6 per group; ANG II, angiotensin II

The development and application of radioimmunoassays for ANG II have made it possible to determine peptide levels over a range of dipsogenic doses of ANG and we have done this in a number of experiments[22] conducted in Dr. Detlev Ganten's laboratory in Heidelberg. Summarized in Table 1 are the plasma levels for ANG II at 15 and 60 minutes after the beginning of infusions of ANG II amide (Hypertensin, CIBA) at doses of 1, 25, 50, and 100 pmol/kg/min into nephrectomized (unanesthetized) rats[22]. The infusion of 25 and 100 pmol/kg/min into unnephrectomized conscious rats produced plasma concentrations after 60 minutes of 204 ± 50.7 fmol/ml; (n = 6) and 848.5 ± 60.3 fmol/ml (n = 7), respectively.

Hsaio and colleagues[21] have demonstrated that it is possible to reliably induce drinking in rats with infusions of Hypertensin at the dose of 25 pmol/kg/min. Thus, it would appear that the intravenous dipsogenic dose of ANG II produces plasma levels of approximately 200 fmol/ml. Given the relationship between the amount of infused exogenous ANG II at the dipsogenic threshold and plasma concentration generated by that dose, one can presume that under experimental situations where circulating levels of ANG II exceed 200 fmol/ml, the RRAS is making a contribution to the observed water intake.

The Capacity of the Renal Renin-Angiotensin System to Generate Angiotensin II to Thirst-Inducing Challenges

Using the rationale stated immediately above, we have conducted a series of experiments examining plasma ANG II levels

Table 2. Conditions and experimental manipulations accompanied by thirst and studied for effects on plasma angiotensin II levels

_____

Physiological
    Moderate Water Deprivation
    - removal of water for 12, 24 and 48h

    Ingestion of a Dry Meal
    - 1h access to laboratory pellets without access to water

Pathophysiological States
    Diabetes Insipidus (DI)
    - Brattleboro rats with ad lib water access and following 14h water deprivation

    Renal Hypertension
    - rats with 2-kidney, 1 clip Goldblatt hypertension characterized as having malignant or benign hypertension

Experimental Thirst Challenges
    Caval Ligation
    - vena cava ligated below the renal veins

    Isoproterenol Treatment
    - subcutaneous injection of 30, 100 and 300 µg/kg

    Polyethylene Glycol Treatment
    - subcutaneous injection of 20% polyethylene glycol (1.5 ml/100g body weight)

_____

under a variety of conditions known to be accompanied by increased water intake[22,32]. Table 2 summarizes the types of situations under which the animals were studied and provides some of the more pertinent experimental details. The situations include (a) moderate levels of water deprivation or depletion following ingestion of a dry meal and after 12, 24, and 48 hours of water privation (It should be noted that rats will survive 6 to 15 days of water deprivation[33].), (b) animals under pathological conditions, (i.e., genetic diabetes insipidus and experimentally induced renal hypertension), and (c) thirst challenges (i.e., caval ligation, isoproterenol treatment and extracellular depletion with PEG. It is possible to compare the plasma levels generated by the preceding manipulations with the levels produced

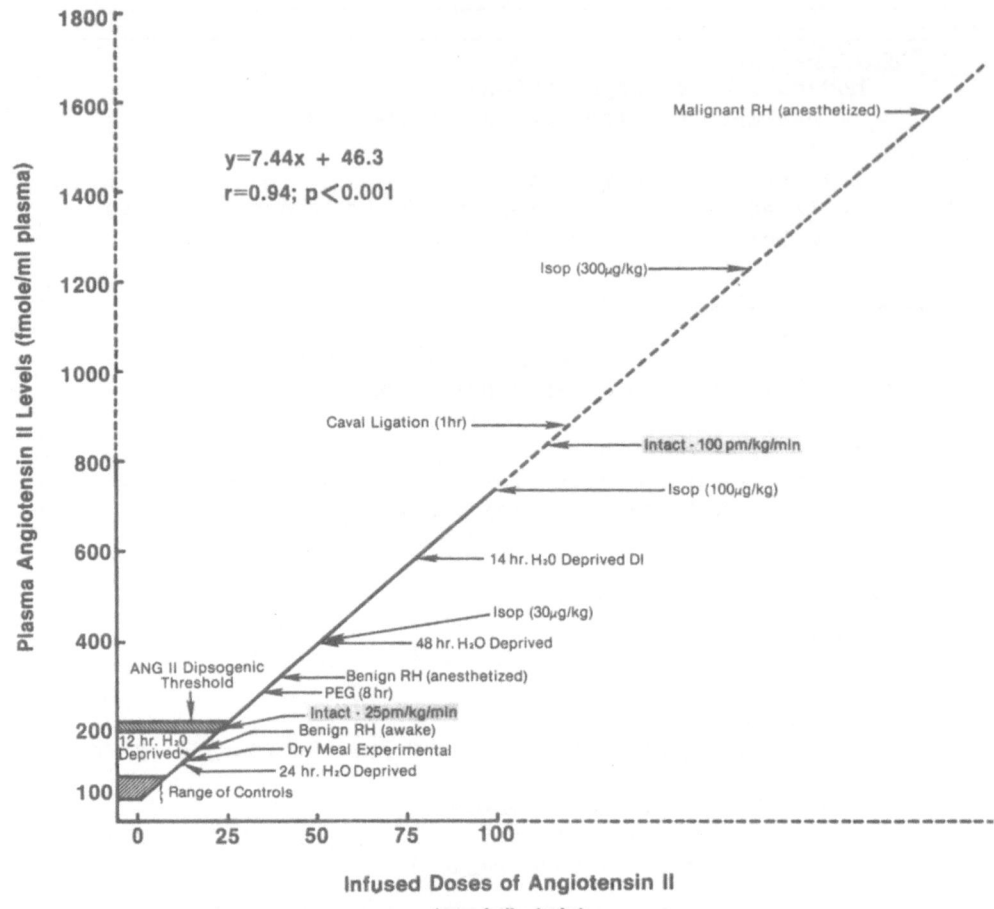

Fig. 1 Regression line of the relationship between infused dose
of ANG II (abscissa) and ANG II concentration measured in
plasma (ordinate). Plasma levels of ANG II produced by
various thirst stimuli are indicated along the regression
line. Abbreviations: DI, diabetes insipidus; RH, renal
hypertension; PEG, polyethylene glycol; ISOP,
isoproterenol. Intact 25 pmol/kg/min and 100 pmol/kg/min
indicates level of circulating ANG II after 60-min
infusion of indicated dose into unanesthetized rats with
kidneys intact. Upper shaded area represents approximate
ANG II plasma level at dipsogenic threshold. [22,32]

by exogenous infusion of ANG II[32]. This type of comparison is provided by Figure 1 which indicates the regression line generated from the data in Table 1 and shows the relationship between dose of ANG II infused and plasma ANG II concentrations produced. The plasma levels of ANG II (ordinate) generated by various thirst challenges are indicated along the regression line. All the experimental manipulations were shown to significantly elevate plasma ANG II levels above control levels. Furthermore, following longer periods of water deprivation (14 hours and 48 hours in diabetes insipidus and in normal rats), caval ligation, isoproterenol treatment, and extracellular fluid depletion with PEG, plasma levels of ANG II were elevated above the experimentally defined dipsogenic threshold. Thus, it seems reasonable to conclude that under virtually all of these latter manipulations much of the induced water intake is due to the dipsogenic action of the octapeptide.

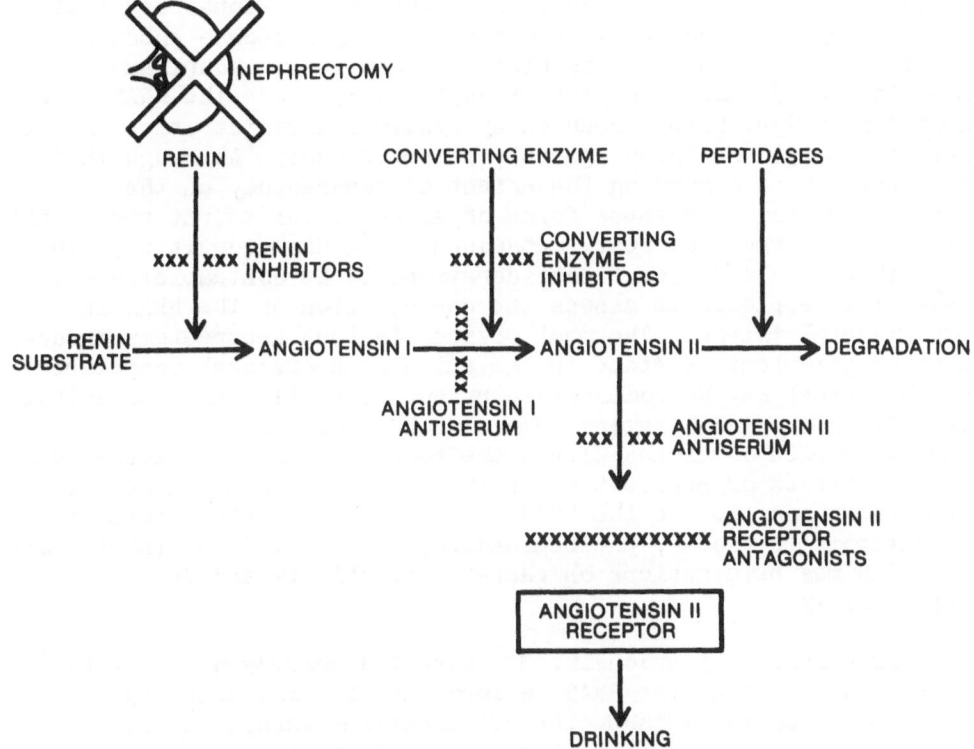

Fig. 2. Synthetic cascade of the renin-angiotensin system showing the sites where the synthesis or action of ANG II can be blocked. Nephrectomy removes the predominant source of renin and xxx marks the sites where inhibition or receptor antagonism can take place.

Disabling the Renal Renin-Angiotensin System

Illustrated in Figure 2 is the metabolic cascade of the RRAS. Also indicated are points of vulnerability where the "flow" between renin and ANG II can be blocked. Major points of interruption are (a) removal of renin at its source by nephrectomy; (b) competitive antagonism of renin by inhibitors; (c) blockage of the conversion of ANG I to ANG II by using converting enzyme inhibitors; (d) competition for the ANG II receptor by employing receptor antagonists; and (e) removing either ANG I or II from availability by using specific antibodies for each peptide. Logically it would seem that each of the preceding methods should be effective in eliminating or markedly reducing the availability of ANG II to interact with receptors mediating the dipsogenic response. However, whereas this may be true in some cases, there are many extenuating limitations of each method that warrant special consideration.

Nephrectomy. In his classic experiments implicating the RRAS in thirst, Fitzsimons[13,34] used nephrectomy to demonstrate that in the absence of a humoral kidney the drinking response produced by ligation of the vena cava is blocked. Similarly, Houpt and Epstein used removal of the kidneys to implicate the RRAS as a mediator of the thirst produced by systemic administration of the beta-adrenergic receptor agonist, isoproterenol. Although their interpretations regarding the effect of nephrectomy on the drinking response to these forms of experimental thirst are in all likelihood largely correct, subsequent work has made it apparent that there are additional considerations to be contemplated when using this approach to assess the contribution of the RRAS in experimental thirst. The most obvious is that nephrectomy induces a major pathological state in the animal. Behavioral competence of the animal may be compromised by both specific and nonspecific effects of the accompanying uremia such as malaise and redistribution of fluids within the body. Another consideration of the effects of nephrectomy is that it alters renal hormonal systems in addition to the RRAS. These include alterations in erythropoietic factor, prostaglandins, kallikrein and others, many of which may have actions on vascular reactivity and/or permeability.

As described previously, Stricker and colleagues[27,28,30,31] have suggested that the RRAS is important for the drinking response produced by isoproterenol treatment because of its pressor action which allows the animals to be behaviorally competent. However, in other studies employing a different research strategy[36] it has been possible to demonstrate that the RRAS system actually provides a first line defense as the mediator in the drinking response produced by isoproterenol treatment.

When isoproterenol is delivered by intravenous infusion, it is possible to control both dose and degree of hypotension produced by the beta-adrenergic receptor agonist. Such studies clearly demonstrate that rats with endocrine kidneys (i.e., after ureteric ligation) drink more readily than nephrectomized animals in response to intravenous isoproterenol.

Renin Inhibitors. Although in principle activation of the RRAS can be blocked by employing renin antagonists such as pepstatin, this approach has not widely been used. Reasons for this are a lack of general availability of such drugs and the relatively low solubility of these inhibitors in biological fluids.

Angiotensin II Receptor Antagonists. After several years of attempting to design peptides that were antagonists for the ANG II receptor, groups began to meet with success in the early 1970's. One of the first ANG II receptor antagonists with significant activity in vivo was sar[1] ala[8] ANG II or saralasin. Initial work demonstrated that saralasin (originally designated as P-113), had an antagonistic action on the ANG II-induced pressor response[37]. Later experiments showed that it was also an effective antagonist to the ANG II mediated dipsogenic response produced by either centrally injected or circulating peptide[38,39].

Armed with a new pharmacological tool, investigators set out to determine the contribution made by ANG II in various forms of experimentally induced thirst. A role for ANG II in the drinking response to 30 hr water deprivation was demonstrated by showing that centrally applied saralasin would delay the onset of the drinking response when water was provided[40]. Since water deprivation includes both cellular and extracellular depletion, it would be expected that drinking should not be totally abolished by the ANG antagonist and that the observed delay in drinking was an appropriate experimental outcome.

However, evidence for a role of the RRAS in mediating thirst produced by caval ligation could not be found by Rolls and Wood[41]. Intravenous infusions of saralasin did not attenuate this form of experimental thirst in spite of the fact that the infusions of antagonists were effective in attenuating drinking to large doses of intravenously administered ANG II. This failure to block drinking to caval ligation, a putative ANG-mediated form of experimental thirst, can in all likelihood be attributed to two aspects associated with the use of this ANG II receptor antagonist.

The first concern is that saralasin has marked agonistic action when used in high concentrations. Mann and colleagues[42] have reported that the agonistic action of the peptide receptor

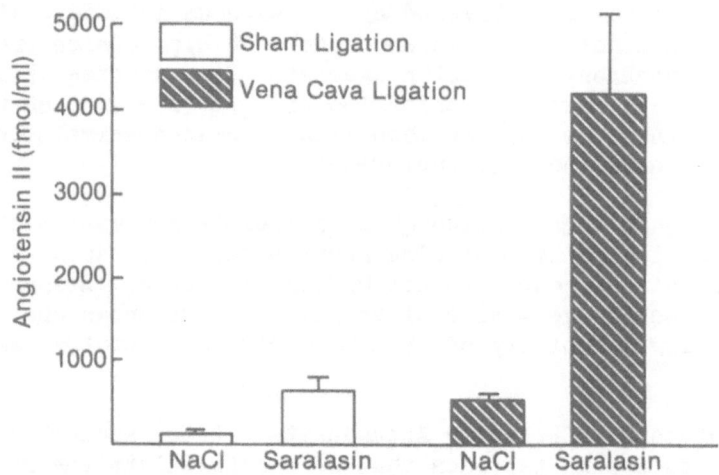

Fig. 3. Plasma ANG II concentrations in rats with ligation of
inferior vena cava above the renal veins (n = 10), and in
controls (n = 12), 195 min after surgery. Half of the
rats of each group received 50 μg/kg/min of saralasin IV
for 30 min before sacrificed, the other half received
0.9% saline. Values are means ± SEM.[43]

antagonist can be minimized with intravenous administration
of successive increases in concentrations of the peptide
antagonist. Using this strategy in thirst experiments in the rat
it has been possible to demonstrate that stepwise increases of
saralasin concentration over the course of 175 min will
significantly attenuate the drinking response following ligation
of the vena cava[43].

    A second concern is related to the extraordinary capacity of
the RRAS to generate high levels of circulating ANG in the face of
blockage of the metabolic cascade between renin and the ANG II
receptor. This consideration is of special concern when severe
hypotension is induced by the thirst generating manipulation. It
must be realized that recovery from the hypotension is interfered
with as well as the dipsogenic response when the systemic action
of ANG II is prevented. Mann and colleagues[43] utilized an
antibody specific for ANG II to test the effects of administration

172

of saralasin following caval ligation on the capacity of the RRAS to generate ANG II. The extraordinarily high concentrations of ANG II seen after these treatments are apparent in Figure 3. Comparing the titers of ANG II which exceeded 4,000 fmol/ml with the range of circulating ANG II levels induced by the variety of different physiological, pathophysiological, and pharmacological challenges summarized in Figure 1, shows the incredible capacity of the RRAS to overcome a functional blockade by an antagonist.

Considering these limitations of saralasin the likely reasons why Rolls and Wood[41] failed to find inhibition of caval ligation-induced drinking become evident. With low doses of antagonist the endogenous RRAS has the capacity to generate sufficient ANG II to overcome the blockade. With high doses, the agonistic activity of saralasin is prominent and actually contributes to the observed drinking.

Converting Enzyme Inhibitors. Conversion of ANG I to the biologically active peptide, ANG II, can be blocked by a group of drugs referred to as ANG-converting enzyme inhibitors. Blockade of the RRAS by these competitive inhibitors has several advantages over nephrectomy and the ANG II antagonists. On the one hand, they circumvent the disadvantages of nephrectomy such as irreversibility and the complications of surgical trauma and anuria. On the other hand, they have no agonist activity of their own as do ANG II antagonists such as saralasin. Nevertheless, initial experiments using the converting enzyme inhibitor SQ 20,881 were not very helpful in determining the role of the RRAS in the drinking response to various thirst stimuli. Lehr et al.[44] and Summy-Long and Severs[45] found that the subcutaneous or intramuscular injection of the converting enzyme inhibitor, SQ 20,881, increased rather than decreased the drinking response to thirst challenges which were commonly thought to involve the RRAS (e.g., isoproterenol treatment, caval ligation, hypovolemia). Lehr et al.[44] suggested that in the presence of converting enzyme inhibitors, the level of ANG I in the blood becomes very high and enters the brain to be converted locally to ANG II. Certainly, potent ANG-converting enzyme activity has been found in the brain[46] but when SQ 20,881 was given intracranially as well as systemically[44,45], only the enhancement effect was prevented and the rats still drank normally to the isoproterenol and hypovolemia.

SQ 20,881 is a nonapeptide which could not be expected to cross readily into brain tissue, but another converting enzyme inhibitor, captopril, is much more potent than SQ 20,881 and does appear to reach the vascular and neural sites of converting enzyme[47]. Using captopril it has been shown that the drinking responses to several putative renin-dependent thirst challenges can be reduced or abolished when ANG II synthesis is prevented in the brain as well as in the circulation. This was accomplished by using either a high subcutaneous dose of captopril (100 mg/kg) which blocks the drinking response to an intracerebroventricular injection of ANG I[48,49], or by giving this same high dose of captopril (100 mg/kg) by gavage (Mann et al., this volume) or by combining a low subcutaneous dose of captopril (0.5 mg/kg) with an additional injection of captopril (20 μg) into the cerebral ventricles[50].

As with the other methods of disabling the RRAS, inhibitors of converting enzyme have been criticized as tools for investigating the role of ANG II in cardiovascular and fluid regulation. The reasons are twofold. First, converting enzyme inhibitors have other effects besides the blockade of ANG II synthesis. For example, they slow the degradation of bradykinin and enkephalin and possibly a number of other peptides[46]. Second, blockade of ANG II synthesis prevents the usual pressor response seen following stimulation of renin release. It has been argued that animals fail to drink to renin-dependent thirst stimuli when treated with captopril not because of the loss of the dipsogenicity of ANG II, but rather because of the loss of the ANG II pressor effect and the debility of the exacerbated hypotension which ensues[27,28].

It seems unlikely that the effects of captopril on other peptides besides ANG could account for its inhibition of drinking. Captopril only reduces drinking to thirst stimuli which are associated with increased renin secretion from the kidney and has no effect on either the drinking to hypertonic saline treatment or to ANG II itself. Nor can the attenuation of drinking be attributed to exacerbated hypotension. Mann et al. (this volume) showed that captopril greatly inhibited the drinking response to PEG treatment even when blood pressure was kept normal with an i.v. infusion of vasopressin.

SUMMARY AND CONCLUSIONS

Although it is well established that ANG II given peripherally or intracerebrally is a potent stimulus for thirst, it remains controversial as to what degree the RRAS contributes to drinking under physiological conditions. The major problem in determining the role of ANG II in thirst is the fact that the RRAS is not the only thirst mechanism involved in defense of extracellular fluid volume. There is evidence that volume receptors in the capacitance vessels play a major role and may compensate for the loss of renin following nephrectomy[51]. It is the multiplicity of mechanisms activated in response to extracellular fluid deficits which makes it difficult to assess the contribution of the RRAS in physiological thirst. Removal of the endogenous source of renin by nephrectomy or inhibition of ANG II synthesis or action may not accurately reflect the quantitiative importance of ANG in the drinking response to a particular thirst stimulus in the intact animal.

The other major stumbling block in the understanding of the role of ANG II in thirst has been the assumption that administration of exogenous ANG II has the same effect on thirst mechanisms as does ANG II generated in response to an extracellular thirst challenge. This assumption fails to consider the possibility of an interaction between the various biological activities of ANG II, or between the ANG II thirst mechanisms and those mediated by the atrial stretch receptors, or between the extracellular and cellular thirst mechanisms. These interactions (Table 3) can be inhibitory or facilitory and must be considered when assessing the drinking response in a particular experimental situation.

Natural thirst stimuli such as water deprivation and hemorrhage lead to various physiological changes which interact in a facilitory way to cause drinking. Water deprivation, for example, is the most common naturally occurring thirst stimulus and one which elicits a robust response. As can be seen in Table 3, water deprivation is associated with all of the physiological changes facilitating water intake and with none of those which are inhibitory.

Experimentally induced thirst on the other hand often operates through only one facilitory mechanism and frequently generates one or more physiological changes which are inhibitory to thirst. Exogenous administration of ANG II in normotensive, water-replete rats, for example, stimulates thirst in the face of greatly elevated blood pressure and in spite of the fact that any water intake causes subsequent dilution of the body fluids and further inhibition. The evidence from studies, comparing the plasma concentrations of ANG II generated by experimental manipulations involving hypovolemia and/or hypotension with plasma levels of ANG II measured during infusions of dipsogenic doses of ANG, clearly indicate that these thirst-inducing treatments elevate plasma levels sufficiently to account for a component of the drinking response. Furthermore, the lack of appreciation for the confounding effects of the inhibitory interactions of exogenous ANG II has probably led to a significant underestimation of the dipsogenicity of plasma concentrations of ANG II.

The maintenance and defense of extracellular fluid volume and composition and the maintenance of normal arterial pressure is central to normal cell function. A review of all the biological actions of ANG II (e.g., pressor action, release of aldosterone,

Table 3. Factors which may accompany thirst-inducing stimuli and which facilitate or inhibit water intake

Thirst Stimuli	Facilitory Factors				Inhibitory Factors			
	Hyper-osmolality	Elevated ANG II	Hypo-tension	Hypo-volemia	Hypo-osmolality	Hyper-osmolality tension	Hyper-volemia
Water Deprivation	yes	yes	yes	yes	no	no	no
Hypertonic Saline	yes	no	no	no	no	slight	?
Hypovolemia	no	yes	slight	yes	yes, later*	no	no
Isoproterenol Treatment	no	yes	yes	no	yes, later*	no	yes, later*
ANG II	no	yes	no	no	yes, later*	yes	yes, later*

* as a result of water intake

etc.) shows them to be, in each case, related to the regulation of the constancy of the extracellular fluid compartment. Equally important, however, is the observation that in no case is ANG II the sole mediator of any of the actions attributed to this peptide. Always other agents can act either independently or in concert with ANG to achieve the ends necessary for control and regulation. Thus, as is true for the other biological actions of ANG, the evidence favors an important, physiological role for the RRAS in the mediation of drinking.

ACKNOWLEDGMENTS

The research reviewed in this paper involved the contributions of many of our colleagues and we wish to thank them. Studies were supported in part by USPHS grants HLP 14558 1R01 HL33796 and 1 K02 MH00064, The Humboldt Foundation, the SFB 90 "Cardiovasculares System", the Deutsche Forshungsgemeinschaft and the Medical Research Council of Canada.

REFERENCES

1. A. Gilman, The relations between blood osmotic pressure, fluid distribution and voluntary water intake, Am. J. Physiol. 120:323 (1937).
2. J. T. Fitzsimons, Drinking by rats depleted of body fluid without increase in osmotic pressure, J. Physiol. (Lond) 159:297 (1961).
3. E. M. Stricker, Extracellular fluid volume and thirst, Am. J. Physiol. 211:232 (1966).
4. E. M. Stricker and J. E. Jalowiec, Restoration of intravascular fluid volume following acute hypovolemia in rats, Am. J. Physiol. 218:191 (1970).
5. E. B. Verney, The antidiuretic hormone and the factors which determine its release. Proc. R. Soc. Lond. 135B:25 (1947).
6. T. N. Thrasher, C. J. Brown, L. C. Keil, and D.J. Ramsay, Thirst and vasopressin release in the dog: an osmoreceptor or sodium receptor mechanism, Am. J. Physiol. 238:R333 (1980).
7. T. N. Thrasher, R. G. Jones, L. C. Keil, C. J. Brown, and D. J. Ramsay, Drinking and vasopressin release during ventricular infusions of hypertonic solutions, Am. J. Physiol. 238:R340 (1980).
8. B. Andersson, Thirst and brain control of water balance, Am. Sci. 59:408 (1971).
9. O. H. Gauer and J. P. Henry, Neurohormonal control of plasma volume. pp. 145-190. in: A. C. Guyton and A. W. Cowley (ed.) International Review of Physiology: Cardiovascular Physiology, Vol. 9, University Park Press, Baltimore (1976).

10. J. Sobocinska, Effect of cervical vagosympathectomy on osmotic reactivity of the thirst mechanism in dogs, Bull. Acad. Pol. Sci. 17:265 (1969).

11. J. Sobocinska, Abolition of the effect of hypovolemia on the thirst threshold after cervical vagosympathectomy in dogs, Bull. Acad. Pol. Sci. 17:341 (1969).

12. J. T. Fitzsimons, Hypovolaemic drinking and renin, J. Physiol. (Lond) 186:130 (1966).

13. J. T. Fitzsimons, The role of the renal thirst factor in drinking induced by extracellular stimuli, J. Physiol. (Lond) 201:349 (1969).

14. S. Kozlowski and E. Szczepanska-Sadowska, Mechanisms of hypovolaemic thirst and interactions between hypovolaemia hyperosmolality and the antiduretic system. pp. 25-35. in: G. Peters, J. T. Fitzsimons and L. Peters-Haefeli (ed.) Control Mechanisms of Drinking. Springer-Verlag, Berlin, Heidelberg and New York (1975).

15. R. L. Hodge, R. D. Lowe, K. K. F. Ng, and J. R. Vane, Role of the vagus nerve in the control of the concentration of angiotensin II in the circulation, Nature 221:177 (1969).

16. J. L. Falk, M. Tang, and R. W. Bryant, Dipsogenic action of diazoxide: a pharmacologic analysis, J. Pharmacol. Exp. Ther. 190:154 (1974).

17. J. M. Linazasoro, C. Jimenez Diaz, and H. Castro Mendoza, The kidney and thirst regulation, Bull. Inst. of Med. Res., Madrid 7:53 (1954).

18. R. C. Nairn, G. M. C. Masson, and A. C. Corcoran, The production of serous effusions in nephrectomized animals by the administration of renal extracts and renin, J. Pathol. Bacteriol. 71:151 (1956).

19. A. W. Ascher and S. G. Anson, A vascular permeability factor of renal origin, Nature 198:1097 (1963).

20. A. N. Epstein and S. Hsaio, Angiotensin as a dipsogen, pp. 108-116. in: G. Peters, J. T. Fitzsimons and L. Peters-Haefeli (ed.) Control Mechanisms of Drinking. Springer-Verlag, Berlin, Heidelberg and New York (1975).

21. S. A. Hsaio, A. N. Epstein, and J. S. Camardo, The dipsogenic potency of peripheral angiotensin II, Horm. Behav. 8:129 (1977).

22. J. F. E. Mann, A. K. Johnson, and D. Ganten, Plasma angiotensin II: dipsogenic levels and angiotensin-generating capacity of renin, Am. J. Physiol. 238:R372 (1980).

23. S. F. Abraham, R. M. Baker, E. H. Blaine, D. A. Denton, and M. J. McKinley, Water drinking induced in sheep by angiotensin - a physiological or pharmacological effect?, J. Comp. Physiol. Psychol. 88:503 (1975).

24. J. T. Fitzsimons, J. Kucharczk, and G. Richards, Systemic angiotensin-induced drinking in the dog: A physiological phenomenon, J. Physiol. (Lond) 276:435 (1978).

25. N. C. Trippodo, R. E. McCaa, and A. C. Guyton, Effects of prolonged angiotensin II infusion on thirst, Am. J. Physiol, 230:1063 (1976).

26. D. Denton, in: The Hunger for Salt An Anthropological, Physiological and Medical Anaysis, Springer-Verlag, Berlin, Heidelberg, New York (1982).

27. E. M. Stricker, The renin-angiotensin system and thirst: a reevaluation. II Drinking elicited in rats by caval ligation and isoproterenol, J. Comp. Physiol. Psychol. 91:1220 (1977).

28. E. M. Stricker, The renin-angiotensin system and thirst: some unanswered questions. Fed. Proc. 37:2704 (1978).

29. J. Atkinson, H.-P. Kaeserman, J. Lambelet, G. Peters and L. Peters-Haefeli, The role of circulating renin in drinking in response to isoprenaline, J. Physiol. (Lond) 291:61 (1979).

30. E. M. Stricker, W. G. Bradshaw, and R. H. McDonald, Jr., The renin-angiotensin system and thirst: A reevaluation, Science 194:1169 (1976).

31. J. A. Hosutt, N. Rowland, and E. M. Stricker, Hypotension and thirst in rats after isoproterenol treatment, Physiol. & Behav., 21:593 (1978).

32. A. K. Johnson, J. F. E. Mann, W. Rascher, J. K. Johnson, and D. Ganten, Plasma angiotensin II concentrations and experimentally induced thirst, Am. J. Physiol. 240:R229 (1981).

33. E. F. Adolph, Do rats thrive when drinking sea water?, Am. J. Physiol. 140:25 (1943).

34. J. T. Fitzsimons, Drinking caused by constriction of the inferior vena cava in the rat, Nature 204:479 (1964).

35. K. A. Houpt, and A. N. Epstein, The complete dependence of beta-adrenergic drinking on the renal dipsogen, Physiol. Behav. 7:897 (1971).

36. R. Rettig, D. Ganten, and A. K. Johnson, Isoproterenol-induced thirst: renal and extrarenal mechanisms, Am. J. Physiol. 241:R152 (1981).

37. D. T. Pals, F. D. Masucci, G. S. Denning, Jr., F. Sipos, and D. Fessler, Role of the pressor action of angiotensin II in experimental hypertension, Circ. Res. 29:673 (1971).

38. E. D. Vaughn, Jr., H. Gavras, J. H. Laragh, and M. N. Koss, Vascular permeability factor: dissociation from the angiotensin II induced pressor and drinking responses, Nature 242:334 (1973).

39. J. T. Fitzsimons, A. N. Epstein, and A. K. Johnson, Peptide antagonists of the renin-angiotensin system in the characterisation of receptors for angiotensin-induced drinking, Brain Res. 153:319 (1978).

40. R. L. Malvin, D. Mouw, and A. J. Vander, Angiotensin: physiological role in water-deprivation-induced thirst of rats, Science 197:171 (1977).

41. B. J. Rolls and R. J. Wood, Role of angiotensin in thirst, Pharmac. Biochem. Behav. 6:245 (1977).

42. J. F. E. Mann, I. Phillips, R. Dietz, H. Haebara, and D. Ganten, Effects of central and peripheral angiotensin blockade in hypertensive rats, Am. J. Physiol. 234:H629 (1978).

43. J. F. E. Mann, A. K. Johnson, W. Rascher, J. Genest, and D. Ganten, Thirst in the rat after ligation of the inferior vena cava: role of angiotensin II, Pharmac. Biochem. Behav. 15:337 (1981).

44. D. Lehr, H. W. Goldman, and P. Casner, Renin-angiotensin role in thirst: paradoxical enhancement of drinking by angiotensin converting enzyme inhibitor, Science 182:1031 (1973).

45. J. Summy-Long and W. B. Severs, Angiotensin and thirst: studies with a converting enzyme inhibitor and a receptor antagonist, Life Sciences 15:569 (1974).

46. M. P. Printz, D. Ganten, T. Unger, and M. I. Phillips, Minireview: The brain renin angiotensin system. pp. 3-52. in: D. Ganten, M. Printz, M. I. Phillips and B. A. Scholkens (ed.) The Renin Angiotensin System in the Brain. Exp. Brain Res. Suppl. 4. Springer-Verlag, Berlin, Heidelberg and New York (1982).

47. M. L. Cohen and K. D. Kurz, Angiotensin converting enzyme inhibition in tissues from spontaneously hypertensive rats after treatment with captopril or MK-421, J. Pharmacol Exp. Ther. 220:63 (1982).

48. M. D. Evered and M. M. Robinson, The renin-angiotensin system in drinking and cardiovascular responses to isoprenaline in the rat, J. Physiol. 316:357 (1981).

49. M. D. Evered and M. M. Robinson, Increased or decreased thirst caused by inhibition of angiotensin-converting enzyme in the rat, J. Physiol. 348:573 (1984).

50. M. M. Robinson and M. D. Evered, Effects of systemic and intracranial inhibition of angiotensin-converting enzyme on isoproterenol-induced drinking in the rat, Eur. J. Pharmacol. 90:343 (1983).

51. J. T. Fitzsimons, The physiology of thirst and sodium appetite. Monographs of the Physiological Society No. 35, pp. 194. Cambridge Univ. Press., Cambridge (1979).

RENIN DEPENDENCE OF INSULIN-INDUCED THIRST

M. Costales, M. Vijande, B. Marín, J.I. Brime and
P. Lopez-Sela
Department of Physiology (Medicine and Biology)
University of Oviedo, Spain

INTRODUCTION

The influence of insulin on food intake has been amply studied.
However there has been little work on the effect of this hormone in
water intake. In 1964 Novin[1] reported increased water intake due to
insulin administration, separated from its effect on food intake.
Booth and Brookover[2] (1968), Spitz[3] (1974) and Waldbillig and Bartness[4] (1981) also found similar results of stimulation of thirst by
insulin. Unpublished results by Fitzsimons reveal that i.p. insulin-
induced drinking (IID) is abolished by nephrectomy. It could favour
the possible participation of the renin-angiotensin system in this
phenomenon. Waldbillig and Bartness[4] found that i.p. insulin produced neither hypovolemia non plasmatic hyperosmolality, both conditions typified as thirst stimuli. On the contrary, animals presented hypervolemia and hypoosmolality.

We have tried to contribute to the controversial knowledge of
the mechanisms implied in the dipsogenic effect of the insulin on
the rat, with special regards to the hypothesis of the participation
of renin-angiotensin system in IID.

METHODS

Male and female Wistar rats (200-400 g) were individually
housed on a 12-12 light-dark cycle and mantained with a standard
laboratory diet and tap water from graduated tubes fitted with glass
spouts and available ad lib. During the experiments food was withdrawn but access to water was mantained. Commercial insulin Actrapid
Novo (vials of 10 ml - 40 U/ml) was used in all experiments. Injections of 1 ml/kg of the different doses used were prepared immedia-

tely before administration, dissolving the appropriate volume of
commercial insulin in saline and administered intraperitoneally.
Before starting with actual experiments, animals were injected at
least twicw 20 U of insulin (i.p.) in order to accustom the rats
to the manipulation. Different injections were separated normally
by 48 hours.

EXPERIMENT 1

To 11 male rats (350-400 g) 0.1, 1.0, 10, 20 and 40 U/kg of
insulin were injected on different days and in a random order. Every
dose was repeated 3 times and the averaged response was computed.
Averaged response to 3 i.p. injections af saline (0.9 %) were used
as control (Fig.1). Factorial analysis of results reveals a significant difference ($p < 0.01$) in water intake by different doses of
insulin, and a significant difference ($p < 0.01$) of water intakes by
different times. Individual comparisons with control data reveals
a significant effect of hormone doses of 10, 20 and 40 insulin U/Kg
after 60' and 120'. In an additional experiment, male rats (250 -
400 g) were injected with saline (n=11) or insulin 20 U/Kg (n=14)
and blood samples were collected from the tail artery by momentary
inmobilization. Hematocrit and blood glucose concentration were determined. Hematocrit fell significantly from preinjection values
(47.9 ± 1.7) after 30' (44.4 ± 2.0 , $p < 0.01$), 60' (41.7 ± 1.9 ,
$p < 0.001$) and 120' (38.7 ± 1.6 , $p < 0.001$). Hematocrit values in
control rats did not change. Glycaemia also fell abruptly and significantly from preoperative values (127.4 ± 7.0) after 30' ($62.8 \pm
7.5$, $p < 0.001$), 60' (57.0 ± 9.1 , $p < 0.001$) and 120' (57.7 ± 13.8,
$p < 0.005$).

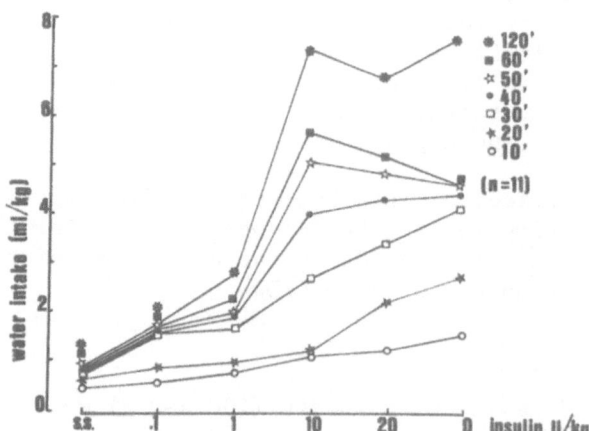

Fig. 1.   Water intakes after insulin injections. All insulin doses
greater than 10 U/KG were statistically significant in
comparison to saline injection, after 1 and 2 hours.

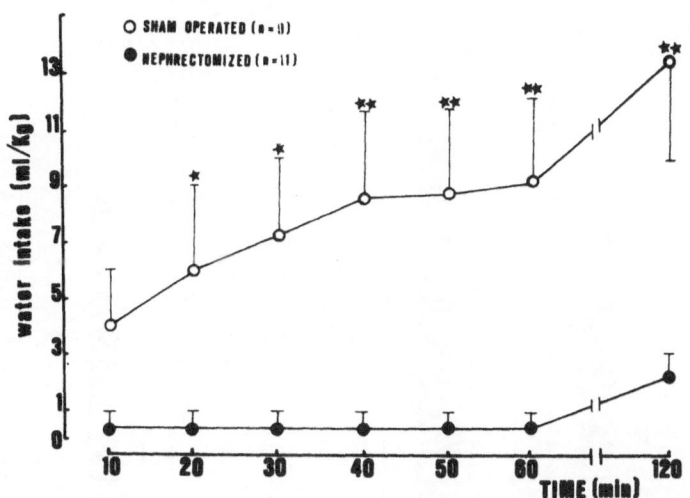

Fig. 2.   The effects of total nephrectomy on insulin-induced drinking, compared with sham operated by unpaired t-test.
* p < 0.05 ,   ** p < 0.005

EXPERIMENT 2 : The effect of bilateral nephrectomy

        Kidneys were removed from 10 male rats (200-350 g) through a
dorsal incision under light ether anaesthesia, and 9 male rats (200-
350 g) suffered a sham operation. After 6 hours 20 U/Kg of insulin
were i.p. administered and the water intake was measured every 10'
during the first 60' and again at 120'. Nephrectomy abolishes absolutely thirst induced by insulin. Similar results were obtained in
a batch of female rats (Fig. 2).

EXPERIMENT 3 : The effect of Converting Enzyme Inhibitors

        The effect of blocking the generation of angiotensin II on
IID was approched treating rats with converting enzyme inhibitors.
Groups of 11 rats (300-400 g) were injected s.c. with Captopril
0.1 mg/Kg, or Captopril 10 mg/Kg. The drug was dissolved in an
appropriate amount of saline and 5 ml/Kg of the solution was injected. The control group received only saline. One hour later all
animals received a i.p. injection of insulin 20 U/Kg. Low dose of
Captopril (Fig. 3) did not alter significantly IID. However, the
administration of 10 mg of Captopril/Kg produces a significant increase of water induced by insulin. The experiment was repeated
using the converting enzyme inhibitor MK-421 on another batch of
male rats (250-400 g). Ten animals received 0.5 mg/Kg of MK-421
s.c. dissolved in 1 ml of saline. Another ten rats received only
saline. One hour later 10 U/Kg of insulin were injected into all
animals. A significant increase of IID was observed after one and
two hours (Fig. 4).

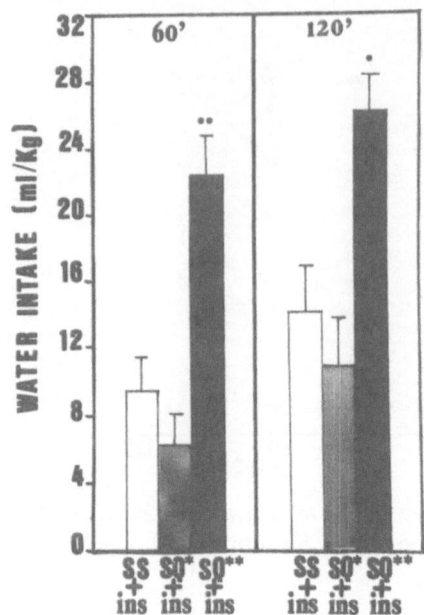 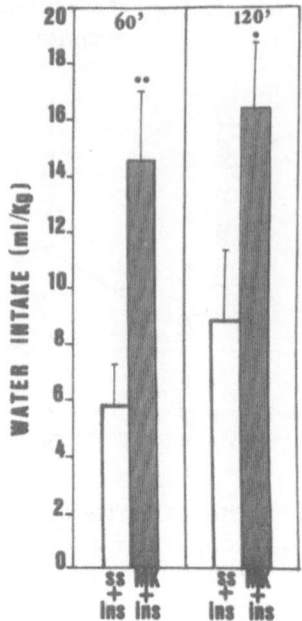

Fig. 3. The effects of s.c. Cap-
toptil (0.1*and 10**mg/Kg)
on water intake induced
by i.p. insulin (20 U/Kg)
in comparison to saline
injection by unpaired t-
test. ● p < 0.005 ,and
●● p < 0.001

Fig. 4. The effects of s.c. MK-
421 (0.5 mg/Kg) on water
intake induced by i.p.
insulin (10 U/Kg) as com-
pared to saline injection
by unpaired t-test.
● p < 0.05 , ●● p < 0.005

EXPERIMENT 4 : The effect of blocking A-II receptors

(Sar^1– Ala^8) angiotensin II was used as an angiotensin II receptor antagonist. To 10 rats (300-350 g), under Nembutal anaesthesia (50 mg/Kg i.p.) a guide cannula (0.7 mm o.d.) was implanted
with its tip aimed to lateral cerebral ventricule, and fixed to the
skull with screws and dental cement. After 3 days of recovery and in
different days, testing of different doses (25 μg , 50 μg and 125
μg) of saralasin began. The drug was dissolved in 5 μl of isotonic
saline. The injection system was composed of a inner cannula (0.4
mm o.d.) atached by PE10 tubing to a 10 μl Hamilton syringe. Control
injection consisted of 5 μl of saline. Fifteen minutes after saralasin administration 20 U/Kg of insulin were injected i.p. dissolved
in 1 ml of saline. Saralasin in doses of 25 μg and 50 μg did not
modify significantly the IID. However, saralasin in doses of 125 μg
caused a substantial decrease one and two hours after insulin administration (Fig. 5).

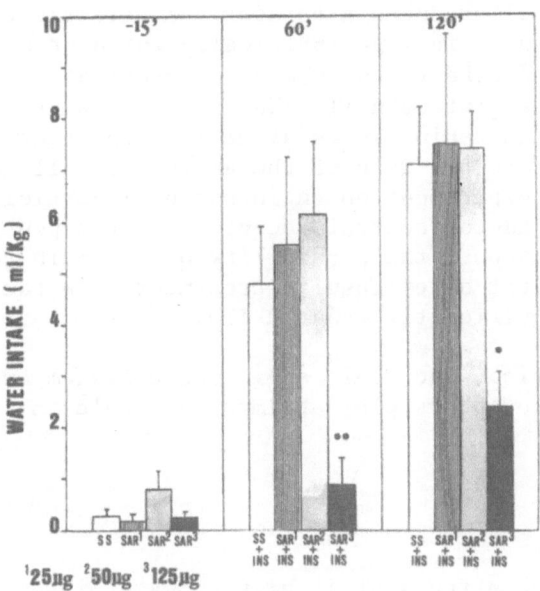

Fig. 5.   Effect of saralasin (Sar[1]-Ala[8] Angiotensin II), 25, 50
or 125 ug i.c.v. on water intake induced by insulin (20
U/Kg) in comparison to saline injection, by paired t-test.
● p < 0.002 , ●● p < 0.001

DISCUSSION

    Insulin shows a weak dipsogenic effect. Our results are in
agreement with those of Novin[1], Spitz[2], and Booth and Pitt[5]. Doses
of 10 , 20 and 40 U/Kg of insulin are very similar in effect and
significantly increased water intake after one hour. These amounts
of insulin are in the pharmacological range and because of that it
is uncertain if insulin could play any role in physiological regulation of water intake. The decrease in hematocrit after insulin
injection reveals the induction of hypervolemia by the hormone. It
would exclude the generation of thirst by mean of signals from
volume receptors. Waldbillig and Bartness[4] also found expansion of
blood volume after insulin treatment, althoug it was smaller than
that induced by only handling the animals.

    The suppression of IID by nephrectomy clearly suggests the
implication of some renal mechanism mediating this behaviour. Circulating renin released from the kidney[6,7] is a potent stimulus to
thirst and very well could be the cause of IID. The use of captopril to abolish IID has several difficulties. Our results from
doses of captopril of 0.1 mg/Kg reveal no affectation by so low
amount of drug. However, 10 mg/Kg of captopril induces an enhancement of IID. Fitzsimons and Elfont[8], demonstrated that situations
in which there was a high renin level, the administration of moderate

doses of captopril produces a paradoxical effect due to the accumulation of angiotensin I peripherically which crosses the blood brain barrier and affects the brain receptors after its transformation there to angiotensin II. Hence, our results favour the hypothesis of high renin levels after insulin treatment in these rats, and therefore the role of the angiotensin II in the IID. The most conclusive experiment on angiotensin II participation on IID is that of blockade of central receptors of the peptide. Our results strongly support the possibility of the main role of angiotensin II in thirst after insulin treatment. In fact, doses 125 µg of saralasin significantly reduced IID.

In conclusion, the renin angiotensin system and the angiotensin II central receptors play an important role in insulin induced thirst.

REFERENCES

1. D. Novin, The effects of insulin on water intake in the rat, in "Thirst"- Proceedings of the 1st International Symposium on Thirst in the Regulation of Body Water, ed., Pergamon Press New York (1964).
2. D. A. Booth and T. Brookover, Hunger elicited in the rat by a single injection of bovine crystalline insulin, Physiology and Behavior, 3:439 (1968).
3. R. Spitz, Induction of drinking by insulin in the rat, European J. Pharmacol. 31:110 (1974).
4. R. J. Waldbillig and T. J. Bartness, Insulin-induced drinking : an analysis of hidrational variables, Physiology and Behavior, 26:787 (1981).
5. D. A. Booth and M.E. Pitt, Role of glucose in insulin-induced feeding and drinking, Physiology and Behavior, 3:447 (1968).
6. J. T. Fitzsimons, The role of a renal thirst factor in drinking induced by extracellular stimuli, Journal of Physiology, 201: 349 (1969).
7. J.T. Fitzsimons, Hypovolemic drinking and renin, Journal of Physiology, 186:130 (1966).
8. B. R. Elfont and J.T. Fitzsimons, Captopril induced drinking depends on and is enhanced by renin, Journal of Physiology, 319:71 (1981).

COMPARISON OF ANGIOTENSIN II AND III

INDUCED DIPSOGENICITY AND PRESSOR ACTION

John W. Wright, Sandra L. Morseth, Margaret J. Sullivan
and Joseph W. Harding

Washington State University
Pullman, WA 99164

INTRODUCTION

Angiotensin III (AIII) may play a greater role in body water balance and blood pressure control than previously envisioned. AIII is the predominant circulating peptide of the rat renin-angiotensin system,[1] and promotes greater neural activity than angiotensin II (AII)[2] when microiontophoretically delivered into the subfornical organ[2]. Intracranial injections of AII at doses of 50 pM and above induce greater drinking and pressor action than AIII in the rat,[3,4] however, doses of AII and AIII below 25 pM induce equivalent dipsogenicity[5]. The present investigation further compared the magnitude of dipsogenic and pressor responses elicited by intracarotid and intracerebroventricular (icv) injections of AII and AIII.

DIPSOGENICITY

Intracarotid infusion of AII was predicted to be more potent than AIII stemming from the observation that circumventricular organs (CVOs) taken from rats revealed nearly equivalent binding for [^{125}I] AII and [^{125}]AIII[5]. Therefore, AII was expected to be more potent because it takes longer for aminopeptidases to metabolize it thus more of it gets to the CVO receptors in an active form, i.e. either AII or converted to AIII. In the case of icv injections we expected equivalency of responding at the doses used since receptors for both ligands are available in close proximity to the injection site.

Materials and Methods

Adult male rats were each prepared with a right brachial artery

187

catheter for angiotensin infusion[6]. Additional rats were prepared
with icv cannula aimed at the right lateral ventricle[5]. During testing the brachial artery catheter was coupled to an infusion system
that permitted reasonably unrestricted movement by the animal. Eight
animals received each dose of AII (vehicle, 1, 10, 100 and 500 pM/
kg/min), another 8 rats received AIII, with a minimum of 48 hr between
treatments. Within each group, 4 rats were infused with ascending
doses of each analogue and the other 4 with descending doses. A 1.5%
solution of gentamycin sulfate in sterile saline served as the vehicle
for all compounds[7] which were infused at a rate of 20 µl/min. Volumes
of water consumed during the 8 hr infusion period were recorded at
1 hr intervals to the nearest 0.1 ml. Eight icv animals were injected with AII at doses of 0, 0.01, 0.1, 1, 10 and 25 pM/2 µl with the
order of the 5 doses and the control injection (2 µl of artificial
CSF) counterbalanced across subjects with all animals receiving
all treatments with 48 hr between treatments. Eight additional icv
animals were treated equivalently except AIII was substituted for AII.

Results

    Total water consumptions for each dose of intra-arterially infused
AII and AIII are presented in Table 1. There was a difference between
analogues with AII infusion yielding greater water consumption than
AIII, $p < 0.05$. AII induced greater consumption than AIII specifically
at the 100 and 500 pM doses. There was an overall dose effect,
$p < 0.0001$. The 500 pM dose differed from the vehicle, 1 and 10 pM; 100
pM differed from the vehicle and the 1 pM dose. There was also an
Analogue X Dose effect, $p < 0.05$. The 500 pM dose of AII resulted in
significantly greater water consumption than all other doses except
100 pM, while the 500 pM dose of AIII was different from the vehicle,
1 and 10 pM of AIII. The water consumption results taken from the
icv rats are shown in Fig. 1. There was an overall difference in favor
of AII due to the significantly greater consumption at 25 pM, $p < 0.05$.
There was a dose effect, $p < 0.001$, with each increment in dose different from the others except between 0.01 and 0.1 pM which did not
differ from the CSF injection. And the Analogues X Dose interaction
was significant, $p < 0.001$. The 25 pM dose resulted in more water
consumption following AII injection as compared with AIII, with no
differences between analogues at other doses.

Table 1. Mean (± SE) Water Intake and Change in Arterial Blood
         Pressure During Intracarotid infusion (pM/kg/min)

Response	Vehicle	1 pM	10 pM	100 pM	500 pM
ml/8 hr					
AII	1.1±0.5	1.9±1.0	3.1±0.9	9.5±2.7	13.5±3.2
AIII	1.2±0.7	1.6±0.8	2.0±0.8	3.4±0.7	6.6±1.2
mm Hg					
AII	3.5±1.5	4.3±1.1	9.8±1.1	42.3±3.6	53.3±2.9
AIII	3.1±1.2	3.0±0.7	9.6±1.3	18.9±2.6	30.0±3.2

PRESSOR RESPONSES

Haywood et al.[6] have reported that maximal pressor action of blood-borne AII occurs with infusion via the carotid artery in rats since CVOs of forebrain structures appear to be the central site for angiotensin receptors in this species. We predicted that AII would be a more potent pressor agent than AIII in agreement with previous reports[8,9]. Icv AII injection also results in a pressor response[10], and at 50 pM and above has been reported to be considerably more potent than AIII[4]. Given the equivalency of AII and AIII binding in rats we predicted that they would yield comparable pressor responses.

Materials and Methods

Adult male rats were each prepared with a right brachial artery catheter for infusion and a femoral artery catheter for blood pressure measurements. Additional rats were prepared with icv cannula and a femoral artery catheter. The brachial artery infusion rats received doses of 0, 1, 10, 100 and 500 pM/kg/min of AII and AIII, 50 μl/min for 10 min, in a counterbalanced design. Four animals received AII doses the first day and AIII the second; the other 4 animals AIII then AII. Thirty min were allowed between treatments. A 5 min baseline mean arterial blood pressure (MABP) was measured prior to each infusion and additional time was provided if the pressure had not returned to original baseline. The icv injection animals were administered doses of 0, 0.1, 1, 10 and 100 pM/2 μl CSF, counterbalanced for ascending and descending order. Four animals received doses of AII on the first test day and AIII the second day. The other 4 animals received AIII followed by AII the next day.

Results

Maximum changes in MABP during each intracarotid infusion dose are provided in Table 1. Again there was an overall difference between analogues with AII infusion resulting in greater pressor responding than AIII, $p < 0.0001$. As expected there was a dose effect, $p < 0.0001$, with each dose different from the other except for the vehicle and 1 pM which did not differ. And there was an Analogue X Dose effect, $p < 0.0001$, with AII yielding greater MABP at 100 and 500 pM. Maximum changes in MABP following icv injection of AII and AIII are presented in Fig. 2. There was no difference between AII and AIII but there was a Dose effect, $p < 0.0001$. Each increment in dose was different from the next except between 0.1 and 1 pM.

DISCUSSION

The water consumption induced by intracarotid infusion of AII generally agree with reports utilizing intravenous infusion in rats[10-14]. AIII induced consumption was significantly below that of

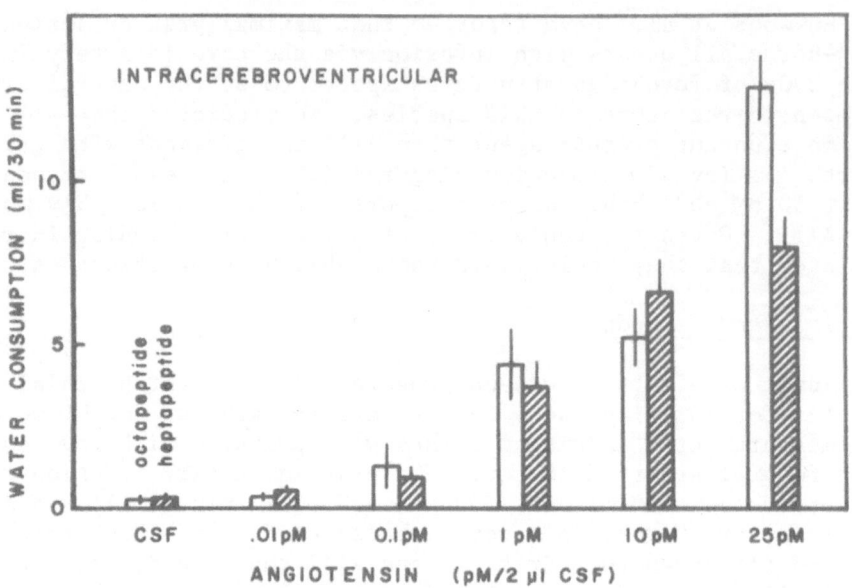

Fig. 1. Mean (± SE) water consumption following icv injections of
octapeptide (AII) or heptapeptide (AIII) according to dose.

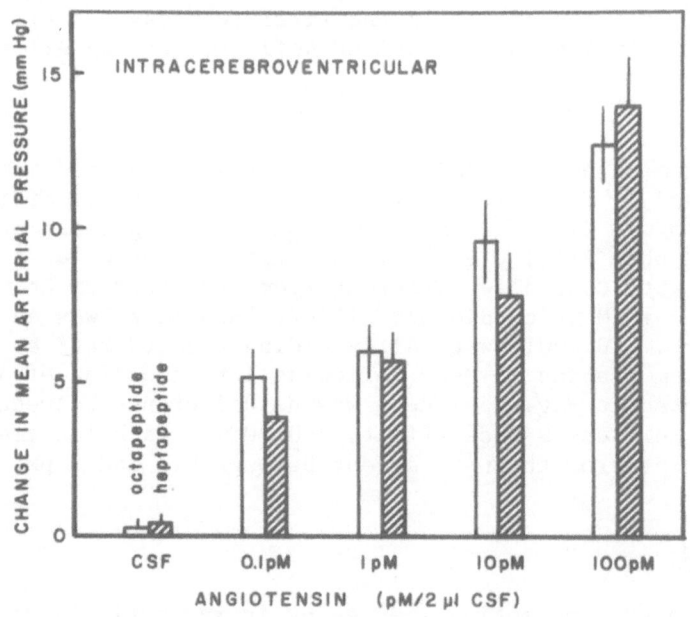

Fig. 2. Mean (± SE) change in arterial pressure following icv
injections of AII or AIII as a function of dose.

AII at each of the two highest doses. Icv injection of AII and AIII
elicited drinking in water replete rats. The present volumes of water
consumed following AII injection agree with those previously reported[15,16] as did the threshold dose of 0.1 pM[14]. AIII injection elicited
equivalent water consumption with respect to AII at 4 of the 5 doses
presently utilized. Only at the highest dose did AII result in greater
consumption. Fitzsimons[3] compared dipsogenicity following AII and
AIII diencephalic injections in rats and found AIII to be about 50% as
dipsogenic as AII at doses of 100 pM and above. This agrees with the
present results at 25 pM. Tonnaer and coworkers[4] found that icv injections of 50 pM AII resulted in a water consumption of about 9.5 ml
during the 20 min post-injection period. The present animals indicated
this level of drinking at the 25 pM dose.

     The observation that intracarotid AII was considerably more potent
than AIII as a pressor agent agrees with earlier reports,[8,9] and supports the hypothesis that there is a common central receptor site for
AII and AIII in the rat. Therefore, AII should be more effective than
AIII because it takes longer for tissue and blood-borne aminopeptidases
to degrade it to an inactive form and thus more of it arrives at CVO
receptors as an active ligand. The icv results further support this
hypothesis in that icv injections of AII and AIII yielded equivalent
pressor responses at the doses presently employed. This is in contrast
to the results of Tonnaer[4] who found AII to be much more potent than
AIII. Several differences appear to exist between our procedures and
theirs. First, we gentled our animals for several days prior to surgery which reduces fluctuations in blood pressure levels that accompany
handling, and freezing responses that interfere with drinking. We
minimized handling by introducing our injector into the guide cannula
while the animal was free-moving and unrestrained. Second, we used
an initial injection of angiotensin to test for correct placement as
recommended by Simpson et al.[14]. And finally we tested over a period
of two or more days rather than one day. Any one or combination of
these factors could perhaps account for some of the discrepancy.

ACKNOWLEDGMENTS

     This work was supported by the American Heart Association. Thanks
are due Mike Mills for collecting a portion of the data and Ruth Day
for preparing the manuscript in camera-ready format.

REFERENCES

1.   P. F. Semple and J. J. Morton, Angiotensin II and angio-
          tensin III in rat blood, Circ. Res. 38:122 (1976).
2.   D. Felix and W. Schlegel, Angiotensin receptive neurons in
          the subfornical organ. Structure-activity relations,
          Brain Res. 149:107 (1978).
3.   J. T. Fitzsimons, The effect on drinking of peptide
          precursors and of shorter chain peptide fragments of

angiotensin II injected into the rat's diencephalon, J. Physiol. 214: 295 (1971).

4. J. A. D. M. Tonnaer, V. M. Wiegant, W. De Jong and D. de Wied, Central effects of angiotensins on drinking and blood pressure: Structure-activity relationships, Brain Res. 236:417 (1982).

5. J. W. Wright, S. L. Morseth, M. J. Mana, E. LaCrosse, E. P. Petersen and J. W. Harding, Central angiotensin III-induced dipsogenicity in rats and gerbils, Brain Res. 295: 121 (1984).

6. J. R. Haywood, G. D. Fink, J. Buggy, M. I. Phillips and M. J. Brody, The area postrema plays no role in the pressor action of angiotensin in the rat, Am. J. Physiol. 239: H108 (1980).

7. K. E. Moe, M. L. Weiss and A. N. Epstein, Sodium appetite during captopril blockade of endogenous angiotensin II formation, Am. J. Physiol. (in press).

8. M. J. Peach, C. A. Sarstedt and E. D. Vaughan, Changes in cardiovascular and adrenal cortical responses to angiotensin III induced by sodium deprivation in the rat, Circ. Res. 38:II-117 (1976).

9. W. S. Spielman, J. O. Davis and R. H. Freeman, Des-asp[1]-angiotensin II: Possible role in mediating the renin-angiotensin response in the rat, Proc. Soc. Exp. Biol. Med.151:177 (1976).

10. M. I. Philips, Biological effects of angiotensin in the brain, in: "Enzymatic Release of Vasoactive Peptides," F. Gross and G. Vogel, eds., Raven Press, New York (1980).

11. J. T. Fitzsimons and B. J. Simons, The effect on drinking in the rat of intravenous infusion of angiotensin, given alone or in combination with other stimuli of thirst, J. Physiol. 203:45 (1969).

12. A. N. Epstein and S. Hsiao, Angiotensin as dipsogen, in: "Control Mechanisms of Drinking," G. Peters, J. T. Fitzsimons and L. Peters-Haefeli, eds., Springer-Verlag, Berlin (1975).

13. S. Hsiao, A. N. Epstein and J. S. Camardo, The dipsogenic potency of peripheral angiotensin II, Hormones Behav. 8:239 (1977).

14. J. B. Simpson, A. N. Epstein and J. S. Camardo, Localization of receptors for the dipsogenic action of angiotensin II in the subfornical organ of the rat, J. comp. Physiol. Psychol. 92:581 (1978).

15. A. K. Johnson and A. N. Epstein, The cerebral ventricles as the avenue for the dipsogenic action of intracranial angiotensin, Brain Res. 86:399 (1975).

16. W. E. Hoffman and M. I. Phillips, Evidence for Sar[1]-Ala[8] angiotensin crossing the blood cerebrospinal fluid barrier to antagonize central effects of angiotensin II, Brain Res. 109:541 (1976).

ANGIOTENSIN II AND ARTERIAL PRESSURE IN THE CONTROL OF THIRST

Marilyn M. Robinson and Mark D. Evered

Departments of Physiology
University of Western Ontario
London, Ontario, Canada, N6A 5C1
and University of Saskatchewan
Saskatoon, Saskatchewan, Canada, S7N 0W0

INTRODUCTION

Stimuli such as hypovolemia or hypotension which increase the secretion of renin from the kidneys increase thirst in man and other vertebrates[1]. However, intravenous infusions of renin or angiotensin II (Ang II) at doses which produce circulating concentrations similar to those observed in hypovolemia and hypotension cause relatively little drinking[2,3,4]. This could be interpreted to mean that the role of Ang II in the control of drinking to these stimuli is minor. The conditions are not comparable, however. Intravenous infusions of Ang II cause large increases in arterial pressure in the water-replete animal whereas under physiological conditions these same concentrations of Ang II would occur when blood pressure was at or below normal.

We have investigated the possibility that the pressor response to Ang II suppresses the drinking response. Ang II was infused I.V. in two doses. The higher dose (100 ng/min) is commonly used in investigations of the renin angiotensin system and reliably causes drinking but produces Ang II concentrations in plasma above the physiological range[3,5]. The lower dose (16.7 ng/min) of Ang II produces plasma concentrations equivalent to those produced by 12 h water deprivation[3,5] but is barely dipsogenic. We compared the drinking response of rats given Ang II alone with that of rats also given a vasodilator to return arterial pressure to normal.

METHODS

Male Wistar rats (250 - 400 g) were used. They were housed individually with free access to food (Purina Rodent Laboratory Chow) and tap water.

The right femoral artery and vein were cannulated in rats used for blood pressure recording. If they were to be used in the drinking experiments, only the vein was cannulated. The arterial cannulas were made from polyethylene tubing (PE 50) drawn out at one end to a smaller diameter and inserted into the vessel 4 cm. The venous cannulas were constructed by joining Silastic tubing (I.D., 0.635 mm; O.D., 1.194 mm) to Microline tubing (I.D., 0.51 mm; O.D., 1.56 mm; the Microline tubing was pulled out to a smaller diameter at one end and the Silastic tubing, after expansion in ether, was slipped over it approximately 1 cm). The Silastic end was inserted into the femoral vein and advanced into the abdominal vena cava a total of 4.5 cm from the point of entry. The catheters were brought under the skin and attached to a pedestal at the back of the neck. When not in use the cannulas were filled with heparinized saline (150 units/ml) and sealed. The surgery was done under general anesthesia (equithesin[6]). The rats were allowed at least a week to recover.

For both blood pressure recording and drinking experiments, the rats were placed in individual wooden open-topped test cages. Lengths of coiled polyethylene tubing (PE 50) connected the arterial catheters to a Statham pressure transducer and the venous catheters to a 2 ml glass syringe mounted on an infusion pump (Sage Instruments, Model 355).

All I.V. infusions were made at the rate of 1 ml/h. Ang II was infused at a dose of either 16.7 or 100 ng/min for 90 min. Fifteen minutes after the infusion was started, the pressor response to Ang II was reduced by injecting either isoproterenol (Isuprel HCl, Winthrop Laboratories, Aurora, Ont.), diazoxide (Hyperstat, Schering Canada Inc., Pointe Claire, Que.), or minoxidil (donated by the Upjohn Company of Canada, Don Mills, Ont.). All drugs were made up in 0.15 M saline except minoxidil which was dissolved in propylene glycol. Isoproterenol and diazoxide were injected S.C. and minoxidil was injected I.P.

The angiotensin-converting enzyme inhibitor, captopril (donated by Dr. Horovitz and Mr. Lucania, Squibb Institute for Medical Research, Princeton, NJ) was infused continuously in all experiments to block the endogenous synthesis of Ang II. We verified (Table 1) that this dose was adequate to prevent the drinking response to the hypotensive drugs given alone but not that to I.V. infusion of hypertonic saline (1 M-NaCl, 1 ml/h), a stimulus to drink not mediated by Ang II.

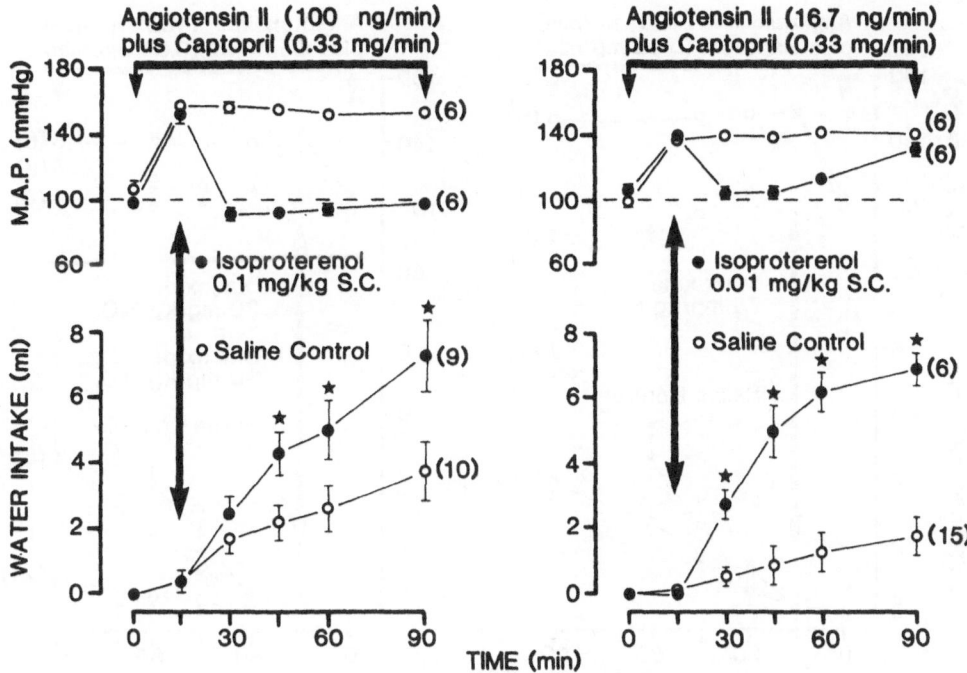

Fig. 1 Mean arterial pressures (M.A.P., Means ± S.E.) and intakes
of rats infused I.V. with Ang II and captopril (number of
rats in parentheses). Isoproterenol or isotonic saline
injected as indicated by the arrow. * Intake significantly
greater than control, p < 0.05.

Data in all cases are presented as means ± S.E.. Differences
between means were compared statistically using either analysis of
variance or, where variances were heterogenous, Mann-Whitney U-test.

RESULTS

Intravenous infusion of Ang II at a rate of either 16.7 or 100
ng/min caused rats to drink (Fig. 1). At the lower dose the amount
drunk was only slightly, but statistically significantly, greater
than that of control rats given only captopril (0.4 ± 0.2; N = 6,
p < 0.05). The higher dose of Ang II caused significantly more
drinking than the lower dose but the average water intake was still
less than 4 ml.

Both doses of Ang II caused large increases in mean arterial
pressure which persisted for the duration of the infusion (Fig. 1).
The difference in the pressor responses to the two doses was statistically significant but even the lower dose increased blood pressure
about 40 mmHg.

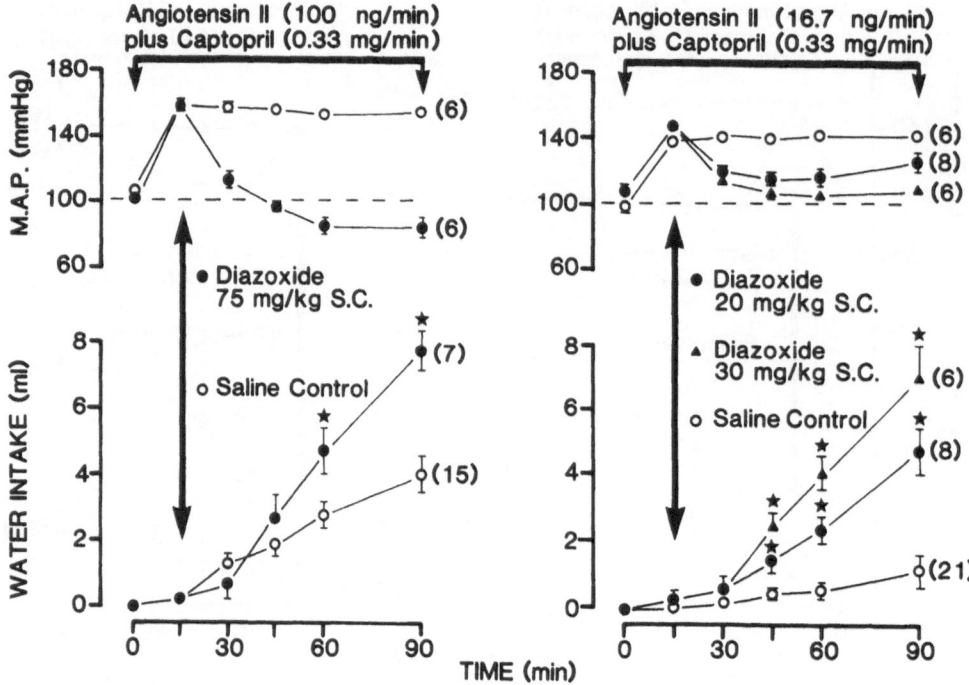

Fig. 2   Mean arterial pressures (M.A.P., Means ± S.E.) and intakes
of rats infused I.V. with Ang II and captopril (number of
rats in parentheses). Diazoxide or isotonic saline
injected as indicated by the arrow. * Intake significantly
greater than control, $p < 0.05$.

When isoproterenol was injected at a dose which reduced the
pressor response the drinking response to each dose of Ang II was
greatly increased (Fig. 1). When mean arterial pressure was closer
to normal there was no longer any significant difference between the
drinking responses to the two doses of Ang II (Fig. 1).

The increased drinking to Ang II caused by reducing the pressor
response was not unique to isoproterenol since another vasodilator,
diazoxide, had the same effect (Fig. 2). The fall in mean arterial
pressure was slower after injection of diazoxide than after isoproterenol; likewise the increase in water intake was delayed. The
results shown in the right-hand panel of Fig. 2, comparing the effects
of 2 doses of diazoxide, demonstrate that the closer mean arterial
pressure was returned to normal, the greater was the increase in the
drinking response. But these results also show that the increased
water intake does not depend on the animal being hypotensive.

Table 1. The effect of captopril (0.33 mg/min, I.V.) on the amount of water drunk following injections of the vasodilators or I.V. infusion of hypertonic saline.

Treatment	Alone ml/90 min	N	Plus Captopril ml/90 min	N	
Isoproterenol (0.1 mg/kg, S.C.)	3.96 ± 1.29	(9)	0.06 ± 0.02	(5)	*
Diazoxide (75 mg/kg, S.C.)	5.84 ± 1.24	(7)	0.06 ± 0.03	(7)	*
(20 mg/kg, S.C.)	2.60 ± 0.63	(8)	0.11 ± 0.09	(8)	*
Minoxidil (10 mg/kg, I.P.)	3.13 ± 0.73	(7)	0.14 ± 0.13	(7)	*
1 M-NaCl (1 ml/h, I.V.)	4.88 ± 0.71	(5)	3.84 ± 1.00	(5)	N.S.

* Intake significantly different from that of rats given the vasodilators alone ($p < 0.05$; Mann-Whitney U-test).

Reducing the pressor response to Ang II with a third vasodilator, minoxidil, again enhanced the drinking response (from 0.58 ± 0.16 to 2.95 ± 0.60, N = 6, $p < 0.001$) although the total amount drunk was less than when either isoproterenol or diazoxide was given.

As shown in Table 1, the dose of captopril chosen was sufficient to block drinking to the vasodilators when they were given alone.

DISCUSSION

The low dipsogenicity of I.V. Ang II, as compared to the much larger drinking responses of hypovolemic or hypotensive rats having similar plasma concentrations of the peptide, has been cited as evidence against Ang II having an important physiological role in the control of thirst[4,7,8]. This argument assumes that the effects of Ang II infused I.V. or generated endogenously are similar. As shown here, however, even low doses of Ang II infused I.V. elevate arterial pressure in the water-replete rat and this pressor response appears to inhibit the drinking.

The pressor action of Ang II injected I.V. also seems to inhibit the secretion of vasopressin. Ang II injected intracranially consistently stimulates vasopressin secretion[9] but has a less reliable effect when given I.V.[10,11]. Mitchell, Barron, Brody and Johnson[12] discovered that Ang II injected I.V. inhibited the magnocellular neurons in the supraoptic nucleus in intact rats but excited them in

197

animals with deafferented sino-aortic baroreceptors. Phenylephrine, which also elevated blood pressure, also inhibited the magnocellular neurons in intact rats and had no effect on them in the debuffered animals. They concluded from these results that the stimulation of vasopressin secretion by Ang II injected I.V. is inhibited by the hypertension.

In summary, increasing the plasma concentration of Ang II without increasing arterial pressure more closely mimics the physiological conditions of extracellular fluid volume depletion. In this situation Ang II, given I.V., is a very effective stimulus to drink and probably also to vasopressin secretion.

REFERENCES

1. J. T. Fitzsimons, "The Physiology of Thirst and Sodium Appetite", Monographs of the Physiological Society No. 35, Cambridge (1979).
2. S. A. Hsaio, A. N. Epstein, and J. S. Camardo, The dipsogenic potency of peripheral angiotensin II, Horm. Behav. 8:129 (1977).
3. A. K. Johnson, J. F. E. Mann, W. Rascher, J. K. Johnson, and D. Ganten, Plasma angiotensin II concentrations and experimentally induced thirst, Am. J. Physiol. 240:R229 (1981).
4. E. M. Stricker, The renin-angiotensin system and thirst: A reevaluation. II Drinking elicited in rats by caval ligation and isoproterenol, J. Comp. Physiol. Psychol. 91:1220 (1977).
5. J. F. E. Mann, A. K. Johnson, and D. Ganten, Plasma angiotensin II: Dipsogenic levels and angiotensin-generating capacity of renin, Am. J. Physiol. 238:R372 (1980).
6. C. P. Gandal, Avian anesthesia, Fed. Proc. 28:1533 (1969).
7. E. M. Stricker, The renin-angiotensin system and thirst: Some unanswered questions, Fed. Proc. 37:2704 (1978).
8. J. A. Hosutt, N. Rowland, and E. M. Stricker, Hypotension and thirst in rats after isoproterenol treatment, Physiol. Behav. 21:593 (1978).
9. L. C. Keil, J. Summy-Long, and W. B. Severs, Release of vasopressin by angiotensin II, Endocrinology 96:1063 (1975).
10. J. P. Bonjour, and R. L. Malvin, Stimulation of ADH release by the renin-angiotensin system, Am. J. Physiol. 218:1555 (1970).
11. J. R. Claybaugh, L. Share, and K. Shimizu, The inability of infusions of angiotensin to elevate the plasma vasopressin concentrations in the anesthetized dog, Endocrinology 90:1647
12. L. D. Mitchell, K. Barron, M. J. Brody, and A. K. Johnson, Two possible actions for circulating angiotensin II in the control of vasopressin release, Peptides 3(3):503 (1982).

ANGIOTENSIN DEPENDENT THIRST FOLLOWING POLYETHYLENEGLYCOL TREATMENT IN THE RAT

Johannes F.E. Mann Siegfried Eisele, Detlev Ganten
A. Kim Johnson, Rainer Rettig, Eberhard Ritz,
and Thomas Unger

Department of Medicine and Pharmacology
University of Heidelberg, 6900, FRG

INTRODUCTION

We have observed that angiotensin II plasma concentrations in polyethyleneglycol (PEG) treated rats were closely correlated with water intake by the animals (1). The angiotensin II plasma levels were in the range previously determined to be dipsogenic (2). The aim of the present investigation was to establish whether the correlation between angiotensin II plasma levels and the amount of water intake in PEG treated rats was based on a causal relationship. As an investigative tool we used captopril - an angiotensin I converting enzyme inhibitor - in order to block the formation of angiotensin II from angiotensin I. We approached the problem by this pharmacological means because it was shown in isoproterenol induced thirst that captopril may blunt water intake (3).

EFFECTS OF CAPTOPRIL ON PEG INDUCED THIRST

Adult male rats were used for the following experi-

ments (n = 8/group). In the morning, they received 325mg
PEG (MW 20,000, Serva Heidelberg, FRG) subcutaneously
(s.c.) (time = 0 hours). Captopril was given at 1 and
100 mg/kg by gastric gavage in glucose solution which was
also given as control. The converting enzyme inhibitor
was given 4 hrs after PEG injection because we had previously observed that angiotensin II plasma levels as
well as water intake in this particular thirst paradigme
will substantially increase only after 4 hrs. Water
intake was measured with graded burettes.

As is evident from fig. 1, high dose captopril
effectively blunted water intake ($p < 0.05 - < 0.01$, ANOVA)
wheras low dose captopril rather enhanced PEG induced
drinking. It is of note that we had determined that the
dose of 1 mg/kg captopril greatly suppressed the conversion of intravenously (i.v.) injected angiotensin I
for about 2 - 3 hrs (4).

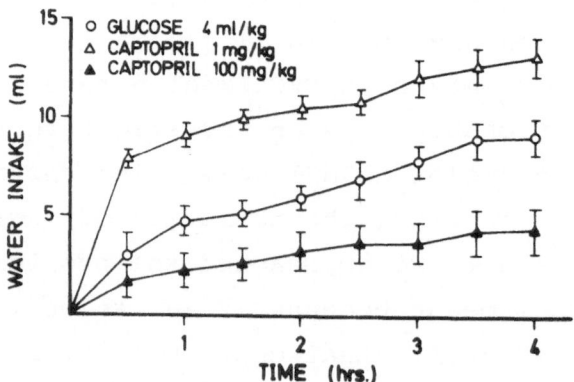

Fig. 1 Cumulative water intake following s.c. PEG
treatment. PEG was given 5 hrs prior to water access.
Time indicates hrs after water was returned to rats.

CAPTOPRIL, PEG INDUCED THIRST AND BLOOD PRESSURE

Low blood pressure in the arterial tree may stimulate water intake or - less frequently - debilitate animals, thus rendering them physically incapable for ingestive behavior (Johnson A. K., this volume; Stricker E. M., see Johnson paper). We therefore recorded blood pressure fluctuations follwing PEG injection in 3 groups of rats (n= 6 each): 1) PEG treatment (s.c.) and oral glucose solution; 2) PEG treatment and oral captopril (100 mg/kg) 4 hrs following PEG; 3) PEG treatment with oral captopril as in group 2, and i.v. vasopressin infusions by a variable speed infusion pump (total volume of infused volume: < 1ml/rat). The speed of the infusion pump was minute by minute adjusted to maintain blood pressure in the individual animal at the level recorded before giving captopril. To measure blood pressure, rats were implanted several days (> 2) prior to the actual experiment with chronic intraarterial catheters.

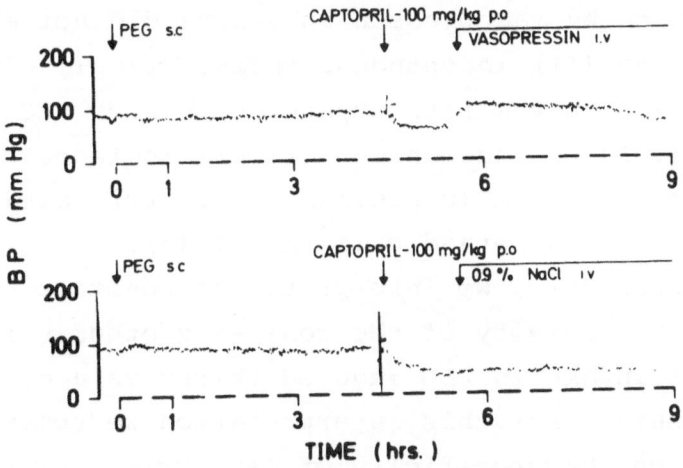

Fig. 2 Two individual blood pressure (BP) recordings (mean arterial blood pressure). At time "0", PEG was s.c. administered. At 4.5 hrs, captopril was given by gavage, and blood pressure was returned to normal by i.v. infusions of vasopressin (top panel). Equal amounts of NaCl were infused as control (lower panel).

The experiments showed that vasopressin infusions could indeed stabilize blood pressure (typical examples see fig. 2; a short blood pressure fall after captopril was allowed before vasopressin was started to ascertain the effectiveness of captopril). However, stabilization of blood pressure did not interfere with the antidipsogenic effect of captopril in PEG induced thirst. Water intake 2.5 hrs following captopril administration was 10 ± 3, 1 ± 1, and 2 ± 1 ml in groups 1, 2, and 3 respectively.

COMMENT

The results described show that high doses of a converting enzyme inhibitor effectively inhibited thirst due to a hypovolemic stimulus associated and correlated with high circulating concentrations of angiotensin II (1). This antidipsogenic effect was not a"non-specific" consequence of the blood pressure lowering by the converting enzyme inhibitor because (I) maintenance of normotension by vasopressin infusions did not alter outcome, and (II) independent thirst stimuli elicited water intake in the rats treated with s.c. PEG and oral captopril. These independent stimuli included hypertonic saline and dry food (unpublished results). Similar data were obtained in betaadrenic thirst (3).

Consequently, we interprete the positive correlation between the activity of the renin-angiotensin system and water intake in PEG induced thirst as a cause-effect relationship. For this interpretation we somewhat rely, however, on the specificity of captopril, our pharmacological "tool". Present evidence indicates that captopril, at the high dose used, infact inhibits conversion of angiotensin I peripherally and in the brain (Fitzsimons, this volume). The low dose (1mg/kg) which

was not antidipsogenic - and, interestingly, not hypotensive (unpublished) - did probably not reach effective
inhibitory concentrations in brain. In the absence of
a blood pressure lowering by low dose captopril, it remains to be established whether peripheral conversion
of angiotensin I was blocked under these conditions.

REFERENCES

1) A.K. Johnson, J.F.E. Mann, W. Rascher, J. Johnson, and D. Ganten, Plasma angiotensin II concentrations
and experimentally induced thirst. Am. J. Physiol., 240:
R229-R234, 1981.

2) J.F.E. Mann, A.K. Johnson, and D. Ganten, Plasma
angiotensin II: dipsogenic levels and the angiotensin
generating capacity of renin. Am. J. Physiol., 238:
R372-R377, 1980.

3) M.D. Evered, and M.M. Robinson, The renin-angiotensin system in drinking and cardiovascular responses
to isoprenaline in the rat, J. Physiol., 316: 357-367,
1981.

4) J.F.E. Mann, W. Rascher, R. Dietz, A. Schömig,
and D. Ganten, Effects of an orally active converting
enzyme inhibitor, SQ 14225, on pressor responses to
angiotensin administered into the brain ventricles of
spontaneously hypertensive rats, Clin. Sci., 56: 585-
591, 1979.

A GENERAL REVIEW OF THE EFFECTS OF ANGIOTENSIN-PEPTIDES

AND 8-SUBSTITUTED ANALOGS OF ANGIOTENSIN II

R. Kazim Türker

Department of Pharmacology
Faculty of Medicine, University of Ankara
Ankara, Turkey

In this short review special attention has been focused on some pharmacological effects of angiotensin peptides which may contribute to its direct physiological function in water and sodium intake.

From the pharmacological point of view angiotensin II (A II) belongs to the polypeptides that affect smooth muscle and blood vessels. Since it has the most pronounced vasoconstrictor and hypertensive effect, this has generally been considered in relation to its possible role in the pathogenesis of hypertension which is reflected in a number of reviews (1).

A II has a primary function in the control of aldosterone secretion (2). In conscious unstressed rat A II produces an increase in aldosterone secretion (3). In zona glomerulosa cells from rat A II at the concentrations as low as 10^{-11} mol/liter increases aldosterone secretion (4). Both in vivo and in vitro conditions, (des-Aspartic acid)[1]-A II, heptapeptide angiotensin III (A III) has a potent aldosterone-stimulating activity in various species which is more potent than A II in rabbit and rat (3,5,6). Other studies, however, indicate that A II is more active than A III and the low activity of heptapeptide is attributed to its rapid degradation in adrenals.

One of the most important physiological role of renin-angiotensin system is the control of water and sodium intake. It has been shown that A II when infused intravenously (7) or injected intracerebroventricularly (8) causes the rat to drink. Further studies elucidated the central sites of action of A II inducing thirst. Subfornical organ, dorso third ventricle, antero-ventral third ventricle are the main brain areas which contain specific sites for the effect of A II (9).

A II produces an enhancement of vascular response to sympathetic nerve stimulation which has been attributed to an increase in release of norepinephrine (NE) from sympathetic nerve endings (10,11,12). A cholinergic link for the effect of A II in the smooth muscle has also been described (10). A II accelerates the β-hydroxylation of dopamine, increasing the formation of new proteins as well as axonal uptake of dopamine in the heart muscle (13). Moreover, A II has been shown to evoke the release of dopamine from tyrosine (14), and to cause an increase in tyrosine hydroxylase activity in adrenergic neurons and brain (15). These effect should be taken into account for the explanation of the action of octapeptide in water and sodium intake.

Like other vasoconstrictors, A II shifts Na^+ in and K^+ out of vascular smooth muscle cells and shows no specifity in this respect (16). In contrast, NE which increases the K^+ efflux from rat aortic strips, A II causes only a rapid, transient rise in K^+ efflux (17). In isolated artery strips and rat uterus muscle an increase in Na^+ efflux was demonstrated under the influence of A II (18). The stimulating effect of A II on Na^+-pump has been suggested as a common pathway for the effect of octapeptide on vascular smooth muscle (18). The contractile effect of A II in the isolated vascular smooth muscle correlated with the increased Na^+ efflux and contractile response to the peptide (18). This mechanism may also explain the potent antiarrhythmic effect of A II agaist ouabain-induced cardiac arrhythmias (19).

Much more interest has been focused on A II tachyphylaxis, presumably because of the potential physiological importance of reactivity of the vessels to the octapeptide. It has been described that A II tachyphylaxis represents saturation of receptor sites which are normally freed of A II by local angiotensinases (20). Other studies, However, indicated that total tissue angiotensinases are not the critical determinant in angiotensin tachyphylaxis (21). Using 1-substituted analog of A II, Sar^1-Phe^8-A II, we have previously found that a rapid tachyphylaxis occurs to this compound in the isolated rabbit aortic strip which is normally resistant to the development of angiotensin tachyphylaxis (22). A rapid tachyphylaxis also develops to the vaconstrictor effect of Sar^1-Phe^8-A II but not Asp^1-Phe^8-A II in renal vasculature supporting our earlier speculation that angiotensinases probably lack aminopeptidases in the vascular wall are very important for the development of tachyphylaxis to A II (22). The same receptors are are responsible for the response to both peptites since a cross-tachyphylaxis is observed between Sar^1-Phe^8-A II and Asp^1-Phe^8-A II, moreover, the specific competitive antagonist of A II, Sar^1-Ile^8-A II, could block the effects of both peptides in vascular smooth muscle. Working on a particular effect of A II, tachyphylaxis phenomenon should always been taken into consideration.

Evidence has been presented indicating that A II increases the synthesis of non-mast cell histamine "Induced Histamine" (28) in the vascular wall (24). This effect is probably one of the various indirect actions of A II which may be contributed its physiological actions.

Angiotensin-peptides also interact with endogenous prostaglandins (PGs). It has been shown that A II induces a release of PGE-like substance from various isolated and isolated perfused tissues (25). A III has been found to be the most potent stimulator for the release of PGE-like material in rabbit mesenterial vasculature (26). in rat stomach fundal strips and isolated canin tracheal muscle (25,27). Recently evidence has been presented indicating that A II also causes an increase in the synthesis of vascular PGI_2 in kidney, lung and atria (28,29). A III has less potent activity in vascular smooth muscle as well as autonomic nervous system (30) but higher agonistic activity in steroidogenesis and PG release.

A II also interacts with several endogenously occurring polypeptides. It is well established that angiotensin converting enzyme is identical to kininase II which causes degradation of kinin-peptides (31). It is therefore obvious that the inhibition of converting enzyme prevents the production of A II but increases the level of kinin-peptides in the biological enviroment. A series of studies has previously been published indicating the effect of various endogenous peptides on drinking behavior in different species. In ducks and pigeons A II, bombesin, eledoisin elicit a dipsogenic effect when injected intracerebroventricularly. Substance P, however, was found to be ineffective. Among all these endogenous peptides only A II has a greater dipsogenic effect in birds and mammals, since bombesin and tachykinins inhibit water intake in mammals (32,33).

COMPETITIVE INHIBITORS OF ANGIOTENSIN PEPTIDES:

The first report has been published in 1970 dealing with the pharmacological effects of 8-substituted analogs of A II. Asp^1-Phe^4--Tyr^8-A II has been shown to have an antagonistic effect against the parent peptide on the isolated rat uterus and rat blood pressure (34). This analog, however, was not a pure A II antagonist since it has some myotropic and pressor activities in the investigated preparations. Extensive series of studies have been carried out in Cleveland Clinic (U.S.A.) dealing about different 8-substitued analogs of A II. Number of 8-substituted analogs of A II, 8-Valine, 8-Alanine,8-Isoleucine,8-Leucine,8-Aminophenylisobutiric acid,8--Dimethyl Glycine (35,36) and several others were tested and found to be the competitive antagonists of A II both in vivo and in vitro (37). All these synthetic peptides have primary structure similar to A II and may be degradated by enzyme(s) which metabolise A II.

On this basis, the first position of the peptides (Asp^1) was replaced with sarcosine, thus Sar^1-Ala^8-A II (Saralasin) was synthetized. This new compound has been shown to be a potent and long--lasting competitive antagonist of A II (38). Similar structural change was later made in Asp^1-Ile^8-A II then the compound Sar^1-Ile^8--A II was synthetized. This analog was the most potent specific and long-lasting competitive antagonist of A II so far described (39). The antagonistic potencies of 8-substituted analogs of A II have been determined by several authors measuring the pA$_2$ values against A II in various preparations (35). The pA$_2$ value of Sar^1-Ile^8-A II was found to be 9.48 while the range of other analogs was between 5.5-8.6 which were significantly lower than that of Sar^1-Ile^8-A II (35,39). No agonistic property of Sar^1-Ile^8-A II was observed in the isolated vascular tissues and isolated perfused organs. This compound also competitively inhibits the effects of A III in blood pressure and intestinal motility indicating that both A II and A III act on the same receptors (32). Some Studies, however, indicate that A II and A III have their own agonistic properties in adrenal glands based upon the observations that 8-substituted synthetic analogs of A II are potent competitive inhibitors of A II-induced smooth muscle contraction and pressor response but these compounds are poor inhibitors for both A II and A III-induced steroidogenesis (40). However, 8-substituted analog of A III, Ile^7-A III, is potent competitive inhibitor of steroidogenetic effect of both A II and A III but ineffective inhibitor A II-induced elevation in blood pressure (40). A III also has more potent agonistic activity than A II in stimulating tyrosine hydroxylase in adrenergic nerve and brain (15). Thus it is assumed that there are different receptor systems for angiotensin-peptides at least in adrenal cortex and smooth muscle. However, even having more potent agonistic activity on PGE release from kidney A III acts through A II receptors as evidenced by equipotent blocking activity of Ile^7-A III and Sar^1-Ile^8-A II (41)..A hexapeptide (Asp^1--Arg^2)-Ile^8-A II has also been synthetized having a potent short--acting antagonist effect agaist A II (42). A lipophilic A II antagonist, Octanoyl (Leu^8)-A II in oil solution has also been defined having 6 to 24 hours inhibition of the effect of A II after intramuscular injection (43).

In conclusion, much work has still to be done for the identification of A II receptors in various organs and systems. The stimulation of central nervous system and catecholamine synthesis, transport of water and sodium across the cell membrane, interaction with endogenous PGs and other endogenously occurring bioactive substances should be investigated further, in order to extend our knowledge of the physiological action of angiotensin-peptides in water and sodium intake.

REFERENCES

1. A.P. Smlyo and A.V. Somlyo, Vascular smooth muscle II. Pharmacolo-

gy of normal and hypertensive vessels, Pharmacolo. Rev., 22:249 (1970).

2. W.B. Campbell, S.N. Brooks and W.A. Pettinger, Angiotensin II and angiotensin III-induced aldosterone release in vivo in the rat, Science, 184:994 (1974).

3. M.J. Peach, C.A. Sarsted and E.D. Vaughan, Changes in cardiovascular and adrenocortical responses to angiotensin III induced by sodium deprivation in rat, Cir. Res. (Suppl. 2) 38:117 (1976).

4. R.F. Bing and D. Schulster, Steroidogenesis in isolated rat adrenal glomerulosa cells: response to physiological concentrations of angiotensin II and effects of potassium, serotonin and (Sar[1]--Ala[8])-angiotensin II, J. Endocrinol., 74:261 (1977).

5. J. Douglas, G. Aguilera, T. Kondo and K. Catt, Angiotensin II receptors and aldosterone production in rat adrenal glomerulosa cells, Endocrinol., 102:685 (1978).

6. F.A.O. Mendelsohn and C.D. Kachel, Action of angiotensins I, II and III on aldosterone production by isolated rat adrenal zona glomerulosa cells: Importance of metabolism and conversion peptides in vitro, Endocrinol., 106:1760 (1980).

7. J.T. Fitzsimons and B.J. Simons, The effect on drinking in the rat of intravenous infusion of angiotensin given alone or in combination with other stimuli of thirst, J. Physiol., 203:45 (1969).

8. A.N. Epstein, J.T. Fitzsimons and B.J. Rolls, Drinking induced by injection of angiotensin into the brain of the rat, J. Physiol., 210:457 (1970).

9. J.B. Simpson, A.N. Epstein and J.S. Camardo Jr., The localization of dipsogenic receptors for angiotensin II in the subfornical organ of rat, J. Comp. Physiol. Psychol., 92:581 (1978).

10. R.K. Türker, Effect of angiotensin on response to norepinephrine and periarterial stimulation of the isolated perfused cat terminal ileum, Eur. J. Pharmacol., 21:171 (1973).

11. R.K. Türker and Z.S. Ercan, Effect of prostaglandin E_2 on the pressor response to periarterial stimulation and norepinephrine of the isolated perfused rabbit kidney, Prostaglandins, 9:695 (1975).

12. Z.S. Ercan, Changes induced by angiotensins and prostaglandin E_2 on the release of transmitter from isolated perfused rabbit kidney during periarterial stimulation, Ar. int. Physiol. Biochim., 83:799 (1975).

13. C. Chevillard, N. Duchene and J.M. Alexandre, How angiotensin II increase cardiac dopamine- -hydroxylation, J. Pharm. Pharmacol., 27:193 (1975).

14. G. Simonnet and M.F. Giorguieff-Chesselet, Stimulating effect of angiotensin II on the spontaneous release of newly synthetized (^{3}H)-dopamine in rat striatal slices, Neuroscience Letters, 15::153 (1979).

15. M.C. Boadle-Biber, V.H. III, Morgenroth and R.H. Roth, Activation of tyrosine hydroxylase by angiotensin II and des-(Asp[1])-angiotensin II heptapeptide (Abstr.), Pharmacologist, 16:213 (1974).

16. S.M. Friedman and C.L. Friedman, Ionic basis of vascular response

to vasoactive substances, Canad. Med. Ass. J., 90:167 (1964).

17. G. Rorive and F. Hagemeijer, Influence de la noradrenaline et de l'angiotensine sur la composition ionique de la fibre musculaire de l'aorte de rat, Ann. Endocr. (Paris), 27:521 (1966).

18. R.K. Türker, I.H. Page and P.A. Khairallah, Angiotensin alteration of sodium fluxes in smooth muscle, Arch. int. Pharmacodyn., 165:394 (1967).

19. R.K. Türker, Effect of angiotensin on the cardiac arrhythmias induced by G-Strophantin, Experientia, 20:707 (1965).

20. P.A. Khairallah, I.H. Page, F.M. Bumpus and R.K. Türker, Angiotensin tachyphylaxis and its reversal, Circ. Res., 19:247 (1966).

21. N.K. Hollenberg, L.R. Barbero and K.P.J. Hinrichs, Angiotensin tachyphylasis and vascular angiotensinase activity, Experientia, 27:1032 (1971).

22. R.K. Türker and Z.S. Ercan, Change of activity in rabbit aorta of three analogs of angiotensin substituted in NH_2-terminal position, Pharmacology, 13:155 (1975).

23. R.W. Schayer, Histamine and circulatory homestasis, Proc. Soc. Exp. Biol Med., 24:1295 (1965).

24. Z.S. Ercan, F.F. Ersoy, T.A. Bökesoy and R.K. Türker, Evidence for the histamine-mediated myotropic effect of angiotensin II in the rabbit aorta, Pharmacology, 18:276 (1979).

25. Z.S. Ercan and R.K. Türker, A comparison between prostaglandi releasing effect of angiotensin II and angiotensin III, Agents and Actions, 7:569 (1977).

26. A. Blumerg, S. Denny, K. Nishikawa, E. Pure, G.R. Marshall and P. Needleman, Angiotensin III-induced prostaglandin(PG)release, Prostaglandins, 11:195 (1976).

27. Z.S. Ercan, F.F. Ersoy and R.K. Türker, The Relaxing effect of angiotensin II and angiotensin III on canin isolated contracted tracheal muscle, J. Pharm. Pharmacol., 30:452 (1978.

28. Z.S. Ercan, H. Zengil, F. Akar and R.K. Türker, Possible prostacyclin mediated effect of angiotensin II in the isolated perfused rat lung, Pros. Leuk. Med., 12:77 (1983).

29. R.K. Türker, Evidence for a prostacyclin-mediated chronotropic effect of angiotensin II in the isolated cat right atria, Eur. J. Pharmacol., 83:271 (1982).

30. R.K. Türker and Z.S. Ercan, A comparative study with angiotensin II and angiotensin III in the anesthetized cats, Res. Commun. Chem. Pathol. Pharmacol., 21:15 (1978).

31. R. Igic, E.G. Erdös, H.S.Y. Yeh, K. Sorella and T. Nakajima, Angiotensin I converting enzyme of the lung, Cir. Re., 21-22 (Suppl):51 (1972).

32. G. de Caro, M. Massi and L.G. Micossi, Effect of bombesin on drinking induced by angiotensin II, carbachol and water deprivation in the rat, Pharmacol. Res. Commun., 12:657 (1980).

33. G. de Caro, M. Mariotti, M. Massi and L.G. Micossi, Dipsogenic effect of angiotensin II, bombesin and tachykinins in the duck, Pharmacol. Biochem. Behav., 13:229 (1980).

34. G.R. Marshall, W. Vine and P. Needleman, Specific competitive

inhibitor of angiotensin II, Proc. Nat. Acad. Sci. (Wash.), 67:
:1624 (1970).

35. R.K. Türker, I.H. Page and F.M. Bumps, Antagonists of angiotensin II, Handbook Exp. Pharmacol., Eds. I.H. Page and F.M. Bumpus, Springer-Verlag, Berlin, New York, V. 37, p. 162 (1974).

36. R.K. Türker and N.Ü. Gündogan, A comparative study of two structurally related analogs of Asp^1-Ile^5-angiotensin II, Naunyn--Schmiedeberg's Arch. Pharmacol., 282:411 (1974).

37. R.K. Türker, M.M. Hall and F.M. Bumps, Competitive inhbition of Asp^1-β-amide-Val^5-angiotensin II by Sar^1-Ile^8-angiotensin II in the cat isolated cardiac muscle and coronary vessels, J. Pharm. Pharmacol., 26:582 (1974).

38. D.T. Pals, F.D. Masucci, F. Sipos and G.S. Denning Jr., Specific competitive antagonist of the vascular action of angiotensin II, Cir. Res., 29:664 (1971).

39. R.K. Türker, M.M. Hall, M.Yamamoto, C.S. Sweet and F.M. Bumpus, A new long-lasting competitive inhibitor of angiotensin, Science, 177:1203 (1972).

40. J.R. Blair-West, J.P. Coghan, D.A. Denton, J.W. Funder, B.A. Scoggins and R.D. Wright, Effect of heptapeptide (2-8) and hexapeptide (3-8) fragments of angiotensin II on aldosterone secretion, J. Clin. Endocrinol. Metab., 32:575 (1971).

41. A.L. Blumberg, K. Nishikawa, S.E. Denny, G.R. Marshall and P. Needleman, Angiotensin (AI,AII,AIII) receptor characterization, Cir. Res., 41:154 (1977).

42. M. Hirata, K. Watanabe, M. Nara, E. Hattori and K.Arakawa, A hexapeptide angiotensin antagonist, des-(Asp^1-Arg^2)-Ile^8-angiotensin II, Endocrinol. Japon., 22:567 (1975).

43. R.C. Abdulkader, H.B. Silva, M. Marcondes, E.A. Vasconcelos and A.C.M. Paiva, Long-lasting inhibition of angiotensin response in rats by depot administration of octanoyl-Leu^8-angiotensin II, J. Pharm. Pharmacol., 31:20 (1979).

EFFECTS OF PEPTIDES OF THE "GUT-BRAIN-SKIN TRIANGLE" ON DRINKING BEHAVIOUR OF RATS AND BIRDS

Giuseppe de Caro

Institute of Pharmacology, University of Camerino

Camerino, Italy

Amphibian skin is an extraordinarily rich source of active peptides. Erspamer and co-workers undertook a systematic biological and chromatographic screening of the skin extracts of more than 500 different species of amphibians and discovered and characterized more than 40 active peptides (1, 2). Most of these peptides contain aminoacid sequence in their "active regions" analogous to mammalian peptides which act as neurotransmitters, autacoids, releasing factors or gastrointestinal hormones. Several of them, or at least substances chemically and biologically resembling them, have been identified in mammalian tissues, including the brain. Thus, the interest of experts in these peptides (particularly in those which have been found in mammalian brain and which exert on it intense and specific effects at extremely low concentrations) is now progressively shifting from their pharmacological to their physiological aspects.

Taken as a class, these peptides of amphibian origin have a wide spectrum of biological activities and at least four different groups of them potently and specifically affect ingestive behaviour of mammals and birds. They are (fig. 1):
1) Tachykinins, which include physalaemin, from the skin of the South American frog Physalaemus biligonigerus and kassinin, from the skin of the African frog Kassina senegalensis. These peptides chemically and biologically resemble substance P as well as eledoisin, the active undecapeptide of the Mediterranian octopus Eledone moschata.
2) Bombesins, which include bombesin, litorin and ranatensin. These peptides have been found in the skin, respectively, of the European dyscoglossid frog Bombina bombina, of the Australian leptodactylid frog Litoria aurea and of the South American frog Rana pipiens.

213

Tachykinins

EL	Pyr-Pro-Ser-Lys-Asp-Ala-Phe-Ile-Gly-Leu-Met-NH$_2$
PH	Pyr-Ala-Asp-Pro-Asn-Lys-Phe-Tyr-Gly-Leu-Met-NH$_2$
SP	Arg-Pro-Lys-Pro-Gln-Gln-Phe-Phe-Gly-Leu-Met-NH$_2$
KS	Asp-Val-Pro-Lys-Ser-Asp-Gln-Phe-Val-Gly-Leu-Met-NH$_2$

Bombesins

BS	Pyr-Gln-Arg-Leu-Gly-Asn-Gln-Trp- -Ala-Val-Gly-His-Leu-Met-NH$_2$
LT	Pyr-Gln-Trp-Ala-Val-Gly-His-Phe-Met-NH$_2$
RT	Pyr-Val-Pro-Gln-Trp-Ala-Val-Gly-His-Phe-Met-NH$_2$

Opioid Peptides

DM	Tyr-D-Ala-Phe-Gly-Tyr-Pro-Ser-NH$_2$
HD	Tyr-D-Ala-Phe-Gly-Tyr-Hyp-Ser-NH$_2$
LE	Tyr-Gly-Gly-Phe-Leu
ME	Tyr-Gly-Gly-Phe-Met

Angiotensins

CA	Ala-Pro-Gly-Asp-Arg-Ile-Tyr-Val-His-Pro-Phe
IA	Asp-Arg-Val-Tyr-Ile-His-Pro-Phe
VA	Asp-Arg-Val-Tyr-Val-His-Pro-Phe

Fig. 1. Aminoacid sequence of tachykinins, bombesins, opioid peptides and angiotensins of nonmammalian and mammalian origin which modify drinking behaviour of rats and pigeons.

EL, eledoisin	DM, dermorphin
PH, physalaemin	HD, hydroxy proline-dermorphin
SP, substance P	LE, leucine-enkephalin
KS, kassinin	ME, methionine-enkephalin
BS, bombesin	CA, crinia-angiotensin II
LT, litorin	IA, isoleucine-angiotensin II
RT, ranatensin	VA, valine-angiotensin II

3) <u>Dermorphins</u>, which include dermorphin and hyp[6]-dermorphin, two heptapeptides with exceptionally intense peripheral and central opioid activities, recently found in the skin of South American frogs belonging to the <u>Phyllomedusae</u> family (<u>sauvagei</u>, <u>rhodei</u>, <u>burmeisteri</u>). Dermorphins closely resemble, from a biological point of view, the endogenous opioid peptides Met- and Leu-enkephalin.

4) <u>Angiotensins</u>, which include crinia-angiotensin, an active undecapeptide of the skin of the Australian frog <u>Crinia georgiana</u>. Crinia-angiotensin is closely related, from a chemical and biological point of view, to the classical angiotensins isoleucine and valine-angiotensin II.

The substances of the last group, the angiotensins, are potent dipsogens both in rats and in birds. All the other peptides, although belonging to chemically and biologically different groups, are potent inhibitors of water intake in rats but, with the exception of opioid peptides, elicit copious drinking in pigeons and duks. At least in rats, some of them also modify feeding behaviour.

These peptides occur, or at least have their counterpart, in mammalian tissues, including the brain (3, 4, 5). Some of them have been identified in brain regions which are considered to be involved in the control of water intake (6).

DRINKING ALTERATIONS IN RATS

1 - Tachykinins
Eledoisin, physalaemin and substance P, in this order, injected into the lateral ventricles of Wistar, Long Evans, Sprague-Dawley and Brattleboro (hetero- and homozygous) rats, inhibit in a dose dependent fashion drinking elicited by ICV angiotensin II, water deprivation and SC NaCl load. Doses of eledoisin ranging between 10 and 100 ng per rat are needed to inhibit drinking induced by ICV angiotensin II, 100 ng/rat, but larger doses of the peptide are necessary to inhibit water intake elicited by water deprivation or dehydration.

The potency ratio between eledoisin and the other tachykinins is approximately 20:1:0.5 (fig. 2), but in cell-dehydrated rats physalaemin is less effective than the other peptides and the potency ratio is about 20:0.5:0.5.

Kassinin is unique, in that it potently inhibits drinking elicited by SC NaCl load but does not affect that induced by other dipsogenic determinants.

In baby rats the antidipsogenic effect of eledoisin and physalaemin appears very early and develops at the same pace as the dipsogenic effect of angiotensin II (7). They inhibit milk intake, too. However, while the inhibitory effect on water intake does not modify or actually increases with age, that on milk intake progressively decreases with age and completely disappears when the rats are 12-15 days old.

The antidipsogenic effect elicited by tachykinins both in adult and baby rats seems to be specific and neither related nor due to other specific or aspecific behavioural alterations. In fact, these peptides never produce discomfort or behavioural alterations and do not affect food intake, even at doses up to 100 times larger than the antidipsogenic ones. Eledoisin and physa-

215

laemin do reduce explorative activity, while substance P increases spontaneous motor activity. However, not only are these effects induced by doses up to 10.000 times larger than those antidipsogenic, but they also last far less than the effect on water intake: for example, after 20 μg of eledoisin the rats are depressed for 30-60 min, but 2 hr later they are still completely adipsic (8).

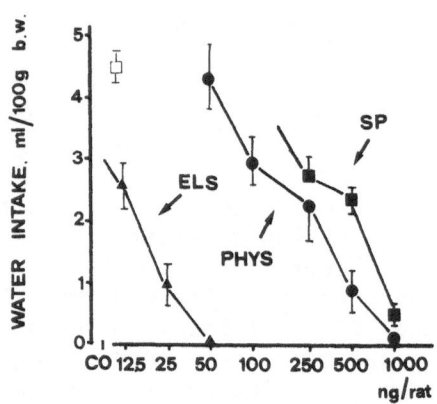

Fig. 2. Dose-antidipsogenic effect of pICV eledoisin (ELS), phy-salaemin (PHYS) or substance P (SP) in rats made thirsty by pICV angiotensin II.
Values are means of 10 to 12 data ± s.e.m.

The effect of tachykinins on water intake involves peculiar hypothalamic areas, as demonstrated by the fact that doses up to 10-100 times inferior to the intraventricular ones are needed to inhibit drinking when the peptides are injected into the anterior hypothalamus or the area preoptica medialis.

The subfornical (SFO) organ is probably only in part involved in the antidipsogenic effect of tachykinins. In fact, these peptides (which do not affect water intake when injected by intraperitoneal route) are normally effective after electrolytic lesioning of the SFO.

The mechanism of the antidipsogenic action of these peptides is probably neurogenic and is not related to the potent vasoplegic effect that they elicit. In fact, they inhibit drinking at doses which do not lower blood pressure. Moreover, injected into the brain ventricles, they produce hypertension, not hypotension, and this response is at least in part due to vasopressin release. They

216

could modify drinking behaviour simply acting as vasoplegic drugs, but they inhibit also carbachol induced drinking, which is not inhibited by vasoplegic drugs (8, 9, 10).

Numerous findings indicate that tachykinins are neurotropic drugs. In fact, a role as neurotransmitter has been suggested for substance P; eledoisin and physalaemin produce electrocortical modifications in the rat (11) and in vitro they potently stimulate isolated neurons of Aplisia punctata, Helix pomatia and Achantina fulica.

For these reasons, we think that the antidipsogenic effect of tachykinins could be more probably explained in terms of neurogenic rather than haemodynamic alterations.

However, experimental data suggest that tachykinins influence fluid homeostasis in a complex way. In fact, not only do they inhibit water intake, but also produce vasopressin release, antidiuresis and consequently water conservation (12). Moreover, pilot experiments suggest that in the range of the intracranial antidipsogenic doses these substances modify sodium appetite and renal transport of electrolytes.

2 - Bombesins

ICV bombesin, ranatensin and litorin inhibit water intake elicited by different dipsogenic determinants. Moreover, they produce, to different extents, other alterations such as grooming, modification of spontaneous motor activity and inhibition of food intake.

The inhibition of angiotensin induced drinking is attained at pulse ICV doses of bombesin as low as 10 ng per rat or less (fig. 3). Larger doses are needed to inhibit drinking induced by water deprivation and cell deydration or by ICV injection of carbachol. Food intake inhibition is attained at doses of at least 1 µg per rat (13).

The antidipsogenic and anorexigenic activity of bombesins is influenced by the modalities of their administration. In fact, administered by pulse ICV injection, bombesin and ranatensin in this order, but not litorin, inhibit angiotensin-induced drinking; bombesin and litorin in this order, but not ranatensin, inhibit water deprivation thirst; food intake is inhibited by bombesin, but it is affected only by extremely large doses of ranatensin and is not modified at all by litorin. When administered by continuous ICV infusion, litorin is more effective than bombesin in inhibiting angiotensin-induced drinking, while ranatensin is equally as active as bombesin in inhibiting drinking induced by water deprivation. The relative effectiveness of bombesins in inhibiting food intake is not affected by the modalities of their administration,

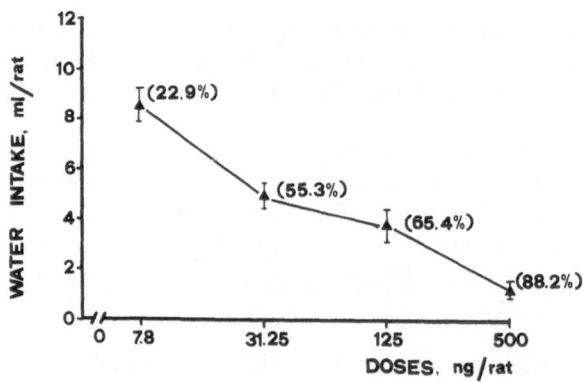

Fig. 3. Dose-antidipsogenic effect curves of pICV bombesin in rats made thirsty by pICV angiotensin II. Values are means of 8 to 10 data ± s.e.m. In parentheses: per cent inhibitory effect.

but when infused these peptides are less effective than when injected into the lateral ventricles (14).

The ICV injection of bombesin to rats, besides drinking inhibition, produces feeding inhibition, grooming, alteration of explorative behaviour and of spontaneous motor activity, but neither neurological nor autonomic alterations. These behavioural alterations become statistically significant 5-10 min after the treatment, and progressively increase thereafter, while drinking inhibition is completely evident already in the first 5 min of observation, even at the minimum effective doses. Moreover, they are induced by doses of bombesin far larger than the antidipsogenic ones (15). On the other hand, litorin and ranatensin produce these behavioural effects only at doses far larger than those effective on water intake. This suggests that the antidipsogenic effect elicited by bombesins is specific and neither due nor related to the other specific or aspecific behavioural alterations that the ICV injection of the peptides can produce.

The mechanism of the antidipsogenic effect of bombesins is unknown, at present. Bombesin displays important effects in the rat brain, including hypothermia and feeding inhibition. In the rat ranatensin is approximately as active as angiotensin II in eliciting hypertension, but bombesin gives erratic vascular responses. Moreover, administered by IP route these peptides do not affect drinking at all. Thus, we think that bombesins, as a class, probably affect the drinking behaviour of rats through a neuro-

218

genic rather than a vascular mechanism of action.

3 - Opioid peptides

The brain opioid peptides leucine- and methionine-enkephalin inhibit water intake at ICV doses ranging between 50 and 200 μg per rat. The synthetic derivative D-Ala^2,D-Leu^5-enkephalin, which is more resistant to enzymatic hydrolysis, is about 2000 times more effective than the naturally occurring brain opioid peptides, while dermorphin is up to 40.000 times more effective, its threshold ICV dose ranging between 0.5 and 1 ng per rat (fig. 4). On a weight to weight basis, dermorphin is the most active of the antidipsogenic peptides at present known (16).

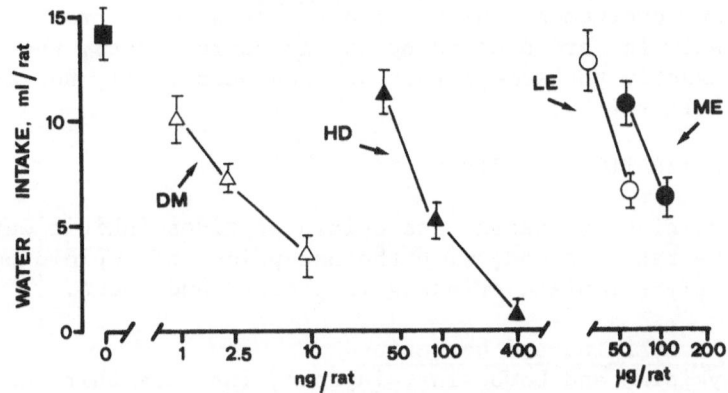

Fig. 4. Dose-antidipsogenic effect curves of pICV dermorphin (DM), hyp^6-dermorphin (HD), leu-enkephalin (LE) or met-enkephalin (ME) in rats made thirsty by pICV angiotensin II. Values are means of 10 to 14 data ± s.e.m.

At antidipsogenic doses enkephalins release vasopressin and produce a long lasting increase in plasma renin activity which is apparently mediated by a renal β-adrenergic stimulation of central origin and which can explain the delayed polydipsia which follows the inhibition of water intake that they produce (17).

The antidipsogenic effect of these peptides is not simply a consequence of central depression or of other behavioural alterations. In fact, it lasts longer than the central depression that brain opioid peptides produce. Moreover, the effect of dermorphin is elicited at doses which do not induce depression or any behavioural alteration at all.

Radioimmunological and immunohistochemical studies have shown that enkephalins are widely distributed in the rat brain, while dermorphin-like immunoreactivity occurs only in a restricted number of brain regions, namely the arcuate nucleus, the OVLT and the SFO. Moreover, we have demonstrated that, when administered directly into the SFO, dermorphin is much more potent than after ICV injection, the inhibitory effect of 1 ng of the peptide being 86% vs the 42% observed after the injection of the same dose into the lateral ventricles. These data suggest that the SFO may be a target site for the antidipsogenic effect of endogenous brain dermorphins. However, at present we have no data to explain the mechanism of the antidipsogenic effect of this as well as that of the other brain opioid peptides.

4 - Angiotensins
The ICV injection of crinia angiotensin, the active undecapeptide of the Australian frog <u>Crinia georgiana</u>, produces in the rat a strong excitement which prevents the animals from drinking. When the animals are sedated by an IP barbiturate, the peptide produces exactly the same effect as angiotensin II, but at doses 10 times larger (18).

DRINKING ALTERATIONS IN BIRDS

Tachykinins, bombesins and opioid peptides inhibit water intake in the rat. Instead, with the exception of opioid peptides, they potently stimulate drinking in pigeons and ducks.

1 - Tachykinins and bombesins
Tachykinins and bombesins, injected into the third brain ventricle, produce a copious and absolutely normal intake of water. Latency to drinking is short and the intake of water is practically complete in 5 min. In these 5 min the pigeons may take as much water as they usually drink during an entire day. The response is repeatable and dose dependent. There is no sign of discomfort or of behavioural alterations. Food intake is never affected (19, 20, 21).

ICV tachykinins and bombesins elicit practically the same effect as angiotensin II but they are 10 times less effective than angiotensin II itself, with the exception of substance P which is about 100 times less active (fig. 5).

Injected by IP route bombesins and tachykinins are still dipsogenic, but at doses at least 1000 times larger than the ICV ones.
The mechanism of the dipsogenic effect of these peptides is not known at present.

Time course and dose response curves of drinking induced by tachykinins and bombesins are identical to those of drinking eli-

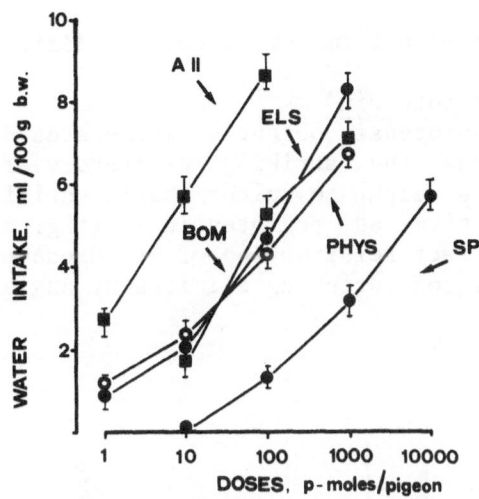

Fig. 5. Dose-dipsogenic effect curves of pICV angiotensin II
         (AII), eledoisin (ELS), physalaemin (PHYS), bombesin
         (BOM) or substance P (SP) in the pigeon.
         Values are means of 10 to 14 data ± s.e.m.

cited by angiotensin II. However, these substances cannot act
through activation of angiotensin II receptors or by activation of
the central isorenin-angiotensin system, since the receptor for
angiotensin II induced drinking is highly specific. Moreover, competitive antagonists of angiotensin II potently inhibit drinking
induced by angiotensin itself but not that elicited by tachykinins
or bombesins (22).

A vascular mechanism of action seems to be a rather unlikely
explanation, not only because tachykinins and bombesins have opposing vascular effects in pigeons, but also because SFO-lesioned
pigeons are exactly as sensitive as non lesioned birds to ICV
tachykinins as well as to ICV or IP bombesins (23).

Thus, on the analogy of the hypothesis put forward for rats,
we think that it is possible to hypothesize a neurogenic mechanism of the dipsogenic action of tachykinins and bombesins in the
pigeons.

Tachykinins and bombesins stimulate water intake in pigeons:
why do they inhibit drinking in rats? On the basis of available
data this opposing effect cannot be explained, neither in terms of

brain alterations nor in terms of vascular modifications.

2 - Crinia-angiotensin
ICV crinia-angiotensin potently stimulates drinking without producing in pigeons the striking excitatory effect that it induces in rats. On a weight to weight basis crinia-angiotensin is exactly as effective as angiotensin II (fig. 6) and the time course and dose-effect relationship of its dipsogenic response are identical to those of drinking elicited by angiotensin II itself (18).

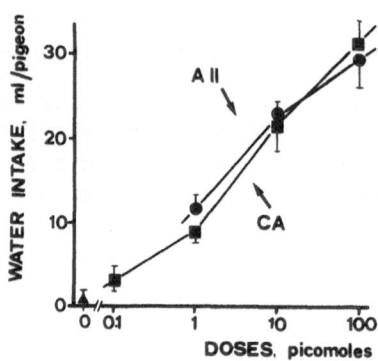

Fig. 6.   Dose-dipsogenic effect curves of pICV angiotensin II (AII) or crinia-angiotensin in the pigeon.
Values are means of 10 to 12 data ± s.e.m.

These data confirm that receptors for angiotensin induced drinking in rats and pigeons, although similar, are different and indicate that different molecular requirements must be satisfied to activate angiotensin receptors in these animals.

CONCLUSIONS

Tachykinins, bombesins and opioid peptides modify ingestive behaviour in Wistar, Sprague-Dawley, Long Evans and Brattleboro rats, as well as in rabbits, pigeons and ducks. What is the biological significance of their effects?

Data at present available suggest that, at least in rats and pigeons, endogenous brain peptides which affect drinking behaviour play a physiological role in the control of body fluid homeostasis and belong to a peptidergic system in the bosom of which each peptide plays a peculiar role. This is suggested by several data.

First of all, the distribution of these peptides, which are not spread in the brain but are concentrated in areas, some of which are considered to be involved in the control of drinking.

Secondly, the level of their active doses, which may be considered in many cases to be in the range of physiological concentrations. This is the case of substance P which, injected into the hypothalamic pre-optic area, inhibits water intake at doses of 1 to 5 ng, that is at doses which are well within an order of magnitude of a "physiological concentration", since in the hypothalamus there are about 100 to 700 ng of immunoreactive substance P (24, 25) per g of tissue.

Third: there are brain areas selectively sensitive to the injection of these peptides. This is the case of substance P; this in also the case of physalaemin, which almost completely inhibits angiotensin II induced drinking when at doses of 1 to 5 ng it is injected into the medial pre-optic and into the anterior hypothalamic areas, but which is far less effective when it is injected into other brain areas; this is the case of dermorphin, which is particularly effective when injected into the subfornical organ (9, 26, 27).

The fourth reason for considering these peptides of physiological importance in body fluid control is the fact that, with the exception of kassinin and bombesin, at fully antidipsogenic doses they never produce discomfort, excitation or depression, neurological or autonomic compromission, other specific or aspecific behavioural alterations, modifications of food intake. Moreover, while neurotropic drugs which inhibit drinking, like morphin and apomorphin, produce taste aversion, these peptides do not have this effect, not even opioid peptides, despite the close biological affinities they have with morphin itself.

The fifth element in favour of a physiological role, at least as far as tachykinins are concerned, comes from ontogeny studies. In fact, the antidipsogenic effect of tachykinins and the dipsogenic one of angiotensin II in the earliest stages of post-natal life involve, at the same time, water and milk intake, but progressively become selective and are specific for water at the age of 8-12 days. That is, the ontogeny of the effect elicited by tachykinins is exactly the same as that of angiotensin II, a peptide whose physiological role in water intake control is beyond any argument.

Tachykinins, bombesins and opioid peptides, on the one hand, tachykinins, bombesins and angiotensins, on the other hand, have completely different chemical and biological properties but produce the same effect, respectively antidipsogenic and dipsogenic, in rats and in pigeons. If they play a physiological role, why are so many and so very different peptides needed to reach the single goal of inhibiting, or stimulating, water intake? As far as birds are concerned, we cannot reply to this question. As far as rats are concerned, we can only put forward a hypothesis, that is that these peptides belong to a "cerebral peptidergic system" which participates in the control of water intake and body fluid homeostasis. In fact, all these peptides inhibit water intake, but this effect is achieved in different ways by the different peptides. Substance P inhibits angiotensin II, water deprivation and cell-dehydration drinking, kassinin inhibits only cell-dehydration thirst. None of them affects food intake. Bombesins have different effects on drinking behaviour, according also to the modalities of their administration, and may also affect food intake. Physalaemin and enkephalins, but not substance P, release vasopressin and produce antidiuresis, while tachykinins, but not opioid peptides, probably directly influence renal transport of electrolytes.

All these data, taken together, strongly suggest that the endogenous peptides of rat brain that we have studied, and with them probably other endogenous brain peptides, are part of a brain peptidergic system and that in the different kinds of thirst one rather than any other peptide of this system is called to intervene according to the peculiar effect that it can elicit on water intake and body fluid homeostasis.

REFERENCES

1. G. Bertaccini, Active polypeptides of nonmammalian origin, Pharmacol. Rev. 28:127 (1976).
2. V. Erspamer, P. Melchiorri, M. Broccardo, G. Falconieri Erspamer, P. Falaschi, G. Improta, L. Negri and T. Ronda, The brain-gut-skin triangle: new peptides, Peptides 2:7 suppl. 2 (1981).
3. A. J. Harmar, Three tachykinins in mammalian brain, Trends NeuroSci. 7:57 (1984).
4. L. H. Lazarus, R. I. Linnoila, O. Hernandez, R. P. DiAugustine, A neuropeptide in mammalian tissues with physalaemin-like immunoreactivity, Nature 287:555 (1980).
5. J. H. Walsh and G. H. Dokray, Localization of bombesin-like immunoreactivity (BLI) in the gut and brain of rat, Gastroenterology 74:1108 (1978).
6. R. Buffa, E. Solcia, E. Magnoni, G. Rindi, L. Negri and P. Melchiorri, Immunohistochemical demonstration of a dermorphin-like peptide in the rat brain, Histochemistry 76:273 (1982).
7. F. Cantalamessa, G. de Caro, A. N. Epstein and M. Perfumi, Effects of the tachykinins eledoisin and physalaemin on drinking behaviour in baby rats, This Symposium.
8. G. de Caro, L. G. Micossi and G. Piccinin, Antidipsogenic effect of intraventricular administration of eledoisin to rats, Pharmacol. Res. Comm. 9:489 (1977).

9. G. de Caro, M. Massi and L. G. Micossi, Antidipsogenic effect of intracranial injections of substance P to rats, J. Physiol. 279:133 (1978).
10. G. de Caro, M. Massi, L. G. Micossi and F. Venturi, Physalaemin, a new potent antidipsogen in the rat, Neuropharmacology 17:925 (1978).
11. G. de Caro, L. G. Micossi, F. Venturi, A. Brancati and E. Scarnati, Behavioural and electrocortical modifications induced in the rat by intraventricular injection of physalaemin and eledoisin, Psychopharmacol. 38:211 (1974).
12. F. Cantalamessa, G. de Caro, M. Massi and M. Perfumi, Possible influence of tachykinins on body fluid homeostasis in the rat, J. Physiol., Paris 79:524 (1984).
13. G. de Caro, M. Massi and L. G. Micossi, Effect of bombesin on drinking induced by angiotensin II, carbachol and water deprivation in the rat, Pharmacol. Res. Comm. 12:657 (1980).
14. G. de Caro, M. Massi, L. G. Micossi and M. Perfumi, Drinking and feeding inhibition by ICV pulse injection or infusion of bombesin, ranatensin and litorin to rats, Peptides 5:607 (1984).
15. F. Cantalamessa, G. de Caro, M. Massi and L. G. Micossi, A study on behavioural alterations induced by intracerebroventricular administration of bombesin to rats, Pharmacol. Res. Comm. 14:163 (1982).
16. G. de Caro, M. Massi, L. G. Micossi and M. Perfumi, Effect of dermorphin and related peptides on drinking behaviour of the rat, in: Central and peripheral endorphins: basic and clinical aspects, E. E. Müller and A. R. Genazzani, ed., Raven Press, New York (1984).
17. F. Cantalamessa, G. de Caro, M. Massi and L. G. Micossi, Stimulation of drinking behaviour and of renin release induced by intracerebroventricular injection of D-Ala^2-D-Leu^5-enkephalin to rats, Pharmacol. Res. Comm. 14:141 (1982).
18. F. Cantalamessa, G. de Caro, M. Massi and L. G. Micossi, Drinking stimulation by a new angiotensin, crinia-angiotensin II, in rats and pigeons, Pharmacol. Biochem. Behav. 17:741 (1982).
19. M. D. Evered, J. T. Fitzsimons and G. de Caro, Drinking behaviour induced by intracranial injections of eledoisin and substance P in the pigeon, Nature 268:332 (1977).
20. G. de Caro, M. Massi and M. Perfumi, Potent dipsogenic effect of physalaemin in the pigeon, Pharmacol. Res. Comm. 10:861 (1978).
21. G. de Caro, M. Massi and L.G. Micossi, Bombesin potently stimulates water intake in the pigeon, Neuropharmacol. 19: 867 (1980).
22. G. de Caro, M. Massi, L. G. Micossi and M. Perfumi, Angiotensin II antagonists versus drinking induced by bombesin or eledoisin in pigeons, Peptides 3:631 (1982).
23. M. Massi, G. de Caro, L. Mazzarella and A. N. Epstein, The

role of the subfornical organ in the drinking behaviour of the pigeon, Brain Research in press.

24. G. W. Treager, H. D. Niall, J. T. Potts Jr., S. E. Leeman and M. M. Chang, Synthesis of substance P, Nature, New Biol. 232:87 (1971).

25. D. Powell, S. E. Leeman, G. W. Treager, H. D. Niall and J. T. Potts, Radioimmunoassay for substance P, Nature, New Biol. 241:252 (1973).

26. G. de Caro, M. Massi, M. Perfumi and F. Venturi, Sensitivity of different nuclei of rat brain to the anti-dipsogenic effect of tachykinins, Appetite 4:198 (1983).

27. M. Perfumi, G. de Caro, M. massi and F. Venturi, Inhibition of ANG II-induced drinking by dermorphin given into the SFO or into the lateral ventricle of intact or of SFO lesioned rats, This Symposium.

SUPPRESSION OF WATER INTAKE BY THE E PROSTAGLANDINS

Nancy J. Kenney

University of Washington

Seattle, WA 98195

The prostaglandins (PGs) are unsaturated fatty acids produced and released by virtually every organ of the body including the brain, the kidney and the vasculature. Since their discovery in the 1930s and synthesis in the 1960s, the list of PGs and PG-like substances, as well as the physiological systems known to be affected by these substances, has grown to immense proportions.

The PGs particularly those of the E series (PGE), are involved in many aspects of the central and peripheral control of body-fluid homeostasis including blood-pressure regulation, vasopressin secretion and the renal response to vasopressin, water and electrolyte excretion and water ingestion. This paper reviews only the effects of the E prostaglandins on water ingestion.

EFFECTS OF ADMINISTRATION OF PROSTAGLANDIN E ON WATER INTAKE

Administration of prostaglandin E_1 (PGE_1) or prostaglandin E_2 (PGE_2) to either the brain or the periphery of the rat results in reduced water intake in response to a variety of dipsogenic stimuli. The exact effect of PGE administration on water ingestion is dependent upon both the dipsogen involved in stimulating the drinking and the site of administration of the PG.

Intracerebroventricular Administration of PGE

Injection of PGE_1 or PGE_2, at doses of 10 ng or more, into the lateral cerebral ventricle of the rat reduces water intake in response to either intracerebroventricular (icv)[1,2,3,4] or intravenous (iv)[5] administration of angiotensin II (A II, Table 1).

Table 1. Percent Reduction of Water Intake (Mean ± SEM) Following Intracerebroventricular (icv) or Intraperitoneal (ip) Administration of Prostaglandin E (PGE).

| Dipsogen | % Suppression of Water Intake | | | | | |
| | icv PGE Dose (ng) | | | ip PGE Dose (µg/kg) | | |
	10	100	1000	10	50	100
icv Angiotensin II	34±12	48±14	44±16	6±16	42±12	74±6
iv Angiotensin II	46±16	74±26	94±12	NT*	NT	NT
Polyethylene Glycol	-12±19	48±20	87±8	20±25	4±21	86±4
Water Deprivation	7±5	12±7	52±5	NT	7±7	49±8
Hypertonic Saline	-14±15	-10±9	1±5	21±12	63±12	77±11

* NT = not tested

Latency to the onset of drinking is not affected by icv PGE[1]. Drinking induced by icv A II also is reduced if the rat is pretreated with arachidonic acid[4], the precursor of PGE_2, or bradykinin[5], a peptide which stimulates endogenous PGE synthesis.

The antidipsogenic action of centrally-administered PGE is not unique to A II. Although the effective dose levels are 10 to 100 times greater than in the case of A II-stimulated drinking, water ingestion in response to polyethylene-glycol induced hypovolemia and to 20-hr water deprivation is attenuated following icv injection of PGE[1] (Table 1).

The antidipsogenesis of centrally-administered PGE does not extend to all drinking stimuli, however. Drinking in response to cellular dehydration is unaffected by treatment with even very high doses of icv PGE[1] (Table 1). The significance of this finding is twofold. First, it suggests that central PGE must act to suppress water intake at a site relatively specific to the dipsogenic stimulus and not at the final pathways of drinking behavior. Second, the failure of icv PGE to reduce intake to cellular dehydration indicates that the cessation of drinking induced by the PGE treatment is not due to malaise. This point is further emphasized by the failure of icv PGE to reduce food intake of food-deprived rats when the rats are hydrated by gavage prior to access to food[1], a situation which reduces the dependency of the ingestion of food on that of water.

228

Table 2. Correlations Between the Percent Suppression of Water Intake and Core-Temperature Increase Induced by Intraventricular Prostaglandin E.

Dipsogen	N	r^2	
		Post-Test Temperature	Temperature Change
icv Angiotensin II (PGE_1)	30	.183	−.084
icv Angiotensin II (PGE_2)	31	.111	.274
iv Angiotensin II (PGE_1)	26	.251	.269
Polyethylene Glycol (PGE_1)	31	.232	.385[*]
Water Deprivation (PGE_1)	20	.602[***]	.504[**]
Hypertonic Saline (PGE_1)	17	.039	.124

[*] $p < .05$ [**] $p < .02$ [***] $p < .01$

In addition to its antidipsogenic action, icv injection of PGE increases core temperature[6] and blood pressure[7,8]. Fluharty[4] has reported a significant correlation between the core temperature and the percent reduction of icv A II-induced water intake following icv PGE_2 administration. He suggested that the pyrogenic action of elevated brain levels of PGE may contribute to its antidipsogenic actions particularly when drinking is stimulated by A II. In my own laboratory, no relationship between either the change of core temperature or core temperature per se subsequent to icv PGE treatment and suppression of A II-induced drinking has been found (Table 2). On the other hand, icv-PGE-induced reductions of water intake stimulated by hypovolemia or water deprivation, for which higher PGE doses are required, are significantly correlated with one or both of these temperature measures (Table 2). This suggests that the antidipsogenesis of high doses of icv PGE may be related to the increased temperature, but that the antidipsogenic effect of icv PGE may be separable from its thermogenic action at lower doses.

The antidipsogenic, thermogenic and pressor actions of

centrally administered PGE can be dissociated under specific experimental conditions. When ovariectomized rats are treated with either estradiol or progesterone, the increase of arterial pressure induced by central PGE_1 is attenuated[9]. The thermogenic effect is augmented and the antidipsogenic activity of the PGE is left unchanged for rats treated with either ovarian steroid[10]. This differential action of ovarian-hormone treatment on the pressor, thermogenic and antidipsogenic actions of icv PGE suggests that the neural mechanisms underlying these three effects of icv PGE treatment are independent.

Intraperitoneal Administration of PGE

Peripherally administered PGE, like PGE administered into the brain, is antidipsogenic. The dipsogens against which peripheral PGE is effective are not identical to those affected by centrally-administered PGE, however.

The greatest similarities in the effectiveness of peripheral and central PGE in suppressing water intake are observed when drinking is stimulated by either icv A II or water deprivation (Table 1). A II-induced water intake is reduced following intraperitoneal (ip) injection of PGE_2 at doses of 50 μg/kg or more[11]. Water-deprivation induced drinking is only affected by ip injection of high doses of the PGE[11] which parallels the finding that only high doses of icv PGE will reduce drinking to this complicated stimulus[1].

Intraperitoneal administration of PGE is less effective than intracranial administration of PGE in reducing water intake due to polyethylene-glycol induced hypovolemia (Table 1). Drinking due to this stimulus is only reduced by a high, generally antidipsogenic, dose of ip PGE_2[11].

The most striking difference in the antidipsogenesis induced by central or peripheral administration of PGE is noted when drinking is elicited by hypertonic-saline treatment (Table 1). While even very high doses of PGE administered into the brain fail to affect water intake to this stimulus[1], relatively low doses of the PGE suppress cell-dehydration induced drinking when the PG is administered into the peritoneal cavity[11,12].

ROLE OF ENDOGENOUS PGE IN THE CONTROL OF WATER INTAKE

A variety of inhibitors of PG synthesis are available. These include such non-steroidal anti-inflammatory agents as indomethacin (Indo), meclofenamate, and acetylsalicylic acid (ASA). All of these drugs act to prevent the conversion of the free-fatty acid precursors of the PGs to the PG endoperoxides through blockade of the actions of an enzyme complex referred to as PG synthetase.

These drugs interfere, not only with the production of PGE, but of all PGs and PG-like substances. While this generality of effect may make interpretation of data resulting from the use of PG-synthetase inhibitors (PGSIs) difficult, studies involving the administration of exogenous PGs have indicated that only the E series PGs are antidipsogenic[1].

Inhibition of PG Synthesis in the Brain

Only the relationship between PG synthesis in the brain and drinking in response to icv A II has been examined. Phillips and Hoffman[13] found that when icv injection of A II was paired with icv treatment with the PGSI, meclofenamate, the length of the drinking bouts of rats was extended and total water consumption was increased. Perez Guaita and Chiaraviglio[2], using ip injections of the inhibitor, Indo, replicated this finding. On the other hand, Kenney and Moe[14], using long-term, low-dose oral Indo treatment, failed to find any effect of PG-synthetase inhibition on drinking to icv A II. Fluharty[4] used a bio-behavioral assay to indicate that his synthetase-inhibitor treatment did block the production of PGE in brain. When the PGE_2 precursor, arachidonic acid, is injected into the cerebral ventricles, an increase of core temperature, which has been demonstrated to be due to the conversion of the arachidonic acid to PGE_2[15], ensues. After determining the effectiveness of icv Indo in reducing or eliminating the thermogenic action of arachidonic acid, Fluharty paired the Indo with icv A II treatment. No change of A II induced water intake was noted with any dose of Indo tested[4].

While the data relating endogenous brain PGE to drinking induced by icv A II are contradictory, the potential for PGE acting within the brain as a limitor of water intake remains. Recently, Malet et al.[16] have reported high levels of specific binding of PGE_2 in areas of brain traditionally regarded as being involved in body-fluid homeostasis including the hypothalamus, amygdala, septum and nucleus of the solitary tract. As techniques of inhibition of PG synthesis in specific areas of brain are refined, more definitive work on the role of endogenous brain PGE in the control of drinking can be conducted.

Inhibition of Peripheral PG Synthesis

Even when investigating the role of endogenous peripheral PGE in the control of drinking, the development of techniques for reduction of PG production which are not debilitating to the animal is difficult. We have found that the least invasive technique for administering PG-synthetase inhibitors to rats is to provide them in the animals' drinking water[14]. Thus, while the animal engages in its spontaneous ingestion of fluids, it self-administers the drug in low doses over extended periods of time. In spite of its

Table 3: Average Percent Change From Baseline ($\pm$ SEM) of Plasma $KH_2\text{-}PGE_2$ Levels of Jugular-Vein Catheterized Rats Over 4 Days of Low-Dose, Oral Prostaglandin-Synthetase Inhibitor Treatment.

Day	% Change of Plasma $KH_2\text{-}PGE_2$ Synthetase-Inhibitor Treatment		
	Control	Indomethacin 1.0 mg/ml	Acetylsalicylic Acid 2.25 mg/ml
1	46 ± 28	-17 ± 15	-35 ± 11
2	31 ± 36	-34 ± 11	-58 ± 6
3	19 ± 41	-54 ± 8	-56 ± 1
4	-8 ± 24	-64 ± 13	-64 ± 8

obvious potential drawbacks, this technique results in a significant and reliable reduction of plasma PGE levels without inducing malaise. After 4 days of exposure to drinking fluids containing either 1.0 mg/ml Indo or 2.25 mg/ml ASA, plasma levels of the PGE_2 metabolite, 15-keto, 13,14 dihydro PGE_2 ($KH_2\text{-}PGE_2$), are reduced by 64% (Table 3). Levels of this metabolite have been found to be directly correlated with those of PGE_2 itself[17]. Although administration of high doses of these PG-synthetase inhibitors has been reported to decrease plasma renin activity[18], this low dose, long-term treatment with Indo or ASA has no effect on plasma A II levels[19].

Whereas the data resulting from attempts to manipulate PG production in brain are contradictory, results of studies involving suppression of peripheral PG synthesis complement those of studies involving ip administration of PGE. Peripherally administered PGE is very effective in reducing water intake in response to either A II or hypertonic-saline treatment[11]. Correspondingly, when endogenous PG synthesis is suppressed, water intake in response to both of these stimuli is increased (Table 4)[14,19]. Total water intake in response to polyethylene-glycol induced hypovolemia is only affected when very high, generally antidipsogenic doses of PGE are administered into the peritoneal cavity[11]. The amount of water consumed in response to such hypovolemia is unchanged when endogenous PG synthesis is inhibited[19] (Table 4).

In addition to augmenting total water consumption to A II and

Table 4. Total Average Total Water Intake and Average Latency to the Onset of Drinking (+ SEM) of Rats Pretreated with Inhibitors of Prostaglandin Synthesis (PGSI) and of Controls Which Were Not Exposed to the PGSI.

Dipsogen	Total Intake (ml)		Latency (sec)	
	Control	PGSI	Control	PGSI
iv Angiotensin II	2.7 + .4	4.4 + .8	642 + 66	456 + 54
Hypertonic Saline	5.6 + .7	9.2 + .6	390 + 38	349 + 84
Polyethylene Glycol	7.4 + .8	8.6 + 1.0	473 + 230	87 + 38

cellular dehydration, suppression of PG synthesis has marked effects on the latency to the onset of drinking (Table 4). Rats which are pretreated with PG-synthetase inhibitors initiate drinking sooner than untreated rats during iv infusion of A II[14] or when made hypovolemic through polyethylene-glycol treatment[19]. No change in the latency to drink following hypertonic saline injections is noted[19] (Table 4). The failure to affect the rate of onset of drinking to cellular dehydration may indicate a differential role of endogenous PG in controlling drinking to this stimulus. The failure to affect latency to drink to cellular dehydration when endogenous PG synthesis is suppressed may, however, be an artifact of the testing technique used with this dipsogen which introduces a 15-min delay between the hypertonic-saline injection and water access.

PROSTAGLANDIN-SYNTHESIS INHIBTION AS A DIPSOGENIC STIMULUS

While the method of long-term, low-dose exposure to PG synthesis inhibitors discussed above has no effect on total daily spontaneous water intake[14], bolus injections of high doses of Indo stimulate drinking by water-sated rats. During the first 4-6 hrs following ip injection of Indo, sated rats drink excessively (Table 5)[20]. This effect is dependent upon the kidney. Bilateral nephrectomy, but not bilateral ureteric ligation, eliminates the Indo induced increase of water intake[20].

While the immediate effect of such Indo treatment is an increase of water intake, over a more extended time period (24-48 hr), water intake is reduced (Table 5). Whether this hypodipsia is due to rebound renal PGE production or to a malaise remains to be determined.

Table 5.   Water intakes (Mean ± SEM) of nondeprived rats following
intraperitoneal injection of 10 mg/kg indomethacin.

| Day | Treatment | Water Intake (ml) | |
		First 4 hours	Total 24 hour
1	vehicle	0.4 ± 0.2	26.9 ± 3.0
2	Indo	5.9 ± 1.5	16.9 ± 3.1
3	vehicle	0.7 ± 0.2	17.7 ± 3.1
4	none		28.8 ± 2.8

DISCUSSION

The E prostaglandins are antidipsogenic when administered into
the cerebrospinal fluid or the peritoneal cavity of rats. Delivery
of PGE into the lateral ventricle of the rat brain is most
effective in reducing water ingestion stimulated by icv or iv
administration of A II.   At somewhat higher dose levels, icv PGE
also reduces drinking elicited by polyethylene-glycol induced
hypovolemia or water deprivation.   Administration of PGE into the
CSF has no effect on drinking in response to cellular dehydration.

Intraperitoneal injection of PGE results in a reduction of
water intake to both A II and cell dehydration.   Inhibition of PG
synthesis augments total water intake to both of these stimuli.   No
reliable effect of either peripheral PGE administration or PG-
synthesis inhibition on water intake to hypovolemia is found.

At the very least, these data indicate that PGE may act
pharmacologically to reduce drinking.   This, in and of itself, is
important to the experimental analysis of the controls of drinking.
Many of the techniques commonly used to investigate the controls of
fluid ingestion may modify PGE production.   Catheterization of the
jugular vein of rats results in a significant increase of plasma
PGE levels which is maintained for up to 2 days after surgery
(Table 3).   This increase is most likely due to vascular
irritation and damage during surgery.   Since the majority of
investigators begin their studies of the dipsogenic effectiveness
of infused substances within 48-hr after surgery, this increase of
plasma PGE may reduce responsiveness to such stimuli as A II or
hypertonic saline.   More reliable drinking which is less likely to
be confounded by the elevated PGE levels may be obtained if testing
is postponed until the third or fourth day after catheterization
when plasma PGE levels have decreased to baseline levels.

Angiotensin II administration has complex effects on PGE synthesis. Elevation of renin-A II levels stimulates PGE release[21]. Increased PGE levels may in turn stimulate further production and release of renin[21]. Given that increased peripheral PGE levels suppress A II induced water consumption, these interrelationships could have an impact upon the study of the role of A II in the control of drinking.

Administration of captopril specifically increases plasma levels of KH_2-PGE_2 which may underly the hypotensive effect of this drug[22]. The possibility that changes of fluid consumption following captopril treatment is due, in part or in whole, to modifications of PGE_2 synthesis and release rather than to a reduction of A II production needs to be examined.

In addition to these pharmacological actions, these data also suggest that endogenous PGE may be involved in the physiological control of fluid ingestion. The role of endogenous brain PGE in the control of water ingestion remains unclear. Techniques for reducing brain PG levels which have been effective for researchers in non-behavioral fields frequently induce malaise and generalized failure to behave making their use in the study of fluid ingestion inappropriate. The results of studies in which less debilitating doses or modes of delivery of PG-synthesis inhibitors have been employed, are contradictory. Such results may be due to the inadequacy of the techniques involved in manipulating brain PG synthesis and/or the the possibility that a given PG may have opposite effects on a given physiological event dependent upon the area of brain affected.

Until recently, data on the occurence and levels of various PGs in brain offered little support for the possibility that the E prostaglandins might play a role in the CNS control of any behavior. PGE had been reported to be present in brain but in limited quantities compared to other PGs such as PGD_2 and PGF_{2a}. The report of both a greater degree of binding for PGE_2 than for either PGD_2 or PGF_{2a} in rat brain and the prevalence of such binding in areas of the brain which have been linked to body-fluid homeostasis, has changed this picture[16]. Research involving the direct application of specific PGE-receptor blockers to those areas of brain found both to have a high density of PGE binding and which are purported to be involved in the control of drinking can now be conducted to more adequately evaluate the role of PGE endogenous to brain in this behavior.

When our work on the role of the PGs in the control of water intake was initiated, our emphasis, like that of most researchers of the physiological basis of drinking, was on the involvement of the central nervous system. The powerful and consistent effects of manipulations of peripheral PGE levels have, if not redirected our focus, at least increased our awareness of a potential peripheral control of water ingestion.

Since a great deal of emphasis has been placed on peripherally produced stimuli which act within the CNS, it must be emphasized that peripherally administered or produced PGE cannot be acting centrally to produce the effects observed. While administration of PGE into the cerebrospinal fluid or into the peritoneal cavity results in antidipsogenesis, the situations under which drinking is reduced vary with the administration site. Most notable of these discrepancies is the marked decrease of cell-dehydration induced drinking by peripherally injected PGE but failure to affect such drinking with icv injections of the PG. In addition, degradation of PGE in the periphery is rapid making it unlikely that circulating PGE reaches the brain even when plasma levels are greatly increased. While no direct measures of brain PGE levels following ip injection of PGE are available, the effects of peripheral versus central PGE administration on core temperature support the idea that the peripherally injected PG is not acting within the brain. As noted above, increased brain levels of PGE reliably induce elevations of core temperature. Peripherally injected PGE is not thermogenic. Rather such treatment results in a reliable decrease of core temperature[11].

Based on the data relating to both administration of exogenous PGE and manipulation of endogenous PG levels, two roles for peripherally produced PGE in the control of water intake are postulated[14,19]. First, PGE may act as a satiety signal, limiting the amount of water ingested in response to either elevated circulating A II levels or to hyperosmotic stimuli. Water consumption in response to these stimuli is decreased following ip administration of PGE and is augmented when PG production is inhibited. Thus, PGE could act alone, or in conjunction with other factors, to prevent overhydration during a given ingestive bout for which the initiation of drinking is related to either of these stimuli. Second, given the marked decrease of latency to drink to either iv A II or hypovolemia when PG production is inhibited, PGE may act as a tonic inhibitor of drinking. As such, it might act to prevent ingestion of water during mild perturbations of fluid balance and reduce the number of drinking bouts initiated.

In addition to the multiple factors purported to be involved in the initiation of drinking, the E prostaglandins, as well as other naturally-occurring substances, may play an active role in the cessation or delay of a drinking bout. In conjunction with the detailed analyses of the stimuli of thirst, the study of the role of the E prostaglandins and other potential antidipsogenic agents may provide a more accurate appraisal of the mechanisms underlying the ingestive aspects of body-fluid regulation.

REFERENCES

1. N. J. Kenney and A. N. Epstein, Antidipsogenic role of the E-prostaglandins, J. comp. physiol. Psychol. 92:383 (1978).
2. M. F. Perez Guaita and E. Chiaraviglio, Effect of prostaglandin E_1 and its biosynthesis inhibitor indomethacin on drinking in the rat, Pharmac. Biochem. Behav. 13:787 (1980).
3. S. Nicolaidis and J. T. Fitzsimons, La dépendance de la prise d'eau induite par l'angiotensine II envers la fonction vasomotrice cérébrale locale chez le rat, C. r. hebd. Seanc. Acad. Sci., Paris, Serie D, 281:1417 (1975).
4. S. J. Fluharty, Cerebral prostaglandin biosynthesis and angiotensin-induced drinking in rats, J. comp. physiol. Psychol. 95:915 (1981).
5. N. J. Kenney, A case study of the neuroendocrine control of goal-directed behavior: The interaction between angiotensin II and prostaglandin E_1 in the control of water intake, in:"Neural Mechanisms of Goal-Directed Behavior and Learning,"R. F. Thompson, L. H. Hicks and V. B. Shvyrkov, eds., Academic Press, New York (1980).
6. W. Felberg, Body temperature and fever: Changes in our views over the last decade, Proc. R. Soc., Series B 191:199 (1975).
7. N. J. Kenney and E. Perara, Pressor action of centrally administered prostaglandin E_1, Adv. Prostaglandin Thromboxane Res. 8:1225 (1980).
8. W. E. Hoffman and P. G. Schmid, Cardiovascular and antidiuretic effects of central prostaglandin E_2, J. Physiol. 288:159 (1979).
9. K. M. Skoog and N. J. Kenney, Pressor responses to central angiotensin II, prostaglandin E_1 and prostaglandin E_2: Effects of ovarian steroids, Neuroendo. 36:144 (1983).
10. K. M. Skoog and N. J. Kenney, Influences of ovarian steroids on angiotensin-induced dipsogenesis and prostaglandin E_1-induced antidipsogenesis and thermogenesis, Soc. Neurosci. Absts. 7:639 (1981).
11. N. J. Kenney, K. E. Moe and K. M. Skoog, The antidipsogenic action of peripheral prostaglandin E_2, Pharmac. Biochem. Behav. 15:263 (1981).
12. D. J. Goldstein, D. J. Marante Perez, J. P. Gunst and J. A Halperin, Prostaglandin E_1 inhibits acute cell dehydration thirst, Pharmac. Biochem. Behav. 10:895 (1979).
13. M. I. Phillips and W. E. Hoffman, Sensitive sites in the brain for the blood pressure and drinking responses to angiotensin II, in:"Central Actions of Angiotensin and Related Hormones," J. P. Buckley and C. Ferrario, eds., Pergamon, New York (1977).
14. N. J. Kenney and K. E. Moe, The role of endogenous prostaglandin E in angiotensin-II-induced drinking in rats, J. comp. physiol. Psychol. 95:383 (1981).

15. J. A. Splawinski, K. Reichenberg, J. Vetulani, J. Marchaj and J. Kaluza, Hyperthermic effect of intraventricular injections of arachidonic acid and prostaglandin E_2 in the rat, Pol. J. Pharmacol. Pharm. 26:101 (1974).

16. C. Malet, H. Scherrer, J. M. Saavedra and F. Dray, Specific binding of [^{3}H]prostaglandin E_2 to rat brain membranes and synaptosomes, Brain Res. 236:227 (1982).

17. S. A. Metz, M. G. Rice and R. P. Robertson, Applications and limitations of measurement of 15-keto-13,14-dihydro prostaglandin E_2 in human blood by radioimmunoassay, Prostaglandins 17:839 (1979).

18. J. C. Romero, C. L. Dunlap and C. G. Strong, The effect of indomethacin and other anti-inflamatory drugs on the renin-angiotensin system, J. clin. Invest. 58:282 (1976).

19. N. J. Kenney and K. E. Moe, Cellular dehydration and hypovolemia: Effect of acetylsalicylic acid on drinking, Pharmac. Biochem. Behav. 17:73 (1982).

20. G. Feuerstein, M. Krausz and Y. Gutman, Effect of indomethacin on water intake of the rat, Pharmac. Biochem. Behav. 9:893 (1978).

21. P. C. Weber and W. Siess, Interactions of renal prostaglandins with the renin-angiotensin system, Pharmac. Ther. 15:321 (1982).

22. S. L. Swartz, G. H. Williams, N. K. Hollenberg, L. Levine, R. G. Dluhy and T. J. Moore, Captopril-induced changes in prostaglandin production: Relationship to vascular response in normal men, J. Clin. Invest. 65:1257 (1980).

BENZODIAZEPINE AND ENDORPHINERGIC MECHANISMS IN RELATION TO SALT

AND WATER INTAKE

Steven J. Cooper

Department of Psychology
University of Birmingham
Birmingham B15 2TT, U.K.

INTRODUCTION

Pharmacological treatments can be used to assess the involvement of specified neurochemical mechanisms in the control of drinking responses. The benzodiazepines are an interesting group of drugs, for which specific, high-affinity binding sites have been identified within the central nervous system[1,2,3]. A little more than a decade ago, Maickel and Maloney[4] showed that in rats which had been adapted to a 23 h water-deprivation schedule, acute treatments with the benzodiazepine receptor agonists, diazepam or chlordiazepoxide, led to significant increases in water consumption. In addition, Falk and Burnidge[5] showed that chlordiazepoxide significantly increased the level of consumption of a 1.5% NaCl solution in rehydrating rats. These studies helped to initiate an examination of the effects of drugs active at benzodiazepine receptors in relation to the controls of water and saline consumption.

Research with drugs which act as agonists at opiate receptors provided the impetus for an investigation of the involvement of endogenous opioid peptides in the mediation of drinking responses[6]. It is thought that the blockade of endogenous opioid peptide activity may be responsible for the typical antidipsogenic effects of drugs like naloxone and naltrexone[7,8]. The present paper provides a brief assessment of the probable involvement of benzodiazepine-related and endorphinergic mechanisms in the control of drinking responses. More tentatively, it suggests that interactions may occur between the two[9].

BENZODIAZEPINES AND WATER INTAKE

Hyperdipsia has been reported to follow the administration of chlordiazepoxide, diazepam, flurazepam, lorazepam, midazolam and oxazepam[10,11]. The effect has generally been reported using water-deprived animals (rats and mice; apparently other species have not been studied). It is mediated by drug action at benzodiazepine receptors, since it has been shown that midazolam-induced hyperdipsia was reversed by the receptor antagonist, Ro15-1788[12]. Time-course studies show that the hyperdipsia results from an extension in the duration of the drinking response in the rehydrating animal, and not from an increase in the local rate of fluid consumption[10]. Water consumption aroused by either osmotic or hypovolemic challenge is also enhanced by chlordiazepoxide treatment[13].

The benzodiazepine receptor antagonists, Ro15-1788 and CGS 8216[14], have no intrinsic effects on water intake in rehydrating rats[15]. These observations indicate that putative endogenous benzodiazepine ligands are not effective during drinking to restore body water deficit. The novel compound, FG 7142, described as an inverse agonist active at benzodiazepine receptors[16], did reduce deprivation-induced water consumption[15].

BENZODIAZEPINES AND SALT INTAKE

Experiments have been carried out in rats adapted to a water-deprivation schedule and allowed to rehydrate with access to salt solutions using a single-stimulus method. The consumption of hypertonic NaCl solution increases following the administration of chlordiazepoxide[5], diazepam[17], or midazolam[18]. The rise in NaCl consumption has been attributed to the punishment-attenuating effects of benzodiazepine treatments[5]. This is unlikely to be the case, however, since chlordiazepoxide has also been shown to increase the ingestion of a highly palatable 0.9% NaCl solution[19,20]. Instead, salt taste may potentiate the hyperdipsic effect of benzodiazepine treatments[20]. Unexpectedly, the benzodiazepine receptor antagonist Ro15-1788 also increased hypertonic saline intake, thus revealing partial agonist properties[18,21]. In contrast, the antagonist CGS 8216 did not affect hypertonic saline consumption[21].

ENDORPHINS AND WATER INTAKE

Drugs which act as opiate receptor antagonists decrease water consumption in water-deprived and nondeprived animals[6,7,8,22]. The effect is not due to a general suppression of drinking responses, since schedule-induced polydipsia remains unaffected[23]. The sites of action of opiate receptor antagonists in blocking deprivation-induced drinking are thought to be located within the central nervous system[24,25]. The pattern of drinking in rehydrating animals following naloxone or naltrexone administration suggests that drinking satiety may be enhanced by the drug treatments[23].

These data suggest that endogenous opioid peptides may be involved in at least some categories of drinking response. Experiments with opiate receptor agonists, however, produce ambiguous results. For example, after treatments with the kappa opiate receptor agonist, ethylketocyclazocine, drinking was dose-dependently inhibited in water-deprived rats[26], although some increases in drinking occurred in nondeprived animals. Similarly, the injection of enkephalins into the cerebral ventricles attenuated drinking in thirsty animals[27], but produced a delayed dipsogenic action when injected into nondeprived animals[28]. The involvement of endogenous opioid peptides in drinking responses remains, therefore, to be better characterized.

ENDORPHINS AND SALT INTAKE

In male mice, the stimulation of sodium appetite by immobilization stress, food deprivation or angiotensin is antagonized by the opiate receptor antagonist, naloxone[27]. It has been suggested that endorphin release may be involved in sodium appetite[29]. The heightened drinking response to a palatable 0.9% NaCl solution is also reversed by naloxone[30]. In two-bottle preference tests carried out using water-deprived rats, naloxone treatments diminish the consumption of preferred hypotonic NaCl solutions[31]. It is difficult, at the present time, to distinguish a possible specific effect of naloxone on salt preference, or appetite, from a more general antidipsogenic effect. Hence, it remains unclear to what extent endogenous opioid peptides play a part in specific behavioural responses to salt solutions.

BENZODIAZEPINE-OPIATE ANTAGONIST INTERACTIONS

Evidence from binding studies and pharmacological experiments suggests effects of benzodiazepine treatments on endogenous opioid mechanisms[32,33]. In drinking experiments, opiate receptor antagonists have been shown to block the hyperdipsia induced by chlordiazepoxide[34,35]. It is possible, therefore, that benzo-diazepine-induced hyperdipsia may depend, in some degree at least, on endogenous opioid peptides. This issue, however, awaits further clarification before the nature of the possible interrelationship can be judged.

REFERENCES

1. C. Braestrup and R. F. Squires, Specific benzodiazepine receptors in rat brain characterized by high-affinity ^{3}H-diazepam binding, Proc. natn. Acad. Sci., U. S. A. 74:3805 (1977).
2. H. Möhler and T. Okada, Benzodiazepine receptors: demonstration in the central nervous system, Science. 198:849 (1977).
3. W. S. Young and M. J. Kuhar, Radiohistochemical localization of benzodiazepine receptors in rat brain, J. Pharmacol. Exp. Ther. 212:337 (1980).
4. R. P. Maickel and G. J. Maloney, Effects of various depressant drugs on deprivation-induced water consumption, Neuropharmacology. 12:777 (1973).
5. J. L. Falk and G. K. Burnidge, Fluid intake and punishment-attenuating drugs, Physiol. Behav. 5:199 (1970)
6. S. J. Cooper and D. J. Sanger, Endorphinergic mechanisms in food, salt and water intake: an overview, Appetite (1984) in press.
7. D. R. Brown and S. G. Holtzman, Suppression of deprivation-induced food and water intake in rats and mice by naloxone, Pharmacol. Biochem. Behav. 11:567 (1979).
8. D. J. Sanger, Opiates and ingestive behaviour, in: "Theory in Psychopharmacology, Vol. 2", S. J. Cooper, ed., Academic Press, London (1983).
9. S. J. Cooper, Minireview. Benzodiazepine-opiate antagonist interactions in relation to feeding and drinking behavior, Life Sci. 32:1043 (1983).
10. S. J. Cooper, Benzodiazepines, barbiturates and drinking, in: "Theory in Psychopharmacology, Vol. 2", S. J. Cooper, ed., Academic Press, London (1983).
11. J. D. Leander, Effects of punishment-attenuating drugs on deprivation-induced drinking: implications for conflict procedures, Drug Dev. Res. 3:185 (1983).
12. S. J. Cooper, Specific benzodiazepine antagonist Ro15-1788 and thirst-induced drinking in the rat, Neuropharmacology. 21:775 (1982).
13. S. J. Cooper, Effects of chlordiazepoxide on drinking compared in rats challenged with hypertonic saline, isoproterenol or polyethylene glycol, Life Sci. 32:2453 (1983).
14. C. A. Boast, P. S. Bernard, B. S. Barbaz, and K. M. Bergen, The neuropharmacology of various diazepam antagonists, Neuropharmacology. 22:1511 (1983)
15. S. J. Cooper and L. B. Estall, Behavioral pharmacology of food, water and salt intake in relation to drug actions at benzo-diazepine receptors, Neurosci. Biobehav. Rev. (1985) in press.
16. C. Braestrup, M. Nielsen, T. Honore, L. H. Jensen, and E. N. Petersen, Benzodiazepine receptor ligands with positive and negative efficacy, Neuropharmacology. 22:1451 (1983).

17. M. Tang, C. Brown, D. Maier, and J. L. Falk, Diazepam-induced NaCl solution intake: independence from renal factors, Pharmacol. Biochem. Behav. 18:983 (1983).

18. M. Tang, S. Soroka, and J. L. Falk, Agonistic action of a benzo-diazepine antagonist: effects of Ro15-1788 and midazolam on hypertonic NaCl intake, Pharmacol. Biochem. Behav. 18:953 (1983).

19. S. Turkish and S. J. Cooper, Enhancement of saline consumption by chlordiazepoxide in thirsty rats: antagonism by Ro15-1788, Pharmacol. Biochem. Behav. (1984) in press.

20. J. L. Falk and M. Tang, Chlordiazepoxide elevates the NaCl solution acceptance-rejection function, Pharmacol. Biochem. Behav. (1984) in press.

21. J. L. Falk and M. Tang, Midazolam-induced increases in NaCl solution ingestion: differential effect of the benzodiazepine antagonists Ro15-1788 and CGS 8216, Pharmacol. Biochem. Behav. (1984) in press.

22. J. D. Leander and M. D. Hynes, Opioid antagonists and drinking: evidence of kappa receptor involvement, Europ. J. Pharmacol. 87:481 (1983).

23. S. J. Cooper and S. G. Holtzman, Patterns of drinking in the rat following the administration of opiate antagonists, Pharmacol. Biochem. Behav. 19:505 (1983).

24. D. R. Brown and S. G. Holtzman, Opiate antagonists: central sites of action in suppressing water intake of the rat, Brain Res. 221:432 (1981).

25. D. A. Czech, E. A. Stein, and M. J. Blake, Naloxone-induced hypodipsia: a CNS mapping study. Life Sci. 33:797 (1983).

26. S. Turkish and S. J. Cooper, Effects of a kappa receptor agonist, ethylketocyclazocine, on water consumption in water-deprived and nondeprived rats in diurnal and nocturnal tests, Pharmacol. Biochem. Behav. (1984) in press.

27. G. de Caro, L. G. Micossi, and F. Venturi, Drinking behaviour induced by intracerebroventricular administration of enkephalins to rats, Nature. 277:51 (1979).

28. F. Cantalamessa, G. de Caro, M. Massi, and L. G. Micossi, Stimulation of drinking behaviour and of renin release induced by intracerebroventricular injections of D-Ala^2, D-Leu^5-enkephalin to rats, Pharmacol. Res. Commun. 14:141 (1982).

29. C. G. Kuta, H. U. Bryant, J. E. Zabik, and G. K. W. Yim, Stress, endogenous opioids and salt intake, Appetite (1984) in press.

30. S. J. Cooper and S. Turkish, Effects of naloxone and its quaternary analogue on fluid consumption in water-deprived rats, Neuropharmacology. 22:797 (1983).

31. S. J. Cooper and D. B. Gilbert, Naloxone suppresses fluid consumption in tests of choice between sodium chloride solutions and water in male and female water-deprived rats, Psychopharmacology (1985) in press.

32. Y. Watanabe, T. Shibuya, B. Salafsky, and H. F. Hill, Prenatal and postnatal exposure to diazepam: effects on opioid receptor binding in rat brain cortex, Europ. J. Pharmacol. 96:141 (1983).
33. T. Duka, M. Wuster, and A. Herz, Benzodiazepines modulate striatal enkephalin levels via gabanergic mechanism, Life Sci. 26:771 (1980).
34. S. J. Cooper, Effects of opiate antagonists and of morphine on chlordiazepoxide-induced hyperdipsia in the water-deprived rat, Neuropharmacology. 21:1013 (1982).
35. S. J. Cooper, Enhancement of osmotic- and hypovolemic-induced drinking by chlordiazepoxide in rats is blocked by naltrexone, Pharmacol. Biochem. Behav. 17:921 (1982).

SELECTIVE ANTIDIPSOGENIC EFFECT OF KASSININ IN WISTAR RATS

G. de Caro and L. G. Micossi

Inst. of Pharmacology, Univ. of Camerino

Camerino, Italy

Kassinin is a dodecapeptide of the tachykinin family (fig. 1) which has been found in the skin of the African frog Kassina senegalensis. It displays the same biological effects of the other tachykinins. However, it has a poor action on blood pressure and salivary secretion, but potently affects smooth muscles of the intestinal and genito-urinary tract, as well as ileal transport of electrolites (1).

KS	Asp-Val-Pro-Lys-Ser-Asp-Glu-Phe-Val-Gly-Leu-Met-NH$_2$
EL	Pyr-Pro-Ser-Lys-Asp-Ala-Phe-Ile-Gly-Leu-Met-NH$_2$
PH	Pyr-Ala-Asp-Pro-Asn-Lys-Phe-Tyr-Gly-Leu-Met-NH$_2$
SP	Arg-Pro-Lys-Pro-Gln-Gln-Phe-Phe-Gly-Leu-Met-NH$_2$
Nα	His-Lys-Thr-Asp-Ser-Phe-Val-Gly-Leu-Met-NH$_2$

Fig. 1.   Aminoacid sequence of the tachykinins kassinin (KS), eledoisin (EL), physalaemin (PH), substance P (SP) and neurokinin α (Nα).

Since kassinin has the same affinity as eledoisin (the most active antidipsogenic tachykinin found up to now) for the SP-E type receptors for substance P and since a kassinin-like peptide has been recently found in mammalian brain (2), we considered it interesting to check whether kassinin shared with eledoisin the same antidipsogenic activity.

Our experiments demonstrated that this tachykinin inhibits water intake induced by cell dehydration, but not that evoked by other dipsogenic determinants and that it produces an intense ex-

245

citement which at least in part can explain the alteration of drinking behaviour.

METHODS

Male albino rats (Wistar, Charles River) bearing permanent cannulae into the lateral brain ventricles were made thirsty by a) intracerebroventricular (ICV) pulse injection of angiotensin II (100 ng/rat in 1 µl 0.9% NaCl solution), b) 16 hr of water deprivation and c) cell dehydration (NaCl, 3 mOsmoles/100 g b.w. subcutaneously, 15 min prior to water presentation).

In some experiments the rats were trained either to drink water or to eat food (pellets) only from 9.30 to 11.30 a.m., while in others they were made hungry by 24 hr of food deprivation.

Water intake was determined to the nearest 0.1 ml from graduated drinking tubes; food intake was determined to the nearest 0.1 g by weighing the pellets of food. Spontaneous motor activity was studied in a open field.

ICV injections of kassinin (Peninsula Laboratories inc., Belmont, USA, dissolved in 1 µl of 0.9% NaCl solution) took place 60 sec before water or food presentation or before testing the rats in the open field.

For details concerning the techniques employed, see (3, 4).

RESULTS

Rats tested in an open field, without access to water and receiving kassinin ICV, 50-400 ng/rat, do not explore the field but run along its wall continuously, resting from time to time for a few sec. The effect has a latency of a few sec, lasts 5-15 min, according to the dose, and when it disappears it leaves the animals perfectly normal. If the rats receive ICV kassinin, 50-200 ng/rat and angiotensin II, 100 ng/rat at the same time, the excitement is much more evident; the rats move rapidly and jerkily and exhibit intense grooming, stratching and rearing activity. In both instances, if the rats have free access to water they either do not drink at all or take water only in rare and brief bouts of drinking. However, when the excitement disappears the animals, if thirsty, drink normally.

Rats made thirsty by ICV angiotensin II or by water deprivation, as well as those trained to drink only from 9.30 to 11.30 a.m., do not drink at all in response to ICV kassinin, 50-200, 100-800 and 400-1600 ng per rat respectively. However, the inhibitory effect lasts as long as the excitement lasts and when it disappears the rats, if thirsty, drink normally and take the same amount of water as untreated animals. In these experiments neither water intake inhibition nor latency to drinking are ever related to the dose of kassinin. Moreover, in angiotensin II experiments if the treatment is repeated two or more times on alternative days, the excitation progressively decreases and with it water intake inhibition decreases, too.

Rats made hungry by food deprivation or trained to eat only from 9.30 to 11.30 a.m. eat normally even if receiving doses of kassinin as large as 1600 ng per rat. In these experimental conditions not even food associated drinking is affected by kassinin treatment.

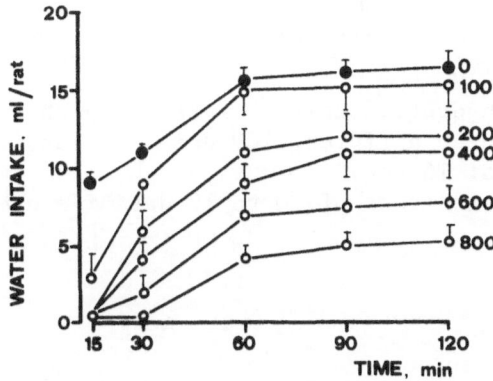

Fig. 2. Cell dehydration thirst: inhibitory effect elicited by different ICV doses of kassinin.
Values are means of 10 to 12 data ± s.e.m.

In animals made thirsty by cell dehydration, kassinin at doses of 100 to 800 ng per rat produces a long-lasting, dose dependent inhibition of drinking which is statistically significant at all the levels of the doses employed (fig. 2). Latency to drinking is significantly related to the dose, too (fig. 3). In these experiments an evident excitement, lasting no more than 10 to 15 min, is evoked by doses of kassinin as large as 800 ng per rat or more. However, at these levels of doses the rats are adipsic not only while being excited but also for a long time well after the excitement is over.

If these experiments are carried out on rats which previously had received the peptide into the brain two or more times on alternative days, the sensitivity of rats to kassinin does not change.

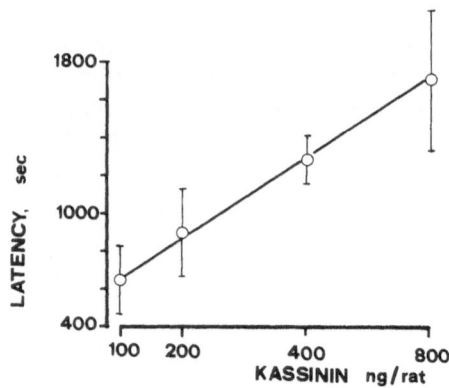

Fig. 3. Cell dehydration thirst: latency to the inhibitory effect elicited by different ICV doses of kassinin. Latency of controls: 16 sec.
Values are means of 10 to 12 data ± s.e.m.

DISCUSSION

The results of our experiments demonstrate that kassinin possesses an antidipsogenic effect which is specific and highly selective. In fact, kassinin does not inhibit food intake, food associated drinking, angiotensin II- or water deprivation-induced drinking, but inhibits water intake induced by cell dehydration.

This effect is easily repeatable, depends on the dose and continues long after excitement has subsided. Moreover, after repeated treatments with kassinin, while the excitatory effect disappears, that on cell dehydration thirst remains unmodified.

The antidipsogenic effect of kassinin is highly selective, since only cell dehydration thirst is inhibited by the peptide. The inhibition of drinking elicited by other dipsogenic determinants is always clearly due to the excitement that kassinin produces.

We consider extremely important the fact that kassinin possesses this selective antidipsogenic effect.

In fact, in the mammalian brain there are at least nine different antidipsogenic peptides belonging to at least three different peptide families. Some of them have a wide spectrum of antidipsogenic effect and inhibit drinking elicited by numerous different challenges, some have a limited spectrum of antidipsogenic effects, but only kassinin affects only one single kind of thirst, that is cell dehydration thirst.

Kassinin is of amphibian origin, but it has its counterpart in the brain: substance K, also called neurokinin α. Moreover, it has been shown that of the two precursor molecules to substance P, α- and β-pre-pro-tachykinin, the latter contains in its chain the aminoacid sequences not only of substance P but also of the kassinin-like substance K (2).

It has been emphasized that the existence of two tachykinin precursors can indicate that in the brain there are "substance P neurons" and "substance P + substance K neurons" or, alternatively, single neurons which produce either or both the peptides in response to different stimuli (2). If it is demonstrated that substance K shares with kassinin not only the same immunochemical properties but also the same specific and highly selective antidipsogenic effects, the occurrence in the brain of such a peptide in close connection with another antidipsogenic peptide will give further, conclusive support to the hypothesis that the different antidipsogenic peptides are members of a peptidergic system and that they participate in the control of water intake and body fluid homeostasis according to the peculiar effect they elicit on drinking behaviour.

REFERENCES

1. V. Erspamer, The tachykinin peptide family, Trends NeuroSci. 4:267 (1981).
2. A. J. Harmar, Three tachykinins in mammalian brain, Trends NeuroSci. 7:57 (1984).
3. G. de Caro, M. Massi, L.G. Micossi, Antidipsogenic effect of intracranial injection of substance P in rats, J. Physiol. 279:133 (1978).
4. G. de Caro, L. G. Micossi, F. Venturi, A. Brancati and E. Scarnati, Behavioural and electrocortical modifications induced in the rat by intraventricular injection of physalaemin and eledoisin, Psychopharmacologia 38:211 (1974).

SENSITIVITY TO DIPSOGENIC PEPTIDES OF PIGEONS BEARING LESIONS

DIRECTED TO THE SUBFORNICAL ORGAN (SFO)

M. Massi, G. de Caro, °A.N. Epstein and [+]L. Mazzarella

Inst. of Pharmacology, Univ. of Camerino, Camerino-Italy
°J. Leidy Lab., Univ. of Pennsylvania, Philadelphia-USA
[+]Inst. of Pathology, EM Center, Univ. of Bari, Bari-Italy

INTRODUCTION

The tachykinins eledoisin and physalaemin, as well as the bombesin peptides litorin, ranatensin and bombesin itself, exert in the pigeon a powerful dipsogenic effect which is similar, as far as drinking behaviour, time course and dose-response relationship are concerned, to that evoked by angiotensin II (Ang II)[1-5].

In mammals (rat, dog, opossum)[6-10], peripheral Ang II induces thirst by acting on the SFO and lesioning of the SFO completely blocks its dipsogenic effect. On the other hand, icv Ang II acts mainly on receptors outside the SFO as shown by the fact that SFO ablation only slightly reduces the dipsogenic response[11].

In the present study we investigated the role of the SFO in drinking induced by Ang II and, comparatively, by tachykinins and bombesins in the pigeon.

MATERIAL AND METHODS

Twenty four White King pigeons of both sexes were employed. After being tested twice for drinking in response to ip Ang II, 100 µg/pigeon, they received either an electrolytic lesion aimed at the SFO (16 animals) or a sham lesion (8 animals). In the same surgery a guide cannula was implanted into their third cerebroventricle.

Statistics are Student's t test.

RESULTS

    As shown in Fig. 1, while sham lesioned animals showed essen-
tially the same response as pre-surgery, 9 of the 16 lesioned pigeons became completely refractory to ip Ang II. No sign of recovery in sensitivity was detected over the 60 days period of observation.

    On the other hand, the 9 pigeons refractory to ip Ang II, were
still sensitive to the dipsogenic action of icv Ang II (Fig. 2).
In response to 10 ng/pigeon, they drank approximately the same
amount of water of sham lesioned animals. At the higher dose of 100
ng/pigeon, their dipsogenic response was clearly present, but was
slightly lower than that of sham lesioned birds. However, 60 days
after surgery their water intake in response to icv Ang II was
not different from that of sham lesioned pigeons.

    The bombesin peptides, including bombesin itself, when given
either icv or ip evoked short latency, marked and equal dipsogenic
responses both in sham and in SFO lesioned pigeons (Fig. 3).
Moreover, no significant difference was observed between sham and
lesioned animals in response to both icv and ip litorin or ranatensin.

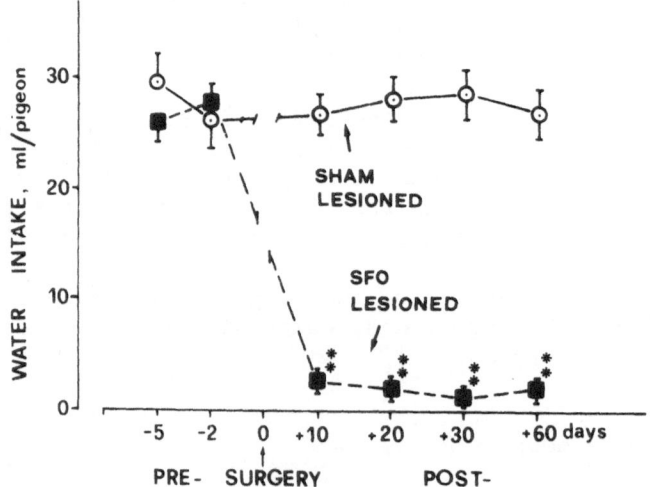

Fig. 1. Water intake of sham and SFO lesioned pigeons in 60 min
         after ip administration of 100 μg of Ang II. Values deter-
         mined at several times pre- and post-operatively are means
         ±SEM. ✱ P < 0.01; where not indicated, difference between
         sham and SFO lesioned pigeons was not statistically sig-
         nificant.

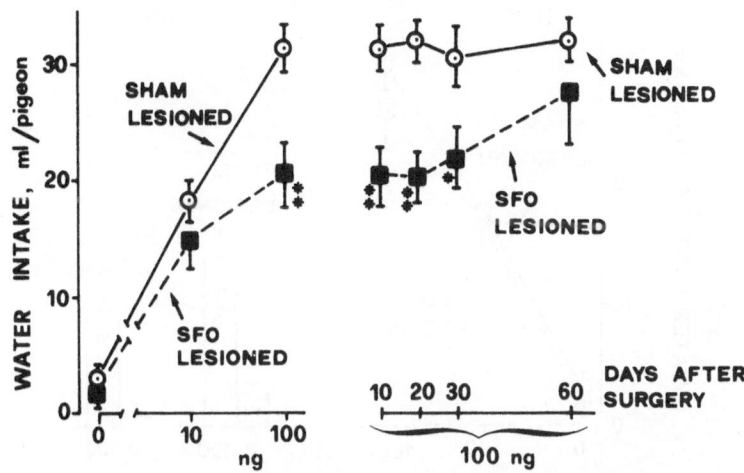

Fig. 2. Water intake (means±SEM) of sham and SFO lesioned pigeons
in 60 min after different icv doses of Ang II. Drinking to
100 ng of Ang II was determined at several times after
surgery. ✻P < 0.01, ✳P < 0.05; where not indicated, difference between the two groups of animals was not significant.

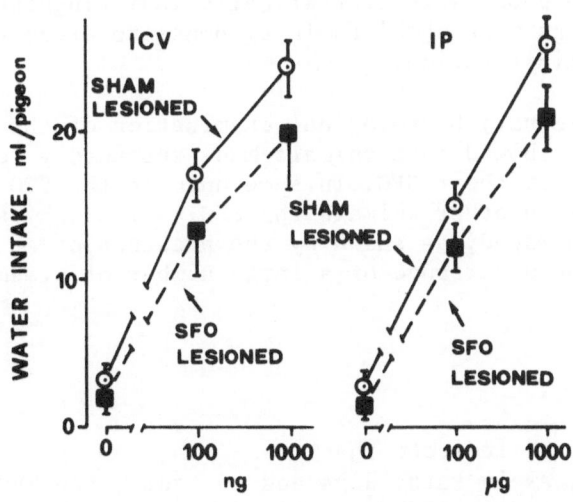

Fig. 3. Water intake (means±SEM) of sham and SFO lesioned pigeons
in 60 min after icv or ip injection of different doses of
bombesin. The drinking response of the two groups of animals was never statistically different.

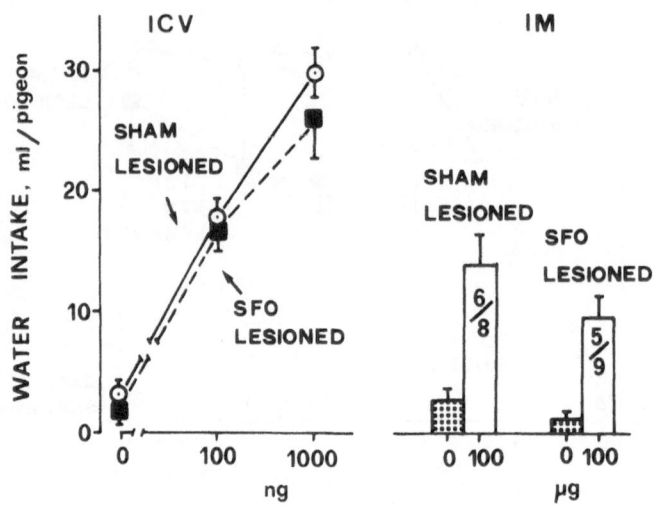

Fig. 4. Water intake (means±SEM) of sham and SFO lesioned pigeons in 60 min after icv or ip injection of different doses of physalaemin. Drinking responses of the two groups of animals were never statistically different.

In response to icv physalaemin, both lesioned and sham lesioned pigeons showed short latency and dose-dependent dipsogenic responses, which were not significantly different from each other. When given by im injection at the dose of 100 μg, only about half the animals responded. Nevertheless, the water intakes of sham and SFO lesioned pigeons were statistically indistinguishable (Fig. 4). Similar findings were obtained in response to eledoisin both after icv and im administration.

The preliminary histological examination of the brain of lesioned pigeons showed that animals made refractory to ip Ang II had marked lesions of their SFO. In some animals the SFO was completely ablated, while in other animals the SFO, although still present, was markedly damaged, as shown by the presence of an intense gliosis and by the occurrence of a large number of plasmacells in its body.

CONCLUSIONS

Our findings indicate that:
1) in pigeons, as in rats, dogs and opossums, the SFO is essential for the dipsogenic effect of blood-borne Ang II ;
2) lesioning of the SFO only slightly reduces the dipsogenic effect of icv Ang II. Moreover, this reduction is observed only at high doses of Ang II and is not permanent;

3) finally, the SFO appears not to be involved in drinking elicited by peripheral or icv administration of bombesins or of **tachy-kinins**.

Previous data from this laboratory have ruled out the possibility that drinking to bombesins and tachykinins is due to activation of Ang II receptors for drinking[12]. The findings of the present study indicate that different neuroanatomical structures are responsible for drinking to Ang II, on one hand, and to the other dipsogenic peptides, on the other hand, and give further support to the hypothesis of the existence of different peptidergic systems involved in the regulation of drinking behaviour in the pigeon.

REFERENCES

1. M. D. Evered, J.T. Fitzsimons and G. de Caro, Drinking behaviour induced by intracranial injections of eledoisin and substance P in the pigeon, Nature 268:332 (1977).
2. G. de Caro, M. Massi and L.G. Micossi, Potent dipsogenic effect of physalaemin in the pigeon, Pharmac. Res. Commun. 10:861 (1978).
3. G. de Caro, M. Massi and L.G. Micossi, Bombesin potently stimulates water intake in the pigeon, Neuropharmacology 19:867 (1978).
4. G. de Caro, M. Mariotti, M. Massi and L. G. Micossi, Relative dipsogenic potency of some partial sequences of bombesin in pigeons and ducks, Pharmac. Res. Commun. 12:483 (1980).
5. M. D. Evered and J. T. Fitzsimons, Drinking and changes in blood pressure in response to angiotensin II in the pigeon Columba livia, J. Physiol. 310:337 (1981).
6. J. B. Simpson and A. Routtenberg, Subfornical organ lesions reduce intravenous angiotensin-induced drinking, Brain Res. 88: :154 (1975).
7. A. E. Abdelaal, S. Y. Assaf, J. Kucharczyk and G. J. Mogenson, Effect of ablation of the subfornical organ on water intake elicited by systemically administered angiotensin II, Can. J. Physiol. Pharmacol. 52:1217 (1974).
8. J. B. Simpson, A. N. Epstein and J. S. Camardo, Jr., Localization of receptors for the dipsogenic action of angiotensin II in the subfornical organ of rat, J. Comp. Physiol. Psychol. 92:581 (1978).
9. T. N. Thrasher, J. B. Simpson and D. J. Ramsay, Lesions of the subfornical organ block angiotensin-induced drinking in the dog. Neuroendocrinology 35:68 (1982).
10. A. L. R. Findlay, R. M. Elfont and A. N. Epstein, The site of the dipsogenic action of angiotensin II in the north American Opossum, Brain Res. 198:85 (1980).
11. R. L. Thunhorst, R. W. Lind and A. K. Johnson, Lesions of the

subfornical organ (SFO) block drinking to peripheral but not central angiotensin, Neurosci. Abs. 7:638 (1981).

12. G. de Caro, M. Massi, L. G. Micossi and M. Perfumi, Angiotensin II antagonists versus drinking induced by bombesin and eledoisin in pigeons, Peptides 3:631 (1982).

INHIBITION OF ANG II-INDUCED DRINKING BY DERMORPHIN GIVEN INTO THE SFO OR INTO THE LATERAL VENTRICLE OF INTACT OR OF SFO LESIONED RATS

M. Perfumi, G. de Caro, M. Massi and F. Venturi

Institute of Pharmacology, University of Camerino

Via Scalzino 5, 62032 Camerino (MC), Italy

INTRODUCTION

Dermorphin is an opioid heptapeptide found in the skin of South American frogs of the Phyllomedusa genus[1,2]. Afterwards, dermorphin-like immunoreactivity has been detected in the central nervous system of rats, pigs and frogs[3].

In addition to its potent antinociceptive action[4], dermorphin has been shown to possess numerous other central actions in rats, including prolactin release, LH and FSH release, central regulation of gastric acid secretion and of gastric emptying[5], and inhibition of drinking induced by angiotensin II (Ang II)[6].

Recently, Buffa et al.[7] reported that, unlike enkephalins and endorphins which are widely distributed in the central nervous system, dermorphin-like immunoreactivity occurs in a restricted number of regions of the rat brain, and particularly in the arcuate nucleus, the organum vascolosum laminae terminalis (OVLT) and the subfornical organ (SFO).

The occurrence of dermorphin-like immunoreactivity in the SFO which is a sensitive site to the dipsogenic action of Ang II[8], suggested to us the opportunity of evaluating whether this organ plays an important role in the inhibitory effect of dermorphin on Ang II--induced drinking. On the other hand, dermorphin was also tested on drinking in response to cell dehydration and to water deprivation in order to assess whether this peptide is a general antidipsogen or whether its action may be considered selective for drinking induced by Ang. II.

MATERIAL AND METHODS

Male Wistar rats (Charles River, Calco, Italy) weighing 250--300 g were employed. They were kept in individual cages in a room in which the temperature was 19±0.5°C and were maintained on food in pellets (Morini, Reggio Emilia, Italy).

Under equithesin anaesthesia, two indwelling guide cannulae (o.d. 600 µ), aimed either at the lateral ventricle (LV) or at the SFO, were stereotaxically implanted in non lesioned animals, according to the technique described by Epstein et al.[9]. Stereotaxic coordinates for the SFO cannula were calculated according to Konig and Klippel atlas of the rat brain[10]. The employed coordinates were: AP = 6.6, L = 0 at 0° in the coronal plane and V = 5.7; to avoid bleeding the sagittal sinus was slightly retracted when introducing the indwelling cannula.

Twelve rats received an electrolytic lesion aimed at the SFO. In same surgery a guide cannula was implanted in their LV, with the same technique employed in intact animals.

Intracranial injections were made in conscious rats through a stainless-steel injector (o.d. 300 µ) temporarily inserted into the guide cannula. The tip of both the LV and SFO injectors extended 1 mm beyond the end of the cannula tip. Substances given intracranially were dissolved in isotonic saline and given in a finale volume of 1 µl into the LV and of 0.1 µl into the SFO. Injections into the SFO were made by means of a 1 µl Hamilton precision syringe.

Drinking was induced by icv injection of Ang II (100 ng/rat), by sc administration of hypertonic NaCl (8.76 g/100 ml; 1 ml/100 g b.w.) given 15 min before dermorphin administration or by a 16 hr period of water deprivation. Water intake was measured to the nearest 0.1 ml by means of graduated drinking tubes. Statistics are Student's t test.

RESULTS

The effect of dermorphin given into the SFO or into the LV on drinking induced by icv Ang II (100 ng/rat) is reported in Fig. 1. When injected into the SFO dermorphin elicited an inhibitory effect clearly more potent than after LV injection. In fact, after SFO injection, dermorphin elicited a highly statistically significant inhibition even at the dose of 0.5 ng/rat (% inhibition = 64.6, $P < 0.01$) and at the dose of 1 ng/rat the inhibition was 86.2%. In the same rats receiving dermorphin into the LV, the peptide did not elicit a statistically significant inhibition at the dose of 0.5 ng, while at the dose of 1 ng/rat the inhibition was 42.9% vs that of 86.2% observed after the SFO administration.

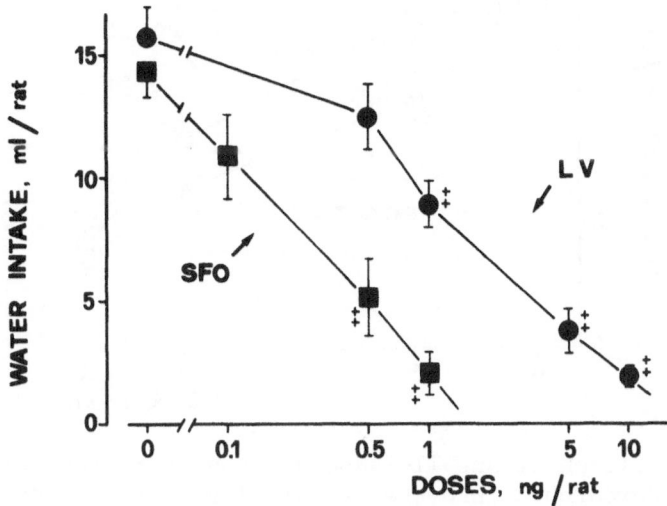

Fig. 1. Inhibition of Ang II-induced drinking following injection
of dermorphin into the LV or into the SFO. Values determined 15 min after the icv injection of Ang II, are means±
±SEM of 8 data. ‡ $P < 0.01$; where not indicated, difference from controls was not statistically significant.

The inhibitory effect of dermorphin (again on drinking induced
by icv Ang II) administered into the LV of intact or of SFO lesioned rats is reported in Fig. 2. At the dose of 1 ng per rat,
lesioned animals showed no significant inhibition of drinking to
Ang II, while intact rats showed a marked and significant inhibition ($P < 0.01$). Increasing the dose of dermorphin, however, progressively decreased the difference in sensitivity to the antidipsogenic action of the peptide between intact and SFO lesioned
rats. In the animals bearing lesions directed to the SFO, the
histological analysis revealed the presence of marked or total lesions of the organ. Moreover, evidence in support of the functional
impairment of the SFO was also given by the observation that lesioned animals were made completely refractory to the dipsogenic
action of ip Ang II, while being clearly sensitive to it preoperatively.

When tested on drinking induced by sc hypertonic NaCl, a statistically highly significant inhibition was elicited by dermorphin
only at doses clearly larger (20 ng or more per rat) than those effective on Ang II-induced drinking (Fig. 3).

Finally, in water deprived rats, dermorphin produced again an
antidipsogenic effect, but only at doses larger than those effective on drinking in response to Ang II. In fact, the doses of 1 and

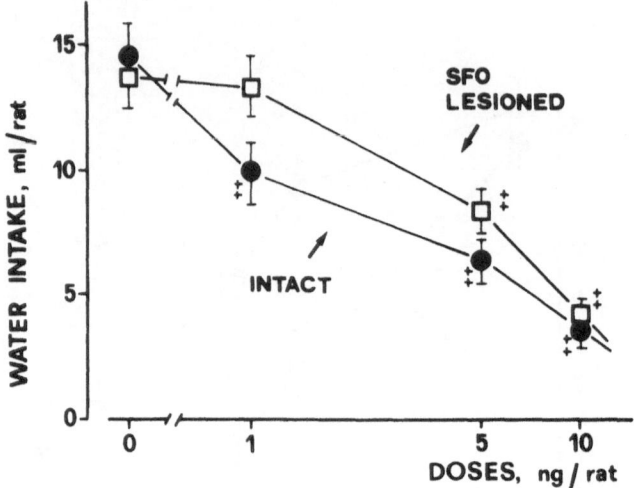

Fig. 2. Inhibition of Ang II-induced drinking following injection of dermorphin into the LV of intact or of SFO lesioned rats. Values, determined 15 min after Ang II injection are means±SEM of 12 data. ‡P < 0.01; where not indicated, difference from controls was not statistically significant.

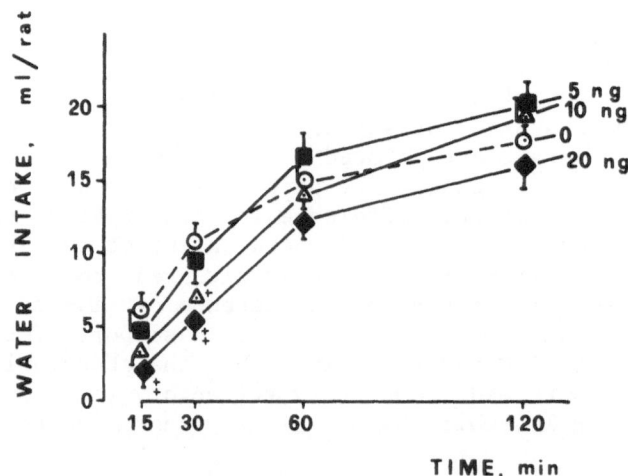

Fig. 3. Time course of drinking following sc hypertonic NaCl (8.76 g/100 ml; 1 ml/100 g b.w.) in controls (0) or in rats receiving different icv doses of dermorphin. Each point is mean±SEM of 15 data. ‡P < 0.01, +P < 0.05; where not indicated, difference from controls was not significant.

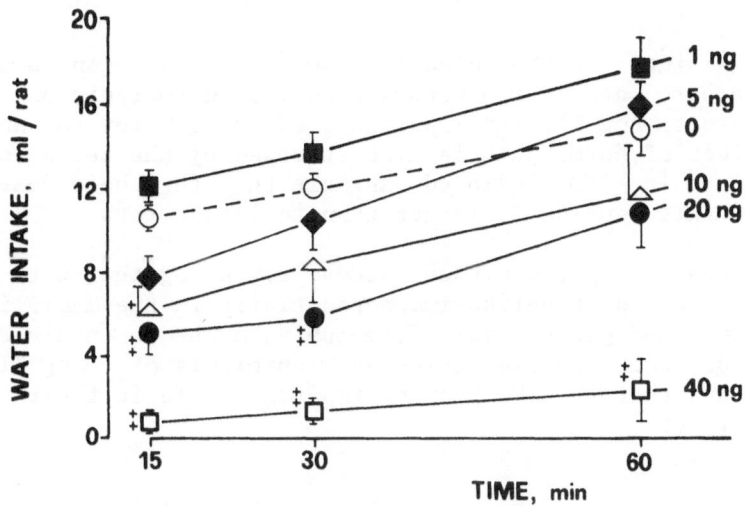

Fig. 4. Time course of drinking in response to water deprivation in controls (O) or in animals treated with different icv doses of dermorphin. Values are means±SEM of 8 data. ‡P < < 0.01, + P < 0.05; where not indicated, difference from controls was not statistically significant.

5 ng/rat, which produced a significant inhibition of drinking induced by Ang II, did not affect at all drinking in response to water deprivation and a highly significant inhibition was obtained only at the dose of 20 ng/rat (Fig. 4).

DISCUSSION

Present results confirm the potent inhibitory effect of icv dermorphin on drinking induced by icv Ang II in the rat, as already observed in a previous study[6]. Moreover, they indicate that this effect is specific at least up to 5 ng/rat, since at these dose levels drinking in response to other dipsogenic determinants (cell dehydration and water deprivation) was not affected and other behavioural alterations were not observed. On the other hand, cell dehydration- and water deprivation-induced drinking were inhibited by dermorphin only at high doses, at which the animals appeared to be sedated and at which the peptide is known to exert marked depressant effects . Therefore it seems likely that they were inhibited only as a consequence of other central depressant effects. These observations argue in favour of a selective antidipsogenic effect of dermorphin for drinking in response to Ang II.

Moreover, the results obtained following direct injection of

dermorphin into the SFO clearly show that this organ is a highly
sensitive site for the inhibitory effect of dermorphin on drinking
in response to Ang II. The importance of the SFO for the antidipsogenic effect of dermorphin is also stressed by the lesioning study,
showing that in SFO lesioned animals the threshold dose for the
effect of this peptide is larger than in intact rats.

The results of the present study, taken together with the occurrence of dermorphin-like immunoreactivity in the mammalian brain,
suggest that endogenous brain dermorphins might be involved in the
control of water intake acting as inhibitors of Ang II-induced
drinking and that the SFO play an important role in their antidipsogenic effect.

REFERENCES

1. P. C. Montecucchi, R. de Castiglione, S. Piani, L. Gozzini and
   V. Erspamer, Aminoacid composition and sequence of dermorphin, a novel opiate-like peptide from the skin of Phyllomedusa sauvagei, Int. J. Peptide Protein Res. 17:275 (1981).
2. P. C. Montecucchi, R. de Castiglione and V. Erspamer, identification of dermorphin and Hyp[6]-dermorphin in skin extracts of
   the Brazilian frog Phyllomedusa rhodei, Int. J. Peptide Protein Res. 17: 316 (1981).
3. L. Negri, P. Melchiorri, G. Falconieri Erspamer, V. Erspamer,
   Radioimmunoassay of dermorphin-like peptides in mammalian
   and nonmammalian tissues, Peptides 2:(suppl. 2) 45 (1981).
4. M. Broccardo, V. Erspamer, G. Falconieri Erspamer, G. Improta,
   G. Linari, P. Melchiorri and P. C. Montecucchi, Pharmacological data on dermorphins, a new class of potent opioid
   peptides from amphibian skin, Br. J. Pharmacol. 73:625 (1981).
5. V. Erspamer and P. Melchiorri, Actions of amphibian skin peptides on the central nervous system and the anterior pituitary, in:"Neuroendocrine Perspectives," E.E. Muller and R.
   M. MacLeod, eds., Elsevier Science Publishers B. V., Amsterdam (1983).
6. G. de Caro, M. Massi, L. G. Micossi and M. Perfumi, Effect of
   dermorphin and related peptides on drinking behaviour of the
   rat, in:"Central and peripheral Endorphins: basic and clinical aspects," E. E. Muller and A. R. Genazzani, eds., Raven
   Press, New York (1984).
7. R. Buffa, E. Solcia, E. Magnoni, G. Rindi, L. Negri and P. Melchiorri, immunohistochemical demonstration of dermorphin-
   -like peptide in the rat brain, Histochemistry 76:273 (1982).
8. J. B. Simpson, A. N. Epstein and J. S. Camardo, Jr, Localization of receptors for the dipsogenic action of Angiotensin
   II in the subfornical organ of rat, J. Comp. Physiol. Psychol. 92:581 (1978).
9. A. N. Epstein, J. T. Fitzsimons and B. J. Rolls, Drinking in-

duced by injection of angiotensin into the brain of the rat. J. Physiol. 210:457 (1970).

10. F.R. Konig and R.A. Klippel, "The rat brain: a stereotaxic atlas of the forebrain and lower parts of the brain stem", The Williams and Wilkins Company, Baltimore (1963).

PERIPHERAL MECHANISMS FOR THE MAINTENANCE AND TERMINATION OF

DRINKING IN THE RAT

Gerard P. Smith

Department of Psychiatry
The New York Hospital-Cornell Medical Center
White Plains, New York 10605

INTRODUCTION

The peripheral, physiological mechanisms for the maintenance
and termination of drinking have been difficult to investigate
because ingested water activates both mechanisms during a bout of
drinking. Thus, when drinking occurs, the time course and degree
of activation of the mechanism for maintaining drinking, and the
onset, time course, and degree of activation of the mechanism for
terminating drinking have not been measured separately. This is a
major obstacle to the development of an adequate theory of drinking behavior. The extent of this problem is made clear by asking,
"Does the administration of a dipsogen increase water intake by
stimulating the maintenance mechanism, inhibiting the termination
mechanism or both?" The answer, of course, is that no one knows.
Despite this ignorance, the work of the last decade has provided
techniques, observations, and ideas that I believe can be used to
answer these fundamental questions. I shall review the available
evidence for these peripheral mechanisms in the rat. See the paper
by Gibbs et al (this volume) and the books by Denton[1] and by Rolls
and Rolls[2] for relevant evidence for other species.

PERIPHERAL MECHANISMS FOR MAINTAINING DRINKING

When sham drinking is made possible by letting ingested water
drain out of an esophageal[3] or gastric fistula,[4] rats drink much
more than normal. Rats sham drank about 4 times as much water as
they really drank after the same interval of water deprivation.[4,5]
Thus, sham drinking activates the peripheral mechanism for maintaining drinking much more than the peripheral mechanism for terminating drinking. From this it is reasonable to conclude that

sensory feedback from the mouth and, possibly, the esophagus are important for maintaining drinking.

Despite their obvious importance, pregastric sites are not necessary for drinking to be maintained in its normal behavioral form of distinct bouts. Rats "ingest" water in bouts intravenously.[6] This intravenous "drinking" is mostly periprandial and is characterized by smaller bouts and by smaller total intakes over 24 hours than normal.[7] Rats also drink intragastrically,[8] but this is apparently very difficult for them to learn to do.[8,9,10] Thus, intravenous and intragastric "drinking" demonstrate the primacy of pregastric sites for maintaining normal drinking.

The adequate stimuli at pregastric sites are taste[11,12,13] and temperature.[14] Their potency is enhanced by water deprivation.[5] These peripheral stimuli are mediated by afferent fibers of the fifth, seventh, ninth and tenth cranial nerves. Selective lesions of the afferent fibers of the fifth nerve[15] or of the chorda tympani, glossopharyngeal, and pharyngeal branch of the vagus[16] decrease real drinking. Presumably, these lesions would also decrease sham drinking, but this has not been demonstrated. Thus, the relative importance of the afferent fibers of these four cranial nerves for maintaining sham drinking is not known. It is clear, however, that sham drinking does not depend on a peripheral cholinergic mechanism because atropine methylnitrate had no significant effect on it.[17]

The hyperdipsia of sham drinking not only indicates that the mouth is the primary site of the mechanism that maintains drinking, but the hyperdipsia also indicates that the normal termination of drinking in the rat depends upon the accumulation of water in the gut and/or its absorption into the circulation.

Two stimuli for terminating drinking decrease sham drinking. Sham drinking is inhibited by a preabsorptive gastric mechanism and by a postabsorptive osmosensitive mechanism.[18] Sham drinking, however, does not appear to be inhibited by the postabsorptive volume expansion produced by a preload of isotonic saline.[18]

PREABSORPTIVE MECHANISMS FOR TERMINATING DRINKING

The stomach is a preabsorptive site for terminating drinking. The best evidence for this is that obtained by Hall and Blass.[18] They used Hall's pyloric noose[19] to test the effectiveness of gastric stimulation by water to terminate sham drinking. When the noose was closed, the satiating effects of water at preabsorptive sites in the intestine or at postabsorptive sites were excluded because the noose prevented ingested water from emptying into the duodenum. When rats drank while the noose was closed, they drank much less than when they sham drank. This is evidence that the accumulation of water in the stomach activates a termination mechanism. But activation of this gastric mechanism is not sufficient to produce normal termination of drinking because rats drank significantly more when the pyloric noose was closed than when the pyloric noose was open.

Gastric vagal afferent fibers may mediate this inhibitory effect, but no one has done the experiment. I emphasize the lack of experimental evidence because in an apparently relevant situation, abdominal vagotomy had no effect[20] or only a slight effect[21] on the termination of eating when liquid food accumulated in the stomach because the pylorus was experimentally closed.

Hall and Blass observed that when the pylorus was closed, the length of fluid deprivation had no effect on water intake.[18] This is clearly abnormal and suggests that the noose is doing more than activating a normal gastric distention mechanism for terminating drinking. This would not be surprising because the pyloric sphincter is densely innervated and vascularized and, thus, the pressure of the noose could produce a number of side effects. The extent to which these side effects contribute to the termination of drinking needs to be clarified.

There is another caveat about these pyloric noose experiments. The termination of drinking when the pyloric noose is closed is not the result of gastric stimuli acting alone. It is an interaction of gastric and oral termination mechanisms. Although I stressed the effect of pregastric stimulation by water to maintain drinking, there is evidence that pregastric stimulation also activates an inhibitory mechanism. This follows from the fact that oral preloads of water inhibit drinking more than intragastric preloads of water.[22] Thus, it is important to note that no one has measured the relative contribution of the pregastric and gastric mechanisms to the termination of drinking that occurs when the pyloric noose is closed.

It is possible that water in the small intestine activates another preabsorptive mechanism that terminates drinking. This idea has not been tested in rats.

POSTABSORPTIVE MECHANISMS FOR THE TERMINATION OF DRINKING

Liver

The liver is the first organ ingested water contacts after it is absorbed into the portal circulation. There is evidence for a hepatic osmoreceptor mechanism that is vagally innervated.[23] In addition to the evidence that this osmoreceptor mechanism is involved in the neuroendocrine control of the renal excretion of water,[24] it may also be involved in the termination of drinking in the rat. After selective hepatic vagotomy, rats drank significantly more water after 17 hours of water deprivation than sham operated rats did.[25] This was a specific effect in the sense that selective gastric or ceoliac vagotomy had no effect on water intake after water deprivation.

But if this effect of hepatic vagotomy were the result of removing the afferent fibers that mediate the hepatic osmoreceptor activity, hepatic vagotomized rats should overdrink to other osmotic challenges such as hypertonic saline or ingested food. But this did not occur (see Section on Abdominal Vagotomy, Table 1). Thus, the importance of the hepatic osmoreceptor mechanism for the termination of drinking is not clear.

Heart

The right atrium is the other putative postabsorptive site for the termination of drinking in the rat. Distention of a small balloon at the confluence of the right superior vena cava and the right atrium inhibited spontaneous water intake, and water intake after 24 hours deprivation, polyethylene glycol (PEG), and isoprenaline, but not after hypertonic saline.[26]

If these effects of balloon distention are mediated through vagal stretch receptors, then cervical vagotomy should block the effect of volume expansion. Vance was the first to investigate the effect of cervical vagotomy on thirst in the rat. In 1970, he reported that right or left cervical vagotomy had no effect on water intake in response to PEG, or on water intake in the presence or absence of food.[27] After left cervical vagotomy, Zimmer et al[28] observed decreased drinking of water in response to hypertonic saline, but increased drinking of saline 24 hours after PEG in nephrectomized rats. The intake of saline was normal at 3.5 and 8 hours after PEG.

In 1980, Moore-Gillon[29] reported a more extensive investigation of the effects of unilateral cervical vagotomy on drinking. When rats were tested within a couple of days after left or right cervical vagotomy, the inhibitory effect of volume expansion by dextran on water intake after water deprivation was reduced. Left or right cervical vagotomy also increased water intake after hypertonic saline, isoprenaline, PEG, and sodium nitrite. Cervical vagotomy had no effect, however, on drinking of water elicited by PEG in nephrectomized rats or after ligation of the inferior vena cava or after diazoxide. Saline intake after these hypovolemic dipsogens was not tested.

Although the effect of cervical vagotomy on the drinking response to hypertonic saline is not clear from these studies, the results are converging evidence for a right atrial mechanism, sensitive to distention, that terminates drinking.

ABDOMINAL VAGOTOMY

It is possible that transection of abdominal vagal fibers travelling in the cervical vagus contributed to the increased drinking of water or saline observed after unilateral cervical vagotomy. Vance,[27] however, found no significant effect of unilateral abdominal vagotomy on drinking after hypertonic saline or on drinking in the presence or absence of food. But Zimmer et al[28] replicated the effect of left cervical vagotomy to increase saline intake 24 hours after PEG by bilateral abdominal vagotomy. In fact, saline intake after bilateral abdominal vagotomy was almost twice that observed after unilateral abdominal vagotomy. Moore-Gillon[29] failed to replicate any of his results with unilateral cervical vagotomy by bilateral abdominal vagotomy.

But given the presumed importance of abdominal vagal afferent fibers for hepatic and gastric mechanisms for terminating drinking (see above), it is puzzling that bilateral abdominal vagotomy had so little effect on drinking in these studies. In considering these mostly negative results of abdominal vagotomy, it is important to note that the pattern of branching of the abdominal vagal system is variable, the vagus regenerates, and all functional tests of completeness of vagotomy evaluate loss of efferent fibers to the stomach. Unfortunately, these fibers account for less than 10% of all the abdominal vagal fibers.[30] Thus, it is possible that the failure to see effects of abdominal vagotomy is due to the difficulty of making the lesion complete.

We became aware of these problems while trying to evaluate the effect of abdominal vagotomy on the satiety effect of gut peptides.[31,32] When we did the vagotomy under microscopic control and ligated the ends of the cut trunks to prevent regeneration and to facilitate anatomical verification, we discovered that rats had marked impairments in drinking after a number of dipsogens.[33] Bilateral abdominal vagotomy that left the hepatic branch intact decreased water intake in rats after hypertonic saline, isoproterenol, and PEG, and decreased 24-hour water intake in the presence and absence of food. These results were surprising. Given the traditional idea that the abdominal vagus was responsible for inhibition of drinking by water distending the stomach or perfusing the hepatic osmoreceptors, an increase of water intake was expected, not a decrease. Kraly et al[33] interpreted the results as evidence that abdominal vagotomy produced a motivational deficit for drinking because its effects transcended specific categories of dipsogens. The multiplicity of deficits, however, could be the result of non-specific debilitation.

Despite these different interpretations, the deficits were sufficiently large and unexpected to attract experimental attention. The result is that most of the deficits in drinking have been confirmed[25] and extended to the drinking response to angiotensin II.[34,35]

Given that bilateral abdominal vagotomy removes afferent and efferent fibers from all the abdominal organs, we reasoned that more selective changes in drinking might be obtained after selective lesions of the main branches of the abdominal vagal system. Under microscopic guidance, we were able to section the hepatic branch, the gastric branches, or the coeliac and accessory coeliac branches selectively and then test the drinking response to a variety of dipsogens. The results showed numerous changes in drinking that were a function of the dipsogenic condition and the vagal branch lesioned.[25,34,36,37]

After 17 hours of water deprivation, only rats with hepatic vagotomy drank different than normal -- they overdrank (Table 1). As discussed in the section on the Liver (see above), the increased drinking after hepatic vagotomy supports the existence of a hepatic osmoreceptor mechanism for inhibiting drinking that is vagally innervated. The normal drinking response to deprivation after total (bilateral) abdominal vagotomy or gastric and coeliac vagotomies undermines the suggestion that deficits in drinking after bilateral abdominal vagotomy are due to non-specific debilitation or mechanical abnormalities of swallowing or gastric emptying.

Note that when both the coeliac and hepatic vagal branches were cut, rats drank normally to water deprivation (Table 1). This suggests a functional interaction between the effects of ingested water mediated through the hepatic and coeliac branches. If the overdrinking produced by hepatic vagotomy is a result of disinhibition, then the normalization of drinking when the coeliac branch is also sectioned suggests a loss of excitation. The decreased drinking in response to angiotensin II and hypertonic saline after coeliac vagotomy is consistent with the loss of an excitatory function of the coeliac branch. But an excitatory function of the coeliac branch would not explain the increased water intake by coeliac and hepatic vagotomized rats during 2 hours of eating liquid food, but not solid food (Table 1).

It is clear that selective vagotomies produced less deficits than total abdominal vagotomy did (Table 1). The apparent exception is the significant decrease in water intake during 24 hours without food in gastric vagotomized rats, but not in total vagotomized rats. This exception should be interpreted cautiously because Kraly et al[33] reported that total vagotomized rats drank significantly less in this condition. When we attempted to replicate this effect, total vagotomized rats drank less than sham, but the difference was not statistically significant.[25]

Table 1. Effect of Dipsogens After Total or Selective Abdominal Vagotomies

Vagal lesion	24 h water without food	Hypertonic saline	Angiotensin	Water deprivation	24 h water to solid food ratio	2 h water to liquid food ratio
Total	Normal	↓	↓	Normal	↓	Normal
Gastric	↓	↓	↓	Normal	↓	Normal
Coeliac	Normal	↓	↓	Normal	Normal	Normal
Coeliac-hepatic	Normal	Normal	Not tested	Normal	Normal	↑
Hepatic	Normal	Normal	Normal	↑	Normal	Normal

↓ significantly decreased; ↑ significantly increased compared to sham vagotomized rats (from reference 25 with permission of the publisher)

Given the importance of hypertonic saline and angiotensin II to the development of current theories of thirst, the decreased drinking to these dipsogens after gastric or coeliac vagotomy is significant. Although both vagotomies produced less drinking after both dipsogens, gastric vagotomy produced a larger drinking deficit in response to hypertonic saline than to angiotensin II.[34,36] There also appears to be a different neurological mechanism underlying the deficient drinking in response to hypertonic saline and to angiotensin II because the deficit to angiotensin II appears within the first 48 hours after total vagotomy,[38] but the deficit in drinking to hypertonic saline takes at least 1 week to develop.[39]

Thus, with the exception of hepatic vagotomized rats drinking after water deprivation, and combined coeliac and hepatic vagotomized rats drinking in response to eating liquid food, the most common abnormality in drinking produced by total or selective abdominal vagotomy is decreased drinking (Table 1). Note that gastric vagotomy which removes the hypothetical negative feedback from the stomach never resulted in overdrinking. The varied and large effects of selective vagotomies did not clarify the effects of total abdominal vagotomy. Instead, they multiplied the phenomena to be explained. Whatever their meaning, they certainly indicate the importance of the abdominal vagal system for normal drinking.

Four aspects of these results with abdominal vagotomy require comment. First, the relative importance of the loss of afferent and efferent fibers after the total or selective vagal lesions is not known. The failure of peripheral and anticholinergic blockade to reproduce these deficits[40,41] suggests that the loss of afferent fibers is the critical lesion. The availability of a new surgical technique[42] for selectively lesioning afferent or efferent vagal rootlets will facilitate the analysis of this problem.

Second, the effects of selective abdominal vagotomies complicate the interpretation of the effects of cervical vagotomies. For example, left cervical vagotomy lesions afferent and efferent fibers distributed in the hepatic, right gastric, and accessory coeliac abdominal branches.[43] Right cervical vagotomy lesions afferent and efferent fibers distributed in the left gastric and coeliac abdominal branches.[43] The fact that either cervical vagotomy leads to overdrinking[29] probably indicates that the effect of denervating the atrial inhibitory mechanism dominates the effects of the abdominal vagal lesions. But it is clear that a more specific denervation of the atrial mechanism is required to test this inference.

Third, the atrial distention mechanism could be responsible, at least in part, for the decreased drinking frequently observed after total or selective abdominal vagotomy. Abdominal vagotomy removes abdominal afferent fibers, but leaves the thoracic afferents from the atrium unopposed. If at least some of the abdominal vagal

272

afferents stimulated drinking, their removal would produce a net increase in inhibition. This increase in inhibition would be further exaggerated in the central vagal terminal fields in the nucleus tractus solitarius if the vagal terminal fields underwent reorganization to enhance the effect of surviving (thoracic) afferent terminals. Such a phenomenon has been observed in the dorsal horn after section of the sciatic nerve with a time course that is similar to the development of the decreased drinking after hypertonic saline.[44]

Fourth, although the increases and decreases of water intake after total or selective vagotomies have been interpreted according to Jacksonian principles as loss of inhibition or excitation respectively, this has not been rigorously proven. To interpret an increase of water intake as the result of loss of inhibition is reasonable, but a decrease in water intake can result from a decrease in the potency of maintenance mechanisms and/or an increase in the potency of termination mechanisms. For example, Kraly[45] suggested that the decreased water intake elicited by hypertonic saline after abdominal vagotomy that left the hepatic branch intact is due to increased potency of termination mechanisms. Much more work is required to evaluate this particular suggestion and the problem in general.

SUMMARY

The mouth is the major site for the maintenance of drinking. The large and prolonged intakes of water during sham drinking is strong evidence for this. The mechanisms that are activated by water in the mouth, pharynx, and esophagus probably involve afferent fibers in the fifth, seventh, ninth, and tenth cranial nerves, but there is no experimental evidence for the relative importance of these nerves to sham drinking.

Since drinking terminates prior to the removal of systemic osmotic or volemic dipsogenic stimuli, the receptor sites for the stimuli that initiate thirst are not the sites of the mechanisms for terminating drinking. There are two preabsorptive sites for terminating drinking -- the mouth and the stomach. Two postabsorptive sites may also contribute to the termination of drinking; they are the liver and the atria of the heart. But I emphasize the lack of evidence for the relative potency of these sites for terminating drinking under any dipsogenic condition.

Vagal afferent fibers have been suggested to mediate the terminating stimuli from the stomach, liver, and atria. There is some evidence for vagal mediation of hepatic and atrial termination of drinking, but none for the role of the vagus in the gastric termination of drinking. Selective gastric vagotomy decreases drinking under some conditions or has no effect, but gastric vagotomy has never produced excessive drinking.

273

Despite repeated suggestions in the literature that the role of the vagus nerve is simply to inhibit drinking, the recent results of total or selective abdominal vagotomies reveal a number of experimental situations where abdominal vagotomy produced a decrease in drinking. The effects of vagotomy are not only more varied than expected, but they are also more specific because the change in drinking produced is a function of the branch of the abdominal vagus lesioned and the dipsogenic stimulus that is operating.

The main impression obtained from reviewing the relevant literature is ignorance. There is only sketchy information about the sites of the peripheral mechanisms for the maintenance and termination of drinking and less than that about the mechanisms themselves. This extensive ignorance about mechanisms that are activated every time drinking occurs is a major impediment to an adequate theory of drinking behavior.

ACKNOWLEDGEMENTS

I thank Mrs. Marion Jacobson and Mrs. Jane Magnetti for typing this manuscript. The writing of the manuscript and the work from my laboratory cited in it were supported by Research Scientist Award MH00149 and Research Award MH15455 from the National Institute of Mental Health.

REFERENCES

1. D. A. Denton, The Hunger for Salt, Springer-Verlag, New York,(1984).
2. B. J. Rolls and E. T. Rolls, Thirst, Cambridge University Press, New York,(1982).
3. D. G. Mook, Oral and postingestional determinants of the intake of various solutions in rats with oesophageal fistulas. J. comp. physiol. Psychol. 56:645-659 (1963).
4. E. M. Blass, R. Jobaris, and W. G. Hall, Oropharyngeal control of drinking in rats. J. comp. physiol. Psychol. 90:909-916 (1976).
5. E. M. Blass and W. G. Hall, Drinking termination: interactions among hydrational, orogastric, and behavioral controls in rats. Psychol. Rev. 83:356-374 (1976).
6. S. Nicolaïdis and N. Rowland, Long-term self-intravenous 'drinking' in the rat. J. comp. physiol. Psychol. 87: 1-15 (1974).
7. N. Rowland and S. Nicolaïdis, Periprandial self-intravenous 'drinking' in the rat. J. comp. physiol. Psychol. 87: 16-25 (1974).
8. A. N. Epstein, Water intake without the act of drinking. Science 131:497-498 (1960).

9. G. L. Holman, Intragastric reinforcement effect. _J. comp. physiol. Psychol._ 69:432-441 (1968).

10. A. Altar and H.J. Carlisle, Intragastric drinking in the rat: evidence for a role of oropharyngeal stimulation. _Physiol. Behav._ 22:1221-1225 (1979).

11. T. Ernits and J. D. Corbit, Taste as a dipsogenic stimulus. _J. comp. physiol. Psychol._ 83:27-31 (1973).

12. D. G. Mook and N. J. Kenney, Taste modulation of fluid intake, _in_: Drinking Behavior: Oral Stimulation, Reinforcement, and Preference. J.A.W.M. Weijnen and J. Mendelson, eds., Plenum Press, New York (1977).

13. L. M. Bartoshuk, Water taste in mammals, _in_: Drinking Behavior: Oral Stimulation, Reinforcement and Preference. J.A.W.M. Weijnen and J. Mendelson, eds., Plenum Press, New York (1977).

14. R. M. Gold, G. Kapatos, J. Prowse, P. M. Quackenbush, and T. W. Oxford, Role of water temperature in the regulation of water intake. _J. comp. physiol. Psychol._ 85: 52-63 (1973).

15. M. F. Jacquin and H.P. Zeigler, Trigeminal orosensation and ingestive behavior in the rat. _Behav. Neuroscience_, 97:62-97 (1983).

16. M. F. Jacquin, Gustation and ingestive behavior in the rat. _Behav. Neuroscience_, 97:98-109 (1983).

17. D. Lorenz, P. Nardi and G.P. Smith, Atropine methyl nitrate inhibits sham feeding in the rat, _Pharmacol. Biochem. Behav._ 8:405-407 (1978).

18. W. G. Hall and E. M. Blass, Orogastric determinants of drinking in rats: interaction between absorptive and peripheral controls. _J. comp. physiol. Psychol._ 91: 365-373 (1977).

19. W. G. Hall, A remote stomach clamp to evaluate oral and gastric controls of drinking in the rat. _Physiol. Behav._ 11:897-901 (1973).

20. F. S. Kraly and J. Gibbs, Vagotomy fails to block the satiating effect of food in the stomach. _Physiol. Behav._ 24:1007-1010 (1980).

21. M. F. Gonzalez and J. A. Deutsch, Vagotomy abolishes cues of satiety produced by gastric distention. _Science._ 212:1283-1284 (1981).

22. N. E. Miller, R. I. Sampliner and P. Woodrow, Thirst reducing effects of water by stomach fistula versus water by mouth, measured by both a consummatory and an instrumental response. _J. comp. physiol. Psychol._ 50: 1-5 (1957).

23. A. Adachi, A. Niijima and H. L. Jacobs, An hepatic osmo-receptor mechanism in the rat: electrophysiological and behavioral studies. Amer. J. Physiol. 231:1043-1049 (1976).

24. R. C. Rogers and D. Novin, The neurological aspects of hepatic osmoregulation, in: The Kidney in Liver Disease, 2nd Edition, M. Epstien ed., Elsevier, New York (1983).

25. G. P. Smith and C. Jerome, Effects of total and selective abdominal vagotomies on water intake in rats. J. Auton. Nerv. System. 9:259-271 (1983).

26. S. Kaufman, Role of right atrial receptors in the control of drinking in the rat. J. Physiol. (London) 349: 389-396 (1984).

27. W. B. Vance, The effects of vagotomy on the water intake of the white rat. Psychonomic Sci. 20:21-22 (1970).

28. L. J. Zimmer, L. Meliza and S. Hsiao, Effects of cervical and subdiaphragmatic vagotomy on osmotic and volemic thirst. Physiol. Behav. 16:665-670 (1976).

29. M. J. Moore-Gillon, Effects of vagotomy on drinking in the rat. J. Physiol. 308:417-426 (1980).

30. E. Agostoni, J. E. Chinnock, M. DeBurgh Daly and J. G. Murray, Functional and histological studies of the vagus nerve and its branches to the heart, lungs and abdominal viscera in the cat. J. Physiol. 135: 182-205 (1957).

31. G. P. Smith, C. Jerome, B.J. Cushin, R. Eterno and K. J. Simansky, Abdominal vagotomy blocks the satiety effect of cholecystokinin in the rat. Science 213: 1036-1037 (1981).

32. G. P. Smith, C. Jerome and J. Gibbs, Abdominal vagotomy does not block the satiety effect of bombesin in the rat. Peptides 2:409-411 (1981).

33. F. S. Kraly, J. Gibbs and G. P. Smith, Disordered drinking after abdominal vagotomy in rats. Nature 258:226-228 (1975).

34. C. Jerome and G.P. Smith, Gastric or coeliac vagotomy decreases drinking after peripheral angiotensin II. Physiol. Behav. 29:533-536 (1982).

35. N. Rowland, Impaired drinking to angiotensin II after subdiaphragmatic vagotomy in rats. Physiol. Behav. 24:1177-1180 (1980).

36. C. Jerome and G.P. Smith, Gastric vagotomy inhibits drinking after hypertonic saline. Physiol. Behav. 28:371-374 (1982).

37. M. G. Tordoff and D. Novin, Coeliac vagotomy attenuates the ingestive responses to epinephrine and hypertonic saline but not insulin, 2-deoxy-D-glucose or poly-ethylene glycol. Physiol. Behav. 29:605-613 (1982).

276

38. K. J. Simansky and G. P. Smith, Acute abdominal vagotomy reduces drinking to peripheral but not central angiotensin II. Peptides 4:159-163 (1983).

39. C. Jerome and G.P. Smith, Development of the drinking deficit to hypertonic saline in rats after abdominal vagotomy. Physiol. Behav. 32:819-821 (1984).

40. E. M. Blass and H.W. Chapman, An evaluation of the contribution of cholinergic mechanisms to thirst. Physiol. Behav. 7:679-686 (1971).

41. D. DeWied, Effect of autonomic blocking agents and structurally related substances on the 'salt arousal of drinking.' Physiol. Behav. 1:193-197 (1966).

42. G. P. Smith, C. Jerome and R. Norgren, Vagal afferent axons mediate the satiety effect of CCK-8. Soc. Neurosci. Abstracts 9:902 (1983).

43. R. Norgren and G. P. Smith, The central distribution of vagal subdiaphragmatic branches in the rat. Soc. Neurosci. Abstracts 9:611 (1983).

44. P. D. Wall, Alterations in the central nervous system after deafferentation: connectivity control. in: Proc. of the Third World Congress on Pain, Adv. in Pain Research and Therapy, Vol. 5, J. J. Bonica, U. Lindblom and A. Iggo eds., Raven Press, New York (1983).

45. F. S. Kraly, Abdominal vagotomy inhibits osmotically induced drinking in the rat. J. comp. physiol. Psychol. 92:999-1013 (1978).

DISTURBANCES IN WATER BALANCE CONTROLS FOLLOWING LESIONS TO THE

AREA POSTREMA AND ADJACENT SOLITARY NUCLEUS

Richard R. Miselis, Thomas M. Hyde and Robert E. Shapiro

Animal Biology, Institute of Neurological Sciences
School of Veterinary Medicine and School of Medicine
University of Pennsylvania, Philadelphia, Pa. 19104

THE AP/cmNTS LESION AND SYNDROME

Lesions which remove the area postrema (AP) and the subjacent portions of the nucleus of the solitary tract (cmNTS) which lie in the caudal brainstem close to the dorsal spinomedullary junction cause dramatic and apparently permanent alterations in energy and fluid balance[1,2]. There is a well characterized syndrome of transient hypophagia and accompanying weight loss. Two to three weeks into this syndrome normophagia resumes with eventual stabilization of body weight but at a lower level. In addition there is a mild hyperdipsia. See figure 1.

The lesion removes the area postrema which is a circumventricular organ (CVO) of the brain and as such lacks a blood brain barrier. For this reason many investigators have suggested that CVOs are sites for putative central receptors. We suspect that they must subserve visceral function providing the brain with information on the status of the internal milieu derived from parameters circulating in the blood. In addition the neighboring portions of the nucleus of the solitary tract are involved in the lesion. More specifically, this includes portions of the commissural and medial subnuclei of the NTS which are relay nuclei for peripheral sensory afferents from the viscera.

NEURAL CONNECTIVITY OF THE AP/cmNTS

When one considers the neural connections of this area of the brain, it is immediately apparent that the lesion has much wider ranging effects than initially indicated by the very small size of

279

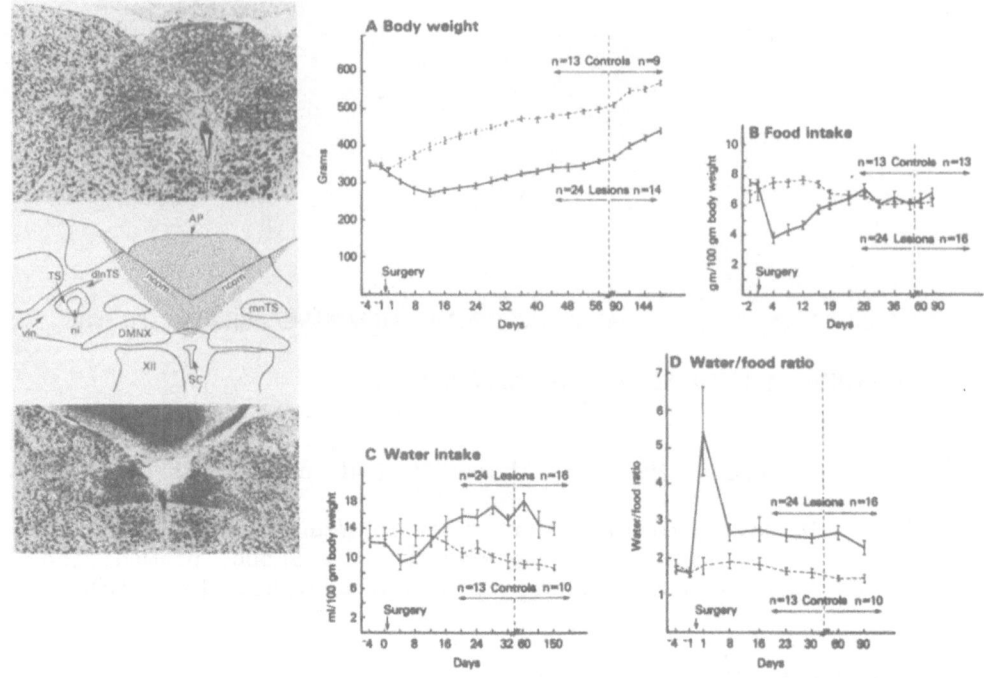

Fig. 1.   Left side. The AP/cmNTS lesion. (A) Photomicrograph of
the normal dorsal medulla of the rat.   (B) Labelled
schematic drawing of the same area.   (C) Photomicrograph of the dorsal medulla of a lesioned rat.   The
shaded area in (B) indicates the missing tissue.
Abbreviations: AP = area postrema, TS = tractus solitarius, DMNX = dorsal motor nucleus of the vagus, XII =
Hypoglossal nucleus, mnNTS = medial subnucleus of the
nucleus of the solitary tract, ncom = commissural subnucleus of NTS, dlnTS = dorsal lateral subnucleus of
NTS, vln = ventral lateral subnucleus of NTS, ni =
subnucleus intersitialis of NTS, and sc = spinal canal.
Right side.   Alterations in body weight control and
ingestive behavior as a result of the AP/cmNTS lesion
in male rats.   (A) Body weight. (B) Food intake. (C)
Water intake. (D) Water/food ratios.

the damaged area. First, we know from tracing studies in which the
entire subdiaphragmatic vagus nerve is dipped into free horseradish
peroxidase (HRP) that this portion of the NTS recives a heavy input
from abdomenal, visceral, sensory fibers. A small portion of these
fibers terminate within the area postrema as well[3-5].   The terminal fields are especially heavy in the commissural subnucleus
underlying the area postrema and in the subnucleus gelatinosus
which is just anterior and slightly lateral to the area postrema.

Abdominal viscera (HRP)   Stomach (CT-HRP)

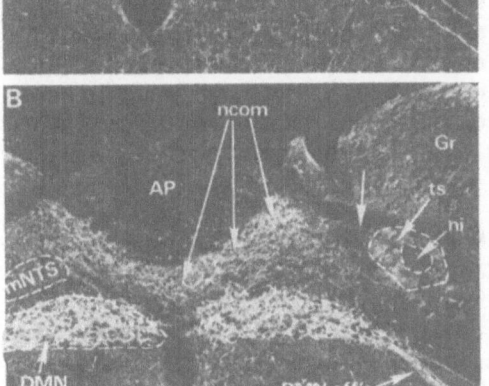
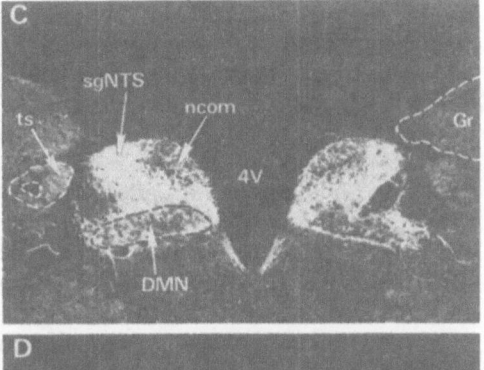
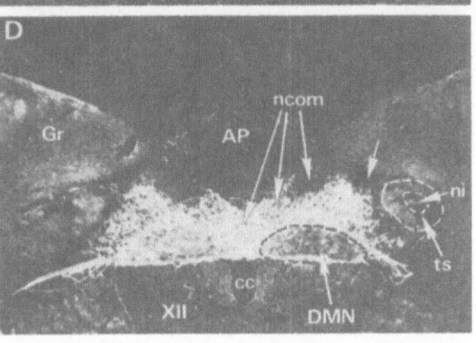

Fig. 2.   Organization of the representation of the abdominal
           viscera in the dorsal medulla.   (A) and (B) represent
           the sensory and motor labelling seen in the dorsal
           medulla when the subdiaphragmatic vagus nerve branches
           are dipped in free horseradish peroxidase (HRP).   (A)
           is just anterior to the AP. (B) is at the level of the
           AP.   (C) and (D) represent labelling seen in the same
           areas of the dorsal medulla when only the stomach is
           injected with the tracer cholera toxin conjugated to
           horseradish peroxidase (CT–HRP).   The greatest density
           of afferent terminals is in sgNTS. Note that much of
           the fiber labelling seen outside of the DMN in C and D
           is of dendrites of motor neurons of the DMN.   The
           unlabelled arrows in B and C point to the dorsal
           lateral subnucleus which is unlabelled and is likely to
           get its input from the thorax. Abbreviations: AP = area
           postrema, cc = central canal, DMN = dorsal motor
           nucleus of the vagus, DMN eff's = efferent axons of the
           DMN motor neurons, Gr =.gracile nucleus, mNTS = the
           medial subnucleus of the nucleus of the solitary tract,
           ncom = commissural subnucleus of NTS, ni = subnucleus
           interstitialis, ts = tractus solitarius, 4V = fourth
           ventricle, XII = hypoglossal nucleus.

See figure 2. When only the stomach is injected with a tracer (in this case, a conjugate of HRP with cholera toxin), the terminal fields are now primarily in the gelatinosus subnucleus[6-8]. Therefore, the terminals in the area postrema and subjacent commissural subnucleus are primarily from abdomenal viscera other than the stomach. We also observed that the motorneurons of the dorsal motor nucleus have extensive dendrites which extend up into particular regions of the overlying NTS[7,8]. See figure 2.

We know that neurons of the area postrema project heavily into the neighboring NTS and lightly into the dorsal motor nucleus of the vagus. In addition the area postrema and subjacent NTS send a strong efferent projection up to specific subnuclei of the lateral parabrachial nucleus which is a second order relay nucleus for ascending visceral input. There are very few afferent projections into the area postrema from within the brain with one major exception. This is an afferent projection from parts of particular subnuclei of the paraventricular and dorsomedial nuclei of the hypothalamus. See figure 3.

The obvious conclusion from these tracing studies is that the AP/cmNTS lesion, by destroying the particular connectivity of this small area of the brain, has far reaching effects within the peripheral and central nervous system. Secondly, the area postrema's major efferent projections are to ascending visceral sensory relay nuclei indicating that, whatever the area postrema might monitor in the blood, it influences brain function by modulating entering and ascending peripheral visceral information and autonomic outflow.

FLUID BALANCE DISTURBANCES

As seen in figure 1 the AP/cmNTS lesion has immediate and dramatic effects on ingestive behavior. Further analysis of the fluid balance disturbances indicates that the mild polydipsia is secondary to a primary polyuria because the polyuria still persists when water and food are withheld overnight. During the deprivation period the rats do concentrate their urine and reduce volume losses indicating a capability for volume conservation; however, the urine osmolality is much less and the urine volume lost is greater than in the control animals. This is true regardless of when the deprivation challenge is done following the lesion. See figure 4. These rats have an increased salt appetite which is also secondary to renal losses[2].

The drinking responses to the above long term challenge (24 hrs. deprivation) appear to be normal because the excessive drinking is balanced by renal losses[2]. In reponse to an acute intracellular dehydration challenge the lesioned rats do initially over drink with no concurrent excessive renal losses. After

A·AP injection site (sagittal plane)

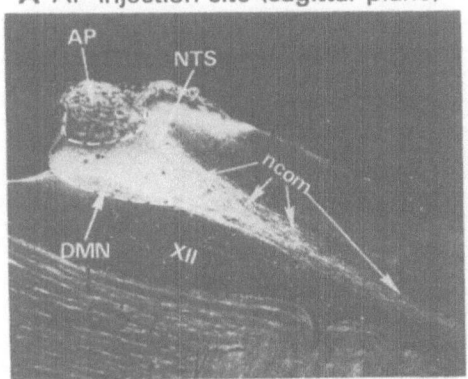

C·PVN (horizontal plane)

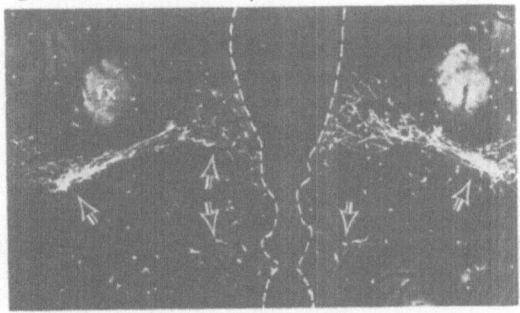

B·Parabrachial (coronal plane)

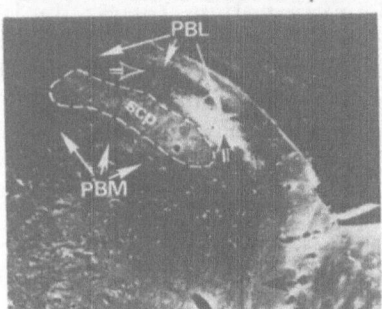

D·Dorsal raphe (sagittal plane)

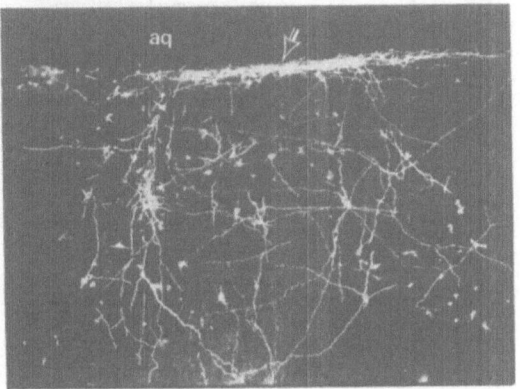

Fig. 3.   The neural connectivity of the area postrema seen with
            injections of CT–HRP into the AP. (A) The injection
            site within the AP and efferent terminal field label-
            ling within the ncomm of NTS and within the DMN. (B)
            Efferent terminal field labelling within the lateral
            parabrachial nucleus (black arrow) at the level of the
            pons. (C) Retrogradely filled neurons (black arrows)
            within the paraventricular nucleus of the hypothalamus.
            (D) Retrogradely filled neurons within the dorsal raphe
            of the midbrain. The black arrow points to the
            labelled axons. It is likely that these neurons were
            labelled by tracer which spilled into the fourth ven-
            tricle. Abbreviations: AP = area postrema, aq =
            aquaduct, DMN = dorsal motor nucleus of the vagus, fx =
            column of the fornix, ncom = commissural subnucleus of
            NTS, NTS = nucleus of the solitary tract, PBL = lateral
            parabrachial nucleus, PBM = medial parabrachial
            nucleus, scp = superior cerebellar peduncle.

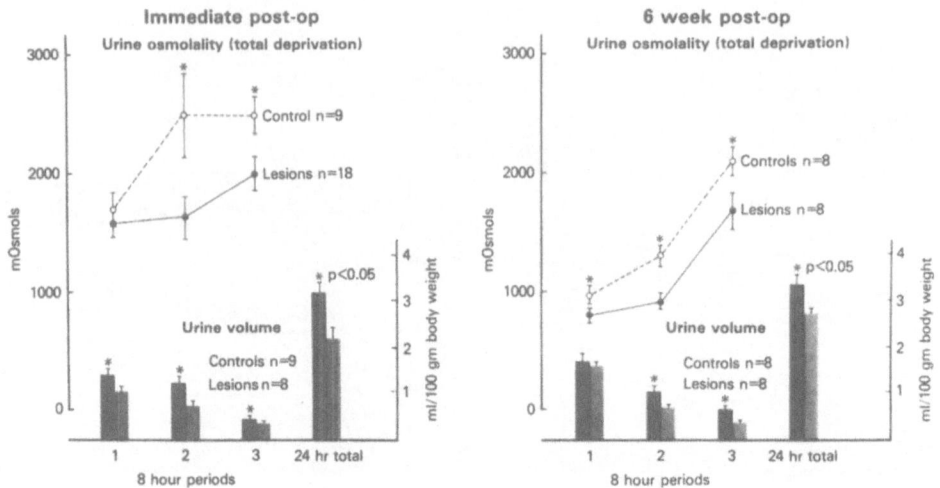

Fig. 4   Urine   osmolalities   and urine volume losses during   24
           hours   of food and water deprivation   done   immediately
           after   the   AP/cmNTS   lesion and again 6   weeks   later.
           Lesioned animals loose more urine and have lower   urine
           osmolalities than sham operated control rats.

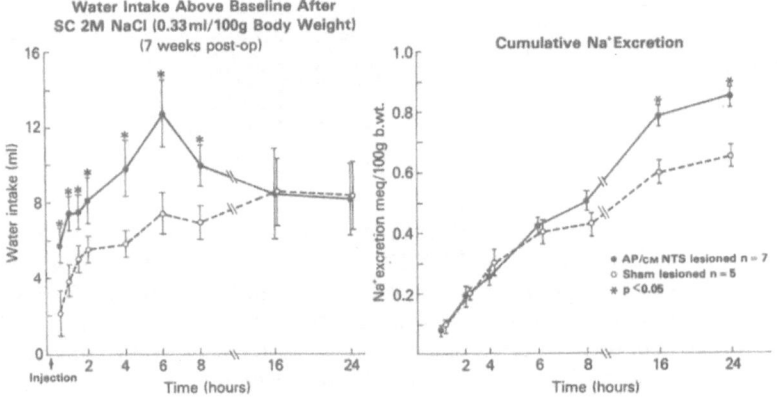

Fig. 5.   Water   intakes   and   renal sodium   excretion   following
            intracellular   dehydration   produced   by   subcutaneous
            (SC)   hypertonic   saline.   The water intakes   are   the
            differences   in drinking between the hypertonic   saline
            injection   day   and   a   baseline   intake   day   for   the
            lesioned and sham rats.

eight hours the lesioned rats begin to compensate for the excessive
drinking by reducing it more than the controls while the renal
volume losses continue. The excessive drinking cannot be explained

by inordinate sodium retention. The rate of sodium excretion is the same as in controls during the first eight hours after which the sodium losses become excessive. See figure figure 5.

It is difficult to interpret the consequences of the AP/cmNTS lesion on fluid balance. The alterations are characterized by excessive behavioral responses in the short term followed by compensation over 24 hours. This is dramatically illustrated in the response to intracellular dehydration. A finer analysis is required in which blood parameters, renal function and drinking patterns are more closely followed over the course of the 24 hour response period. The neuroanatomical studies suggest an alteration in the inflow of visceral sensory information because the lesion removes the area postrema which provides central visceral information as well as modulation of peripheral visceral sensory inflow. In addition there is the loss of peripheral sensory afferents which terminate within the area of the lesion. Other aspects of this lesion remain to be explored. The consequences on cardiovascular and gastrointestinal function need to be investigated in regard to altered sensory input and motor outflow.

REFERENCES

1. T. M. Hyde and R. R. Miselis, Effects of area postrema/caudal medial nucleus of the solitary tract lesions on food intake and body weight, Am. J. Physiol. 244:R577 (1983).
2. T. M. Hyde and R. R. Miselis, Area postrema and adjacent nucleus of the solitary tract in water and sodium balance, Am. J. Physiol. 247:R173 (1984).
3. R. Eng, R. E. Shapiro and R. R. Miselis, Vagal afferents to the area postrema/caudal medial nucleus of the solitary tract important for food intake and body weight: an HRP study, Neurosci. Abstr. 8:273 (1982).
4. R. Norgren and G. P. Smith, The central distribution of vagal subdiagphragmatic branches in the rat, Neurosci. Abstr. 9:611 (1983).
5. M. Kalia and M-M. Mesulam, Brain stem projections of sensory and motor components of the vagus complex in the cat: II. Laryngeal, tracheobronchial, pulmonary, cardiac and gastrointestinal branches, J. Comp. Neurol. 193:467 (1980).
6. D. G. Gwyn, R. A. Leslie and D. A. Hopkins, Gastric afferents to the nucleus of the solitary tract in the cat, Neurosci. Lett. 14:13 (1979).
7. R. E. Shapiro and R. R. Miselis, Organization of gastric efferent and afferent projections within the dorsal medulla oblongatta in the rat, Anat. Rec. 205:182A (1983).
8. R. R. Miselis and R. E. Shapiro, Dorsal motor nucleus neurons have extensive dendrites penetrating the nucleus of the solitary tract, Fed. Proc. 42:1125 (1983).

PREABSORPTIVE AND POSTABSORPTIVE FACTORS IN THE TERMINATION OF

DRINKING IN THE RHESUS MONKEY

James Gibbs*, Barbara J. Rolls**, and Edmund T. Rolls***

*Dept. of Psych., The N.Y. Hospital-Cornell Medical Ctr.
White Plains, N.Y. ** Dept. of Psych., Johns Hopkins Univ.
School of Med. Baltimore, MD ***Dept. of Experimental
Psych., The Univ. of Oxford, Oxford OX1 3UD, UK

INTRODUCTION

The physiological signals which terminate water intake are not well understood. It is possible that animals and humans may employ preabsorptive signals (generated by water acting in the mouth, esophagus, stomach, or small intestine), postabsorptive signals (generated by water acting in the hepatic portal system or in other extracellular or intracellular compartments), or some combination of preabsorptive and postabsorptive signals in order to recognize the appropriate moment to stop drinking. It is possible that the specific signals employed to stop drinking which has been generated by different means (water deprivation, food intake, or hormonal and pharmacological agents) may be different. Finally, of course, it is possible that different species utilize entirely different signals, or place differential emphasis on elements within the same array of physiological signals.[1-9]

We have begun to study the mechanisms which terminate drinking in the rhesus monkey when drinking occurs after a period of total water deprivation. We have paid special attention to the roles of preabsorptive events.

These studies were supported by the National Institute of Mental Health of the United States Public Health Service, and the Medical Research Council of Great Britain. The talents and energies of S. Maddison and R.J. Wood were instrumental in the experiments reported here.

METHODS

Five young adult male rhesus monkeys (<u>Macaca mulatta</u>), 3.1-4.9 kg body weight, were used as subjects. At surgery, each animal was provided with three chronic devices: (1) a gastric cannula, which could be temporarily opened at a test to allow the drainage and complete recovery of all ingested water (gastric sham drinking), (2) a duodenal cannula, positioned 2 cm distal to the pylorus, which could also be opened to allow recovery of ingested water (duodenal sham drinking) or, alternatively, the infusion of fluids directly into the duodenal lumen, and (3) an intravenous catheter, implanted in the external jugular vein, with its distal end advanced to the right atrium; this catheter allowed the intravenous infusion of fluids and frequent blood sampling for measurements of plasma volume and concentration.

After a 22-hour water deprivation during which solid food was available, each monkey was placed in a primate chair within a sound-insulated cubicle for a single daily 60-min test, during which water was presented via a drinking spout. In addition to water intake, and depending on the aims of the particular experiment, we could measure: adequacy of gastric or duodenal drainage, behavioral observations, volumes of fluids infused intraduodenally or intravenously (always at the rate of 5 $ml-min^{-1}$, and always starting at the moment the drinking spout was presented), and plasma osmolality, sodium, and protein concentration. Following the test, each monkey was given a 30-min period of water access, then removed from the primate chair and returned to its individual cage. For further methodological details and a complete account of all aspects of these experiments, see references 10, 11, and 12.

RESULTS AND DISCUSSION

Gastric and Duodenal Sham Drinking

On days when both gastric and duodenal cannulas were closed (essentially normal drinking), monkeys rapidly drank a mean of 119 ml within the first 10 min, at a rate which sharply decreased after the first 5 min (see Figure 1). After the 10-min point, they stopped drinking significant amounts in single bouts, then usually appeared drowsy or asleep. Mean total intake during the 60-min period was 137 ml.

When only gastric cannulas were open, monkeys drank almost continually throughout the test. This type of response occurred the first time gastric cannulas were ever opened, and on all subsequent occasions. Mean total intake was 878 ml ($p < 0.01$ compared to the cannulas closed condition -- see Figure 1). When only the

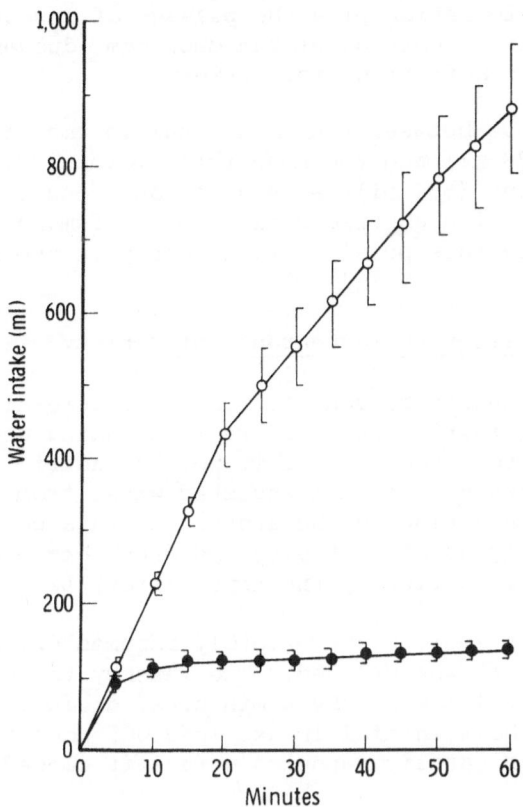

Fig. 1.   Cumulative water intake (mean ± SE) by five monkeys
after water deprivation. When both cannulas were
closed (solid symbols), ingested water remained
within the gut and drinking was normal. When the
gastric cannulas were open (sham drinking -- open
symbols), all ingested water drained immediately to
the outside, was collected and measured.

duodenal cannules were open, sham drinking was very similar to
that seen when only gastric cannulas were open, except that the
rate was not as marked. Mean total intake was 634 ml (p<0.01
compared to the cannulas closed condition).

These results demonstrate that the stimulation of oropharyngeal
and esophageal receptors by ingested water (gastric cannula open
condition) is not sufficient for satiety. Furthermore, oropharyngeal

and esophageal stimulation plus the passage of ingested water through the stomach and first portion of the duodenum (duodenal cannula open condition) are not sufficient for satiety.

The difference between the mean total volume drunk during gastric sham drinking (878 ml) and the mean total volume drunk during duodenal sham drinking (634 ml) was significant, and it suggested the partial operation of a gastric satiety mechanism in the latter condition. We tested this possibility directly in the next experiment.

Gastric Drainage after the Appearance of Satiety

With both cannulas closed, four monkeys were allowed to re-hydrate until the initial phase of rapid drinking ended (8.0 ± 1.1 min after water was presented). At this point, gastric cannulas were opened to allow drainage of all ingested water from the stomach. The mean volume remaining in the stomach at this point was 95 ml; thus, approximately 24 ml of the volume drunk had emptied from stomach to small intestine by the time satiety first appeared.

We found that every monkey rapidly resumed drinking on every occasion the stomach was drained. The latency to resumption of drinking was 2.7 ± 1.1 min. By 8 min after drainage, monkeys had re-drunk 40% of their initial intake ($p < 0.005$ compared to volume drunk on days when gastric cannulas were left closed).

The experiment shows that the presence of water in the stomach is necessary for the termination of drinking under these conditions. The anatomical location of the mechanism for this termination is not made clear by this experiment. The reason monkeys resume drinking following gastric drainage might be the absence of water in the stomach, activating a gastric satiety mechanism directly, or the absence of water emptied from stomach to small intestine, resulting in the secondary failure of an intestinal satiety signal.

To begin to differentiate between these two possibilities, we evaluated the satiating potency of water infused directly into the small intestine during gastric sham drinking.

Intestinal Infusions during Gastric Sham Drinking

The left side of Figure 2 demonstrates the effects of duodenal infusions of isotonic saline or water on gastric sham drinking, and on plasma sodium concentration (as an index of plasma dilution). Duodenal delivery of 100 ml of isotonic saline had little effect on the initial rapid rate of sham drinking (and these results were no different from sham drinking when no infusion was given -- not shown in Figure 2). Thus, a large volume of isotonic saline in the intestine, and any consequent distention, are not sufficient to alter sham drinking.

290

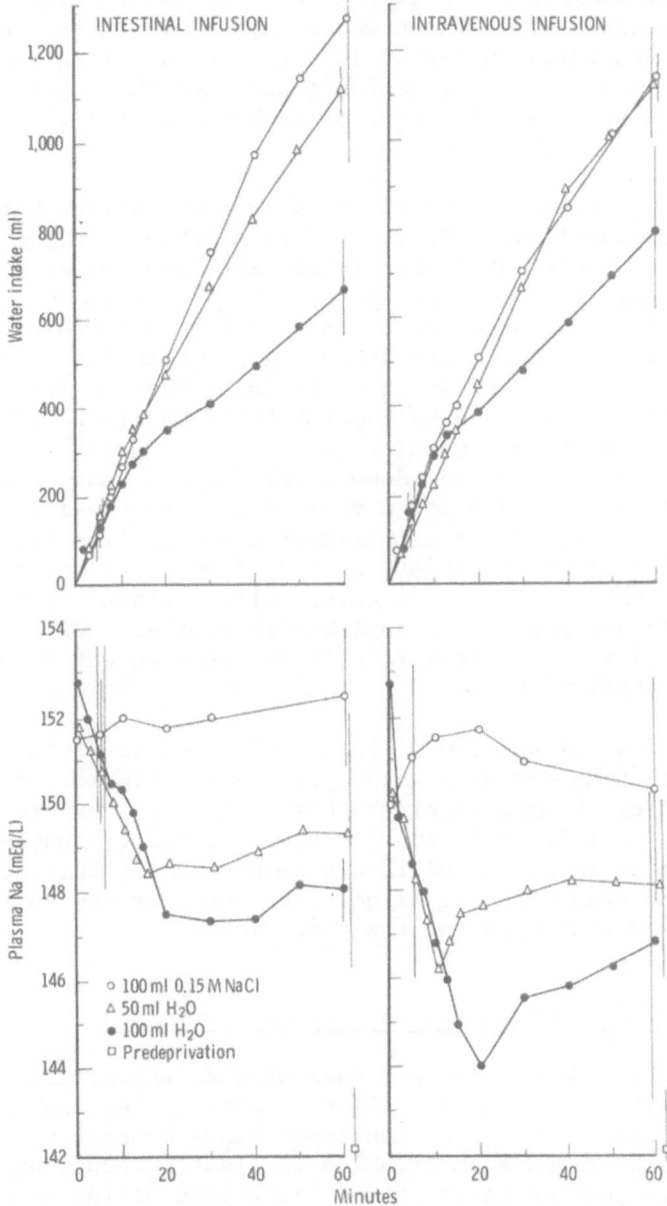

Fig. 2. Cumulative water intakes and concurrent changes in
plasma sodium concentration (mean ± SEM) in four
monkeys sham drinking with open gastric cannulas.
At the moment water was presented (0 min), infusions
of water or 0.15 M NaCl were begun. The left-hand
panels depict the behavioral and physiological changes

seen when infusions were delivered into the duodenum.
The right-hand panels depict the changes seen when
infusions were delivered into·the right atrium.
The rate of infusion was always 5 ml-min^{-1}: the
50 ml infusion lasted 10 min, the 100 ml infusion lasted
20 min. Pre-deprivation plasma sodium concentrations
are noted at the right of each lower panel.

Duodenal delivery of 50 ml of water had a slight but statistically insignificant effect on sham drinking; 100 ml of water
had a larger and statistically significant ($p < 0.05$) effect
(Figure 2, upper left panel). Note, however, that even this large
volume of intraduodenal water did not produce a sustained display
of satiety, such as that seen in the closed cannulas condition of
Figure 1. It would be of interest to determine how large a volume
of water must be delivered to the intestine to elicit such a satiety response. In this regard, recall that the gastric drainage
experiment (described above) demonstrated that, under normal conditions, only 24 ml had emptied from stomach to small intestine
at the time a sustained satiety response began. The striking difference in the powerful satiety action of normally ingested water
and the weak satiety action of water which contacts only the
oropharyngeal and intestinal surfaces is another indication that
the stomach plays a critical role in terminating water intake
under these conditions.

Both the 50 ml and 100 ml infusions of water produced changes
in plasma dilution measures which were rather similar, but they
produced different behavioral changes (Figure 2, compare upper and
lower panels on left-hand side). This discrepancy suggests that
plasma dilution is not a sufficient mechanism for the decrease in
sham drinking seen with the larger infusion. We tested this suggestion more directly in the next experiment.

Intravenous Infusions during Gastric Sham Drinking

The right side of Figure 2 demonstrates the effects of intravenous infusions of isotonic saline or water. The inability of
intravenous saline (and the resulting plasma expansion -- not
shown) to alter rapid sham drinking is clear. Infusions of 50 ml
of water were just as ineffective. Infusions of 100 ml of water
produced a significant suppression of sham drinking; however,
comparison of Figure 2 (upper right panel) with the closed cannulas
condition seen in Figure 1 demonstrates the marked weakness of this
large volume of intravenous water compared to the satiating potency
of water which remains in the gut.

Finally, compare the right-hand side of Figure 2 with the left-
hand side. Note that equivolumetric infusions of water were, for

both the 50 ml and 100 ml volumes, more effective in suppressing
sham drinking when delivered into the intestine than when delivered
into the vein (two upper panels). Moreover, for both volumes, the
greater satiating effectiveness of intestinal water was achieved in
the face of a greater plasma diluting capacity of intravenous water
(two lower panels). These observations strongly suggest the existence of an intestinal (and/or hepatic portal) factor which contributes to satiety under these conditions.

INTERPRETATIONS AND CONCLUSIONS

1. Oropharyngeal stimulation by ingested water is not sufficient to terminate water intake in the water-deprived
   rhesus monkey. Nevertheless, oropharyngeal stimulation
   may be necessary for, or contributory to, the termination
   of drinking.

2. Two observations suggest the existence of a gastric satiety
   factor: (1) the fact that a considerably greater volume
   is consumed when the gastric cannula alone is open than
   when the duodenal cannula alone is open, and (2) the fact
   that a large volume of water infused directly into the
   duodenum during sham drinking has a weak satiating effect,
   but a similar volume of water which is not removed from
   the gut following ingestion has a potent satiating
   effect.

   The large and reliable 'restorative' drinking (which rapidly
   follows the removal of gastric water after normal satiety
   has occurred) is direct evidence that water must accumulate
   in the stomach for satiety to appear.

3. The greater reduction in drinking which follows intraduodenal
   delivery of water, as compared to intravenous delivery of
   the same volume into the systemic circulation, indicates the
   existence of an intestinal or hepatic portal satiety mechanism.

4. The evidence that systemic plasma changes which follow water
   absorption are a sufficient condition for the onset of
   satiety is very weak. Nevertheless, such systemic changes
   may play an increasingly important role in maintaining
   satiety over time.

Generalizing from these findings, we suggest that satiety for
water intake is the end result of a cascade of physiological events
which evolve as water acts sequentially at multiple levels from mouth
to systemic circulation. If this generalization has any validity,
further experiments will reveal the precise identities of these

unknown physiological events, their interplay, and the times at which each is called into action to terminate drinking behavior and maintain satiety for water.

REFERENCES

1. E.F. Adolph, Measurements of water drinking in dogs, Am. J. Physiol. 125:75-86 (1939).
2. R.T. Bellows, Time factors in water drinking in dogs, Am. J. Physiol. 125:87-97 (1939).
3. E.J. Towbin, Gastric distention as a factor in the satiation of thirst in esophagostomized dogs, Am. J. Physiol. 159:533-541 (1949).
4. A.N. Epstein, Water intake without the act of drinking, Science 131:497-498 (1960).
5. E. Bott, D.A. Denton, and A. Weller, Water drinking in sheep with oesophageal fistulae, J. Physiol. 176:323-336 (1965).
6. S. Nicolaïdis and N. Rowland, Long-term self-intravenous drinking in the rat, J. Comp. Physiol. Psychol. 87:1-15 (1974).
7. E.M. Blass and W.G. Hall, Drinking termination: interactions among hydrational, orogastric, and behavioral controls in rats, Psychol. Rev. 83:356-374 (1976).
8. D.J. Ramsay, B.J. Rolls, and R.J. Wood, Thirst following water deprivation in dogs, Am. J. Physiol. 232:R93-R100 (1977).
9. D.J. Ramsay, B.J. Rolls, and R.J. Wood, Body fluid changes which influence drinking in the water-deprived rat, J. Physiol. 266:453-469 (1977).
10. S. Maddison, R.J. Wood, E.T. Rolls, B.J. Rolls, and J. Gibbs, Drinking in the rhesus monkey: peripheral factors, J. Comp. Physiol. Psychol. 94:365-374 (1980).
11. S. Maddison, B.J. Rolls, E.T. Rolls, and R.J. Wood, The role of gastric factors in drinking termination in the monkey, J. Physiol. 305:73P (1980).
12. R.J. Wood, S. Maddison, E.T. Rolls, B.J. Rolls, and J. Gibbs, Drinking in rhesus monkeys: roles of presystemic and systemic factors in control of drinking, J. Comp. Physiol. Psychol. 94:1135-1148 (1980).

294

HISTAMINE PLAYS A ROLE IN DRINKING ELICITED BY EATING IN THE RAT

F. Scott Kraly

Department of Psychology
Colgate University
Hamilton, NY 13346

Eating elicits the vagally-mediated[1] release of histamine from endocrinelike non-mast cells[2] in the gastric mucosa of the rat.[3] Exogenous histamine elicits drinking in the rat.[4,5] To examine whether eating elicits drinking through a histaminergic mechanism, a pharmacological probe was developed to block drinking elicited by systemic histamine in male Sprague-Dawley rats.

PROBE FOR A HISTAMINERGIC COMPONENT

Both the H_1 antagonist dexbrompheniramine (DXB) and the H_2 antagonist cimetidine (C) inhibit drinking elicited by s.c. histamine; the combination of DXB plus C can abolish drinking elicited by all doses of s.c. histamine which are normally dipsogenic in the rat.[6] Selection of the proper doses of these antagonists establishes a probe specific for histamine-elicited drinking. For example, the histamine probe (1 mg/kg DXB plus 16 mg/kg C i.p.) (a) abolishes drinking elicited by s.c. histamine,[6] but (b) fails to inhibit drinking elicited by varying degrees of water deprivation[6,7,8] or by s.c. serotonin[7] in the rat. Thus, this pharmacological treatment specifically abolishes drinking elicited by exogenous systemic histamine.

DRINKING ELICITED BY NORMAL EATING

Assuming this probe also blocks dipsogenic effects of endogenous systemic histamine, the administration of the probe prior to a meal should reveal whether histamine plays a role in drinking elicited by eating.

Eating and Drinking after Food Deprivation

The histamine probe inhibits by about 25 percent drinking elicited by a meal of solid chow for rats eating after 24-h food deprivation.[9] Larger doses of H_2 antagonist alone (i.e., given without DXB) also inhibit drinking by about 25 percent for rats eating either solid chow or liquid food after food deprivation.[10] Even greater percentage inhibition of food-related drinking can be achieved with larger doses of H_1 and/or H_2 antagonists, but because larger doses (unlike the histamine probe) can nonspecifically suppress ingestion of water and food, such results are difficult to interpret.[9,10]

Spontaneous Eating and Drinking

The histamine probe alters the pattern of drinking in nondeprived rats spontaneously taking meals during the dark phase of a 12:12-h light/dark cycle: It (a) abolishes drinking prior to a meal in those few rats which exhibit such drinking; (b) delays the latency to initiate drinking after eating has begun; and (c) inhibits by over 60 percent drinking during and within 30 min after a meal.[8] These effects on drinking occur without effects on meal size, eating rate and latency to rest after eating.

DRINKING ELICITED BY PREGASTRIC FOOD-CONTINGENT STIMULATION

Preabsorptive food-contingent stimulation of the pregastric (oropharynx and esophagus) segment of the gastrointestinal tract appears to be sufficient for the vagally-mediated release of gastric mucosal histamine in the rat.[3,11,12,13] Thus, sham feeding in the rat with open gastric fistula should activate a histaminergic mechanism for food-related drinking.

Sham feeding elicits drinking in the rat.[9] Vagal efferents are important for this phenomenon, because abdominal vagotomy or peripheral cholinergic blockade using atropine methyl nitrate i.p. inhibits such drinking.[14] These effects of vagotomy and cholinergic blockade are separable from their inhibitory effects on sham feeding.[14]

The histamine probe abolishes drinking elicited by sham feeding (Table 1) without inhibiting sham feeding.[9,14] While drinking elicited by sham feeding is abolished by combined H_1 and H_2 antagonism, such drinking appears to be unaffected by H_1 antagonism alone and is merely attenuated by a dose of H_2 antagonist which should abolish H_2-mediated gastric secretory activity.[14] These results suggest synergy between H_1 and H_2 receptors for mediating the dipsogenic effect of endogenous histamine. This is consistent with the synergistic relation of these receptors for mediating the effects of

Table 1. Mean ($\pm$ SE) 60-min Water Intake per Food-
Contingent Stimulation (ml/ml)

	.9% NaCl	DXB + C	p
Pregastric (n=7)	.22 $\pm$ .06	.01 $\pm$ .01	<.02
Post-pregastric (n=16)	.22 $\pm$ .08	.28 $\pm$ .11	>.20

exogenous histamine on drinking[6] and on vasodilation of submucosal arterioles in the stomach of the rat.[15]

Sham feeding also elicits the vagally-mediated[16,17] release of insulin from the pancreas of the rat.[13,18] Should this endogenous insulin play a part in drinking elicited by sham feeding (or normal eating), one would expect insulin's dipsogenic effect to be mediated by histamine receptors, because histaminergic receptor antagonism abolishes drinking elicited by sham feeding which releases both insulin and histamine. Consistent with this notion is the finding that antagonism of peripheral histamine receptors blocks drinking elicited by exogenous insulin in the rat.[7] This latter finding may be explained by the ability of exogenous insulin to elicit the release of gastric mucosal histamine in the rat.[19]

DRINKING ELICITED BY POST-PREGASTRIC FOOD-CONTINGENT STIMULATION

Pregastric food-contingent stimulation elicits drinking which can be abolished by histaminergic antagonism. Whether food-contingent stimulation of post-pregastric sites depends on histamine receptors to elicit drinking has been evaluated by preliminary work in our laboratory: The histamine probe fails to inhibit drinking elicited by intragastric infusion of 5 ml of liquid food in the rat (Table 1). Thus, it appears that post-pregastric stimulation is not necessary while pregastric stimulation is sufficient to fully activate the peripheral histaminergic mechanism for drinking elicited by eating.

CONCLUSION

These results support the following working hypothesis: Preabsorptive stimulation by food in the pregastric segment of the gastrointestinal tract elicits the vagally-mediated release of pancreatic insulin and gastric mucosal histamine. Insulin contributes to drinking either by activating peripheral receptors for histamine or by eliciting further release of histamine from the gastric mucosa. Histamine contributes to drinking by activating gastric vagal afferents[20] and peripheral histamine receptors.

There are three reasons why this working hypothesis considers peripheral rather than brain histamine: First, since intravenous C^{14}-labeled histamine is not found in brain,[21] it is assumed that the dipsogenic effects of systemic exogenous or endogenous histamine depends on activating peripheral and not central histamine receptors. Second, since the H_2-receptor antagonists used in this work should fail to cross blood-brain barrier,[22] their antidipsogenic effects should occur through peripheral receptors. Third, drinking elicited by s.c. histamine is abolished by manipulations that should not block central receptors for histamine: Selective gastric vagotomy abolishes drinking in response to low doses of s.c. histamine, while such vagotomy combined with the angiotensin-converting enzyme inhibitor SQ14,225 (Captopril) abolishes drinking after higher doses of s.c. histamine.[20]

Finally, the identification of a preabsorptive pregastric mechanism as a component of drinking around mealtime, together with the demonstration of this mechanism's dependence upon peripheral histamine receptors, is one step toward a systematic experimental analysis of the peripheral neuroendocrine control of drinking around mealtime. Future work should seek direct evidence for the release of gastric mucosal histamine during sham feeding, examine the relation between release of histamine and drinking behavior, and consider separately the roles of nonhistaminergic gastric and intestinal preabsorptive and postabsorptive mechanisms[23] for drinking elicited by eating.

REFERENCES

1. M.A. Beaven, A. Horakova, W.B. Severs, and B.B. Brodie, Selective labeling of histamine in rat gastric mucosa: Application to measurement of turnover rate. J. Pharmacol. Exper. Ther. 161: 320 (1968)
2. A.H. Soll, K.J. Lewin, and M.A. Beaven, Isolation of histamine-containing cells from rat gastric mucosa: Biochemical and morphologic differences from mast cells. Gastroent. 80: 717 (1981)
3. G. Kahlson, E. Rosengren, and R. Thunberg, Accelerated mobilization and formation of histamine in the gastric mucosa evoked by vagal excitation. J. Physiol. 190: 455 (1967)
4. Y. Gutman and M. Krausz, Drinking induced by dextran and histamine: Relation to kidneys and renin. Eur. J. Pharmacol. 23: 256 (1973)
5. S.F. Leibowitz, Histamine: A stimulatory effect on drinking behavior in the rat. Brain Res. 63: 440 (1973)
6. F.S. Kraly, A probe for a histaminergic component of drinking in the rat. Physiol. Behav. 31: 229 (1983)
7. F.S. Kraly, L.A. Miller, and E.S. Hecht, Histaminergic mechanism for drinking elicited by insulin in the rat. Physiol. Behav. 31: 233 (1983)
8. F.S. Kraly and S.M. Specht, Histamine plays a major role for drinking elicited by spontaneous eating in the rat. Physiol. Behav. 33: in press (1984)

298

9. F.S. Kraly, Histamine plays a part in induction of drinking by food intake. Nature 302: 65 (1983)

10. F.S. Kraly and K.R. June, A vagally mediated histaminergic component of food-related drinking in the rat. J. Comp. Physiol. Psych. 96: 89 (1982)

11. A.K. Ganguly and P. Gopinath, Effect of stimulation of vagus nerves on gastric tissue histamine concentration in albino rats. Experientia 35: 54 (1979)

12. G. Kahlson, E. Rosengren, D. Svahn, and R. Thunberg, Mobilization and formation of histamine in the gastric mucosa as related to acid secretion. J. Physiol. 174: 400 (1964)

13. H.R. Berthoud and A. Jeanrenaud, Sham feeding-induced cephalic phase insulin release in the rat. Am. J. Physiol. 242: E280 (1982)

14. F.S. Kraly, Preabsorptive pregastric vagally mediated histaminergic component of drinking elicited by eating in the rat. Behav. Neurosci. 98: 349 (1984)

15. P.H. Guth, T.L. Moler, and E. Smith, H_1 and H_2 histamine receptors in rat gastric submucosal arterioles. Microvasc. Res. 19: 320 (1980)

16. H.R. Berthoud, A. Niijima, J.F. Sauter, and B. Jeanrenaud, Evidence for a role of the gastric, coeliac and hepatic branches in vagally stimulated insulin secretion in the rat. J. Auton. Nerv. Syst. 7: 97 (1983)

17. E. Ionescu, F. Rohner-Jeanrenaud, H.R. Berthoud, and B. Jeanrenaud, Increases in plasma insulin levels in response to electrified stimulation of the dorsal motor nucleus of the vagus. Endocrinology 112: 904 (1983)

18. H.R. Berthoud, E.R. Trimble, and A.J. Moody, Lack of gastric inhibitory peptide (GIP) response to vagal stimulation in the rat. Peptides 3: 907 (1982)

19. M. Ekelund, R. Hakanson, J. Hedenbro, G. Liedberg, I. Lunquist, J.F. Rehfeld, and F. Sundler, Effects of insulin on serum gastrin concentrations, gastric acid secretion and histamine mobilization in the rat. Acta Physiol. Scand. 114: 17 (1982)

20. F.S. Kraly and L.A. Miller, Histamine-elicited drinking is dependent upon gastric vagal afferents and peripheral angiotensin II in the rat. Physiol. Behav. 28: 841 (1982)

21. S.H. Snyder, J. Axelrod, and H. Bauter, The fate of C^{14}-histamine in animal tissues. J. Pharmacol. Exper. Ther. 144: 373 (1964)

22. R.W. Brimblecombe, W.A.M. Duncan, G.J. Durant, C.R. Ganellin, G.B. Leslie, and M.E. Parsons, Characterization and development of cimetidine as a histamine H_2-receptor antagonist. Gastroent. 74: 339 (1978)

23. F.S. Kraly, The physiology of drinking elicited by eating. Psych. Rev. 91: in press (1984)

SATIETY AND THE EFFECTS OF WATER INTAKE ON

VASOPRESSIN SECRETION

David J. Ramsay and Terry N. Thrasher

Physiology Department
University of California, San Francisco
San Francisco, CA 94143

Two stimuli have been identified in the control of drinking which follows a period of water deprivation. When animals are deprived of water, but not food, extracellular fluid volume decreases and extracellular fluid osmolality rises. In dehydrated dogs, removal of the volume deficit reduces subsequent drinking by 27%, whereas removal of the central osmotic stimulus decreases water intake by 72%.[1] Although the proportions may vary, other species show volume and osmotic components in dehydration induced thirst[2,3]

When dehydrated dogs are allowed access to water, a volume of fluid approximately equal to the deficit is consumed within 5 min[4]. Under these circumstances, there is insufficient time for significant absorption of the ingested fluid. Indeed, Ramsay et al[1] have shown that plasma osmolality and sodium concentration did not show significant dilution in dehydrated dogs until 10 minutes following the onset of drinking, whereas drinking was completed in 2.5 min. Similarly, plasma protein concentration and hemocrit did not fall until 12.5 min. In these experiments, however, the volume and composition of the extracellular fluid returned to normal some 50 minutes later. Thus an adequate quantity of water was consumed in 2.5 min to correct body fluid composition some 50 min later. In the dog, as with other animals who drink rapidly, oropharyngeal and gastric mechanisms have been proposed to explain the phenomenon.[5,6] Permanent satiety can only occur when the volume and composition of the body fluids is returned to normal following complete absorption of the ingested fluid. The mechanisms which account for rapid satiety, and the associated effects on vasopressin secretion, will be examined in this chapter.

OROPHARYNGEAL CUES AND RAPID SATIETY

In experiments carried out in our laboratory, a group of dogs were prepared with gastric fistulae.[7] The fistula was made from nylon, and had a removable plug. With the plug removed, ingested fluid immediately drained from the stomach, and was not absorbed. When the plug was in place, ingested fluid passed through the pyloric sphincter, and was absorbed normally in the intestine. On separate occasions, dehydrated dogs were offered water with either the fistula open or closed. On another occasion, a volume of water equal to that drunk spontaneously was tubed directly into the stomach via the fistula. Access to water was offered again 1 hour later to test the degree of permanent satiety. The experiments were also carried out giving artificial extracellular fluid rather than water.

The results are shown in fig. 1. In confirmation of earlier work, intact dogs drank a volume of water equal to their deficits within 5 mins. Sixty minutes later, satiety was permanent, as shown by the lack of drinking when offered water at that time. When artificial extracellular fluid was offered to intact dehydrated dogs, the volumes drunk were very similar to spontaneous intake of water. Thus rapid satiety did not depend upon water being offered. However, when access to water was allowed 1 hour later, approximately 60% of the spontaneous intake was consumed. Although artificial extracellular fluid was as effective as water in causing rapid satiety, only the volume component of permanent satiety followed absorption of the ingested fluids.

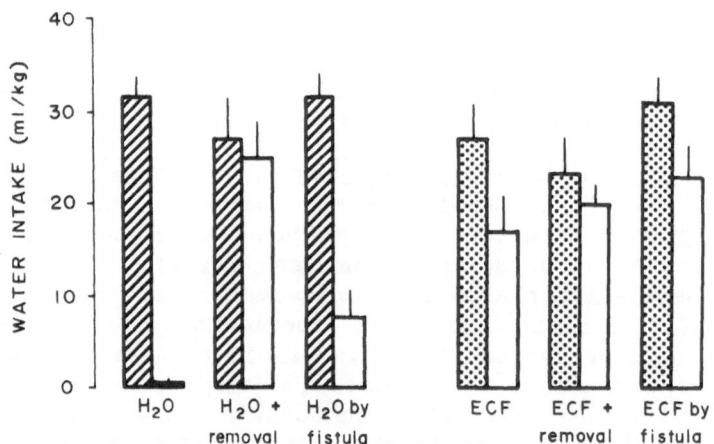

Fig. 1 Fluid consumption after 24-h water deprivation [hatched - water or stippled ECF] and 60 min later (open bars. From Thrasher et al (7) with permission

Of further interest were the results of ingestion of fluid with the gastric fistula open. Independent of the nature of the fluid offered, volumes equal to that of spontaneous intake were ingested. Thus rapid satiety must depend on oropharyngeal cues. However, such rapid satiety is only temporary, as shown by the volumes of water ingested when offered 60 min later. These results are different quantitatively from those reported in dehydrated dogs with esophageal fistulae by Bellows (1939) where twice the calculated fluid deficit was consumed. The nature of the discrepancy is not clear, but may be related to the gastric rather than the esophageal fistula preparation.

These results do not support the participation of hepatic osmoreceptors in satiety [8]. Dogs drank equal volumes of water and artificial extracellular fluid. Moreover, satiety occurred with similar volumes of ingested fluid independent of whether the gastric fistula was open or closed. It is difficult to assign an essential role to hepatic osmoreceptors in the satiety mechanism under these circumstances.

The residual drinking which is present 60 min following placement of the water deficit directly into the stomach is intriguing. Presumably the loss of oropharyngeal cues for satiety can still be detected at a time when permanent satiety is usually present.

RAPID SATIETY AND VASOPRESSIN SECRETION

During dehydration, hypovolaemia and hyperosmolality interact to increase vasopressin secretion, in a similar manner to the stimulation of thirst [9]. However, the results in fig. 2 demonstrate the remarkable similarity between rapid satiety and inhibition of vasopressin secretion in dehydrated dogs. Plasma vsopressin concentration fell dramatically following the onset of drinking, and was significantly reduced at 6 min, whereas plasma osmolality did not fall until 12 min. As the half-life of vasopressin in the dog is approximately 6 min [10], secretion of vasopressin must have ceased during water drinking in order to account for the rate of fall in its plasma concentration.

The similarity between inhibition of vasopressin secretion and rapid satiety are striking. Rapid reductions in plasma vasopressin are present when artificial extracellular fluid rather than water is ingested, or when either fluid is drunk with the gastric fistula open. In contrast, direct gastric loading of fluid has no rapid effect on plasma vasopressin concentrations. Presumably oropharyngeal cues are important in the inhibition of vasopressin secretion, as they are in rapid satiety. Neurophysiological evidence which shows that placing water on the tongue [11] or drinking [12] inhibits firing in some hypothalamic and supraoptic neurones is

ORAL WATER

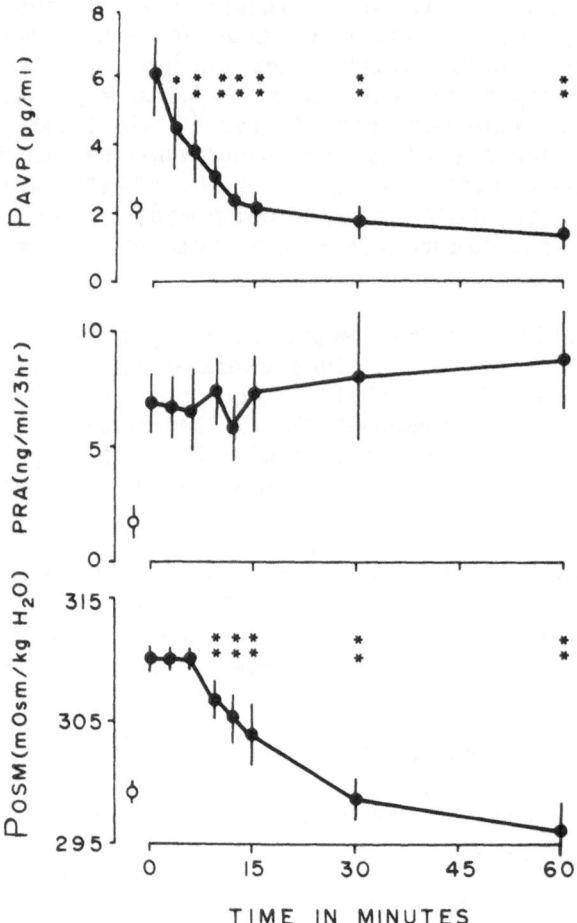

Fig. 2 Voluntary rehydration with water at time 0
in 24-h water deprived dogs. Open circles -
normally hydrated dogs. Closed circles -
dehydrated dogs. (Data from Thrasher et al, 7).

certainly compatible with our findings. For the reasons stated previously, these experiments do not support the involvement of hepatic osmoreceptors in the inhibition of vasopressin secretion.

The effect of drinking seems to be specific to the inhibition of vasopressin. Plasma ACTH levels do not fall following water ingestion by dehydrated dogs . These experiments also showed that plasma renin activity remains constant and drinking causes increased renin secretion in dehydrated sheep [14] . Plasma aldosterone concentrations increase following rehydration in both sheep and dogs. The association between drinking and inhibition of vasopressin secretion has also been reported in monkeys [15] and man. [16]

EFFECT OF BLOOD PRESSURE

The act of drinking is known to be associated with increases in arterial blood pressure.[17] Thus a possible mechanism for both satiety and inhibition of vasopressin could be the increase in blood pressure. This possiblity was examined by testing the effects of increasing blood pressure in dehydrated dogs with phenylephrine on subsequent drinking and plasma vasopressin.

The results of these experiments are shown in fig. 3. Dehydrated dogs, as other species, show a marked rise in arterial blood pressure and heart rate during drinking. When this increase in blood pressure was reproduced with a bolus injection of phenylephrine (between 5-15 microg/Kg iv), vasopressin secretion was not inhibited. Furthermore, when water was offered, there was no inhibition of water intake, and plasma vasopressin concentration fell rapidly.

In a second series of experiments, a soft food mixture was offered to dehydrated dogs, rather than water. The dogs ate the food avidly, approximately matching the swallowing movements associated with drinking. Blood pressure and heart rate increased. However, plasma vasopressin remained constant. Moreover, when water was offered, normal volumes were consumed, and rapid reductions in plasma vasopressin followed. These experiments demonstrate clearly that drinking fluid is necessary to inhibit vasopressin secretion, and that the increase in blood pressure associated with drinking does not provide the inhibitory stimulus.

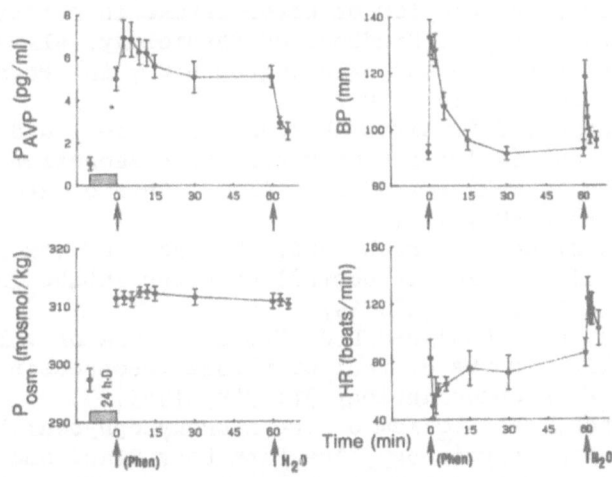

Fig. 3 Effect of Phenylephrine on blood pressure, heart rate and plasma osmolality and vasopressin. Phenylephrine did not inhibit vasopressin secretion but subsequent drinking of water did.

In summary, dehydrated dogs use accurate orophryngeal signals to ensure rapid intake of fluid deficits. Presumably such rapid satiety mechanisms had important survival value when access to water was limited and could not be assured on a continuous basis. Similar powerful inhibitory oropharyngeal cues also inhibit vasopressin secretion such that the hormonal control of water excretion anticipates the ingested fluid load before absorption. Such a mechanism might also have had survival value. Dual protection against fluid overload by preventing excessive fluid intake, and ensuring rapid elimination of excessive intake should that occur, could be achieved through oropharyngeal inhibition of drinking and vasopressin secretion.

REFERENCES

1. D.J. Ramsay, B.J. Rolls and R.J. Wood, Thirst following water deprivation in dogs. Am. J. Physiol 232. (Regulatory, Integrative Comp. Physiol 1) R 93, 1977.
2. D.J. Ramsay, B.J. Rolls and R.J. Wood, Body fluid changes which influence drinking in the water deprived rat. J. Physiol (Lond) 266: 453, 1977.
3. R.J. Wood, E.T. Rolls and R.J. Rolls, Physiological mechanisms for thirst in the non-human primate. Am. J. Physiol 242 (Regulatory, Integrative Comp. Physiol 11) R423.
4. E.F. Adolph, Measurements of water drinking in dogs. Am. J. Physiol 125:75, 1939.
5. R.T. Bellows, Time factors in water drinking in dogs. Am. J. Physiol 125:87, 1939.
6. E.F. Adolph, Regulation of water intake in relation to body water content, In. Handbook of Physiology, Alimentary Canal. Food and Water Intake, Washington D.C., Am. Physiol Soc. sect 6, vol. 1, 163, 1967.
7. T.N. Thrasher, J.F. Nistal-Herrera, L.C. Keil and D.J. Ramsay, Satiety and inhibition of vasopressin secretion after drinking in dehydrated dogs, Am. J. Physiol 240 (Endocrinol. Metab. 3) E 394, 1981.
8. S. Kozlowski and K. Drzewieski, The role of osmoreception in portal circulation in control of water intake in dogs, Acta. Physiol. Pol. 24; 325, 1973.
9. C.E. Wade, L.C. Keil and D.J. Ramsay, Role of volume and osmolality in the control of plasma vasopressin in dehydrated dogs, Neuroendocrinology 37: 349, 1983.
10. H.D. Lauson, Metabolism of the neurohypophyseal hormones, In. Handbook of Physiology, The Pituitary Gland and its Neuro-endocrine Control. Washington, D.C: Am. Physiol Soc. Sect 7, vol IV, Part 1 287, 1974.
11. S. Nicolaidis, Response des unites osmosensibles hypo-thalamiques aux stimulations salines et aqueuses de la langue, Compt. Rend. 263: 2352, 1968.

12. J.D. Vincent, E. Arnauld and B. Biolac, Activity of osmosensitive single cells in the hypothalamus of the behaving
monkey during drinking, Brain Res. 44, 371, 1972.

13. T.N. Thrasher, C.E. Wade, L.C. Keil and D.J. Ramsay, Sodium
balance and aldosterone during dehydration and rehydraton
in the dog, Am.J. Physiol 247 (Regulatory, Integrative
Comp. Physiol 16), 76, 1984.

14. J.R. Blair-West, A.H. Brook, A. Gibson, M. Morris and P.T.
Pullan, Renin, antidiuretic hormone and the kidney in water
restriction and rehydration, J. Physiol (Lond) 294, 181,
1979.

15. E. Arnauld and J. DuPont, Vasopressin release and fixing at
supraoptic neurosecretory neurons during drinking in the
dehydrated monkey, Pflugers Arch. 394, 195, 1982.

16. J.E. Greenleaf, A. Geelen, L.C. Keil, S.E. Kravik, C.E. Wade,
T.N. Thrasher, P.R. Barnes, G. Pyka and C Nesvig, Effects
of drinking on plasma vasopressin, renin and aldosterone in
dehydrated humans, Federation Proceed. 43, 716, 1984.

17. W.E. Hoffman, M.I. Phillips, E. Wilson and P.G. Schmid, A
pressor response associated with drinking in rats. Proc.
Soc. Exper. Biol Med. 154: 121, 1977.

THE NATURE AND LOCALIZATION OF CENTRAL RECEPTOR SYSTEMS

John B. Simpson

Department of Psychology, NI-25
University of Washington
Seattle, WA 98195

The purpose of this paper is to consider the nature and the localization of central receptor systems involved in the control of fluid intake. Inherent in a discussion of these topics is a consideration of how such information describing receptor systems is obtained. Two commonly used techniques which are well suited to such analyses are the experimental brain lesion and its variants, and the intracranial chemical injection. Although other techniques can and often are applied to the questions of localization and nature of central systems, it is truly only in experiments that employ the measurement of specific ingestive behavior that such central mechanisms may be directly understood. Other techniques, which may be described as ancillary, do indeed provide relevant information regarding the organization for central behavioral mechanisms; however, it is only by the use of behavior as the dependent measure that a definitive understanding of those brain systems relevant to behavior can be achieved.

One additional point regarding analyses of central fluid control mechanisms should be stressed: experiments which restrict their scope solely to the use of behavior as the dependent measure(s) may fail to indicate the scope and complexity of central regulatory mechanisms. As will be detailed below, it is now apparent that stimuli indicating alterations in fluid homeostasis often initiate a variety of compensatory mechanisms. These effects may include not only alterations in fluid ingestion, but autonomic, cardiovascular and renal effects as well. Moreover, it appears that in some situations the same tissue site may be responsible for these varied effects of a common stimulus. For example, the hormone angiotensin II, increased during hypovolemia, acts at the subfornical organ to cause synergistic dipsogenic, pressor and antidiuretic effects (Simpson, 1981).

309

While it thus seems that behavior must be studied in order to appreciate central mechanisms of behavior, it is also clear that more than behavior per se should be measured in sophisticated, successful experiments.

I shall now describe experiments which have been designed to establish the nature of and localization of central receptor systems involved in fluid ingestion. This summary will be of necessity idiosyncratic, yet should be illustrative of the methodological and interpretive requirements in this type of analysis.

SUBFORNICAL ORGAN AND ANGIOTENSIN II

A variety of reports have now indicated that the subfornical organ (SFO) is a central site of action of the peptide hormone, angiotensin II (AII). The SFO is one of the circumventricular organs, structures which lack the blood-brain barrier and hence would be accessible to circulating AII. Aryeh Routtenberg and I (Simpson and Routtenberg, 1973) initiated the first series of such studies which were designed both to establish a function for the SFO, and to identify a site of action of AII within the forebrain. The initial approach was a dose-response analysis of the effect on water intake of direct injections of AII into the interstitium of the SFO, and it was found that the SFO was a very sensitive site for this effect. A more thorough dose-response analysis (Simpson, Epstein and Camardo, 1978) was subsequently completed. Injections of 0.5 ul volume were made via a chronic indwelling cannula system of fine gauge (27-34 ga. system) at a constant time (5-12 days) after surgery in order to stabilize and minimize tissue damage from the injection system itself. Remote intracranial injections were performed in each animal's home cage in a test situation where the probability of spontaneous drinking was nil. The power of this dose-response analysis of the behavioral actions of a hormone may be appreciated in Fig. 1. The injection of the hormone directly into the SFO, as well as into control loci, provoked dose-dependent water intake. Indeed, we found that the SFO was the most sensitive site of injection of the hormone for its dipsogenic effect relative to other areas, either in our laboratory or relative to reports in the literature. It was also noted that the latency from injection until the onset of drinking behavior was significantly shorter at the SFO than at the other loci, an observation consistent with a primary site of action of the hormone at that locus. This sensitivity, for both the dipsogenic and for the central pressor effect of AII, has been replicated in more recent work (Mangiapane and Simpson, 1980a). The injection studies, then, suggest an SFO site of action because: i)of the superior dose-response characteristics of the SFO vs. other sites of application; and ii)because of the shorter latency until onset of effects at the SFO vs. other tested loci.

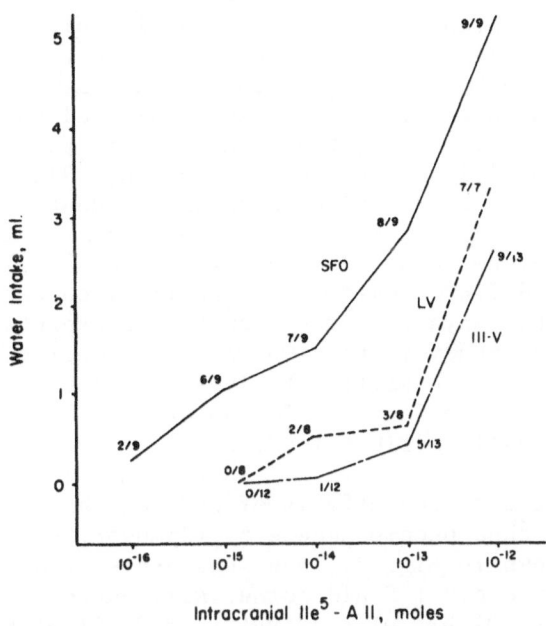

Fig. 1. Water intake after injection of AII to SFO, or to adjacent
lateral (LV) or third (III-V) sites. At each point, num-
erator is number drinking, and denomenator is number
of rats injected. From Simpson, Epstein & Camardo, 1978.

Another intracranial injection technique which can be used in
localization studies of blood-borne stimuli would be the selective
application of an antagonist to the brain during systemic infusion
of the stimulus. This method allows access of hormone to the whole
body during a local and specific antagonism of the hormone within
the brain. This permits measures both of pharmacological as well as
anatomical specificity; in other words, a reversible and specific
pharmacological lesion is produced. Simpson, Epstein and Camardo
(1978) infused the AII antagonist, saralasin, to the SFO or other
brain loci during intravenous infusion of AII. This study found,
first, that intracranial saralasin antagonized, in dose-dependent
fashion, the drinking produced normally by systemic AII infusions,
and second, that the SFO was the most sensitive site of this
antagonism. This study is important in the experimental basis for
the hypothesis that the SFO is a site of action of circulating AII.
However, it suffers, as do all intracranial injection studies, from
the potential problem of diffusion of injectate.

The use of the intracranial injection technique does not
unambiguously specify where a chemical stimulus might act in the
brain. The assumption that the tissue at the tip of a cannula is
the area where the chemical acts is not always correct, especially
since diffusion of injectate is a well-known phenomenon. For
example, in the first thorough study of the dipsogenic effect of

311

intracranial injection of AII, a dose-response analysis indicated
that the preoptic area was the most sensitive site of application,
and hence likely the site of action of AII (Epstein, Fitzsimons and
Rolls, 1970). The SFO was not tested in that study. Subsequent
studies revealed, however, that hormone injected into the preoptic
area likely was diffusing up the cannula shaft for distribution via
the cerebral ventricles (Johnson and Epstein, 1975); such movement
of peptide then led to drinking (Simpson and Routtenberg, 1973;
Buggy, et al., 1975). It is clear, then, that results derived from
this technique alone do not indicate where injectate acts; sites of
application may differ from sites of action (Routtenberg, 1972).
The only available ways to strengthen the interpretation of an intracranial injection study is to compare the latencies and dose-response
characteristics between different sites. Even still, diffusion
could account for such data, if for no other reason that it is not
feasible to test all possible loci.

An assessment of the effects of SFO lesions on drinking to
blood-borne AII thus became necessary (Simpson, Epstein and Camardo,
1978) and is shown in Fig. 2. Note that this study examined blood-
borne, not cerebrospinal fluid-borne, AII. Rats were prepared with
localized lesions of SFO or adjacent tissue, or as operative controls.

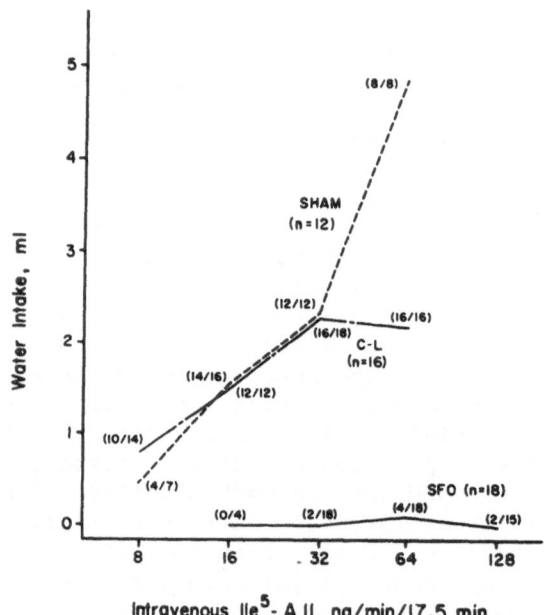

Fig. 2. Water intake during iv AII infusion in rats with lesions
         of SFO, of adjacent tissue (C-L), or operative controls
         (SHAM). At each point, numerator indicates number drinking,
         and denomenator the number of rats infused. Data from
         Simpson, Epstein and Camardo, 1978.

312

Intrajugular infusion of AII was given at the relatively low doses of 8-128 ng/min for a 20 min test period. This treatment shows, in the SHAM group, that drinking reliably provoked by these infusions. Of key importance is the group with lesions of the SFO, in which AII infusion at any dose failed to cause drinking in this testing situation. The same was true in another group of rats when testing was accomplished at 80-90 days after the SFO lesion surgery, which indicates the permanence of this effect. Dogs (Thrasher, Simpson and Ramsay, 1982) and opossums (Elfont, Epstein and Findlay, 1980) also fail to drink to systemic AII infusion, although sheep (McKinley, et al., this volume) still drink following SFO lesions.

Animals bearing the SFO lesion also show other deficiencies in fluid regulation. For example, they show a diminished pressor response to intravenous AII infusion (Mangiapane and Simpson, 1980b) as well as decreased secretion of vasopressin to intraventricular injection of AII. These results agree with the observations that AII injection into the SFO causes increased blood pressure (Mangiapane and Simpson, 1980a) as well as vasopressin secretion (Simpson, et al., 1979). Animals with SFO lesions also show reduced drinking to hyperoncotic colloid dialysis with polyethylene glycol and to peripheral injections of the beta adrenergic agonist, isoproterenol (Simpson, Epstein and Camardo, 1978).

The only logical inference which may be derived from lesion studies is that some particular tissue is necessary for the given behavior of interest. That is, the absence of or enhancement of a particular behavioral capacity following brain damage indicates not necessarily that the damaged area plays a role in the normal behavioral function, but rather that the brain does not function normally for that behavioral capacity in the absence of that tissue. It also follows that lesion studies themselves do not unequivocally indicate the location of receptors or, necessarily, systems involved in behavior. The fact that lesions can destroy vasculature, meninges, ependyma, axons of passage, etc. should caution against the simplistic interpretation of such experiments. As an extreme and perhaps absurd example, transection of the neuraxis at the level of the medulla would prevent AII-induced drinking, as do SFO lesions. The shortcomings in interpretation of lesion experiments should be apparent.

The strongest experimental evidence for localization of function comes from several experiments, using diverse techniques such as the lesion and the intracranial injection, and by demonstrating mutually confirmatory results with the different approaches. This proposition is well illustrated by the experiments discussed above involving circulating AII and the SFO, in which several mutually confirmtory experiments point to the conclusion that the SFO is indeed a receptor for the behavioral action of the hormone.

There are examples of failure of mutual confirmation of experiments on the same localization topic using combined lesion and

injection analysis. One such failure was reported by Fisher and Leavitt (1966), who found that application of atropine to one site in
the "cholinergic thirst circuit" prevented the drinking normally
provoked by application of carbachol to a second point. These data
were interpreted to mean that the muscarinic antagonist prevented
some sore of reverberating activity in the neural circuit responsible
for drinking, hence blocking the carbachol-induced drinking. Somewhat
puzzling to those authors was that lesions failed to duplicate the
effects of local injection of atropine. Indeed, it could be argued
that if the effects of injected atropine were simply to uncouple the
circuit, then lesions should precisely have duplicated this effect.
It was this descrepancy which led Routtenberg (1967) to postulate
that the failure of lesions to disrupt carbachol-induced drinking,
and the success of injected atropine to do so, was due to diffusion
of injected cholinomimetic and/or cholinolytic agents to some site
of action other than the tip of the indwelling cannulae--a site of
action near the cerebral ventricles. One locus both with ventricular proximity and with cholinergic afferents is the SFO.

SUBFORNICAL ORGAN AND CHOLINERGIC COMPOUNDS

     It has long been recognized that the SFO receives a major chol-
inergic input. The source of these afferents in unknown. Simpson
and Routtenberg (1972), in an examination of the "ventricular hypothesis" (Routtenberg, 1967), prepared rats with chronic cannulae
directed at the SFO, the ventricles, or the lateral hypothalamus.
The latter site was particularly noted as a locus at which cholinergic injections provoked drinking. It was found, first, that the
SFO was an extremely sensitive site at which carbachol injection
provoked drinking with short latency, and second, that lesions of
the SFO reduced or prevented drinking typically elicited by carbachol
injections elsewhere within the forebrain. At least one function,
then, of the cholinergic input to the SFO was concerned with water
intake.

     More recent experiments have replicated the intracranial injec-
tion data, and expanded the known effects of cholinergic stimulation
of the SFO. Mangiapane and Simpson (1983) examined the effects of
injection of acetylcholine itself, rather than the synthetic carbachol, on drinking and on blood pressure following injection into the
SFO, the adjacent ventral fornical commissure, or the adjacent third
cerebral ventricle. Rats were prepared with an aortic catheter for
direct measurement of blood pressure, and with a single intracranial
cannula directed at one of the three loci. Fig. 5 shows the magnitude of the dipsogenic and pressor effects following SFO injection
of various doses of ACh. The effects are highly correlated, and
show a consistent temporal relationship such that the onset of
the pressor effect always preceded the onset of drinking behavior.
At low doses, the animals often showed a measurable pressor effect

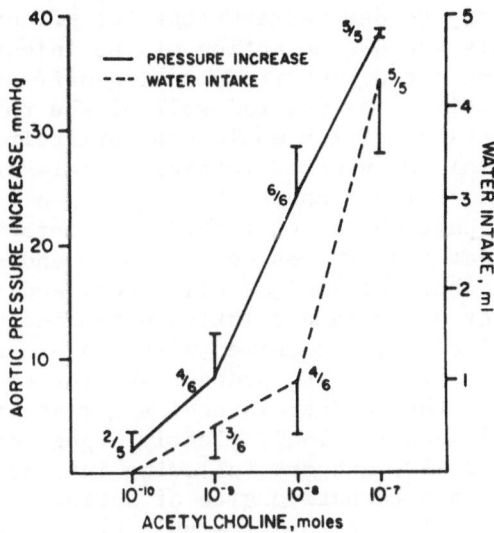

Fig. 3. Dipsogenic and pressor effects of SFO injection of acetylcholine chloride.

in the absence of elicited drinking. Both effects were antagonized by intracranial pretreatment with atropine, and the pressor effect was blunted by systemic phentolamine. The former effect indicates that both pressor and dipsogenic effects occur via specific cholinergic, muscarinic receptors, and the latter effect is suggestive of at least partial sympathetic mediation of the cholinergic pressor effect.

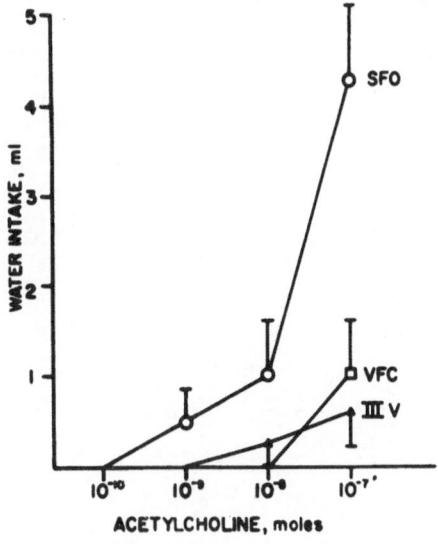

Fig. 4. Dose response of drinking elicited by acetylcholine applied to SFO, fornical commissure (VFC), or third ventricle (IIIV).

315

It is necessary to demonstrate that the effects of SFO injections of ACh likely are due to action of the injected chemical at that structure, and not elsewhere following diffusion. Given the geography of the SFO on the rostral wall of the third ventricle, then control injections of ACh would appropriately be made into the adjacent neuropil of the ventral fornical commissure or into the adjacent ventricle. Fig. 6 shows the comparison for the dipsogenic effect of SFO vs. parenchymal or ventricular injection of ACh. The elicited drinking at the SFO was consistently and significantly greater than the other two loci at all doses, and the two control loci did not differ. The same relative potencies, with SFO greater than the control loci, also occurred with the pressor effect. The dose-response data for both the pressor and for the dipsogenic effects, as well as the shorter latencies for the two effects at the SFO vs. the two control loci, clearly argue for an action of injected ACh at the SFO, and not following interstitial or ventricular diffusion to an alternative site of action. The very rapid hydrolysis of ACh, relative to carbachol, by endogenous cholinesterases further suggests that, in nervous tissue, the material must be applied very near to its site of action.

One additional experiment involving the actions of the cholinergic input to the SFO will be discussed. The connections between the SFO and the supraoptic and paraventricular nuclei have been documented (Miselis, 1981; Lind and Johnson, 1982), and the fact that AII injected into the SFO causes measurable increases in plasma vasopressin (Simpson, et al., 1979) is suggestive of a function of the SFO in the regulation not only of water intake, but of water

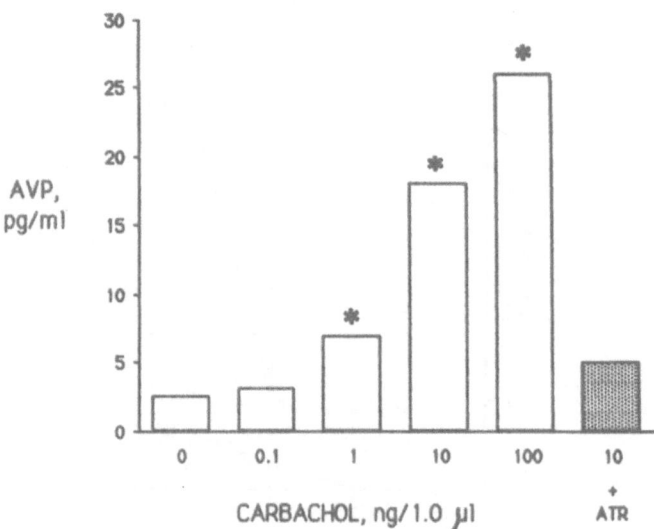

Fig. 5. Dose-response of vasopressin secretion elicited by SFO injection of carbachol, including atripine pretreatment (ATR).

excretion. Animals were prepared with cannulae in the SFO, and the effects of injection of various doses of the cholinomimetic, carbachol, on plasma immunoreactive vasopressin were measured. Animals were killed by decapitation at 5 min after the injection. As can be appreciated in Fig. 5, SFO injection of this agent produced dose-dependent increases in plasma vasopressin concentrations. The effect was antagonized by atropine pretreatment, indicating specific muscarinic receptors in the effect. Moreover, although not shown here, the effect was greater in the SFO than in adjacent ventricle or tissue. The effect of application of cholinomimetics to the SFO, then, precisely duplicates the effects of AII injection there: drinking, a pressor response, and vasopressin secretion all occur following injection of either agent. Of interest is the observation that, at least for drinking, the two effects are in pharmacological parallel and not in series--atropine does not affect AII drinking, and saralasin does not antagonize carbachol drinking (Mangiapane and Simpson, 1979). At the SFO, then, these two classes of input have identical effects; further studies have noted the effects of the two compounds on sodium appetite.

CARBACHOL AND SODIUM APPETITE

A valuable approach to understanding the nature of central receptor mechanisms for fluid intake is the measurement of several fluid homeostatic parameters simultaneously. Recent experiments have examined the effects of intracranial infusions of the cholinomimetic, carbachol, on water and sodium intake and excretion (Fitts, Thunhorst and Simpson, 1985a; 1985b).

The first experiments demonstrated that intracranial carbachol infusions, at 400 or 2000 ng/hr, caused a brisk natriuresis and kaliuresis, with a small antidiuresis and substantial water drinking. Despite the negative sodium balance owing to the natriuresis, sodium ingestion was not provoked. A comparison in our laboratory, as well as to the literature, suggested that intraventricular infusions of AII or of carbachol produce similar effects--drinking, antidiuresis, natriuresis, and a pressor effect. Because of this, carbachol is frequently used as a "control" drug for the effect of AII on salt intake; because carbachol does not cause a salt appetite, the effect of AII on salt appetite is said to be specific. We decided to investigate whether carbachol merely fails to elicit salt intake, or whether it actually suppresses a sodium appetite.

The sodium appetite which occurs following diuretic treatment was chosen as the model to evaluate the effect of carbachol. Lateral ventricular infusion of carbachol (80 or 400 ng/hr) or of AII (87 or 438 ng/hr) were started 3.75 hr after furosemide injection (25 mg/kg, sc). Urinary loss of sodium was measured, and all animals were in substantial negative sodium balance 4 hr later, when access

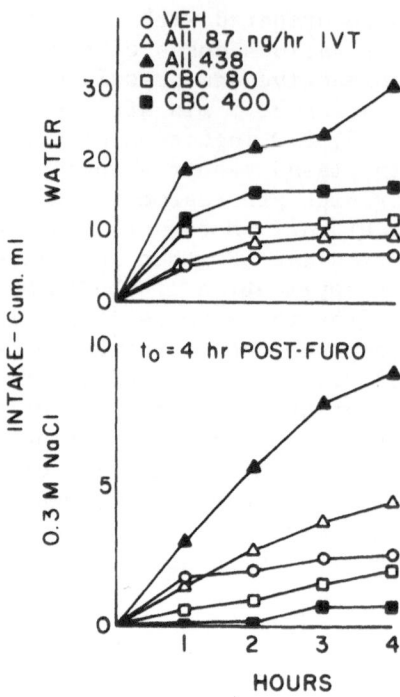

Fig. 6.   Cumulative water and 0.3 M NaCl intake during intracranial
          infusions of AII or carbachol or vehicle. Infusions started
          15 min prior to ingestion measurements, and 3.75 hr after
          furosemide injection.

to ingestive fluids was permitted. Drinking was monitored for 4 hr,
and no food was present.

The ingestion of water and sodium is shown in Fig. 6. The large
increases in water and sodium ingestion in the AII-infused animals
resulted in an enhanced water and sodium balance relative to vehicle-
infused controls; AII hastened recovery from the water and sodium
deficits induced by the diuretic. In contrast, infusion of carbachol
exacerbated the negative sodium balance produced by the diuretic.
The rats receiving the carbachol, especially at the 400 ng/hr dose,
consumed essentially no sodium while continuing to show substantial
water intake. The rats, then, continued to ingest water in the face
of a substantial negative sodium balance. In this model of sodium
appetite, then, carbachol actually suppressed the need-induced
salt appetite.

Several key plasma parameters were measured in another group of
rats, receiving the 400 ng/hr carbachol treatment and allowed to
ingest fluid for 2 hr according to the treatment regimen described
above. Plasma osmolality (282 mOsm/l) and sodium concentration
(128 mEq/l) were reduced by 7% and by 8% relative to vehicle-infused

controls, respectively. The intracranial carbachol clearly enhanced the negative sodium balance following diuretic treatment. The continued water ingestion actually produced a dilutional hyponatremia, again underscoring the antagonistic effect of carbachol on sodium appetite. Similar suppression of sodium appetite following hyperoncotic colloid dialysis was also seen with intracranial carbachol.

One interpretation of such data is that intracranial AII infusions cause behaviors and excretory adjustments consistent with hypovolemia. In this situation, enhanced ingestion of both water and sodium are appropriate to reconstitute fluid volume. In contrast, infusions of carbachol into the cerebral ventricles cause alterations which are consistent with dehydration of the cellular fluid compartment. Hence, ingestion of water coupled with natriuresis and no ingestion of salt are appropriate responses.

It is clear that the ingestion of water provoked both by AII and by carbachol is only a small part of the story of their central effects on fluid balance, and focussing only on water drinking does not indicate the full nature of their effects. It is only by asking further questions--by simultaneously monitoring several dependent measures--that a clear appreciation of these agents can be obtained. Indeed, infusion of carbachol vs. AII are now understood to provoke different constellations of effects, even though both cause water ingestion.

Direct assessment of the nature and localization of central receptor systems for fluid ingestion demands that behavior be used as one dependent measure. Experiments are thus restricted to manipulations of the central nervous system in conscious, behaving animals. However, the techniques useful in such analysis--specifically, intracranial injection and brain lesions--are not without their particular difficulties. This review has indicated some of the problems associated with each. However, as is indicated from some of the described experiments, careful analysis and interpretation of such experiments can indeed tell us about the nature and localization of central receptor systems. Indeed, using varied techniques and multiple dependent measures will be necessary in this task.

REFERENCES

1. J. Buggy, A.E. Fisher, W.E. Hoffman, A.K. Johnson, and M.I. Phillips. Ventricular obstruction: effect on drinking induced by intracranial angiotensin. Science 190: 72 (1975).
2. R.M. Elfont, A.N. Epstein, and A.L.R. Findlay. The role of the subfornical organ in angiotensin-induced drinking in the North American opossum, J. Physiol. (London) 301: 49 (1979).
3. A.N. Epstein, J.T. Fitzsimons, and B.J. Rolls, Drinking induced

by injection of angiotensin into the brain of the rat, J. Physiol. (London), 210: 457 (1970).

4. D.A. Fitts, R.L. Thunhorst, and J.B. Simpson, Fluid intake, distribution, and excretion during lateral ventricular infusions of carbachol in rats, Brain Res. In press.

5. D.A. Fitts, R.L. Thunhorst, and J.B. Simpson, Modulation of salt appetite by lateral ventricular infusions of angiotensin II and carbachol, Brain Res. In press.

6. A.K. Johnson and A.N. Epstein, The cerebral ventricles as the avenue for the dipsogenic action of intracranial angiotensin, Brain Res. 86: 399 (1975).

7. R.A. Leavitt and A.E. Fisher, Anticholinergic blockade of centrally induced thirst, Science, 154: 520 (1966).

8. R.W. Lind and A.K. Johnson, Subfornical organ-median preoptic connections and drinking and pressor responses to angiotensin II, Neuroscience, 2: 1043 (1982).

9. M.L. Mangiapane and J.B. Simpson, Pharmacological independence of subfornical organ receptors mediating drinking, Brain Res. 178: 507 (1979).

10. M.L. Mangiapane and J.B. Simpson, Subfornical organ: forebrain site of pressor and dipsogenic action of angiotensin II, Amer. J. Physiol. 239: R382 (1980a).

11. M.L. Mangiapane and J.B. Simpson, Subfornical organ lesions reduce the pressor effect of systemic angiotensin, Neuroendocrinol. 31: 380 (1980b).

12. M.L. Mangiapane and J.B. Simpson, Drinking and pressor responses after acetylcholine injection into subfornical organ, Amer. J. Physiol. 244: R508 (1983).

13. R.R. Miselis, The efferent projections of the subfornical organ of the rat: a circumventricular organ within a neural network subserving fluid balance, Brain Res. 230: 1 (1981).

14. A. Routtenberg, Drinking induced by carbachol: thirst circuit or ventricular modification? Science 157: 838 (1967).

15. A. Routtenberg, Intracranial chemical injection and behavior: a critical review, Behav. Biol. 7: 601 (1972).

16. J.B. Simpson, The circumventricular organs and the central actions of angiotensin, Neuroendocrinol. 32: 248 (1981).

17. J.B. Simpson, A.N. Epstein and J.S. Camardo, The localization of dipsogenic receptors for angiotensin II in the subfornical organ, J. Comp. Physiol. Psychol. 92: 581 (1978).

18. J.B. Simpson, M. Reed, L.C. Keil, T.N. Thrasher and D.J. Ramsay, Forebrain analysis of vasopressin secretion and water intake induced by angiotensin II, Fed. Proc. 38: 2968 (1979).

19. J.B. Simpson and A. Routtenberg, The subfornical organ and carbachol induced drinking. Brain Res. 45: 135 (1972).

20. J.B. Simpson and A. Routtenberg, Subfornical organ: site of drinking elicitation by angiotensin II, Science 181: 1172 (1973).

21. T.N. Thrasher, J.B. Simpson and D.J. Ramsay, Lesions of the subfornical organ block angiotensin drinking in the dog, Neuroendocrinol. 35: 68 (1982).

ADIPSIA IN SHEEP CAUSED BY CEREBRAL LESIONS

M.J. McKinley, D.A. Denton, M. Leventer, R.R. Miselis, R.G. Park, E. Tarjan, J.B. Simpsom, and R.S. Weisinger

Howard Florey Institute of Experimental Physiology and Medicine, University of Melbourne, Parkville, Victoria, Australia 3052

During the past decade it has been recognized that several structures in the anterior wall of the third ventricle (A3V) are involved in the regulation of water intake. With regard to these structures, the subfornical organ (SFO) is a probable receptor site for angiotensin to induce water drinking[1], while ablation of the organum vasculosum of the lamina terminalis (OVLT) and the antero-ventral third ventricle wall region (AV3V) disrupts osmo-regulatory water intake[2,3]. Destruction of the preoptic medianus nucleus (PMN) also reduces water drinking in rats[4] while in goats ablation of most of the A3V results in adipsia[5]. In view of the likelihood that most of the A3V is involved in the regulation of thirst, we decided to study the effects of either individual or combined ablation of some of these regions in the sheep. The anterior commissure (AC), which courses horizontally across the A3V, provides a convenient boundary for division of the A3V into dorsal and ventral regions. The antero-dorsal wall of the third ventricle (AD3V) contains the SFO, dorsal part of the PMN, and the medial septum, while the anteroventral wall of the third ventricle (AV3V) encompasses the ventral PMN, the OVLT and preoptic periventricular tissue (Figure 1).

Two groups of sheep were studied. The first group were ovariectomized ewes in which the carotid arteries had been surgic-ally enclosed in skin loops in the neck and in which electrodes had been surgically implanted over various parts of the A3V. After recovery from surgery, food and water intake was measured each day and blood samples were regularly obtained from the carotid artery for measurement of plasma [Na], [K] and osmolality. After baseline measurements had been obtained, a lesion was then

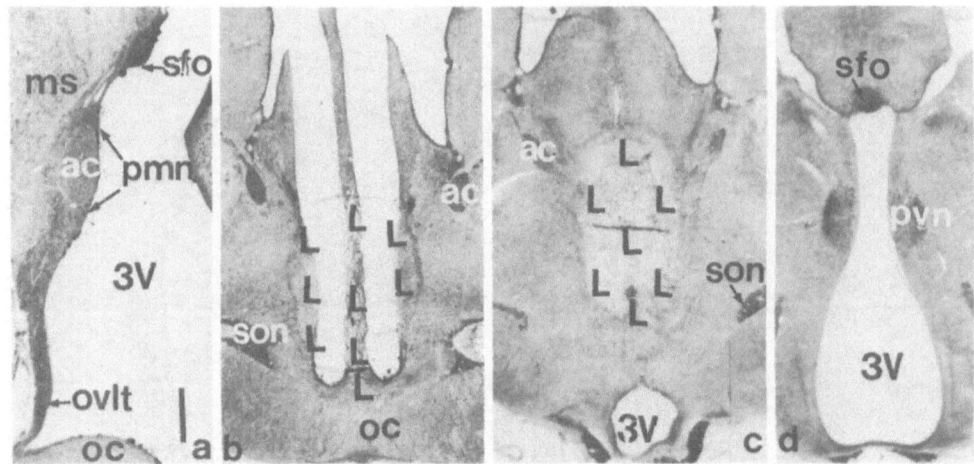

Fig. 1.   Photomicrographs of sheep brains : (a) Mid-sagittal
section of A3V region. (b), (c) and (d) Coronal sections from the brain of a permanently adipsic sheep at
the level of the OVLT in (b), the PMN and preoptic
periventricular region in (c) and showing intact SFO and
PVN in (d). Scale bar = 1mm. Abbreviations : ac =
anterior commissure, ms = medial septum, oc = optic
chiasma, pvn = paraventricular nucleus, son = supraoptic nucleus, 3V = third ventricle, L = lesion.

produced in the A3V by application of radiofrequency current across
the indwelling electrodes. The lesion was made while the sheep
were conscious, and apart from hyperventilation which was probably
due to local brain heating, this procedure did not disturb the
animals or cause any apparent discomfort. This allowed us to
measure the acute effects of the lesion on water intake. After
studying these animals for several months, they were killed, the
brains examined histologically and the site of the lesion
ascertained.

     In 4 of the sheep the lesion was confined to the AV3V and in
another 4 animals to the AD3V region. Although such lesions
cause reduced water drinking in response to systemic infusions of
hypertonic saline[3,6], there is only a transient reduction on the
day to day water intake with ablation of the AV3V[6], and we found
no consistent effect on this parameter with AD3V lesions. In
several other sheep in which there was near complete ablation of
the AV3V and additionally substantial damage to the AD3V (although
the SFO was usually left intact), periods of adipsia from 4-22
days were observed despite severe hypernatremia (Figure 2), and
they were maintained during these periods of adipsia by administration of water into the rumen. Food intake was not initially
reduced by the lesions and only decreased when the sheep became

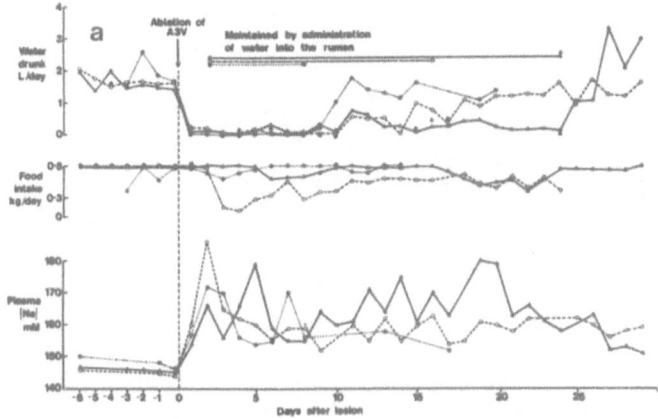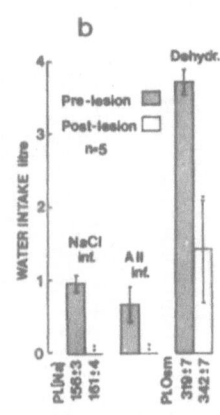

Fig. 2. (a) Metabolic measurements made in 3 of the 5 sheep which
became temporarily adipsic after ablation of the A3V.
(b) Water intake in response to intracarotid infusion for
20 min of 4M NaCl at 1.3 ml/min (NaCl inf), angiotensin II
at 1.6 µg/min (AII inf) and in response to water deprivation (dehydr) for 3 days before and after A3V ablation.
Plasma [Na] in mM and plasma osmolality in mosm/kg at the
end of the infusion or dehydration are given.

severely dehydrated, a state which normally causes reduced food
intake in this species. When the sheep recommenced drinking water,
the volume of water imbibed each day gradually increased to reach
pre-lesion levels. At this time, despite the fact that water
drinking had recommenced, severe deficits in the regulation of the
thirst mechanism still remained. When the response of these
animals to intracarotid infusion of hypertonic saline or angiotensin II was tested, it was found to be completely absent (Fig.2b).
Moreover, water intake in response to water deprivation was greatly
inhibited despite a far greater degree of hyperosmolarity (Fig.2b).
Additionally we have observed impaired vasopressin secretion and
renal Na excretion[6,8] and hyperreninemia when these sheep were
dehydrated, a pattern of effects comparable to those observed in
rats bearing AV3V-lesions[2,7].

Although most of the sheep in this group only ceased drinking
water temporarily, there were two animals that became permanently
adipsic (Fig.3). While water drinking was completely disrupted
despite severe hypernatremia, appetite for their daily food ration
was normal for their hydration state, and the animals remained
alert and in apparent good health as long as they were given
replacement water therapy. Although they never drank water after
the A3V lesion was made, they would readily drink solutions of
concentrated NaCl if presented to them.

Regarding the second group of 7 sheep which we studied, they
underwent similar preparation as the first group but in addition
the parotid duct had been surgically exteriorized to produce a

323

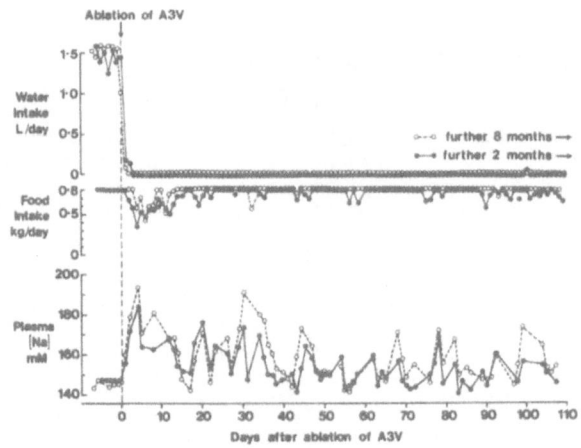

Fig.3.   Metabolic observations made in 2 sheep which became
         permanently adipsic after ablation of tissue in the A3V,
         and were maintained by administration of water into the
         rumen.

permanent parotid fistula. This caused loss of Na from the body
in the saliva lost from the fistula and these sheep were provided
with a solution of 0.6M $NaHCO_3$ to drink for 2 hours each day. They
obtained delivery of this solution as well as their water into
separate cups by pressing pedals situated in the cage. The electrodes were also implanted over the A3V in this group. The post-
mortem histological examination revealed that substantial damage
to both AV3V and AD3V tissue had been incurred in all 7 sheep with
parotid fistulas (e.g. Fig.1). In regard to water intake, a pattern
of effects was observed similar to that seen in the first group.
After ablating tissue in the A3V, 5 sheep ceased drinking water
temporarily for periods ranging from 13-26 days (Fig.4a), while
two became permanently adipsic with regard to water with hypernatremia being observed in these sheep (Fig.4b). However, all 7 sheep
continued to ingest adequate quantities of 0.6M $NaHCO_3$ solution
after ablation of the A3V. Furthermore, we tested the ability of
these sheep to regulate Na intake in response to (i) greater Na
deficit or (ii) intracarotid infusion of hypertonic NaCl solution.
Despite the large disruption of the thirst mechanism caused by
these lesions in the A3V, the sheep were able to make adequate
compensatory adjustment of Na intake in the face of altered Na
balance. Withdrawal of the $NaHCO_3$ solution caused increased intake
of Na when it was made available on the following day in these
sheep. In the opposite sense, intracarotid infusion of hypertonic
saline prior to the period of access to Na solution caused a large
reduction of Na intake, but did not induce water drinking as it had
done when the animals were tested prior to making lesions (Fig.5).

    In summary, these results demonstrate that although the A3V
region is crucial for regulating water intake in sheep, it does not

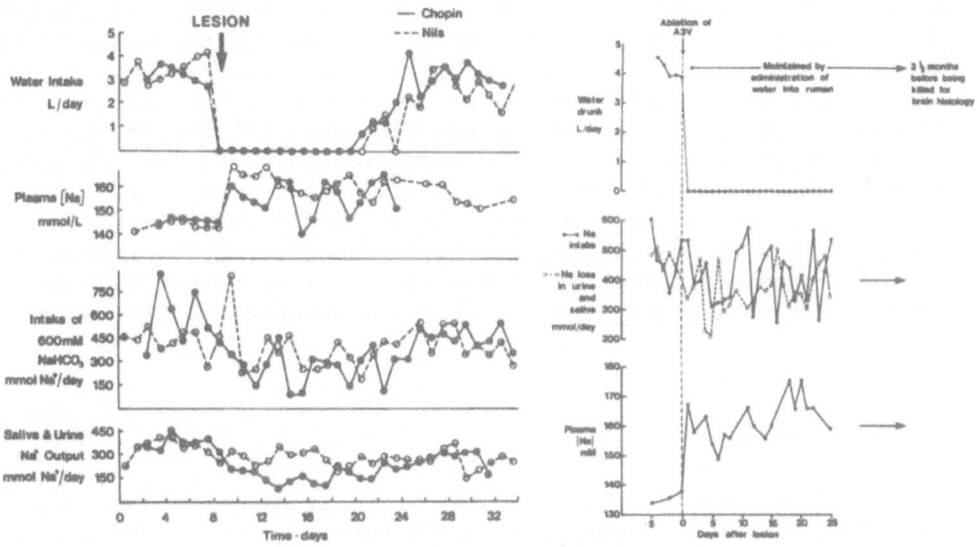

Fig. 4.   (a) Metabolic observations made in 2 of the 5 sheep with
          parotid fistulas which temporarily stopped drinking water
          (given maintenance intrarumenal water) after ablation of
          A3V tissue. (b) Intake of water and 600mM NaHCO3 in 1 of
          the 2 sheep with parotid fistulas which became permanently
          adipsic after ablation of the A3V.

play an essential role in the control of Na appetite. With regard
to the histological analysis, it was necessary to destroy tissue
1.5-2mm on either side of the coronal midline in both ventral and
dorsal aspects of the anterior wall of the third ventricle to cause
adipsia. The ablated tissue common to all 4 sheep which became
permanently adipsic was complete destruction of the preoptic

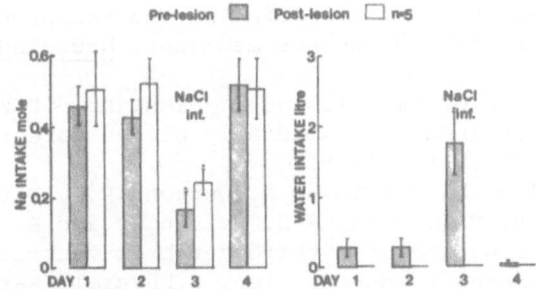

Fig. 5.   Comparison of pre- and post-lesion intake of 600mM
          NaHCO3 solution and water during 2 h of access to Na on
          4 successive days in the 5 sheep with A3V-lesions and a
          parotid fistula which were tested. On day 3, an infusion
          of 4M NaCl at 1.6 ml/min for 30 min was commenced 10 min
          prior to access to Na.

medianus nucleus and preoptic periventricular region, and substantial damage to the OVLT, medial septum, nucleus of the diagonal band and AC. Although the SFO, small parts of the rostral hypothalamus and lateral septum were damaged in one or two of the adipsic animals, destruction of these areas was not essential for the production of the adipsic syndrome. Because there was only marginal differences between the sites and extent of the lesions in permanently adipsic versus temporarily adipsic sheep, it is likely that the recovery observed in some animals was due to the healing of reversible injury and oedema on the periphery of the lesion site, rather than the recruitment of another level of cerebral organization for water drinking.

ACKNOWLEDGEMENTS

This work was supported by the National Health and Medical Research Council of Australia. Brett Purcell, Don Reilly and Susan Eason provided skilled technical assistance.

REFERENCES

1. J.B. Simpson, A.N. Epstein and J.S. Camardo. Localization of receptors for the dipsogenic action of angiotensin II in the subfornical organ of rat. J.Comp. Physiol.Psychol. 92:581(1978)

2. J. Buggy and A.K.Johnson. Preoptic-hypothalamic periventricular lesions : thirst deficits and hypernatremia. Am. J. Physiol. 233 : R44 (1977).

3. M.J. McKinley, D.A.Denton, L.G. Leksell, D.R. Mouw, B.A. Scoggins, M.H. Smith, R.S.Weisinger and R.D. Wright. Osmoregulatory thirst in sheep is disrupted by ablation of the anterior wall of the optic recess. Brain Res. 236 : 210 (1982).

4. M.L. Mangiapane, T.N.Thrasher, L.C. Keil, J.B. Simpson and W.F. Ganong. Deficits in drinking and vasopressin secretion after lesions of the nucleus medianus. Neuroendocrinol. 37 : 73 (1983).

5. B. Anderson, L.G. Leksell and F. Lishajko. Perturbations in fluid balance induced by medially placed forebrain lesions. Brain Res. 99 : 261 (1975).

6. M.J. McKinley, M. Congiu, D.A. Denton, R.G. Park, J.Penschow, J.B. Simpson, E. Tarjan, R.S. Weisinger and R.D. Wright. The anterior wall of the third cerebral ventricle and homeostatic responses to dehydration. J.Physiol (Paris) suppl. In press (1984).

7. M.J. McKinley, D.A. Denton, R.G. Park and R.S. Weisinger. Cerebral involvement in dehydration-induced natriuresis. Brain Res. 263: 340 (1983).

8. M.J. Brody, J.R. Haywood and K.B. Touw. Neural mechanisms in hypertension. Ann.Rev.Physiol. 42: 441 (1980).

THE ORGANUM VASCULOSUM LAMINAE TERMINALIS

AND WATER BALANCE IN DOGS

Terry N. Thrasher and David J. Ramsay

Department of Physiology, School of Medicine
University of California, San Francisco
San Francisco, CA. 94143

INTRODUCTION

Circumventricular organs (CVO) are unique regions of the brain which lack the blood-brain barrier. Thus, neurons, or their processes, located within CVO's are in relatively free communication with blood. This anatomical peculiarity and other evidence which suggests the forebrain contains osmoreceptive elements, have made the subfornical organ (SFO) and the organum vasculosum laminae terminalis (OVLT), prime candidates as receptor areas for the regulation of water balance.

Investigations carried out primarily in rats have provided a large body of evidence which implicates the SFO as a central receptor site for blood-borne angiotensin II (1). The role of the OVLT in water balance is less certain. In the rat, lesions of the antero-ventral region of the third ventricle (AV3V), which includes the OVLT, cause major disruptions in both osmotic and volemic regulation of water balance (2,3). Thus, it is possible that destruction of the OVLT may contribute to some of the deficits observed in rats with AV3V lesions. However, assigning specific functional roles to tissues within the AV3V region is complicated by the fact that the lesion also includes the median preoptic nucleus (MPON), which may also be an important component in the regulation of water balance (4,5).

Studies carried out in dogs suggested to us that the cerebral osmoreceptors involved in regulating drinking and secretion of arginine vasopressin (AVP) are located in a CVO (6 and chapter entitled "Hyperosmotic and hypovolemic thirst",this volume). In our first test of this hypothesis, we found that discrete electrolytic lesions of the SFO eliminated drinking in response to systemic administration of angiotensin II (AII) but did not affect drinking to hypertonic NaCl (7). This report summarizes the effects of lesioning the OVLT on drinking and secretion of AVP in response to hypertonic NaCl, AII, hemorrhage and dehydration.

ACUTE EFFECTS OF OVLT LESIONS

Plasma osmolality rose 18 ± 3 mosmol/kg within 24h after lesioning the OVLT due to spontaneous water diuresis. The rise in plasma osmolality was not accompanied by a compensatory increase in water intake and plasma levels of vasopressin. Rather, water intake was highly variable but did not differ from the control period and plasma vasopressin remained at basal levels. Gradually over 1 to 3 weeks, plasma osmolality returned to levels within the range of control after which time, responses to acute challenges to water balance were re-evaluated.

OSMOTIC RESPONSIVENESS

As has been reported previously (8) ablation of the OVLT increased the latency to drink from 7 ± 1 min to 19 ± 2 min in response to hypertonic NaCl (0.32 mM/kg/min,iv). Furthermore, water intake in response to a NaCl load of 14.6 mM/kg was reduced from 58 ± 6 ml/kg to 21 ± 4 ml/kg (p<0.01) after the lesion. As with drinking, the osmotic threshold required to elicit secretion of AVP was elevated and the magnitude of the response was blunted after the lesion (Table 1). In the control dogs, with lesions adjacent to but not including the OVLT, there were no changes in osmotic responsiveness.

Table 1. OVLT lesion and secretion of AVP in response to hypertonic NaCl, AII, and hemorrhage.

Stimulus	ΔP_{AVP} (pg/ml)	
	Before Lesion	After Lesion
NaCl (14.6 mM/kg body wt)	11.3 ± 1.8	3.9 ± 0.7 *
AII (20 pmol/kg/min)	2.5 ± 0.6	0.1 ± 0.2 *
Hemorrhage (20 ml/kg)	31 ± 9	31 ± 8

Values shown are the means $\pm$1SE of the change in P_{AVP} from control.
*P<0.01 compared to response before lesion.

AII RESPONSIVENESS

Ablation of the OVLT markedly reduced (p<0.01) drinking in response to AII (Fig. 1). Before the lesion all six dogs drank

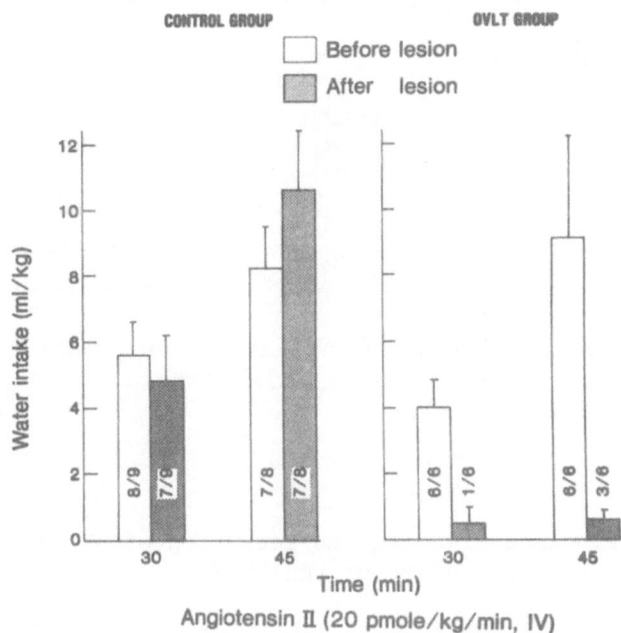

CONTROL GROUP   OVLT GROUP

Fig. 1.   Effect of OVLT lesion on AII-induced drinking

during a 30 min infusion of AII but only one of six drank following
the lesion.   In a second protocol, in which water was offered after
45 min of AII infusion, three of six dogs drank but the volumes consumed were much less compared to the pre-lesion response.   Secretion
of AVP in response to AII was also eliminated following destruction
of the OVLT (Table 1).   Again, there were no deficits in either
drinking or secretion of AVP in the dogs with control lesions.

VOLEMIC RESPONSIVENESS

Before lesioning blood pressure fell from a control level of
96 ± 3 mmHg to an average of 82 ± 6 mmHg after removal of 20 ml/kg of
blood.   After destruction of the OVLT, blood pressure fell from a
control level of 97 ± 4 mmHg to 88 ± 2 mmHg during the hemorrhage period.
Statistical comparison indicated no difference between the responses,
therefore destruction of the OVLT did not compromise the ability to
maintain arterial blood pressure in response to hypovolemia.   In the
control group, the degree of hypotension following hemorrhage was
not affected by the lesion nor was it different from the dogs with
OVLT lesions. The volume of water drunk in response to hemorrhage is
shown in Fig. 2.   In contrast to the deficits in osmotic and AII-
induced drinking, there was no significant effect of the lesion on
drinking in response to hypovolemia.   Furthermore, destruction of
the OVLT did not affect the rise in plasma AVP in response to the
hemorrhage (Table 1).   These results clearly indicate that mechan-

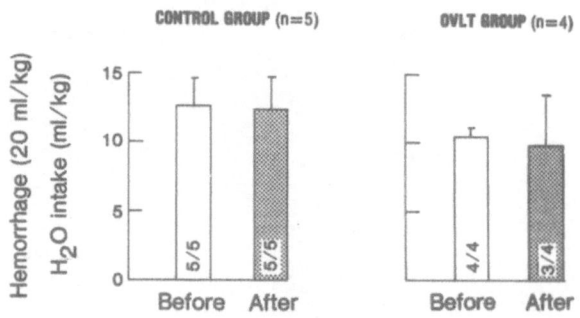

Fig. 2. Effect of OVLT lesion on drinking
in response to hemorrhage.

isms which stimulate drinking and secretion of AVP in response to
hypovolemia, are independent of the OVLT.

RESPONSE TO DEHYDRATION

The changes in plasma osmolality and AVP after 24h of water
deprivation are shown in Table 2. The increase in plasma osmolality
was significantly greater after destruction of the OVLT but the
increase in plasma AVP and water intake were not different. As
plasma AVP and water intake did not increase in proportion to the
rise in plasma osmolality, it suggests that these responses were
attenuated by the lesion. This point can be illustrated by calculat-

Table 2. OVLT lesion and response to 24h dehydration

	Before lesion[a]	After lesion[a]
ΔPosm (mosmol/kg)	8.8 ± 1.5	15.2 ± 1.6 *
ΔP_{AVP} (pg/ml)	4.5 ± 1.1	3.2 ± 1.1
Water Intake (ml/kg)[b]	32 ± 7	39 ± 7
$\Delta P_{AVP}/\Delta$Posm	0.47 ± 0.06	0.20 ± 0.05 *
H_2O/ΔPosm	4.1 ± 0.5	2.7 ± 0.5 *

Values shown are the means $\pm$1SE; n = 5.
a. Change induced by 24h dehydration.
b. Volume consumed in one hour following 24h dehydration.
*P<0.05 compared to response before lesion.

ing the ratio of the change in plasma AVP to the change in plasma osmolality before and after ablation of the OVLT. Comparison of the ratios indicates that the rise in plasma AVP was much less per unit change in plasma osmolality after the lesion (Table 2). A similar conclusion is reached by comparing the ratio of water intake to the change in plasma osmolality. The dogs drank less per unit change in plasma osmolality after destruction of the OVLT.

DISCUSSION

Our results show that ablation of the OVLT leads to disruption of mechanisms regulating thirst and secretion of AVP in response to hypertonicity and AII. In contrast, the lesion does not appear to affect responsiveness to hypovolemia. During water deprivation, which combines osmotic and volemic stimuli, the increase in plasma osmolality was 73% higher after the lesion without a corresponding increase in plasma AVP or water intake. This effect is compatible with a loss of osmotic sensitivity in the presence of normal volume regulation. McKinley et al. (9) have reported similar results in sheep. They found that lesions which involve the OVLT cause disruption of osmotically induced drinking and a greater rise in plasma osmolality during dehydration. Taken together, these observations support the idea that the OVLT is an important link in the regulation of water balance.

The alterations in response to hypertonic NaCl, which include an elevated threshold and a reduction in the magnitude of the response, are compatible with the hypothesis that the OVLT contains osmoreceptors (8,9). However, the OVLT cannot be the only osmoreceptive site as drinking and secretion of AVP in response to hypertonic NaCl are not totally eliminated. But, considering the magnitude of the change in plasma osmolality required to induce these responses, e.g., 7 ± 2 mosmol/kg before vs. 24 ± 2 mosmol/kg after lesioning, it is difficult not to conclude that the physiologically relevant osmoreceptors are contained within the OVLT.

Hypovolemia can stimulate drinking and secretion of the AVP through either peripheral cardiovascular afferents or through increased plasma levels of AII. Because destruction of the OVLT did not affect these responses to hemorrhage, the results indicate that signals arising from cardiovascular afferents are independent of osmoregulatory pathways. However, AII-induced drinking and secretion of AVP do appear to require the integrity of the OVLT. At present we do not have a clear explanation for this effect although there are a number of possibilities. It may be that loss of efferent signals from the OVLT is interpreted as a hypo-osmolar state which, in turn, raises the threshold for AII-induced drinking. Alternatively, important pathways from the SFO may pass in close proximity to the OVLT and are cut by the lesion. Further study will be required to evaluate these possibilities.

One problem with interpretation of lesions is that it is impossible to differentiate between damage to neurons and to fibers of passage. Recently we have begun experiments in collaboration with Dr. R. Miselis to trace the connectivity of forebrain CVO's in the dog. The approach has been to inject horseradish peroxidase (HRP) conjugated to wheat germ agglutinin into the supraoptic nucleus (SON), or into the MPON. In two dogs in which the injectate was localized in the SON, we observed retrograde filling of neurons in the OVLT and the MPON. When the HRP was injected into the MPON, neurons in the OVLT again stained positive for HRP. These results give direct evidence of efferent input to the SON and MPON originating from neurons located within the OVLT.

In conclusion, our results indicate that the OVLT plays an important role in the neural mechanisms responsible for maintenance of normal water balance. Furthermore, it appears highly likely that Verney's osmoreceptors are located within this forebrain CVO.

(This work was supported by NIH grant HL-29714 and HL-01106.)

REFERENCES

1. J. B. Simpson, The circumventricular organs and the central actions of angiotensin, Neuroendocrinology 33: 248 (1981).
2. J. Buggy and A. K. Johnson, Preoptic-hypothalamic periventricular lesions: thirst deficits and hypernatremia, Am.J.Physiol. 233: R44 (1977).
3. A. K. Johnson and J. Buggy, Periventricular preoptic-hypothalamus is vital for thirst and normal water economy, Am.J.Physiol. 234: R122 (1978).
4. M. J. Brody and A. K. Johnson, Role of the anteroventral third ventricle region in fluid and electrolyte balance, arterial pressure regulation, and hypertension, in: "Frontiers in Neuroendocrinology", vol. 6, L. Martini and W. F. Ganong, eds., Raven Press, New York (1980).
5. R. R. Miselis, The efferent projections of the subfornical organ of the rat: a circumventricular organ within a neural network subserving water balance, Brain Res. 230: 1 (1981).
6. T. N. Thrasher, C. J. Brown, L. C. Keil, and D. J. Ramsay. Thirst and vasopressin release in the dog: an osmoreceptor or sodium receptor mechanism? Am.J.Physiol. 238: R333 (1980).
7. T. N. Thrasher, J. B. Simpson, and D. J. Ramsay, Lesions of the subfornical organ block angiotensin-induced drinking in the dog, Neuroendocrinology 35: 68 (1982).
8. T. N. Thrasher, L. C. Keil, and D. J. Ramsay. Lesions of the organum vasculosum of the lamina terminalis (OVLT) attenuate osmotically-induced drinking and vasopressin secretion in the dog, Endocrinology 110: 1837 (1982).
9. M. J. McKinley, D. A. Denton, L. C. Leksell, D. R. Mouw, B. A. Scoggins, M. H. Smith, R. S. Weisinger, and R. D. Wright. Osmoregulatory thirst in sheep is disrupted by ablation of the anterior wall of the optic recess, Brain Res. 236: 210 (1982).

IN VITRO DOWN REGULATION AND POSSIBLE INTERNALIZATION OF CENTRAL

ANGIOTENSIN RECEPTORS

J.W. Harding, J.B. Erickson, J.W. Wright, C.G. Camara
and R.H. Abhold

Department of VCAPP
Washington State University
Pullman, WA 99164-6520

INTRODUCTION

Angiotensin receptors in peripheral tissues have been shown to up or down regulate in response to changes in circulating angiotensin [2/3]. Nevertheless it has been difficult to demonstrate changes in central angiotensin receptors in response to alterations in blood or cerebral spinal fluid (CSF) levels of angiotensins [4/5]. Failure to detect such changes could be the result of a dilution effect whereby only a small percentage of angiotensin receptors are exposed to altered ligand levels or a rapidly recycling population of receptors. This study demonstrates that tachyphalaxis does occur to intracerebro-ventricularly (ICV) applied angiotensins. This change in angiotensin sensitivity may be a result of receptor down regulation and internalization which appear to occur at least in the in vitro preparation used in this study.

MATERIALS AND METHODS

Adult male rats were prepared with ICV cannula aimed at the right lateral ventricle. All animals were first ICV injected with 2μl of artificial CSF, the vehicle for AII and AIII injections. Ten minutes later and at subsequent 20 min. intervals, rats were injected with either 10 pmol of AII or AIII. Water consumption was recorded for each interval.

Additional rats were killed by decapitation, the brains removed, and the hypothalamus, septum, AV3V, and thalamus dissected out as a unit. Tissue was polytron homogenized in 40 volumes cold Krebs-Ringer buffer (KRB) pH 7.4.

The homogenate was spun at 100g for 10 min. The supernatant was spun again at 35,000g for 15 min. The pellet was rehomogenized in KRB and spun again. The pellet was once more rehomogenized in KRB. Tissue homogenate was then preincubated for 30 min. with various angiotensins under 95% O_2: 5% CO_2 at 37°C. After preincubation, the suspension was spun at 35,000g for 15 min. The supernatant was discarded and the pellet rehomogenized in KRB. Postincubation to allow for ligand dissociation was carried out for various times between 120 -180 minutes. The incubate was spun and then rehomogenized in 50mM Tris, 5mM Na_2EDTA, 5mM DTT, and 150mM NaCl (HB). The suspension (200μl) was added to ≃ 200,000 cpm of labeled ligand (200μl) and either buffer (100μl) or 5μM unlabeled ligand (100μl). After 30 min. incubation at 22°C the suspension was filtered on PEI (.3%) soaked 934-AH filters and washed with 4-5ml aliquots of HB. Specific binding was the difference between total binding and binding with 1μM cold ligand.

Dual filter experiments were carried out using 934-AH filters soaked in PEI (3%)[1] or BSA (.1%). Supernatants were separated with 10 min. spins at 13,000 x g. All sonications were for 5 sec.

Label ligands were iodinated by an immobilized glucose oxidase-lactoperoxidase method and purified by HPLC to > 99.5% purity.

Kruskel-Wallis H tests which were used for independent populations to assess significance, were followed by Mann-Whitney U tests for pair wise comparisons.

RESULTS

Water consumption (ml/10 min) following ICV injections of 2μl of artificial CSF followed by repeated injections of 10 pmol of AII or AIII are shown in Figure 1. The initial injections of both AII and AIII produced robust and equivalent drinking responses. Subsequent injections of AIII produced little addditional drinking even though the quantity consumed after the initial injection was well below the maximum capability of the animal[6]. The pattern of drinking was quite different for AII with significant but continually decreasing responsiveness to AII. This decreased sensitivity of the rats to AII was shown in subsequent experiments to last 30-40 min. (data not shown).

The effect of tissue preincubation with various angiotensins on ^{125}I-AII and ^{125}I-Sar[1],Ile[8]-AII binding is shown in Table 1. ^{125}I-AII binding was dramatically diminished after a 30 min. preincubation with AII, AIII, or Sar[1],Ile[8]-AII. The most effective initiator of receptor down regulation was Sar[1], Ile[8]-AII. The effectiveness of AII or Sar[1],Ile[8]-AII at reducing binding was concentration-dependent with statistically significant reductions

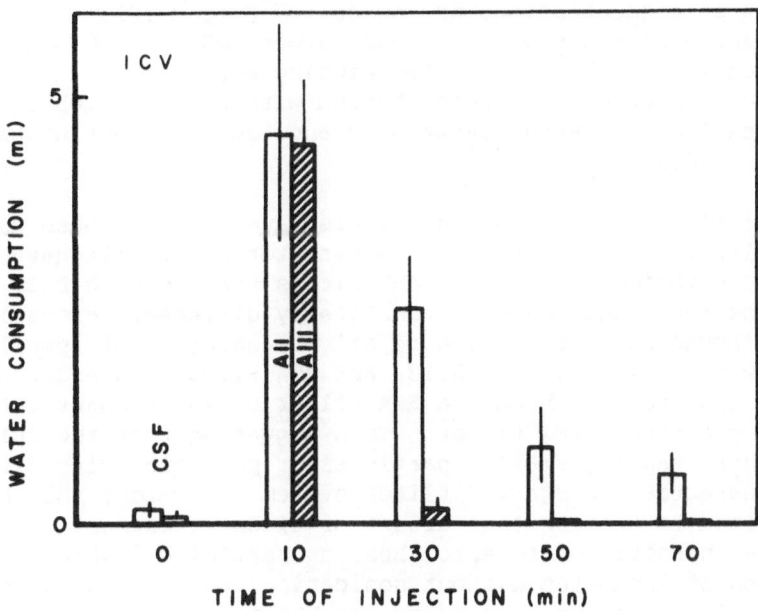

Fig. 1. Mean ± S.E.) water consumption (ml/10 min) following icv injections (2 μl) of artificial CSF and repeated injections of AII or AII (10pM/2μl CSF) (n = 8).

Table 1. Percent inhibition of ^{125}I-AII or ^{125}I-Sar^1,Ile^8-AII binding to rat brain membranes following preincubation with unlabeled angiotensins (mean ± S.D., n = 3)

Label	Angiotensin	Concentration of preincubated angiotensin*					
		60 min post-incu.**					150 min
		10^{-9}	10^{-8}	10^{-7}	10^{-6}	10^{-5}	10^{-5}
^{125}I-AII	AII	16±4	35±7	32±11	44±10	59±12	49±12
^{125}I-AII	AIII	1	3	9	16	58	31
^{125}I-AII	Sar^1,Ile^8-AII	36±1	65±5	71±9	95±5	92±3	82±7
^{125}I-Sar^{1}Ile^8-AII	AII	6±1	13±3	22±4	29±5	42±12	44±5

*All data points were significantly different from control, $p < .01$.
**All comparable 60 min. and 150 min. postincubation points were not significantly different, $p > .05$.

occurring at concentrations as low as 10^{-9}M (p < .01). No
significant difference was observed between 60 and 150 min. postincubation times. ^{125}I-Sar^1,Ile^8 binding was also
diminished subsequent to preincubation with AII (p > .05). Again no
differences were apparent between 60 and 120 min. postincubation
times (p > .05).

Polyethylenimine (PEI)-treated filters have recently been used in a
rapid filtration assay for soluble receptors.[1] Preliminary
experiments indicated that 934-AH filters treated with PEI,
polylysine (PL), and BSA had dramatically different recoveries of
membrane-bound receptors. The relative recoveries of specific
binding were PEI--3.17, PL--3.13, and BSA--1.00. In addition,
material that passed through a BSA filter could be quantitatively
trapped by a single PEI filter. This suggested that the PEI filters
were either trapping smaller particles or particles with different
charge character. A stacked filter system (BSA--top; PEI--bottom)
was utilized to examine that effect of tissue treatment on the size
or charge character of receptor bearing particles (Table 2).
Filtration of incubates without sonication resulted in 75% of the
specific binding being trapped by the BSA filter. This is
considerably more than is seen on a single BSA filter and may be due
to the decrease in flow rate that occurs in the stacked filter
system. When incubates are sonicated prior to incubation, little
change in the distribution of specific binding between the filters is
seen. However, sonication following incubation produces a dramatic
shift of specific binding to the PEI filters. If the incubation
suspension is centrifuged and the supernatant filtered using PEI
filters, the effect of tissue treatment is magnified (Table 3).
Sonication prior to incubation increases supernatant binding over 60%
when compared to no sonication. When sonication is performed after
incubation, a dramatic 355% increase in supernatant binding is
observed.

Table 2. Effect of incubation conditions on ^{125}I-Sar^1,Ile^8-AII
binding to rat brain membranes (% of total specific binding ± S.D., n = 3)

Tissue Treatment	Filter Treatment	
	BSA*	PEI**
No sonication	75 ± 5	25 ± 5
Sonication, incubation	67 ± 7	33 ± 7
Incubation, sonication	49 ± 3	51 ± 3
Sonication, incubation sonication	34	66

*Bovine serum albumin
**Polyethylenimine

Table 3. Effect of incubation conditions on $^{125}I\text{-Sar}^{1}$, Ile^{8}-AII binding on 13,000 x g supernatants prepared from rat brain membranes (% control binding)

Treatment	Binding
No sonication	61
Incubation, sonication	278
Control--sonication, incubation	100

DISCUSSION

Repetitive ICV application of AII or AIII suppressed water intake to subsequent injections of peptide. AIII was much more effective than AII in initiating this tachyphalaxis response. Amazingly, a single 10 pmol application of AIII completely eliminated any additional drinking while multiple injections of AII were needed to achieve similar results. Two explanations seem plausible to account for AIII's superior effectiveness. (1) AII has a higher affinity for the receptor and is able to more effectively stimulate the remaining nondesensitized receptors. Or (2) AII must first be converted to AIII before it can interact with the angiotensin recognition site. If the first explanation were correct, one would expect a larger initial response from AII; however, this is not observed. If the second explanation were correct one might predict the following: (1) Less AIII should be needed to initiate tachyphalaxis than AII, (2) Cross tachyphalaxis to AII should be observed after the tissue is desensitized by AIII, (3) Membrane sites should be available to convert AII to AIII, and (4) the active agent AIII should be more effectively degraded than the precursor, AII. Preliminary data from our laboratory indicates that cross tachyphalaxis does occur, that isolated membranes can convert AII to AIII, that AII is more resistant to degradation in the ventricular space than AIII, and that AIII but not AII is metabolized by CSF.

The explanation for tachyphalaxis may be due to changes in the state of the receptors or an actual down regulation and internalization of the receptor complex [2,3]. The notion of down regulation in this system is attractive, except for the fact that chronic infusions of AII have failed to alter receptor numbers [4,5]. This failure could be due to several factors, including a dilution effect in which only those few receptors exposed to the ventricular system would change, or a rapid recycling of receptors so that rapid sampling would be necessary. Rapid recycling is consistent with the observation that full responsiveness returns in less than 40 minutes after desensitization. Instead of performing the rapid sampling studies which would still be plagued by dilution effects, we chose to determine whether angiotensin receptors could, in fact, down regulate in an in vitro system where all receptors could be exposed to ligand. The results clearly indicate that they can. The relative effectiveness of the various angiotensins in initiating loss of receptors may not be due to receptor affinities but rather degradation resistance. The purpose of the postincubation step was to insure dissociation of preincubation ligand. Multiple postincubation times were used to make certain all ligand had been removed.

The last set of experiments was designed as a preliminary assessment of possible receptor internalization. The dual filter experiment showed a shift in receptor binding from the upper BSA filter to the lower PEI filter when sonication followed incubation. This suggests that receptors are changing compartments subsequent to ligand binding. The purpose of the sonication was to release internalized receptors which are most likely located in coated vesicles. Sonication prior to incubation produced little change in the ratio of filter binding, thus indicating that the shift with postincubation sonciation was not due to mechanical treatment and production of small particles. The idea that receptors are moving to a new compartment after ligand binding is further supported by the enrichment of receptors in the 13,000g supernatant following sonication. Again, this process requires prior exposure to ligand and is not due to sonication alone. In total, these results are consistent with the concept that tachyphalaxis to centrally applied angiotensins is a result of reversible but rapid receptor internalization.

ACKNOWLEDGEMENTS

This study was supported by a grant from the American Heart Association, Inc. and the American Heart Association of Washington.

REFERENCES

1. R.F. Bruno, K. Lawson-Wendling, and T.A. Pugsley, A rapid filtration assay for soluble receptors using polyethenimine treated filters. In press.
2. M.A. Devynck, B. Rouzaine-Dubois, E. Chevillotte, and P. Meyer, Variations in the number of uterine receptors following changes in plasma angiotensin levels. Eur. J. Pharmacol. 40: 27-37 (1976).
3. R.L. Hauger, G. Aguilera, and K.J. Catt, Angiotensin II regulates its receptor sites in the adrenal glomerulosa zone. Nature 271:176-177 (1978).
4. R. Singh, C.M. Ferrario, and R.C. Speth, Effects of intraventricular angiotensin II antagonist infusion on ^{125}I-AngII binding in rat brain. Soc. Neurosci. Abstr. 9:591 (1983).
5. R. Singh, A. Husain, C.M. Ferrario, and R.C. Speth, Rat brain angiotensin II receptors:effects of intraventricular angiotensin II infusion. Brain Res. In press.
6. J.W. Wright, S. Morseth, J.J. Mana, E. LaCrosse, E.P. Petersen, and J.W. Harding, Central angiotensin III-induced dipsiogenicity in rats and gerbils. Brain Res. 295:121-126 (1984).

EFFECT OF INTRACEREBROVENTRICULAR ADMINISTERED VANADATE ON SALT AND WATER INTAKE AND EXCRETION IN THE RAT

Emma Chiaraviglio and Constanza Lozada

Instituto de Investigación Médica
Mercedes y Martìn Ferreyra
Casilla de Correo 389 - 5000 Còrdoba, Argentina

INTRODUCTION

The activity of sensors mediating thirst and ADH release has been correlated to NaCl concentration of the brain CSF. Substances known to inhibit active trans-membrane enzymatic Na transport have been found to inhibit water intake and ADH release (1,2), suggesting that active Na transport may be essential for the excitation of cerebral sensors involved in water balance. Therefore, the discovery that vanadate, a natural occurring substance, is a potent "in vitro" inhibitor of Na,K-ATPase activity (3) made it an interesting "tool" to study water balance. In rats, acute sodium depletion by peritoneal dialysis produced dramatic changes in serum and CSF sodium concentration (4,5). Furthermore infusion of hypertonic artificial CSF into the 3rd ventricle decreased Na intake induced by sodium depletion (5). Therefore it was of interest to study whether an inhibitor of enzymatic Na-transport such as vanadate, would affect sodium and water intake and excretion in conscious rats.

METHODS

The experiments were carried out on adult male albino rats, weighing 250-350 g at the beginning of the experiment. Animals were previously trained for sodium depletion, water deprivation and intake tests, according to the experiments to which they were assigned. A group of rats was depleted of sodium by peritoneal dialysis (pd) against isotonic glucose. The technique was previously described (4,5). After dialysis, rats were caged individually without food and with free access to distilled water. Twenty four hours after pd a two bottle test of 1.8% NaCl and distilled water was run. Urine was collected at the end of the test, measured and aliquots stored frozen for Na and

K determination. Under anesthesia the rats were implanted with a stainless steel cannula 15mm in length and 0.6mm o.d. lowered into the 3rd brain ventricle. After a week of recovery animals were assigned to an experimental procedure. Twenty four hours after pd, rats received a pulse injection (1μl) of vanadate (1.0mM) dissolved in artificial isotonic cerebrospinal fluid (CSF) (5), through implanted cannula. Intake tests were given 10 minutes later. Each rat received a vanadate injection between two CSF (vehicle) injections. Continuous infusion of vanadate (1.0mM) during 24h was carried out by an osmotic (Alzet) minipump (mp) implanted under the loose skin of the back. Infusion starded when mp was connected to the guide cannula, one hour before pd. Twenty four hours later, the infusion was stopped and a two bottle intake test was given. In this experiments vanadate was dissolveld in distilled water since 24h infusion of isotonic CSF decreased sodium intake "per se"(5). Controls were the same rats infused with distilled water. A catheter was inserted into the jugular vein. On the following day a mp filled with vanadate 1.0mM, delivering 1μl/h was implanted and connected to the vein. Twenty four hours later, infusion was discontinued, and a two bottle intake test was run. Same rats received, as control, a mp filled with saline. Vanadate (1.0mM) plus ascorbic acid (AsAc, 1.0mM) dissolved in CSF were icv injected (1μl) in 6 rats depleted of sodium. Control values were taken from the same animals injected with CFS, vanadate 1.0mM, or AsAc, 1.0mM. Ascorbic acid which is capable of reducing vanadate ions, would result in formation of vanadyl a compound which "in vitro" prevents the inhibition of Na,K-ATPase (6).

RESULTS

Intracerebroventricular injection (1μl) of vanadate (1.0mM) significantly decreased sodium intake in sodium depleted rats: (6.9± ±0.86ml) compared with control value (10.5±1.01ml), the difference was significant (P < 0.01) (Fig. 1). The same concentration of vanadate injected into the lateral hypothalamus did not affect Na intake. The urine volume and sodium excreted during the test increased 134% and 280% respectively as is shown by Fig. 2, whereas urine K excretion decreased 76%; vanadate 0.5mM, a dose which did not affect sodium intake increased urine volume and sodium excretion 158% and 320% respectively, while decreasing potassium excretion. Similar effects were observed with the dose of 1.5mM. Continuous infusion of vanadate, 1.0mM at a rate of 1μl/h, during 24h between pd and test decreased intake of 1.8% NaCl solution by 69%. Comparing the mean volumes ingested by controls (9.5±1.3ml) with that of vanadate infused rats (2.9±1.2ml) the difference was significant (P < 0.001). Moreover, when the same concentration of vanadate at the same rate was given by intravenous route the volume of sodium ingested by experimental and control animals did not differ. Injection of vanadate plus ascorbic acid (1μl) did not affect sodium intake; the intake volume did not differ from that observed with injections of CSF or AsAc 1mM. However it differed (P < 0.01) from volumes ingested after vanadate 1.0mM (Fig. 3).

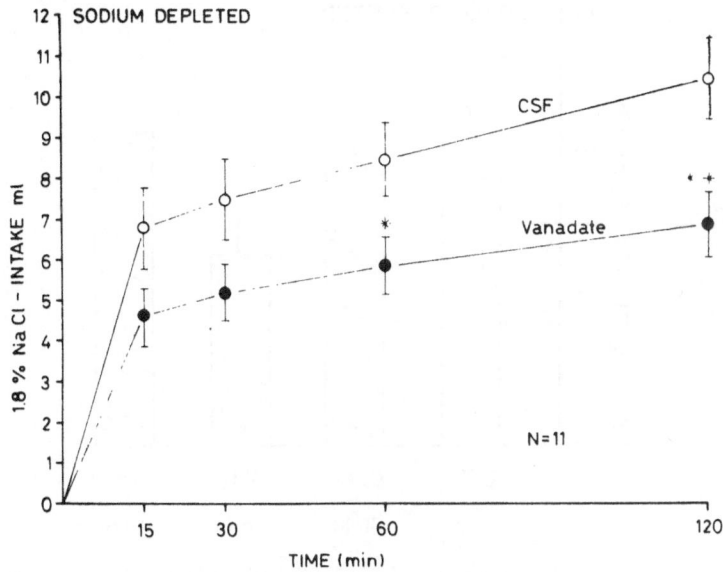

Fig. 1. Effect of icv administration of vanadate (1.0mM) into sodium
        depleted rats on the cumulative intake of 1.8% NaCl solution.
        Mean ± SE of mean. **P < 0.001, * P < 0.01.

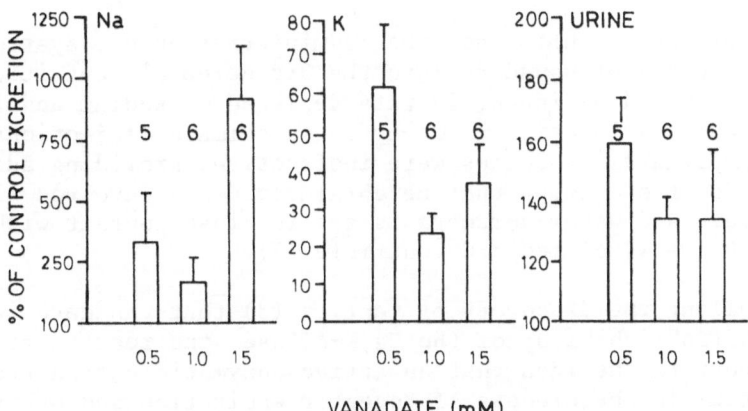

Fig. 2. Urine volume and Na and K excreted during 2 hour drinking
        test by sodium depleted rats after vanadate injection.
        Mean ± SE, percent of control excretion.

Ventricular injection of vanadate 1.0, 1.5 or 1.75mM did not
affect significantly the water intake induced by water deprivation
as compared with intake observed after CSF injection.

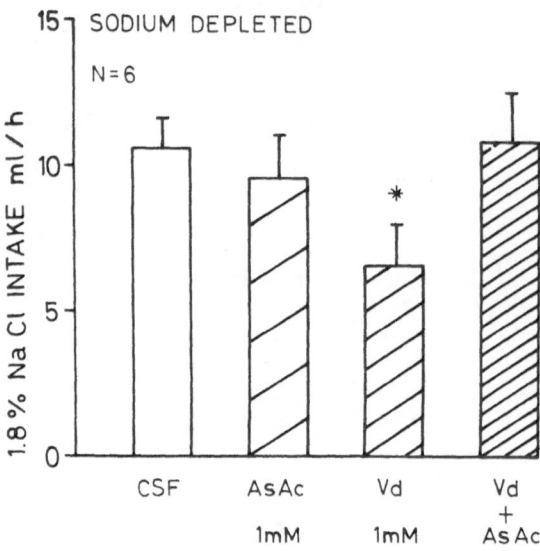

Fig. 3. Sodium chloride intake induced by sodium depletion in rats
icv injected with vehicle (CSF), ascorbic acid (AsAc 1.0mM),
vanadate (Vd 1.0mM) and vanadate plus AsAc.
Mean ± SE of mean. * P < 0.01.

DISCUSSION

     The present results show that administration of systemically
inactive amounts of vanadate into the 3rd cerebral ventricle caused
a decrease in sodium intake in rats depleted of sodium and induced
diuresis and natriuresis in non-hydrated animals. Injections in the
lateral hypothalamic nucleus were ineffective, providing additional
evidence for the concept that cerebral receptors involved in regulation of salt and water homeostasis are in close contact with CSF, on
or near the walls of the 3rd ventricle (7).

     Recalling the discovery by Cantley (3) that vanadate is a power-
ful "in vitro" inhibitor of the Na,K-ATPase, the results reported here
lend support to the idea that an active enzymatic cation transport may
be essential in the process of receptor excitation and/or transduction.
The suggestion that salt appetite may depend on the rate of active sodium transport was first reported by Denton (8) in sheep. Since then,
several reports extended the idea that activity of receptors mediating
water intake and antidiuretic hormone (ADH) release may depend on enzymatic cation transport (1,2,9,10,11). Sodium depleted rats, despite
the fact they were in salt and water imbalance (4), responded to vanadate with copious diuresis and natriuresis, a finding in disagreement
with reports which show that vanadate increased water and salt excretion only in volume expanded rats (10,11). Ventricular injection of

vanadate plus a reducing agent (see Fig. 3) did not decrease sodium intake. This experiment favors the idea that Na-transport enzyme could be linked to receptor activation. Vanadate inhibits "in vitro" Na,K-ATPase (3) whereas vanadil is inactive or a mild inhibitor (6).

Vanadate did not affect water intake induced by water deprivation. It has been reported that ouabain and ethacrinic acid, inhibitors of active sodium transport, inhibited drinking in sheep (1,2). Ouabain is a highly specific inhibitor of Na,K pump, acting at sites accessible from the intracellular medium. Vanadate, on the contrary, acts at sites accessible from the intracellular medium (12) and its actions are potentiated by K ions at the extracellular surface (13). In addition, vanadate is not a specific inhibitor of Na,K pump but inhibits Ca,Mg-ATPases and other enzymatic processes as well. Whatever mechanism of the vanadate effect described here it is evident from the literature that vanadate is a paramount enzymatic inhibitor.

REFERENCES

1. M. Rundgren, M.J. McKinley, L.G. Leksell and B. Andersson, Inhibition of thirst and apparent ADH release by intracerebroventricular ethacrinic acid, Acta physiol. scand. 105:123 (1979).
2. R.S. Weisinger, D.A. Denton and M.J. McKinley, Inhibition of water intake by ouabain administration in sheep, Pharmac.Biochem.Behav. 7:121 (1977).
3. L.C. Cantley, L. Josephson, R. Warner, M. Yanagisawa, C. Lechene and G. Guidotti, Vanadate is a potent (Na,K)-ATPase inhibitor found in ATP derived from muscle, J. Biol. Chem. 252:7421 (1977).
4. M.C. Ferreyra and E. Chiaraviglio, Changes in volemia and natremia and onset of sodium appetite in sodium depleted rats, Physiol. Behav. 19:197 (1977).
5. E. Chiaraviglio and M.F. Pérez Guaita, Effect of cerebroventricular infusion of hypertonic sodium solutions on sodium intake in the rat, This Symposium.
6. J.J. Grantham; The renal sodium pump and vanadate, Am. J. Physiol. 239:F97 (1980).
7. E. Chiaraviglio and M.F. Pérez Guaita, Antero third ventricle (A3V) lesion and homeostasis regulation, Symposium on Body Fluid Homeostasis, Evian (1983).
8. D.A. Denton, F.W. Kaintz, The inhibition of salt appetite of sodium deficient sheep by intracarotid infusion of ouabain, Comm.Behav. Biol. 4:183 (1969).
9. B. Appelgren and S. Ericksson, Apparent inhibition of the ADH secretion induced by intracerebroventricular vanadate, Acta physiol. scand. 108:C18 (1980).
10. W.E. Balfour and A.L.R. Findlay, Intracerebroventricular vanadate and natriuresis in the rat, J. Physiol. 322:35P (1982).
11. W.E. Balfour, J.J. Grantham and I.M. Glynn, Vanadate stimulate natriuresis, Nature, London, 275:768 (1978).

12. L. Beaugé, Vanadate-Potassium interactions in the inhibition of Na,K-ATPase. From: "Na,K-ATPase: Structure and Function". Skou, J.C. and Norby, J. Editors, Academic Press, New York, pp. 373 (1979).
13. L. Beaugé and I.M. Glynn, Commercial ATP containing traces of vanadate alters the response of $(Na^+ + K^+)$ATPase to external potassium, <u>Nature</u>, London, 272:551 (1978).

The authors wish to thank Dr. L. Beaugé for valuable discussion. Supported by grant 148/83 from CONICOR., Córdoba, Argentina.

THE VISCERAL NEURAXIS IN THIRST AND RENAL FUNCTION

Richard R. Miselis

Animal Biology
School of Veterinary Medicine
University of Pennsylvania
Philadelphia, Pa. 19104

INTRODUCTION

Truisms are, of course, self evident. In spite of their obviousness we need to reconsider some in order to appreciate understanding that we take for granted and to remind ourselves of the value of particular approaches we use in our research on brain function. It seems evident to me that understanding behavior and higher order brain function reduces to knowing the rules governing the flow of information in neural networks. Biochemical, molecular, cellular and pharmocological events affect brain function by rising to modulate this flow of infomation in neural circuits. Given these axioms it is, therefore, important to learn the organization of the brain's neural networks.

In the last decade there has been tremendous progress in identifying elements of the neurocircuitry which modulate visceral function. The advance is due to the improvement in neuroanatomical methodology which is now sensitive enough to follow the projections of the finest diameter axonal fibers over long distances both anterogradely and retrogradely. One of the major observations emerging from the new data is that there are long, and much more direct connections than previously known relaying sensory information up to the hypothalamus, limbic structures and the cortex from visceral organs [1]. This is reciprocated by newly discovered long and direct connections from cortex, limbic structures and hypothalamus to the brainstem visceral relay nuclei and motor output nuclei of the autonomic nervous system[2-4]. The network modulates the autonomic nervous system.

Some elements of this network also have strong connections with both the posterior and anterior pituitary gland[5] which brings modulation of endocrine function within its domain. A neurocircuitry encompassing both the autonomic nervous system and the pituitary gland enables the broad function of homeostasis.

Behavior is of great importance in the goals of homeostatic and visceral function. Not surprisingly, the neurocircuitry modulating endocrine and ANS function also relates to regions of the hypothalamus traditionally implicated in the control of ingestive behavior. These areas include the preoptic region and the lateral hypothalamus.

Another important finding emerging from recent neuroanatomical studies is the realization that the circumventricular organs of the brain are likely sites for the central receptors to physiological parameters so long hypothesized to play a role in the behavioral and physiological controls of homeostasis[6-9]. This is reinforced by the fact that they have neural projections to elements of the network of neural circuitry referred to above which modulates the pituitary gland and the autonomic nervous system[7,9,10]. In the remainder of this review I will discuss the neurocircuitry of the two circumventricular organs (CVO) which have a prominent connectivity with other brain structures and a third CVO which is part of the anteroventral third ventricular (AV3V) area of the preoptic region. All are implicated in thirst and renal function. It is this circuitry and those elements which modulate endocrine, autonomic and perhaps behavioral function which I refer to as the visceral neuraxis.

CIRCUMVENTRICULAR ORGANS

Circumventricular organs (CVOs) are small midline structures lying at strategic positions on the surface of the cerebral ventricles. See figure 1. They interface subarachnoid, ventricular and vascular spaces. The capillary density is extremely high and there is a fenestrated endothelial lining to the capillaries indicating a lack of the blood brain barrier in CVOs. This feature suggests that they could be sites of central receptors partricularly for circulating factors that are excluded from the brain by the blood brain barrier. To fulfill this role there would have to be neuronal elements and neural projections to other structures of the brain in order to transfer the information. Indeed, two CVOs meet this requirement quite well. The subfornical organ (SFO) and the area postrema (AP) have many neuronal elements and both have profound efferent projections to very specific sites within the brain. They project heavily to particular components of the visceral neuraxis.

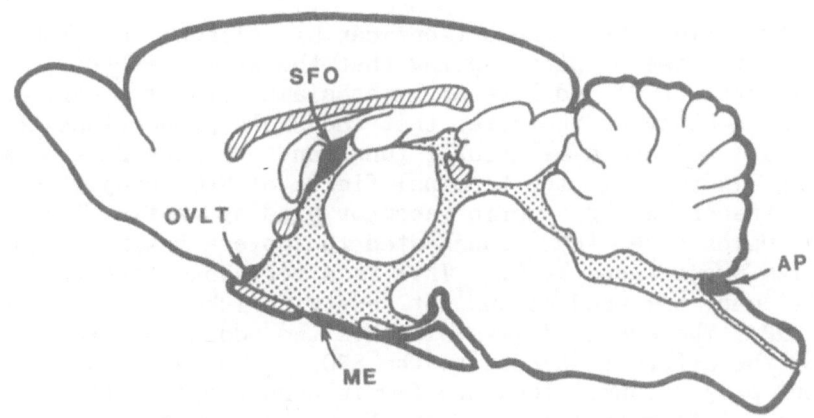

Fig. 1.   Circumventricular organs in the rat. This schematic is
          of a midsagittal section of the rat brain.   It illus-
          trates the location of the subfornical organ (SFO), the
          organum vasculosum of the lamina terminalis (OVLT), the
          median eminence (ME) and the area postrema (AP) along
          the surface of the cerebroventricular system shown by
          the stippled areas.

The Subfornical Organ

     The SFO lies in the dorsal anterior region of the third ven-
tricle.   Its efferent neural projections[7,9] fall into three categories: 1) endocrine; 2) autonomic; and 3) behavioral.   The projections comprising the endocrine category are twofold.   The first
is to the magnocellular neurons of the hypothalamus which, in turn,
form the neurohypophysial tract to the posterior pituitary.   The
second group is a lighter projection to the periventricular, medial
and anterior parvocellular subnuclei of the paraventricular nucleus
(PVN) of the hypothalamus.   This part of the PVN projects to the
median eminence from which emerges the portal system to the anterior pituitary[5].   The autonomic projections are those to other
parvocellular subnuclei of the PVN from which emerge long descending projections to the preganglionic neurons of the entire autonomic nervous system: the parasympathetic motor nuclei of cranial
nerves in the brainstem and the sympathetic intermediolateral cell
column in the spinal cord[3,4].   Since there are no well known sites
which are universally accepted as specific motor output areas for
patterns of ingestive behavior, it is not possible to conclude with

certainty which SFO projections can be classed as behavioral. However, the lesion data showing that the AV3V area[11,12], lateral preoptic area (LPO) and lateral hypothalamus (LH)[13,14] are involved in drinking behavior indicates that the SFO's projections to these sites may mediate a behavioral function. Figure 2 is a summary charting indicating the terminal fields of SFO projections seen when a tracer having specfic receptor binding affinities (horseradish peroxidase [HRP] conjugated to cholera toxin) is used to trace its projections[15,16]. This is a very sensitive tracer which reveals many new projections not seen with labelled amino acids or free HRP. The HRP conjugate also has the added advantage of revealing the afferent sources to the SFO. While the SFO does have afferent projections, they are few in number and scattered forming no obvious influential line of input. The few afferent inputs are overwhelmingly dominated by a potent efferent projection from the SFO. The efferent projections pour into and dominate the AV3V area. There is heavy labelling of the nucleus medianus, the organum vasculosum of the lamina terminalis and the anterior periventricular nucleus and pathway. They penetrate the preoptic region and hypothalamus. There is a very light but specific projection into the medial septal nucleus and a projection to the bed nucleus of the stria terminalis which is light in the anterior ventral area beneath the anterior limb of the anterior commissure and very heavy in the caudal ventromedial region just beneath the anterior commissure. There is a heavy density of terminals in all of the supraoptic nucleus, the nucleus circularis and most of the subnuclei of the PVN. In addition there is a significant set of terminals seen along the dorsal perifornical area, in the LH at the level of the ventromedial nucleus just beneath the zona incerta with some involvement of the zona incerta and in the paraventricular nucleus of the thalamus. Other new projections are much lighter but still very evident. These include projections to suprachiasmatic nucleus, the substantia innominata, the dorsal medial nucleus, more caudal areas of the periventricular nucleus, the arcuate nucleus, the ventral area of the lateral hypothalamus the central and medial nuclei of the amygdala and the lateral parabrachial nucleus of the pons[15,16]. See figure 2 for an illustration of these terminal fields.

The Organum Vasculosum of the Lamina Terminalis (OVLT)

The OVLT is ventral to the SFO lying at the bottom of the anterior wall of the third ventricle on top of the optic chiasm. It is much less neuronal than the SFO and there is no good comprehensive study of its neural projections. Generally there are only incidental reports of retrogradely filled cells in the OVLT indicating projections to the nucleus medianus[17] and to the supraoptic nucleus[7], many reports of fibers terminating in the OVLT or discriptions of cells containing specific peptides with no information

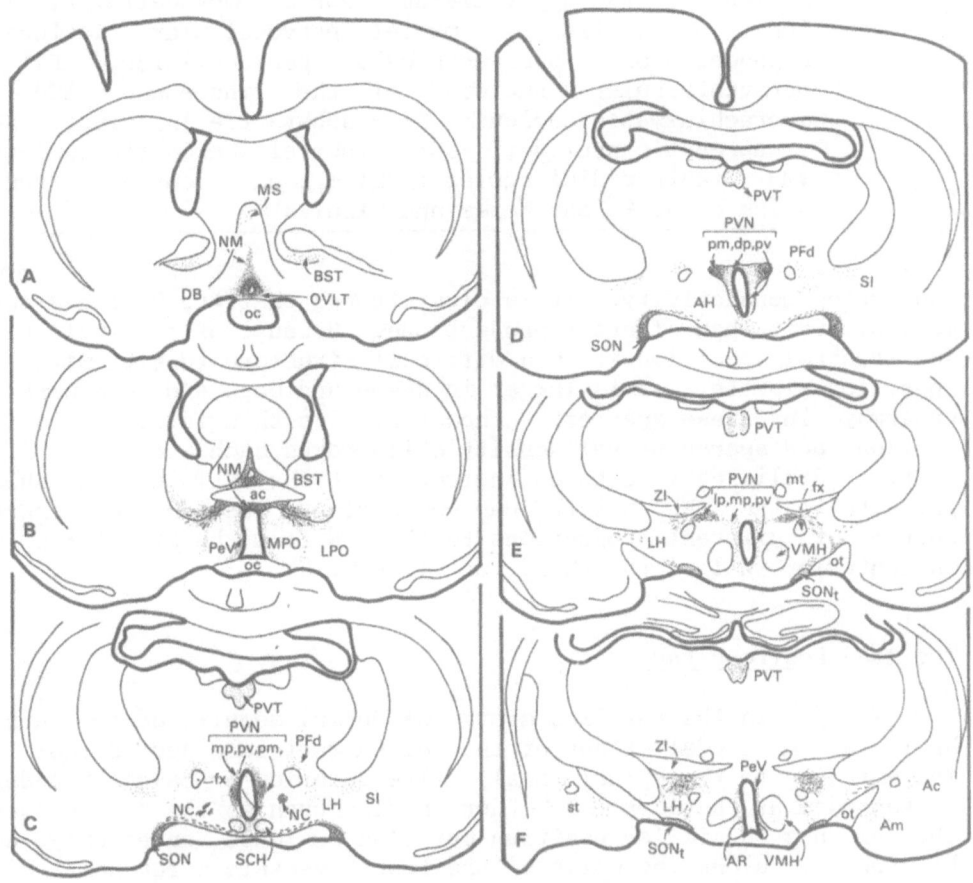

Fig. 2.   Summary charting of the efferent projections from the SFO using cholera toxin–horseradish peroxidase as the neuronal tracer. The fine dots indicate apparent terminal fields and the short dashes indicate fibers of passage or preterminal fibers. A to F are transverse sections from anterior to posterior. Abbreviations: ac – anterior commissure; Ac – central nucleus of the amygdala; AH – anterior hypothalamic nucleus; Am – medial nucleus of the amygdala; Ar – arcuate nucleus; BST - bed nucleus of the stria terminalis; DB – nucleus of the diagonal band; fx – column of the fornix; LH – lateral hypothalamic area; LPO – lateral preoptic area; MPO – medial preoptic area; MS – medial septal nucleus; mt – mammillothalamic tract; NC – nucleus circularis; NM –· nucleus medianus; oc – optic chiasm; ot – optic tract; OVLT – organum vasculosum of the lamina terminalis; PeV – periventricular area of the hypothalamus; PFd – dorsal perifornical area; PVN dp,lp,mp,pm,pv – paraventricular

(continued)

nucleus of the hypothalamus – dorsal parvocellular, lateral parvocellular, medial parvocellular, posterior magnocellular, periventricular parvocellular; PVT – paraventricular nucleus of the thalamus; SCH – suprachiasmatic nucleus; SI – substantia innominata; SON – supraoptic nucleus; SONt – tuberal supraoptic nucleus; VMH – ventromedial nucleus; ZI – zona incerta. From Weiss et al[15], and Weiss and Miselis[16].

---

concerning connectivity. It is clear though that the OVLT has many more afferent than efferent projections. Because of its small size and ventral position it is a difficult structure to hit with an injection pipette. It is larger in sheep and dogs and some work is underway in these species; however, its thin laminar configuration and sparse neural density still makes such a study difficult. Preliminary data do indicate that the OVLT in the sheep projects to the SON, but the nucleus medianus provides the largest source of afferent projections to the SON from the structures of the anterior wall of the third ventricle[18].

The Area Postrema (AP)

The AP in the rat lies along the dorsal medulla of the hindbrain at the caudal extent of the fourth ventricle just dorsal to the entrance to the spinal canal. Its anterior or caudal boundary is regarded as the obex. In other species such as the dog or sheep the AP has a similar position but also extends anteriorly and bilaterally along the walls of the fourth ventricle for some distance. Beneath the AP is the nucleus of the solitary tract (NTS), composed of several subnuclei, and the dorsal motor nucleus (DMN) of the vagus nerve. This area with the AP is often referred to as the vagal complex because of the vagal axonal fibers which either originate here (preganglionic motor neurons within the DMN) or terminate here (visceral sensory fibers ending in the AP, NTS or DMN). It is obviously a major focal point of parasympathetic autonomic nervous system activity and is a target for direct descending projections from the cortex, limbic structures and hypothalamus[2-4].

The AP has specific neural connections to relay nuclei for ascending visceral afferent information entering via the cranial nerves[10]. The first of these nuclei is the NTS. The AP's densest projections are into the immediately adjacent subnuclei of the NTS particularly the dorsolateral, medial and commissural subnuclei. Of these projections the densest terminal field is in the dorsolateral subnucleus which is the likely homologous receiving area in rats for peripheral baroreceptors. A lighter density of fibers project caudaly through the commissural subnucleus and anteriorly

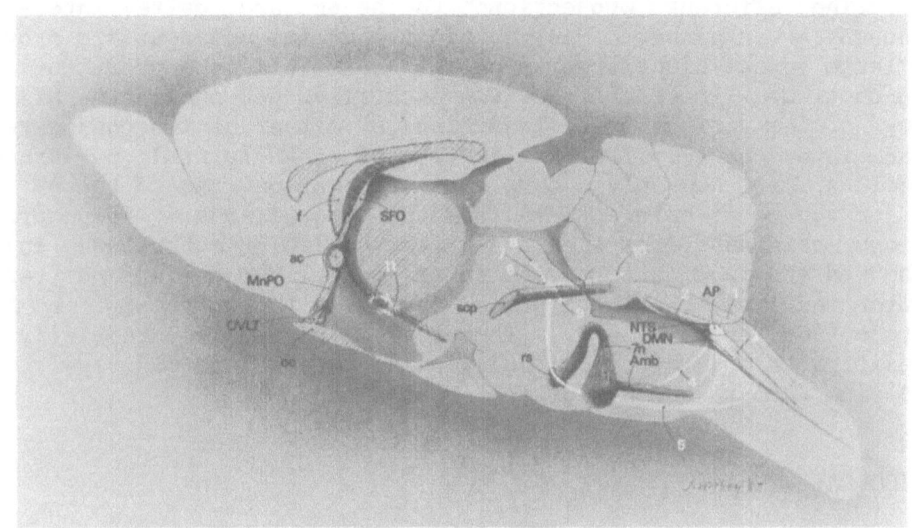

Fig. 3. Summary diagram of the neuronal connectivity of the AP.
1 indicates a light projection to the spinal trigeminal
and paratrigeminal nuclei; 2, the rostral NTS; 3, the
caudal NTS; 4, the nucleus ambiguus; 5, the trajectory of
the pontopetal fibers traversing the ventrolateral catecholaminergic column; 6 and 7, subdivisions of the lateral parabrachial nucleus; 8, the periaqueductal gray;
9, medial parabrachial and mesencephalic nucleus of the
trigeminal; 10, cerebellar vermis; 11, circular plexus of
neurons afferent to the AP from the PVN. Abbreviations:
AP – area postrema; ac – anterior commissure; Amb – nucleus ambiguus; DMN – dorsal motor nucleus of the vagus
nerve; f – fornix; MnPO – nucleus medianus; oc – optic
chiasm; OVLT – organum vasculosum of the lamina terminalis; rs – rubrospinal tract; scp – superior cerebellar
peduncle; SFO – subfornical organ. From Shapiro and
Miselis[10].

into the more cranial parts of the medial division of NTS. This
anterior area is referred to as the gustatory area of NTS because
it receives taste afferent projections. There is one very prominent, long distant projection from the AP to the lateral parabrachial nucleus (PBL) of the pons another major relay for ascending visceral sensory information. In addition there are light
projections to motor outflow nuclei of the vagus nerve. These
projections are to both the DMN and nucleus ambiguus. See figure 3
for the complete summary.

The afferent projections to the AP are quite interesting though few in number. Many consider that the adjacent NTS provides a large source of afferent projections. This is a prominent misconception. There are very few backfilled neurons in the NTS with any of the various sized injections of either of two conjugates of HRP into the AP. The same is true of the lateral parabrachial nucleus; it had only a couple of afferent neurons to the AP. Its major source of afferents is from the hypothalamus. These projections originate from the lateral parvocellular subnucleus (lp) of the PVN and from parts of the dorsal medial nucleus and periventricular nucleus[10]. These areas especially the lp receive a heavy projection from the SFO. Therefore, the SFO is potentially in a position to modulate incoming visceral information and vagal motor outflow through the PVN of the hypothalamus.

VISCERAL NEURAXIS

The CVO's are interconnected enabling coordination of their putative interoreceptive function and their effect on behavior, the endocrine system and the autonomic nervous system. This is provided by certain of the efferent neural projections of the AP and the SFO to particularly important relay nuclei. The AP projects to three major brainstem nuclei which directly relay information forward to hypothalamic, limbic and cortical sites. These are the NTS and PBL which pass visceral afferent information forward and the A1 catecholaminergic cell group of the ventrolateral medulla which provides noradrenergic influence to rostral sites. The SFO projects caudally to the PVN and the LH of the hypothalamus, the AV3V area of the preoptic region and the bed nucleus of the stria terminalis. These terminal areas, in turn, connect to the pituitary gland, to the autonomic nervous system or connect to or receive projections from other CVO's. This forms a complex neural network, the visceral neuraxis, which can modulate behavior, the pituitary gland or the autonomic nervous system. It is the neural substrate for homeostatic function in fluid balance.

The origin of behavior is still not clearly understood from consideration of this neural network. Classical studies point to areas such as the LH and LPO[13,14] and more current work suggests that the AV3V[19] area may be the origin of drinking behavior or thirst. However, none of these models explain how these areas actually lead to complex somatic motor patterns of drinking behavior. The difficulty is that we do not know which of the projections from the above sites could be mediators of drive for motor patterns of ingestive behavior. Considering the complexity of the behavior one might expect involvement of basal ganglia and perhaps cortex. In all likelihood the pattern of projections mediating drive will be complex and will involve the above areas and their projections to several levels of the neural organization for somatic motor behavior rather than a simple descending pathway from the hypothalamus. The latter may exist but is most likely to subserve simple biasing of particular components of a behavioral response.

CONCEPTUAL DRIFT IN CONCEPTS OF FUNCTIONAL ORGANIZATION

With the current emphasis on neural networks there has been a shift in the conceptualization of functional organization. In the past we thought more in terms of hypothalamic centers of function. Now, it is obvious that the areas of the hypothalamus thought to be centers have numerous and prominent connections with more distant sites which are also integrative in function. Secondly, the sites of centers are known to subserve multiple functions making it difficult to accept an area as a specific center. As a result conceptualization of functional organization has drifted from functional centers to functional networks. This alteration in thinking makes it possible to incorporate the prominent neural connectivity of the classical centers with other brain sites and to appreciate how hypothalamic sites subserve more than one function since multiple networks can occupy the same area. It also permits better understanding of how different functions can coordinate with each other because different functional networks come together at several anatomical sites where they can exchange information.

REFERENCES

1. J. A. Ricardo and E. T. Koh, Anatomical evidence of direct projections from the nucleus of the solitary tract to the hypothalamus, amygdala, and other forebrain structures in the rat, Brain Res. 153:1 (1978).
2. C. B. Saper, Convergence of autonomic and limbic connections in the insular cortex of the rat, J. Comp. Neurol. 210:163 (1982).
3. C. B. Saper, A. D. Loewy, L. W. Swanson and W. M. Cowan, Direct hypothalamo-autonomic connections, Brain Res. 117:305 (1976).
4. L. W. Swanson and H. G. J. M. Kuypers, The paraventricular nucleus of the hypothalamus: Cytoarchitectonic subdivisions and organization of projections to the pituitary, dorsal vagal complex, and spinal cord as demonstrated by retrograde fluorescence double-labeling methods, J. Comp. Neurol. 194:555 (1980).
5. S. J. Wiegand and J. L. Price, Cells of origin of the afferent fibers to the median eminence in the rat, J. Comp. Neurol. 192:1 (1980).
6. J. B. Simpson and A. Routtenberg, Subfornical organ: site of drinking elicitation by angiotensin II, Science 181:1172 (1972).

7. R. R. Miselis, R. E. Shapiro and P. J. Hand, Subfornical organ efferents to neural systems for control of body water, Science 205:1022 (1979).

8. R. Eng and R. R. Miselis, Polydipsia and ablation of angiotensin-induced drinking after transections of subfornical organ efferent projections in the rat, Brain Res. 225:200 (1981).

9. R. R. Miselis, The efferent projections of the subfornical organ of the rat: A circumventricular organ within a neural network subserving water balance, Brain Res. 230:1 (1981).

10. R. E. Shapiro and R. R. Miselis, The central neural connections of the area postrema of the rat, J. Comp. Neurol. 234:344 (1985).

11. J. Buggy and A. K. Johnson, Preoptic-hypothalamic periventricular lesions: thirst deficits and hypernatremia, Amer. J. Physiol. 233:R44 (1977).

12. A. K. Johnson and J. Buggy, Periventricular preoptic-hypothalamus is vital for thirst and normal water economy, Amer. J. Physiol. 234:R122 (1978).

13. A. N. Epstein, The lateral hypothalamic syndrome: its implications for the physiological psychology of hunger and thirst, in:"Progress in Physiological Psychology, Vol.4," E. Stellar and J. M. Sprague, eds., Academic Press, New York, (1971).

14. A. N. Epstein, The neuroendocrinology of thirst and salt appetite, in:"Frontiers in Endocrinology, Vol.4," W. F. Ganong and L. Martini, eds., Raven Press, New York, (1978).

15. M. L. Weiss, R. E. Shapiro and R. R. Miselis, Subfornical organ (SFO) connectivity examined using cholera toxin-horseradish peroxidase conjugate, Neurosci. Abstr. 10: 609 (1984).

16. M. L. Weiss and R. R. Miselis, Unpublished observations.

17. C. B. Saper and D. Levisohn, Afferent connections of the median preoptic nucleus in the rat: Anatomical evidence for a cardiovascular integrative mechanism in the anteroventral third ventricular (AV3V) region, Brain Res. 288:21 (1983).

18. R. Miselis, M. J. Mckinley, J. B. Simpson, M. Leventer and D. A. Denton, Projections of structures of the lamina terminalis to the supraoptic nucleus in sheep, Neurosci. Abstr. 10:609 (1984).

19. R. W. Lind and A. K. Johnson, Subfornical organ-median preoptic connections and drinking and pressor responses to angiotensin II, J. Neurosci. 2:1043 (1982).

THE ROLE OF THE ZONA INCERTA IN WATER INTAKE REGULATION

Sebastian P. Grossman

Committee on Biopsychology
University of Chicago
Chicago, IL. 60637

INTRODUCTION

Our understanding of the brain mechanisms that regulate water intake has undergone revolutionary change in the past decade. Where we once thought that the lateral hypothalamus controlled thirst and related behaviors, it now seems clear that we must look to the zona incerta (ZI) and preoptic region. My associates and I have intensively studied the effects of ZI lesions in the past decade. References to specific experiments and a more detailed account of our findings are available (1).

AD LIBITUM WATER INTAKE

The rat with ZI lesions appears to drink only in order to ingest and digest dry food. After recovery from a sometimes severe (40-60% below baseline) post-operative hypodipsia which last approximately 1 week ad libitum water intake recovers to nearly normal levels (a seemingly permanent reduction of 15-30 % of the preoperative baseline remains in most animals but this may be due, at least in part, to the permanently lowered body weight that is typical of the rat with ZI lesions).

However,the rat with ZI lesions ingests no water when food deprived (up to 4 days), and does not compensate for the self-induced body fluid shortage when food is subsequently returned. It is interesting to note that rats with ZI lesions do increase their water intake when dry food (but not water) is available during the deprivation period, suggesting that accumulated prandial and/or digestive needs may provide adequate stimuli for drinking.

Rats with ZI lesions drink essentially all of their daily water ration at night, displaying a clearly exaggerated circadian rhythm.

RESPONSE TO CELLULAR DEHYDRATION

Zona incerta lesions appear to abolish the normal drinking response to cellular dehydration. We have consistently failed to observe drinking 4 - 6 hours after IP, SC, or IV injections of a wide range of volumes and concentrations of hypertonic saline. Some animals do increase their 24 hour intake slightly after very large salt loads (as has been reported for rats with preoptic area lesions) but many of our ZI rats do not. The effects of ZI lesions on cellular dehydration related thirst appears to be permanent although we have observed partial recovery in some animals, 6 months after the surgery.

I should emphasize, at this point, that our control over the size and precise location of the electrolytic lesions that have been used in much of this research is not sufficient to produce a "perfect" ZI syndrome in all animals. In spite of extensive experience, we typically produce a spectrum of effects that range from slight but statistically significant impairments to the complete loss of responsiveness to various experimental treatments. When the data from these animals are subsequently averaged for statistical treatment and pictorial display, the erroneous impression may be created that the rat with ZI lesions retains capabilities which in fact it does not (i.e. in many of our publications it may appear that the rat with ZI lesions does, on the average respond to salt loads albeit at a far lower level than the intact control).

It is interesting to note that the renal response to salt loads appears to be intact after ZI lesions. When hypertonic saline is administered IV, the rat with ZI lesions does not drink but excretes most of the unwanted salt within 6 hours of the injection. When this means of desalination is blocked by nephrectomy, the rat with ZI lesions makes no attempt to dilute its body fluids by drinking after salt loads as high as 5 ml of a 2.0 M NaCl solution.

356

Because the ZI rat drinks nearly all of its daily water ration at night, we compared the effectiveness of salt loads administered 3 hours after the beginning of the light or dark portion of the 12/12 hour cycle. Some rats with ZI lesions did not increase their water intake during 6 or 24 hours after the salt load regardless of the time of administration. However, as a group, the injections given at night proved to be more effective than those administered during the day. This suggests that the increased drinking seen in some ZI rats 24 hours but not 6 hours after injections of hypertonic saline may not merely reflect an extension of the test period but the facilitatory effects of being tested during the dark, normally active part of the day.

Animals with zona incerta lesions do increase their water intake when a large amount of salt (3% w/w) is added to their already quite salty chow. Since these animals do not respond to very large salt loads administered IV, IP, or SC, the response to oral salt would appear to be due to the irritation of oral receptors rather than any post-ingestional consequences.

RESPONSE TO HYPOVOLEMIA

Rats with rostromedial zona incerta lesions respond normally or nearly so to experimental treatments (polyethylene glycol, Formalin) which induce vascular hypovolemia. The fact that ZI lesions abolish water intake during prolonged periods of food deprivation but have little or no effect on the drinking response to hypovolemia provides strong evidence that the latter does not play a significant role in the regulation of drinking under normal circumstances (the extreme thirst seen after extensive blood loss being an example of the conditions which permit hypovolemic influences to operate).

Some ZI lesions do impair the drinking response to polyethylene glycol (PG). Rats with such lesions do not increase their water intake significantly 6 - 8 hours after the SC injections of low doses of PG (5 ml of 5%;10%;15% or 20% solutions) and drink significantly less than controls during the 24 hour period following the treatment. All of these animals do, however, drink as much as controls 6-8 as well as 24 hrs after injections of 5 ml of a 30% PG solution. In these experiments, there was no correlation between the effect of the lesion on cellular and extracellular thirst. Some animals that consistently failed to increase their water intake after hypertonic saline responded normally to PG while others showed impairments in their response to low doses of PG but displayed significant delayed drinking to hypertonic saline. I should emphasize at this point that ZI lesions only rarely result in an impaired water drinking response to PG while severe effects on the drinking response to salt loads are ubiquitous.

OTHER DIPSOGENS

Rats with rostromedial zona incerta lesions do not drink in response to a variety of dipsogens that have been classified as "extracellular", including intra-cerebroventricular or IV injections of angiotensin and IP injections of isoproterenol. These observations indicate that the dipsogenic effects of these treatments are not mediated by the same pathways that are responsible for the drinking response to PG.

SODIUM APPETITE

Rats with ZI lesions do not display the preference for mildly sapid solutions typical of the species. They also drink less of a 0.5 M saline solution than controls following SC injections of Formalin.

When given access to 0.5 M saline 24 hours after polyethylene glycol injections, normal rats consume a significant amount of saline. Rats with ZI lesions (most of which increase their water intake significantly during the first 24 hours after PG) drink little or no saline. The few rats with ZI lesions that do not drink water in response to the PG injections also do not drink saline when it is offered 24 hours after the injection.

SENSORY-MOTOR AND AROUSAL FUNCTIONS

There has been much recent discussion of the fact that lateral hypothalamic (LH) lesions produce such severe sensory-motor dysfunctions that their effects on ingestive behavior are difficult to evaluate. We (2,3) as well as others (4) have reproduced aspects of the LH lesion syndrome by microinjections of the neurotoxin kainic acid into the region. This agent does not affect fibers of passage that are believed to be responsible for the sensory-motor effects of LH lesions and does, in fact, not produce the comatose conditions typically seen after LH lesions. The effects of kainic acid injections into the LH are, unfortunately, far less severe than those of electrolytic lesions and the possibility of diffusion of the neurotoxin to the ZI has not

358

entirely been ruled out. The role of the LH in food-
and water-intake regulation therefore remains an open
question.

This brief digression from our principal topic
highlights the fact that the rostro-medial zona incerta
lesions which severely impair or abolish "regulatory"
drinking do not produce overt sensory or motor dysfunctions as long as the lesions do not invade the
medial lemniscus, lateral aspects of the LH or medial
portions of the internal capsule.

It has nonetheless been suggested (5) that covert
sensory motor impairments might make drinking difficult
enough for the rat with ZI lesions to produce hypodipsia.
This hypothesis rests on the demonstration that some ZI
lesions which produced hypodipsia (but no impairments
in drinking responses to hypertonic saline) reduced the
amount of water obtained per lick. Some of our own
larger lesions have produced excessive food spillage
which may be due to sensory-motor dysfunctions. This is
not, however, found in the vast majority of animals
with rostro-medial ZI lesions that are most effective
in producing the impairments in water intake discussed
above, and we have found no correlation between food
spillage and the severity of the drinking deficits.

Although it seems quite possible that the decreased
efficiency in licking may have contributed to the hypodipsia reported by Evered and Mogenson (5), I do not
believe that it materially contributes to the complex
syndrome observed in my laboratory and it may be important to distinguish clearly between the two types of
lesions. A sensory-motor hypothesis cannot, in my opinion
account for the fact that rats with the rostro-medial
ZI lesions studied in my laboratory do not drink at all
in response to a variety of very potent dipsogenic treatments (including very intense salt loads, central or
systemic angiotensin and isoproterenol) but respond
essentially normally to Formalin or polyethylene glycol.
It also fails to explain why food-deprived ZI rats should
refuse to drink for several days (although less water
is needed during periods of deprivation, obligatory
water loss should produce severe dehydration over a
period of 3 days). We have found, moreover, that rats
with ZI lesions will drink several hundred ML per day
when properly motivated by the addition of sucrose to
the drinking water. The increased intake seen in these
experiments was fully as large as that displayed by
controls.

ANATOMICAL CONSIDERATIONS

Lesions which produce the severe impairments in regulatory drinking discussed above are in the rostro-medial zona incerta, dorso-lateral to the dorso-medial nucleus of the hypothalamus and immediately adjacent to the medial aspects of the medial lemniscus (because the fibers of this prominent bundle react differently to electric current than adjacent cellular components of the area, many of our best lesions assume a tear-like shape which surrounds the medial portions of the medial lemniscus without damaging it. Some of the effective lesions damage the mammillothalamic tract but this appears to be irrelevant. There is no damage to the region of the dorsolateral hypothalamus that forms the triangle between the lateral portions of the medial lemniscus and the medial aspects of the internal capsule. Damage to this area invariably produces aphagia and adipsia and we discard animals which display these effects. Some of our larger lesions destroy portions of the dorsomedial hypothalamus but this too appears to be irrelevant since lesions confined to that area do not reproduce the ZI lesion syndrome.

We have reason to believe that the effects of our lesions are due to the destruction of cell bodies in the rostro-medial ZI since microinjections of kainic acid which selectively destroyed cells in that region without affecting neighboring aspects of the hypo-thalamus or thalamus reproduced the effects of electro-lytic lesions.

REFERENCES

1. S.P.Grossman, A reassessment of the brain mechanisms that control thirst. Neurosci. Biobehav. Rev.8:95-104 (1984).
2. S.P.Grossman, D.Dacey, A.E. Halaris, T.Collier and A. Routtenberg, Aphagia and adipsia after preferential destruction of nerve cells bodies in the hypothalamus, Science 202:557-559 (1978).
3. S.P.Grossman and L.Grossman, Iontophoretic injections of kainic acid into the rat lateral hypothalamus: Effects on ingestive behavior.Physiol.Behav.29:553-559 (1982).
4. E.M.Stricker, A.F. Swerdloff and M.J. Zigmond, Intra-hypothalamic injections of kainic acid produce feeding and drinking deficits in rats. Brain Res. 158:470-473 (1978).
5. M.D.Evered and G.J.Mogenson, Impairment in fluid ingestion in rats with lesions of the zona incerta. Am J.Physiol.233:R53-R58 (1977).

NEURAL CIRCUITS THAT CONTRIBUTE TO PROCUREMENT OF WATER

Gordon J. Mogenson and Blanche Box

Department of Physiology
University of Western Ontario
London, Ontario, Canada   N6A 5C1

INTRODUCTION

It is more than 18 years since Jim Stevenson and I observed a strong and persistent drinking response in rats receiving electrical stimulation of the lateral hypothalamus[1]. We had, according to the Zeitgeist of the mid-1960's, activated the hypothalamic "thirst center". It turned out, after some reflection, that there was a good deal of vagueness about this interpretation. However, the assumption was that by stimulating the so-called hypothalamic "thirst center" we activated a brain site concerned with integrating thirst signals and with generating "command signals" for the procurement and ingestion of water[2].

This view of the integrative activities of the central nervous system was also part of the Zeitgeist - based on the pioneering studies of Sherrington - and considered an appropriate conceptual framework for the neural control of ingestive behaviors[3]. Two fundamental questions were suggested by this conceptual framework. First, what are the neuroanatomical and neurophysiological mechanisms associated with the integration of "thirst signals" by the hypothalamus? Second, how does the "thirst center" interface with the motor system enabling "command signals" resulting from integrative activities of the "thirst center" to become translated into the procurement and ingestion of water.

For a number of years research was directed at the first of these questions and the second largely ignored. Experiments utilizing electrophysiological recording techniques seemed appropriate to elucidate the first question. But the earlier promise of this approach has not been realized[4]. Indeed, as complex sets of neural circuits, which include limbic, basal ganglia and circumventricular structures and preoptic region and zona incerta have been implicated over the years in "thirst mechanisms", it has become necessary to reformulate the question[5].

The second question concerned with the interface of motivational and motor systems in drinking behavior has only begun to receive attention in recent years. A major reason is that axonal tracer techniques have demonstrated neural connections to the motor system from neural structures associated with thirst mechanisms[5].

NEURAL SUBSTRATES OF THE INITIATION, PROCUREMENT AND CONSUMMATORY PHASES OF INGESTIVE BEHAVIOR

The intake side of body water homeostasis depends on behavioral adaptive mechanisms (the output side on renal mechanisms). However, it is only in recent years that the complexity of the behavioral responses has been recognized.

Earlier views having to do with a hypothalamic "thirst center" emphasized the motivational aspects of thirst and the initiation phase of ingestive behavior. The consummatory phase was the subject of only limited investigation and the procurement phase pretty much ignored until recently.

The objective of this presentation is to call attention to the neural mechanisms associated with the procurement phase of ingestive behaviors and to present some preliminary results which suggest that recently demonstrated projections to the mesencephalic locomotor region from the forebrain contribute to the locomotor component of food - and water - seeking behaviors.

NEURAL CORRELATES OF WATER PROCUREMENT

The procurement of water depends on head and eye movements for orientation and on forward progression. These responses are necessary for an animal in the wild to reach a pond or stream and for a laboratory animal to reach and orient to a water spout. The superior colliculus and components of the motor system contribute to head and eye movements[6]. The neural systems for forward locomotion include the mesencephalic locomotor region (MLR) as well as some of the classical motor structures. There has been considerable interest in MLR - spinal cord mechanisms[7], but only recently have forebrain influences on the MLR received attention. These forebrain influences include the contributions of limbic structures associated with behavioral response initiation for ingestive and other adaptive behaviors[5,8].

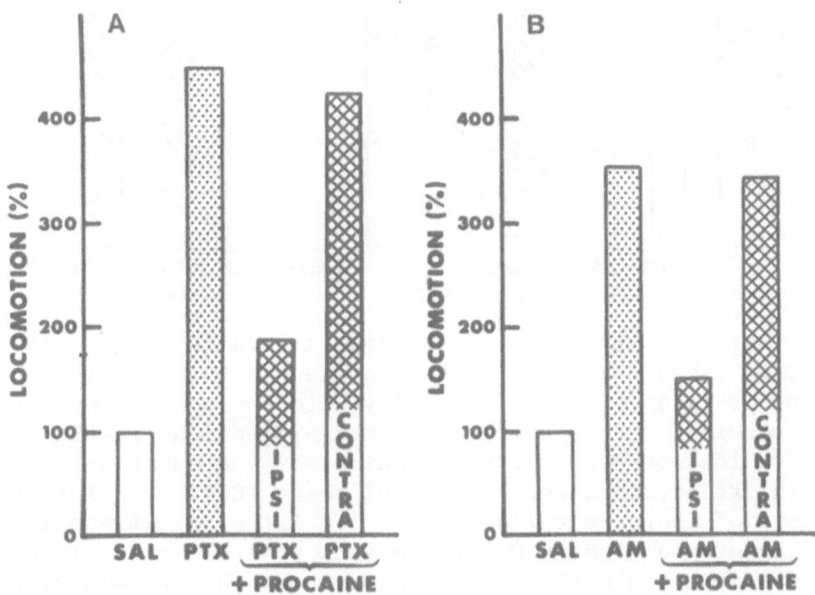

Fig. 1. Unilateral injections of procaine (30 µg in 0.2 µl) into
the zona incerta (panel A) or into the mesencephalic locomotor
region (panel B) of rats blocked reversibly the neural projections
from the subpallidal region to the MLR and reduce locomotor activity. In A locomotor activity was increased more than four-fold by
injecting picrotoxin (PTX) a GABA antagonist, into the substantia
innominata of the subpallidal region (results of a typical animal,
from Mogenson, Swanson & Wu[10]). In B locomotor activity was increased three to four-fold by injecting amphetamine (AM) into the
nucleus accumbens (results of a typical animal, from Brudznski &
Mogenson[12]). Injecting procaine into the contralateral (CONTRA)
zona incerta or MLR had little or no effect on locomotor activity
whereas injecting procaine into the ipsilateral (IPSI) zona incerta or MLR reduced significantly locomotor activity.

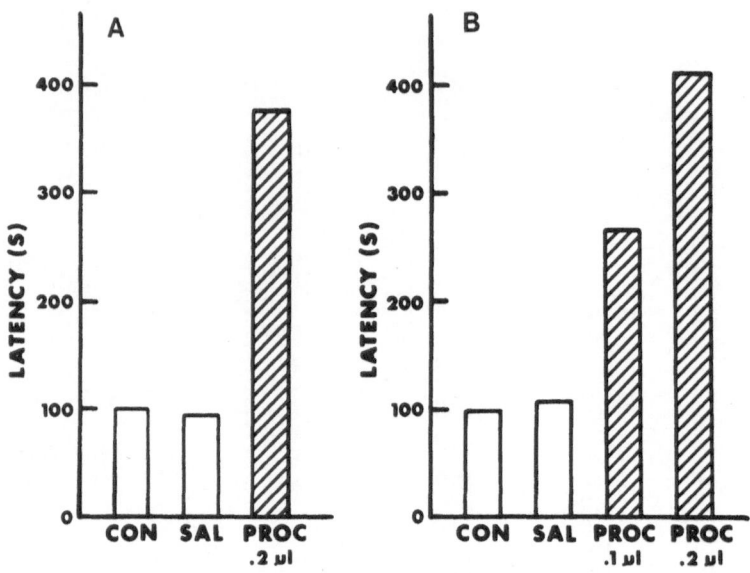

Fig. 2. Rats were trained to run along a straight alley (180 cm
long) to obtain a reward of saccharin-glucose solution (0.25 ml).
Injecting procaine (PROC) bilaterally into the zona incerta to
block reversibly the subpallidal to MLR projections resulted in a
deficit in locomotor activity as indicated by a substantial increase in latency compared to control tests (CON) or bilateral
injections of isotonic saline (SAL). In A the animal was injected
with 30 μg of procaine in 0.2 μl to each zona incerta. In B the
animal was injected with 15 μg of procaine in 0.1 μl and with 30
μg of procaine in 0.2 μl to each zona incerta. (Latency in %).

In a series of experiments using axonal tracer and electrophysiological recording techniques we have investigated neural
projections to the MLR from the substantia innominata, lateral
preoptic region and ventral globus pallidus, designated subpallidal region[9,10]. Since the subpallidal region receives projections
from the amygdala and hippocampus, either directly or indirectly
via the nucleus accumbens, it has been suggested that limbic-
accumbens-subpallidal - MLR connections may have a role in limbic-
motor integration[9]. Behavioral experiments have been undertaken
to investigate this hypothesis. Preliminary results appeared in a
recent article[8], and some of our subsequent observations will now
be presented briefly, which implicate the subpallidal to MLR projection in the locomotor component of the procurement phase of
water intake.

We first recorded locomotor activity of rats in an open-field apparatus and observed the effects of blocking the subpallidal to MLR projections by injecting procaine, a reversible neural blocker, into the zona incerta[10]. Bilateral injections of procaine significantly reduced exploratory locomotor activity in response to novel visual objects[11]. We also observed that locomotor activity from injecting picrotoxin unilaterally into the subpallidal region or from injecting amphetamine (or dopamine) into the nucleus accumbens was reduced by injecting procaine into the ipsilateral zona incerta or into the ipsilateral MLR (see Fig. 1). Procaine injected into the contralateral zona incerta or MLR, as a control for possible non-specific behavioral effects, did not reduce locomotor activity from the picrotoxin, amphetamine or dopamine injections.

We then trained mildly deprived rats to traverse a straight alley to obtain a highly palatable liquid reward (5% saccharin, 95% glucose). Latency to proceed from the start box to the spout to obtain 0.25 ml of the liquid reward (distance of 180 cm) stabilized at 2.0 - 3.0 sec. When procaine was injected bilaterally into the zona incerta there was a four or five-fold increase in the latency (see Fig. 2). Blocking the substantia innominata to MLR pathway with procaine disrupted the procurement phase of the fluid-seeking behavior.

SUMMARY

Neural projections from the subpallidal area to the mesencephalic locomotor region may provide a functional link by which limbic forebrain integrative activities influence locomotor activity. This subpallidal to MLR projection may contribute to the locomotor component of adaptive behaviors including the procurement of water and other fluids. Experiments in which procaine was used to block this projection reversibly provide evidence in support of this suggestion.

ACKNOWLEDGEMENTS

Anne Szpak assisted with the experiments shown in Fig. 2. Supported by NSERC of Canada.

REFERENCES

1. G. J. Mogenson and J. A. F. Stevenson, Drinking and self-stimulation with electrical stimulation of the lateral hypothalamus, Physiol. Behav. 1:251-254 (1966).
2. A. N. Epstein, H. R. Kissileff and E. Stellar, "The Neuropsychology of Thirst", Winston, New York (1973).
3. E. Stellar, Drive and motivation, in: "Handbook of Physiology", Vol. 3, J. Field, ed., Williams and Wilkins, Baltimore (1960).

4. G. J. Mogenson, Electrophysiological studies of the mechanisms that initiate ingestive behaviors with special emphasis on water intake, in: "Neural Integration of Physiological Mechanisms and Behaviour", G. J. Mogenson and F. R. Calaresu, eds., University of Toronto Press, Toronto (1975).

5. L. W. Swanson and G. J. Mogenson, Neural mechanisms for the functional coupling of autonomic, endocrine and somatomotor responses in adaptive behavior, Brain Res. Rev. 3:1-34 (1981).

6. R. J. Wurtz and J. Albons, Visual-motor function of the primate superior colliculus, Ann. Rev. Neurosci. 3:189-226 (1980).

7. S. Grillner and M. I. Shik, On the descending control of the lumbrosacral spinal cord from the "mesencephalic locomotor region", Acta Physiol. Scand. 87:320-333 (1973).

8. G. J. Mogenson, Limbic-motor integration – with emphasis on initiation of exploratory and goal-directed locomotion, in: "Modulation of Sensorimotor Activity During Altered Behavioral States", R. Bandler, ed., New York, Alan R. Liss, Inc. (1984).

9. L. W. Swanson, G. J. Mogenson, C. R. Gerfen and P. Robinson, Evidence for a projection from the lateral preoptic area and substantia innominata to the "mesencephalic locomotor region" in the rat, Brain Res. 295:161-178 (1984).

10. G. J. Mogenson, L. W. Swanson and M. Wu, Evidence that projections from substantia innominata to zona incerta and mesencephalic locomotor region contribute to locomotor activity, Brain Res., 1985 (in press).

11. G. J. Mogenson and M. Nielsen, Neuropharmacological evidence to suggest that the nucleus accumbens and subpallidal region contribute to exploratory locomotion, Behav. Neural Biol., 1984 (in press).

12. S. Brudznski and G. J. Mogenson, Association of the mesencephalic locomotor region with locomotor activity induced by injections of amphetamine into the nucleus accumbens, Brain Res., 1985 (in press).

THE ROLE OF THE SEPTAL AREA IN THE REGULATION OF DRINKING BEHAVIOR AND PLASMA ADH SECRETION

Michele Iovino* and Luca Steardo

Drinking of Neurology, 2nd Medical School
University of Naples
via Pansini 5, 80131 Naples, Italy

INTRODUCTION

Drinking behavior and ADH secretion are normally regulated by two physiological parameters: plasma osmolality and blood volume. A variety of experiments dating back to the classical observations of Verney[9] indicate that signals elicited by increased plasma osmolality and reduced extracellular fluid volume stimulate the release of ADH and initiate the drinking of water. Single neurons in the septal area respond with an increase in activity to only plasma osmolality increases, but not to extracellular hypovolemia[5]. Thus of the two stimuli which normally elicit thirst and ADH secretion, the septal nuclei of the rat forebrain seem to be involved only in the mechanisms regulating intracellular fluid volume. To investigate this hypothesis water intake and circulating levels of ADH were measured in sham- and septal-lesioned rats receiving a subcutaneous injection of hypertonic saline, which reduces intracellular fluid volume, or polyethylen glycol, which reduces extracellular fluid volume. Additionally, rats received total, lateral or medio-ventral destruction of the septal area in an attempt to delineate the critical region involved in drinking behavior and ADH release.

Author for correspondece at: via Arnedi 1, I-84012 Angri, Italy

MATERIALS AND METHODS

Male Wistar rats (initial weight 160 ± 10 g) were used in individual metabolism cages to measure the 24-h water intake and urine output in an animal room maintained at a constant temperature (23 ± 1°C) and humidity (70%) with automatic light control from 06.00 to 20.00 h. Commercial rat chow and water were available ad lib, except where specifically noted. Standard stereotaxic procedures and sodium pentobarbital anesthesia (Nembutal, i.p. 40 mg/kg) were used on all animals to produce an electrolytic lesion or sham lesion. The lesion was performed using a Stoelting Stereotaxic Apparatus; the tip of a thorium electrode was lowered at level of septal area according to the coordinates of Pellegrino and Cushman Atlas[7]. Rats in the total septal (TS) lesion group (n = 20) sustained a bilateral electrolytic lesion in the septal area (2.2, ± 0.5, 5.5), rats in the lateral septal (LS) lesion group (n = 20) sustained two lesions, in the right and left LS areas (2.2, ± 1.2, 5), and rats in the medio-ventral septal (MVS) lesion group (n = 20) sustained a single lesion in the MVS area (2.2, 0.0, 7). Radiofrequency lesions were generated using a Grass LM-4 radiofrequency generator. Lesioning parameters were adjusted for each electrode. As an average, these were of the order of 3 mA/20 sec in the TS lesion group, 0.5 mA/20 sec in the LS lesion group, and 1.2 mA/20 sec in the MVS lesion group. On the basis of the postoperative changes in daily water intake and urine output two groups were formed. In one group (septal drinkers), water intake and urine output were increased by at least 50% during the 7-8 postoperative days compared with the 3 day period immediately preceding surgery. Rats without postoperative increases in water intake and urine output were selected as the septal non-drinker group.

Drinking behavior and plasma ADH concentrations were investigated in the following ways: (1) during the first 7-8 postoperative days daily water intake and urine output were measured, at the same time of the day (09.00 a.m.), with water and food available ad li; (2) the following days each rat received 1 M NaCl (HS) or 20% polyethylen glycol (PEG) subcutaneously (15 ml/kg) at 09.00 a.m. Additionally, some rats received PEG intraperitoneally. Food and water were withheld for 3 h (HS) or 6 h (PEG) after injection and water intake was subsequently measured, by means of graduate tubes, over a 3 h period for HS-treated rats and a 2 h period for PEG-treated rats; (3) at least 8 days after, these protocols were repeated in the same experimental groups in order to determine plasma ADH release.

Rats were killed by decapitation and blood samples were collected from the body trunk in precooled centrifuge tubes containing a solution of 0.25 ml disodium EDTA (15 mg). The plasma was separated by centrifugation (4 C, 2,000 g for 10 min). Plasma ADH concentrations were measured by radioimmunoassay as described elsewhere[6].

All values are given as mean ± SEM. Comparisons among the means were made by Student's paired t-test.

RESULTS

In fig. 1 the results of the histological examination of the lesion of all brains plotted on a schematic frontal section through the septum are shown. The MVS lesion was centered in the medial septal area and tended to extend laterally at ventral level (Fig. 1A). The LS lesion destroyed most of the tissue lying between the ventricles at this level (Fig. 1B). In the TS lesion group the lateral and medial septal areas were completely destroyed in all cases (Fig. 1C).

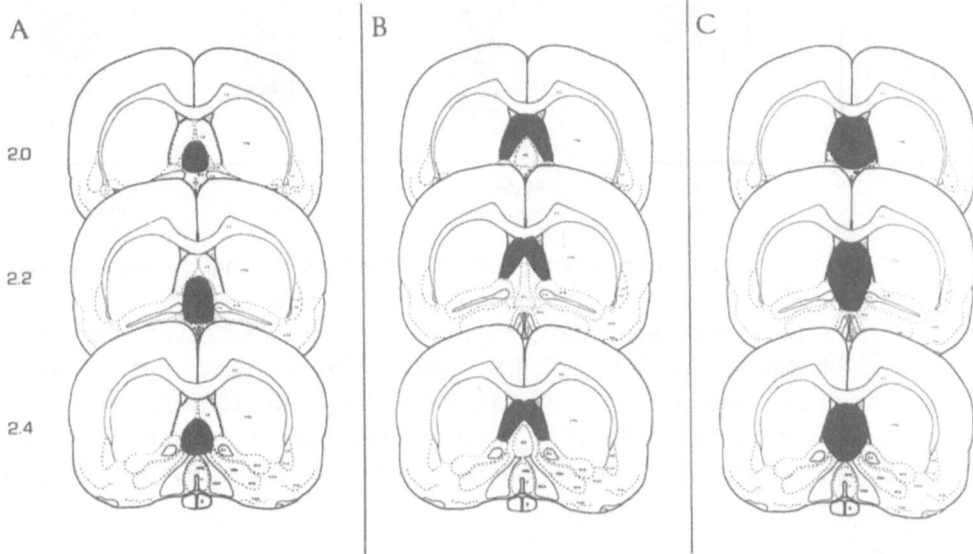

Fig. 1. Reconstruction of the site and extension of tissue destruction following electrolytic lesion. Darkly shaded regions represent the extent of a typical MVS (A), LS (B) or TS (C) lesion. Numbers represent millimeters anterior to bregma (Pellegrino and Cushman Atlas[7]).

Following electrolytic lesions, 12 of 20 rats (60%) bearing MVS lesions became hyperdipsic and polyuric, whilst only 7 of 20 rats (35%) with lesions of the LS areas increased their water intake and urine output postoperatively. Curiously, a reduction of these percenteges was observed in rats with total destruction of the septal complex. In fact, only 5 of 20 rats (25%) became hyper-dipsic and polyuric. However, no differences were observed among the MVS-, LS- and TS-lesioned rats that increased their water intake and urine output (Fig. 2).

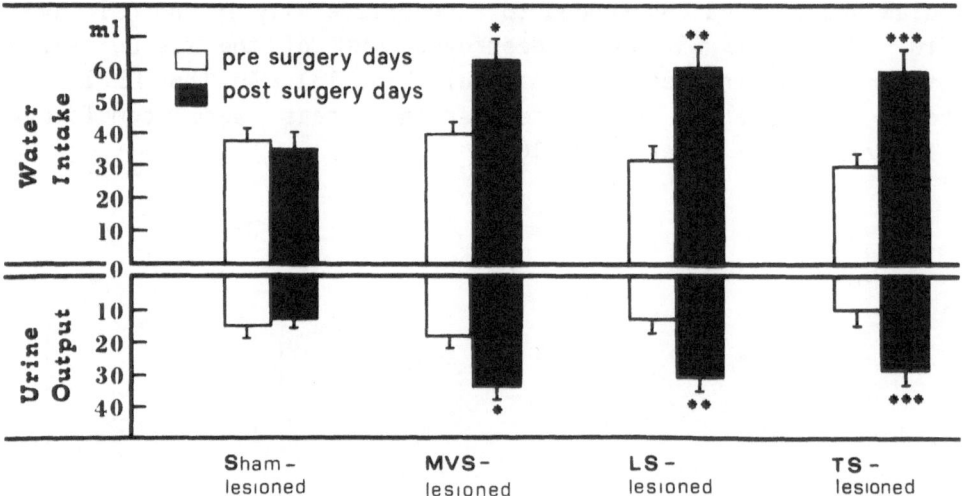

Fig. 2. Water intake and urine output taken over the 3 days prior and the 8 days following surgery in sham-, MVS-, LS- and TS-lesioned rats. Data represent mean ± SEM averaged from 24 sham-, 12 MVS-, 7 LS- and 5 TS-lesioned rats. Water intake and urine output in MVS-, LS- and TS-lesioned rats was statistically ($P < 0.001$) significant if compared to the same rats before surgery and to sham-lesioned rats.

Drinking behavior stimulted by cellular dehydration (HS) or hypovolemia (PEG) is presented in Table 1. Sham-lesioned rats drank 7.4 ± 0.8 mls of water over a 3 h period to HS. Similar data were observed in septal drinkers (8 ± 0.9 ml/3 h) and septal non-drinkers (7.7 ± 0.8 ml/3 h).

After the subcutaneous or intraperitoneal injection of PEG, septal drinker and septal non-drinker rats were found not to differ from sham-lesioned rats in water intake (Table 1). In addition, no differences were observed among the MVS-, LS- and TS-lesioned in either NaCl or PEG treatments.

Table 1. Effect of hypertonic sodium chloride or polyethylen glycol on drinking behavior in sham-lesioned, septal non-drinker and septal drinker rats.

| | Water intake (ml) | | |
| | Hypertonic sodium chloride | Polyethylen glycol | |
		s.c.	i.p.
Sham-lesioned :	7.4 ± 0.8	5.1 ± 0.5	5.5 ± 0.4
Septal non-drinkers:	8 ± 0.9	5.7 ± 0.7	6 ± 0.5
Septal drinkers :	7.7 ± 0.8	5 ± 0.4	6.3 ± 0.6

Hypertonic sodium chloride (1 M NaCl) was administered 3 h prior to water access subcutaneously (s.c.; 15 ml|kg). Polyethylen glycol (20%) was administered subcutaneously or intraperitoneally (i.p.; 15 ml/kg) 6 h prior to water access. The volume of water consumed during a 3 h period for HS-treated rats and a 2 h period for PEG-treated rats was recorded for each rat. Data represent mean ± SEM averaged from 7-8 rats.

ADH release stimulated by cellular dehydration (HS) or hypovolemia (PEG) is presented in Table 2. Plasma ADH concentration was measured by radioimmunoassay in MVS-, LS- and TS-lesioned rats. The ADH release which occurred in response to HS in sham-lesioned rats was markedly blunted in MVS-, LS- and TS-lesioned rats. However, the reduction observed in MVS-lesioned rats was significantly lower than in LS-lesioned rats. On other hand, a normal increase in plasma ADH levels occurred in all experimental groups following PEG treatment (Table 2).

Table 2. Effect of hypertonic sodium chloride or polyethylen glycol on plasma ADH concentration in sham-lesioned, septal non-drinker and septal drinker rats.

	Plasma ADH (μU/ml)	
	Hypertonic sodium chloride	Polyethylen glycol
Sham-lesioned	8.6 ± 1.2 (10)	7.2 ± 0.8 (10)
Septal non-drinkers		
MVS-lesioned:	3.4 ± 0.5* § (6)	6.8 ± 0.4 (6)
LS-lesioned:	5.1 ± 0.6** (6)	7 ± 0.5 (6)
TS-lesioned:	3.5 ± 0.7** (6)	6.5 ± 0.6 (6)
Septal drinkers		
MVS-lesioned:	3.1 ± 0.5* § (6)	6.7 ± 0.4 (6)
LS-lesioned:	4.8 ± 0.6** (4)	6.9 ± 0.2 (3)
TS-lesioned:	4.1 ± 0.6** (3)	7 ± 0.4 (2)

Rats received hypertonic sodium chloride (1 M NaCl) or polyethylen glycol (20%) subcutaneously (15 ml/kg) at 09.00 a.m. Food and water were withheld for 3 h (HS) or 6 h (PEG) after injection, after which rats were sacrificed. ADH extracted from 1 ml plasma was measured by radioimmunoassay. Data represent mean ± SEM, figures in parantheses represent number of animals used.
*$P < 0.001$ vs sham-lesioned rats.
**$P < 0.01$ vs sham-lesioned rats.
§$P < 0.05$ vs LS-lesioned rats.

DISCUSSION

The results of the present study showed that: (i) electrolytic lesions confined to the MVS area yield the highest percentage (60%) of hyperdipsic rats; (ii) hyperdipsic animals did not overdrink in response to HS or PEG treatments, thus showing a dissociation between basal and thirst-evoked conditions; (iii) septal nuclei play an important role in the regulation of ADH response to osmotic stimuli but not to extracellular hypovolemia. Regarding the locus and size of the septal lesion, it has been reported that massive septal lesions enhance daily water intake in the rat, however, as mentioned above, total ablation of the septal complex does not increase daily water intake in all operated animals, in fact a large percentage of rats maintain their water intakes normal. In the Black and Mogenson[2] experiment only 8 of 32 (25%) bearing TS lesions increased their water intakes post-operatively. Furthermore, Besch and Van Dyne[1] have reported that ventrally placed lesions produced the most frequent increase in daily water intake, and Tondat and Almli[8] showed that knife cuts which transect connecting fibers of the septum in the ventral plane produce the highest percentage (55%) of hyperdipsic rats as compared to rats with TS destruction. According to this data, in our experimental conditions, rats with electrolytic lesions confined to the MVS area became hyperdipsic in a higher percentage (60%) in comparison with LS-lesioned (35%) and TS-lesioned (25%) rats. The results of this study also indicate that hyperdipsic rats did not overdrink in response to extracellular hypovolemia (PEG) or cellular dehydration (HS). These findings replicate the results of Black and Mogenson[2] and Tondat and Almli[8] . The experiments of these investigators have resulted in similar failures to observe the removal of inhibition of thirst mediated by hypovolemia or cellular dehydration. On other hand, it has been reported that septal hyperdipsic rats overdrink only to PEG administered intraperitoneally[3,4] . This hypothesis was tested in the present study. However, in our experimetal conditions, identical results between intraperitoneal and subcutaneous administration of PEG were observed. In addition, the work of Tondat and Almli[8] agrees nicely with our data that sham-, septal drinkers and septal non-drinkers drink equal amounts of water following intraperitoneal or subcutaneous injections of PEG.

Finally, plasma ADH concentration was investigated under thirst stimuli that normally elicit ADH release: plasma hyperosmolality (HS) and extracellular hypovolemia (PEG). No difference in plasma ADH concentration was observed in all experimental groups following PEG administration, thus showing that the septal area is not involved in the mechanisms regulating the release of ADH from signals elicited by reduced extracellular fluid volume. On the other hand, septal-lesioned rats were unable to display a normal increase in plasma ADH concentration following HS administration. In addition, when the lesion was confined to the MVS area the ADH release to HS was lower in comparison with LS-lesioned rats. The differences observed in ADH secretion to HS and PEG treatments, indicate that the septal area is involved in ADH releasing mechanisms via osmotic sensitive neurons, located mainly in the MVS area, but would not appear to be responsible for the ADH response to volumetric depletions. According to these data, Bridge and Hatton[5] have reported that septal neurons respond with an increase in activity only to plasma hyperosmolality, but not to extracellular hypovolemia.

REFERENCES

1. N.F. Besch and G.C. Van Dyne, Effects of locus and size of septal lesion on consummatory behavior in the rat, Physiol. Behav. 4: 953 (1969).
2. S.L. Black and G.J. Mogenson, The regulation of serum sodium in septal lesioned rats: a test of two hypotheses, Physiol. Behav. 10: 379 (1973).
3. E.M. Blass and D.G. Hanson, Primary hyperdipsia in the rat following septal lesions, J. Comp. Physiol. Psychol. 70: 87 1970.
4. E.M. Blass, A.I. Nussbaum, and D.G. Hanson, Septal hyperdipsia specific enhancement of drinking to angiotensin in rats, J. Comp. Physiol. Psychol. 87: 422 (1974).
5. J.G. Bridge and J.I. Hatton, Septal unit activity in response to alterations in blood volume and osmotic pressure, Physiol. Behav. 10: 769 (1973).
6. M. Iovino, S. Poenaru, and L. Annunziato, Basal and thirst-evoked vasopressin secretion in rats with electrolytic lesion of the medio-ventral septal area, Brain Res. 258: 123 (1983).
7. L.J. Pellegrino and A.J. Cushman, Stereotaxic Atlas of the Rat Brain, Appleton Century-Crofts, New York, 1967.
8. L.M. Tondat and C.R. Almli, Hyperdipsia produced by severing ventral septal fiber systems, Physiol. Behav. 15: 701 (1975).
9. E.B. Verney, The antidiuretic hormone and factors which determine its release, Proc. Roy. Soc. B 135: 25 (1947).

THE SURFACE MORPHOLOGY OF THE CAT SUBFORNICAL ORGAN

D. Felix° and H. Felix*

°Division of Animal Physiology, Zoology Institute
University of Berne, Berne, Switzerland
*ENT Department, University Hospital, 8091 Zurich
Switzerland

INTRODUCTION

As one of the circumventricular organs, the subfornical organ (SFO) is present in all mammalian and other vertebrate species. Several studies on the ultrastructure of the SFO, as revealed by scanning (SEM) and transmission electron microscopy (TEM), have been carried out (for a review see 1). The fact that this highly vascular organ has intimate contact with the cerebrospinal fluid (CSF) and that it is located outside the blood-brain barrier has led to the hypothesis that this structure is involved in the monitoring of the contents of the CSF and in neurosecretion. The results of recent studies have indicated that the SFO may contain receptors for the octapeptide angiotensin II which are neuronally linked to a drinking circuit. When injected into the ventricles of unanesthetized animals, angiotensin induces thirst (2). Furthermore, neurones in the SFO can be activated by angiotensin ejected microiontophoretically as well as by intravenous injection and direct application on to the surface (3). Since these studies have shown the existence of blood-borne receptors and receptors on the ventricular side, the surface morphology of the SFO is of particular interest.

RESULTS

The SFO is always found at the fusion point between lamina terminalis and tela choroidea. The choroid plexuses of the third and lateral ventricles are attached to the ventricular walls immediately dorsal to the SFO. The most marked characteristic of this organ is its somewhat bulbous appearance (Fig. 1A, B). The shape of the SFO in the cat appears to be different in male and

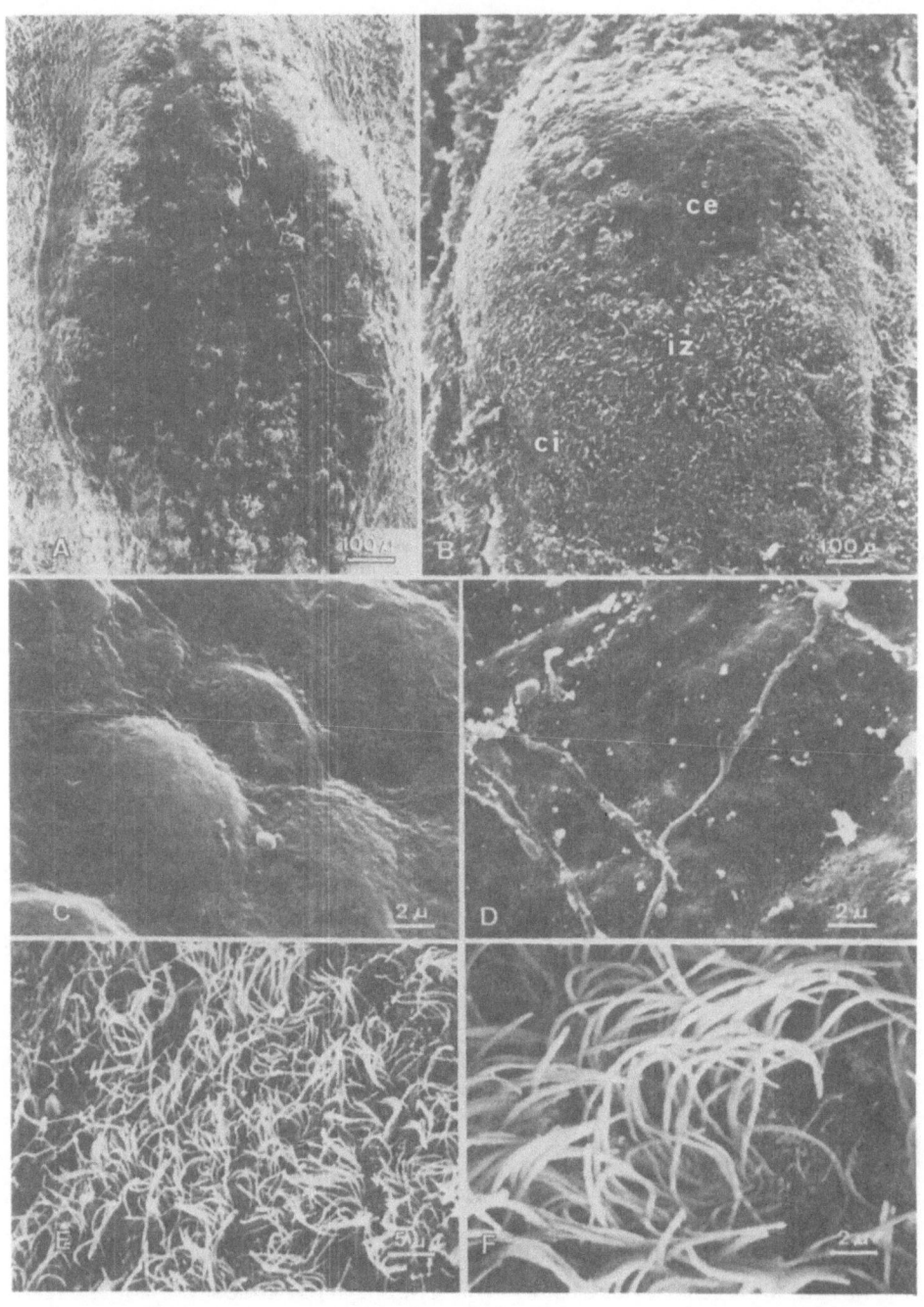

Fig. 1: SEM view of a female (A) and male (B) SFO; central zone (C, D) and ciliated zone (E, F).

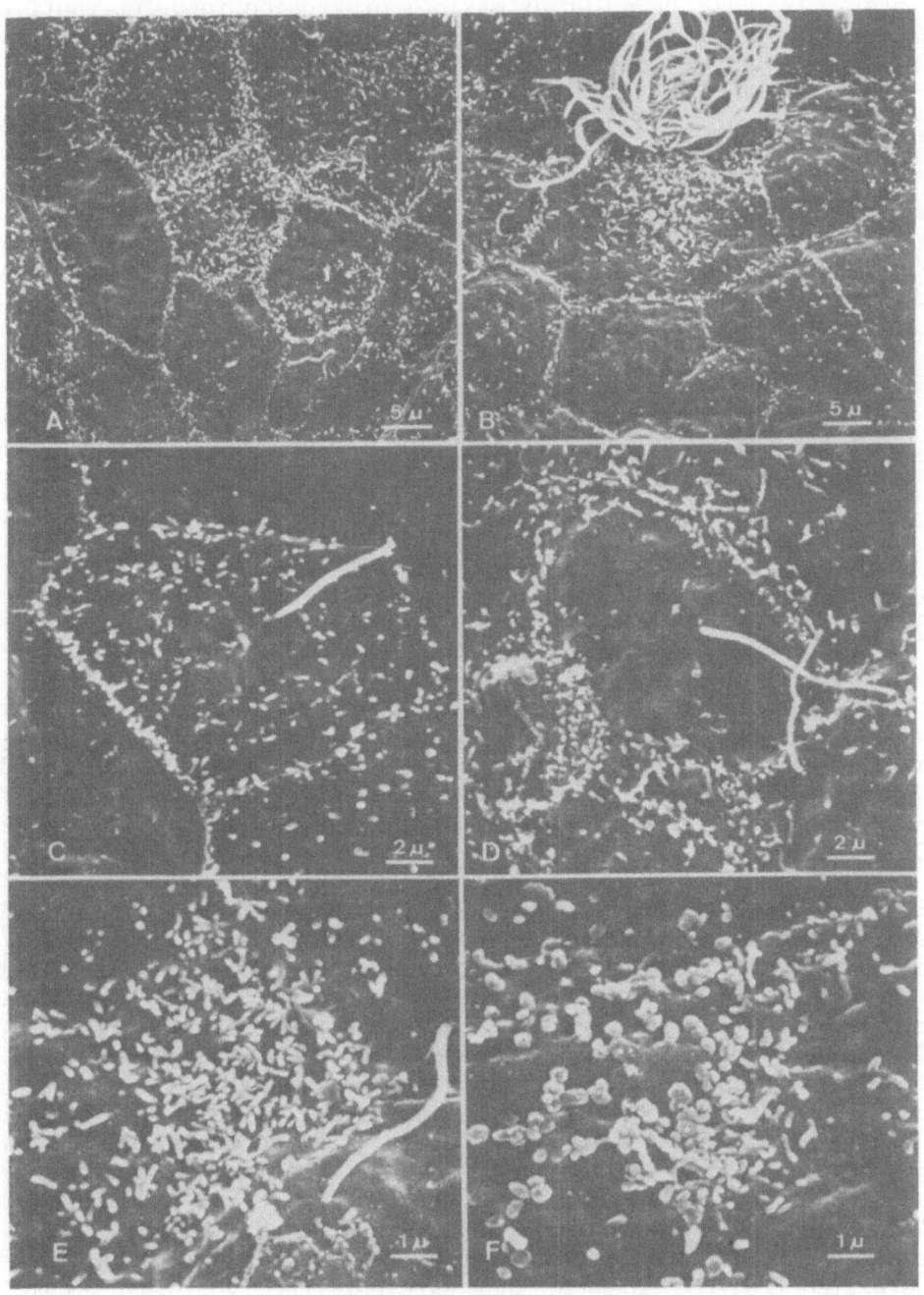

Fig. 2: Ependymal cells of the intermediate zone.

female animals: in the male it shows a pronounced bulging and is relatively round. In the female, however, it appears to be more flattened and elongated, sloping smoothly from the anterior and posterior end to the point of maximum protrusion into the ventricles. On the basis of surface morphology results obtained by SEM studies, the SFO can be divided into three distinct regions.

The central zone (Fig. 1B, ce) is identical to the non-ciliated part of the corpus. In the male cat this zone constitutes about one third of the crest, whereas in the female it covers almost the whole SFO body. The ependymal cells are mostly flat without any specialized surface structures such as cilia and microvilli (Fig. 1C). In many SFO's so far investigated the ependyma in the central zone forms a homogeneous monolayer in which individual cell borders cannot be distinguished. However, in a few cases these cells display rectilinear and slightly depressed boundaries. Across the surface numerous processes of various length ranging from 10 to 50 µm are visible. These long processes branch and have occasional varicosities (Fig. 1D). The fibres descend beneath the surface into the subependymal region.

Moving towards the lateral edges and the anterior or posterior stalks the surface of the SFO becomes more heterogeneous. This intermediate zone (Fig. 1B, iz) is characterized by the presence of specialized surface protrusions such as kinocilia and microvilli. One of the distinguishing features of the ependymal cells is the special arrangement of cilia and microvilli (Fig. 2A, B). Ependymal cells are occasionally encircled by a line of microvilli but more often the microvilli cover the entire cell. Many of the cells may have a solitary cilium protruding from the centre (Fig. 2C, D). In addition many cells are covered by small bleb-like protrusions, clearly distinct from microvilli by their size and shape (Fig. 2 E, F).

Surface variations become most prominent towards the posterior stalk which may correspond to zone 3 in the rat (4). However, in the cat this region includes part of the SFO corpus as well. In this zone bulging cells were apparent. The group of cilia were located either on the surface of the so-called pluriciliated cells (Fig. 3D) or between ependymal cells (Fig. 3 E, F). TEM sections show that the latter correspond to the aperture of cilia-carrying channels penetrating the ependymal surface. The openings of these funnel-shaped holes, being formed by several ependymal cells, can extend to a diameter of approximately 10 to 50 µm. In addition thin sections reveal small canaliculi entering the ependymal layer. These canaliculi are encircled by a single ependymal cell. Since the dense cover of cilia and microvilli sometimes prevents their detection from the surface, they were not readily visible in the SEM study. Besides the single cells with uneven surfaces and the two kinds of channels, the posterior part of the SFO possesses

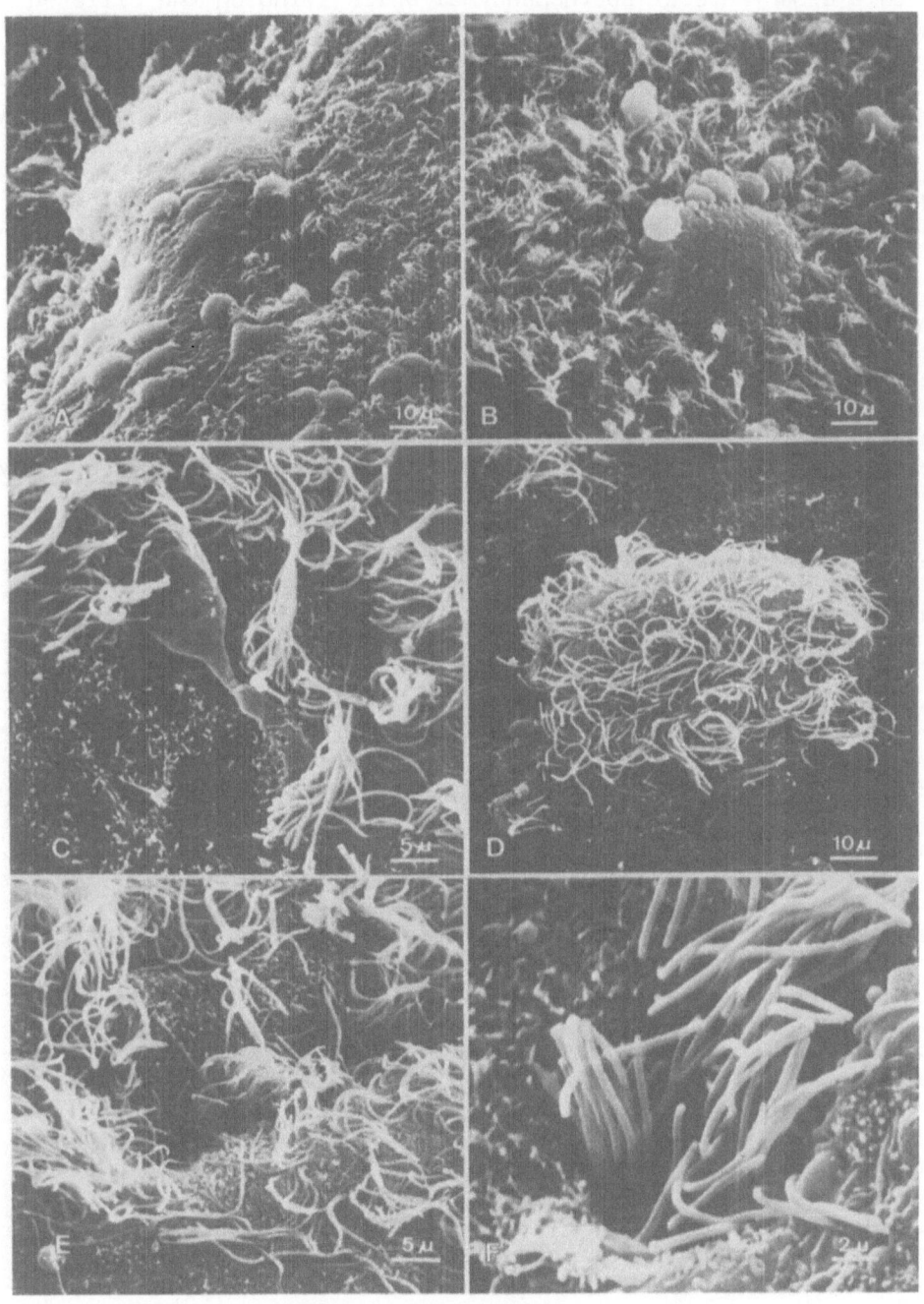

Fig. 3: Surface variations of the intermediate zone.

large protrusions which contain several ependymal elements (Fig. 3 A, B). TEM sections showed the true nature of these bumps as cellular caverns beneath the surface of the SFO. In the same region of this organ several supraependymal cells lying on the ciliated surface of the organ were observed (Fig. 3 C). These cells were pleomorphic and had long pseudopodial processes.

The ciliated zone (Fig. 1B, ci) resembles the surface covering of the ventricular wall. Kinocilia and microvilli are more frequently found and evenly distributed on the slightly sloping stalks and edges of the SFO. In contrast to the intermediate zone this ependymal surface was not interrupted by crypts or channels (Fig. 1E, F).

Attention has been directed towards the structural variation among ependymal cells lining the third ventricle. Furthermore, in this investigation we tried to examine the three-dimensional structure of processes and their vascular terminations beneath the ependymal layer by splitting the SFO in half. We were able to observe tanycytes unbranched throughout their course from the ventricular wall into the subependymal layers of the SFO. The tanycytic processes often terminated on capillary walls, where they divided into several slender branches (Fig. 4). Our results are similar to those obtained in other ventricular structures (5) and lend support to the involvement of ependymal tanycytes as a pathway in neuroendocrine integration.

DISCUSSION

Our studies revealed similarities and differences between the SFO of the cat and other animals. In agreement with the findings in rats (4) distinct regional differences must be taken into consideration, suggesting corresponding functional differentiation. The major difference between the SFO in the cat and the rabbit is the overall bumpy cellular surface seen in the latter (6). The cat has a large non-ciliated zone and many ependymal cells with solitary cilia were observed in the intermediate zone. It is interesting to note that there was a distinct difference between the SFO of male and female cats. Whether this sexual differentiation reflects possible relations with hydric balance, as reported in polyethylenglycol-induced thirst (7), needs to be further elucidated.

From morphological investigations it has been concluded that the SFO cells may detect physical and chemical gradients between CSF and blood. Experimental interventions led to the suggestion firstly that the SFO plays a role in neurosecretion, including both secretory activity into the capillaries and intraventricular ependymosecretion, and secondly that the organ is highly specialized in monitoring reception. The sites of release were found to be either in the pericapillary spaces or through interstitial vacuoles

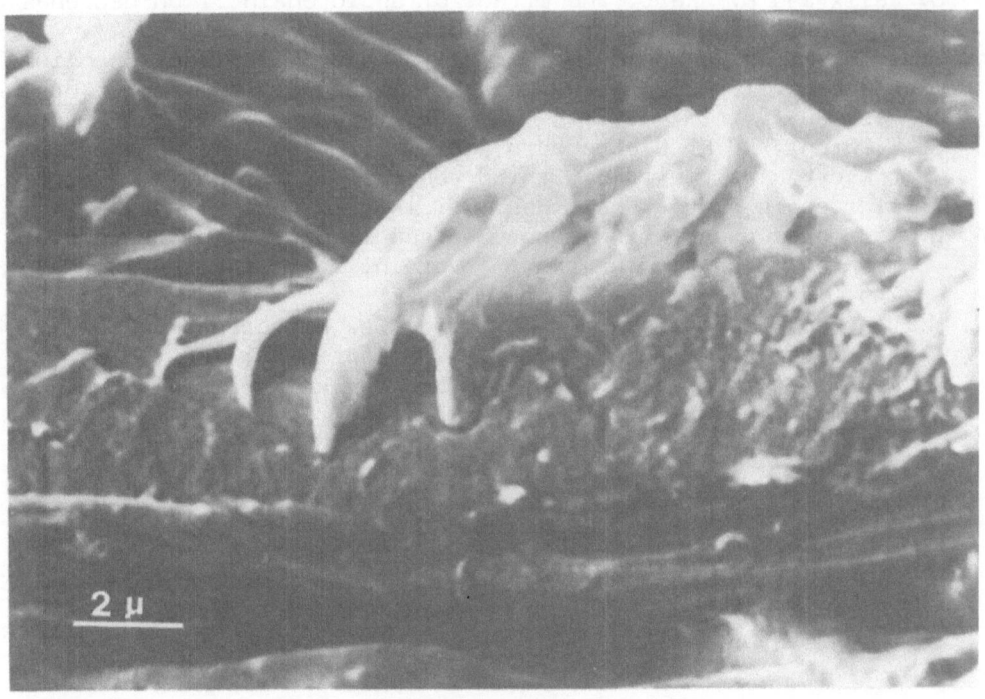

Fig. 4: Tanycyte process dividing at its termination on the capillary wall.

into the ventricular cavity (6). The evidence for CSF reabsorption is based partly on the detection by Andres (8) of miniature ciliated cannulae penetrating the SFO from the ventricular surface in the dog. In contrast to the rat and bird SFO both canaliculi in single ependymal cells and funnel-shaped channels encircled by several cells were found in the SFO of the cat.

To summarize, the ependymal lining of the SFO merits further investigation. The very distinct regional differences in the ependymal cells might be a morphological expression of a higher absorption activity of these cells (1). The role of this activity in extracellular fluid regulation will undoubtedly stimulate further research into the function of this fascinating small organ.

This work was supported by grant 3.017.81 from the Swiss National Science Foundation.

REFERENCES

1. H.-D. Dellmann and J. B. Simpson, The subfornical organ, Int.Rev. Cytol. 58: 333 (1979).
2. J. T. Fitzsimons, Thirst, Physiol.Rev. 52: 468 (1972).

3. D. Felix and K. Akert, The effect of angiotensin II on neurones of the cat subfornical organ, Brain Res. 76: 350 (1974).
4. M. I. Phillips, L. Balhorn, M. Leavitt and W. Hoffman, Scanning electron microscope study of the rat subfornical organ. Brain Res. 80: 95 (1974).
5. J. E. Bruni, R. E. Clattenburg and D. G. Montemurro, Ependymal tanycytes of the rabbit third ventricle: a scanning electron microcopic study, Brain Res. 73: 145 (1974).
6. H. Leonhardt and B. Lindemann, Surface morphology of the subfornical organ in the rabbit's brain, Z.Zellforsch. 146: 243 (1973).
7. M. Vijande, M. Costales and B. Marin, Sex difference in polyethylenglycol-induced thirst, Experientia 34: 742 (1978).
8. K. H. Andres, Ependymkanälchen im Subfornikalorgan vom Hund, Naturwissenschaften 52, 433 (1965).

ENDOGENOUS ANGIOTENSIN AND SODIUM APPETITE

James T. Fitzsimons

The Physiological Laboratory
Downing Street
Cambridge CB2 3EG, England

INTRODUCTION

Intracerebroventricular administration of angiotensin II or its precursors causes an increase in sodium appetite that is independent of the thirst also aroused by these substances. After renin, increased intakes of water and NaCl persist for many hours, sometimes for days, owing to continuing generation of angiotensin II from locally available components of the cerebral renin-angiotensin system[1]. The initial increase in sodium appetite is not secondary to sodium loss[2], though angiotensin also causes natriuresis, because the animals may develop a substantial positive sodium balance. Nor is the increased NaCl intake the consequence of the accompanying increase in water intake since it still occurs when access to water is prevented. The appetite is specific for the sodium ion. A pre-existing hypovolaemia (Fuller & Fitzsimons, this workshop), or treatment with mineralocorticoids[3] augments the effect of angiotensin on sodium appetite.

Though there is good evidence that increases in circulating angiotensin II stimulate thirst[4], it has been less easy to show that increases in circulating angiotensin contribute to sodium appetite. Intravenous infusion of angiotensin can induce rats to drink hypertonic NaCl as well as water, but it is difficult to demonstrate that this is the result of angiotensin acting directly on the central nervous system, since effective rates of infusion are also natriuretic[5]. Beta adrenergic-induced increases in endogenous renin secretion, produced by injecting isoprenaline or phentolamine, are also relatively ineffective at causing increased sodium appetite in the short term[6], though some increase occurs with repeated injections[7].

The difficulty of demonstrating a role for circulating angiotensin in sodium appetite could be partly explained by the stimulus to the appetite being multifactorial. As is the case for hypovolaemic thirst, angiotensin is probably just one of a number of factors that cause increases in sodium appetite, and which normally act together[4]. Peripheral administration may not be the most effective way of demonstrating a role for angiotensin. Circulating angiotensin might not gain access to responsive brain structures, a real possibility in view of the likely existence but unknown function of a brain renin-angiotensin system. The angiotensin pressor response could be inhibitory. The taste of the highly concentrated NaCl solutions which are generally used in experiments on sodium appetite could be sufficiently aversive to interfere with the response. Indeed, to make an animal drink salty water in order to satisfy its hunger for salt is to oblige it to perform an act that it would rarely have to perform in nature.

The question of the relevance of such experiments in understanding sodium appetite in nature is therefore fraught with difficulty. However, although we must have reservations about the physiological significance of some results, the fact that a rat can be made to drink hypertonic NaCl by factors which reproduce those present in sodium depletion may mean that these factors normally have a role to play. Experiments described here in which increases in hypertonic NaCl intake were obtained by manipulating the renal renin-angiotensin system, surgically and by using the angiotensin converting enzyme inhibitor captopril, show that there are circumstances in which one of these factors, namely peripherally generated angiotensin II, may stimulate sodium appetite.

CAPTOPRIL IN RATS WITH LIGATED URETERS

Bilateral ligation of the ureters results in an increase in circulating renin[8]. When offered water only, rats with ligated ureters drank more than control rats subjected to a sham operation, though the amounts of water taken were small (see below) and it took some hours for the values to become significantly different from the control intakes. This was not an effect of anuria since bilaterally nephrectomized rats showed no such increase in water intake. In view of other evidence[9] it is reasonable to conclude that this is an example of renin-induced thirst.

The rat with ligated ureters is particularly responsive to the dipsogenic effect of captopril[10]. Captopril alone causes some drinking and it enhances drinking in response to some other stimuli to thirst[11,12]. By preventing conversion of angiotensin I to angiotensin II, captopril releases renin secretion from the inhibitory feedback of angiotensin II, resulting in a further increase in renin secretion in this preparation (Fig. 1).

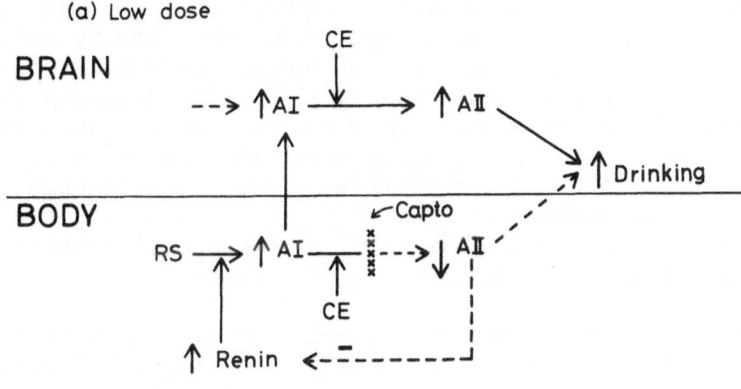

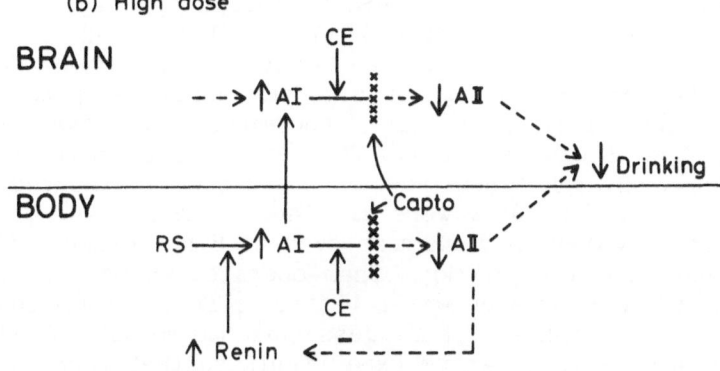

Fig. 1.   Interpretation of the effects of low dosage and high
           dosage of systemically administered captopril on drinking.
           RS = renin substrate, A I and A II = angiotensin I and II,
           CE = angiotensin converting enzyme, Capto = captopril.
           From reference 10.

        After subcutaneous injection of moderate doses of captopril,
rats with ligated ureters drank vigorously, taking more water than
similarly injected sham-operated rats. The 3 h cumulative water
intakes of rats with ligated ureters (ml/kg body weight, $\pm$ SEM,
numbers of rats in parentheses) were 27.3 $\pm$ 4.3 (9) after the
lowest (0.5 mg/kg) dose and 25.0 $\pm$ 3.9 (9) after the middle (5.0
mg/kg) dose, both values being significantly (<0.01) greater than
the control intake of 9.8 $\pm$ 1.6 (8). Intakes of sham-operated rats
after these two doses were less at 7.1 $\pm$ 2.8 (7) and 12.0 $\pm$ 2.2 (9)
respectively; only the latter (middle dose) value was significantly
(<0.01) greater than the control of 3.7 $\pm$ 1.6 (8). After the
highest (50.0 mg/kg) dose, the 3 h intakes were 8.1 $\pm$ 3.0 (8) by
rats with ligated ureters and 4.3 $\pm$ 2.0 (7) by sham-operated rats,
neither value being different from the corresponding controls.

The fact that in response to captopril larger amounts of water were drunk by rats with ligated ureters than by animals able to excrete urine, shows that such drinking was not secondary to increased renal fluid loss. Captopril-induced drinking is angiotensin-dependent because it did not occur in rats made anuric by nephrectomy. Nor did it occur after the highest dose of captopril used in these experiments, a dose large enough to prevent significant angiotensin II formation in the brain as well as in the periphery, but not large enough to interfere with drinking caused by non-angiotensin stimuli.

Since moderate doses of captopril, associated with increased angiotensin II formation in the brain, were highly effective at causing increased water intake, it seemed that the rat with ureteric ligation might be an ideal preparation in which to demonstrate a role for renal renin in sodium appetite. However, rats with ligated ureters offered 2.7 % NaCl to drink in addition to water after captopril (0.5, 5.0 or 50.0 mg/kg, 8, 6 and 3 rats respectively), showed no sign of a sodium appetite up to 8 h after operation (they cannot be followed for much longer than this), even though the body fluid osmolality fell as a result of increased water intake. By 3 h, when most of the captopril-induced drinking was over, intakes of water were similar to the intakes of the groups offered water only whereas even at 8 h, intakes of 2.7 % NaCl did not exceed 1.0 ml/kg. Sham-operated animals behaved similarly, with intakes of water little different from the water-only groups and intakes of NaCl less than 1.0 ml/kg. Food was not available in these short-term experiments so that there was no other source of sodium available to the rat.

CAPTOPRIL IN NORMAL RATS

Although in the short term captopril did not stimulate sodium appetite in the intact rat, it did when it was given mixed in the food over a period of several days[13]. Similarly, when captopril was administered in the drinking water at a concentration of 1 mg/ml, which meant a dose of about 40 mg per day for the rat drinking an average amount of water, the intake of hypertonic NaCl to which the rats also had access began to increase after 1-2 days and reached a plateau by about 4-5 days (Fig. 2)[14]. Water intake was unaffected by captopril when both NaCl and water were available to drink, though it increased when only water was available. Food intake was also unaffected. The pattern of increased NaCl intake with no change in water intake was also seen when captopril was given in the food or by subcutaneous injection. Therefore, the rats were not trying to increase their fluid intake by drinking NaCl instead of water because the water tasted of captopril. Stimulation of sodium appetite by captopril seemed to be a specific effect and a large one.

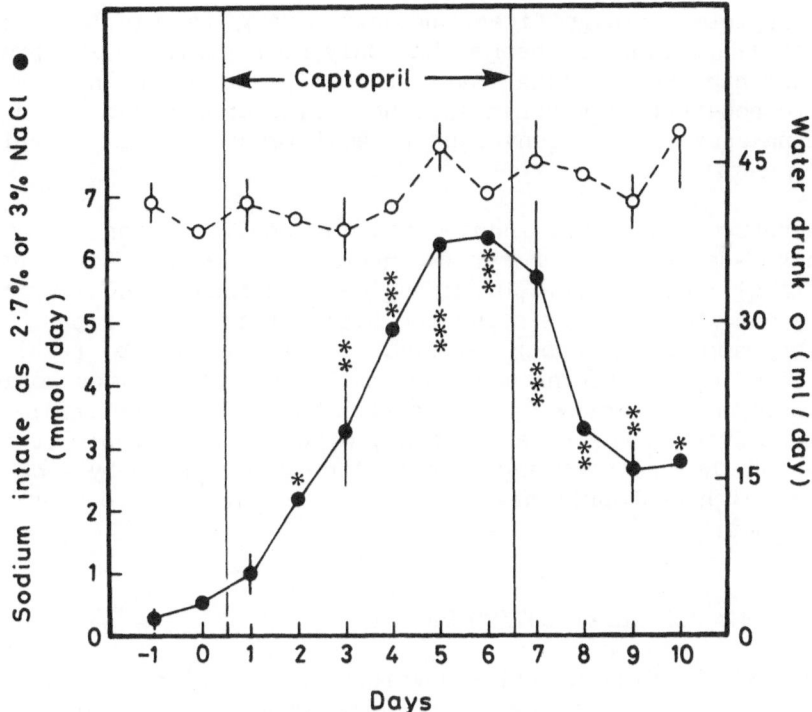

Fig. 2. Mean 24 h intake (± SEM) of sodium in mmol taken as 2.7 % or 3 % NaCl (filled symbols, left-hand ordinate), and water in ml (open symbols, right-hand ordinate), of rats with access to both fluids and receiving captopril in the drinking water (1 mg/ml) on days 1 to 6 inclusive. Sodium intake excludes sodium in the food. n = 16, up to and including day 4, n = 11 thereafter. * p<0.05, ** p<0.01, *** p<0.001, compared with day 0 by paired t test. From reference 14.

The long-term stimulation of sodium appetite by captopril was not secondary to increased urinary sodium loss. The rats remained in sodium and fluid balance and the haematocrit did not alter. Sodium appetite did not occur when captopril reached the brain from the periphery through a blood-brain barrier disrupted by trauma, nor when captopril was injected directly into the brain. It required daily but intermittent administration of captopril such as occurs when the drug is given by single injections, or when it is contained in the food or water and the animal is showing its normal nycthemeral rhythmicity of eating and drinking behaviour. It did not occur when captopril was given continuously by intravenous infusion at a constant rate over a wide range of rates, but it did occur as the infusion rate was reduced to a low value. With continuous administration of captopril, converting enzyme blockade is too complete for significant amounts of angiotensin II to be formed in the brain. It seems that only when peripheral captopril levels are high but falling that circulating angiotensin I is high enough to penetrate the brain and there find enough unblocked converting enzyme to be converted to angiotensin II and stimulate sodium appetite.

Therefore, captopril-induced sodium appetite is an angiotensin-dependent phenomenon. Moderate levels of circulating captopril which block peripheral angiotensin converting enzyme activity but which do not cross the blood-brain barrier in sufficient amounts to block cerebral converting enzyme, result in an increase in circulating angiotensin I which stimulates sodium appetite when it is converted to angiotensin II in the brain. It seems that even in the absence of any other stimulus, prolonged exposure of the brain to angiotensin such as produced by long-term treatment with captopril may be all that is required to arouse sodium appetite.

ADRENALECTOMY AND DEOXYCORTICOSTERONE

Increased sodium appetite develops in the rat made sodium-deficient by removal of the adrenal glands and, paradoxically, when there is retention of sodium produced by excessive amounts of mineralocorticoids[15]. The adrenalectomized rat is hyper-reninaemic whereas renin secretion is suppressed during treatment with deoxycorticosterone (DOC). Therefore, it seemed that the effect of captopril in these two preparations could throw light on the extent to which angiotensin is involved in sodium appetite. Captopril was administered as follows. At the start of the experiment, in mid-morning, food was removed and captopril, 0.5 mg/kg (low dosage, L) or 50.0 mg/kg (high dosage, H) in 0.9 % NaCl (5.0 ml/kg), or 0.9 % NaCl alone (C), was injected subcutaneously. At 6 h, each rat was given a second injection of the same amount of 0.9 % NaCl or captopril as it had already received. Food was returned and the drinking water was replaced with water containing captopril (0.01 mg/ml for the low dosage schedule and 1 mg/ml for the high), and the experiment was continued until 24 h had lapsed since the first injection.

The chronically adrenalectomized rat showed a further increase in intake of 2.7 % NaCl above its already increased baseline intake after the low dosage schedule of captopril, and a fall in intake after the high (Fig. 3). The differences were evident within a few hours of injection and water intake was also increased after the low dosage schedule. The effect of captopril on sodium appetite in DOC-treated rats (35 mg /kg daily for 3-5 days and then every other day) was quite different, with neither dosage affecting intake of NaCl. DOC on its own caused polydipsia as well as increased intake

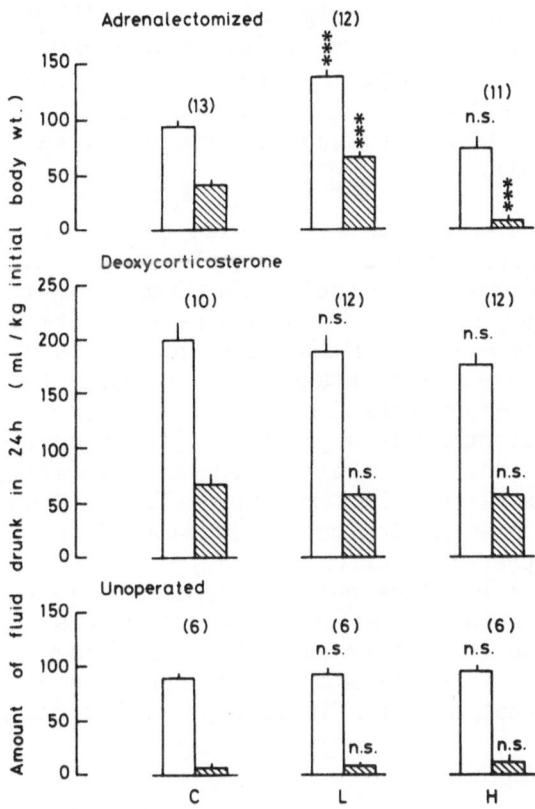

Fig. 3. Mean 24 h intake (with one SEM, number of observations in
parentheses) of water (open columns) and 2.7 % NaCl
(hatched columns) of adrenalectomized, DOC-treated and
unoperated rats offered both fluids, in response to 0.9 %
NaCl (C), low dosage (L) or high dosage (H) schedules of
captopril. For details of the dosage schedule, see the
text. *** p<0.001, compared with the corresponding C
value by unpaired t test. n.s. = not significant. Based
on reference 16.

of NaCl which was also unaffected by captopril. Captopril had
little effect on the NaCl intake of unoperated rats over this
period, but had administration been continued for longer than 24 h
in these animals the phenomenon of increasing NaCl intake already
described would have occurred.

The hyper-reninaemia of the adrenalectomized rat was further
increased by captopril. Plasma renin concentrations (in pmol
angiotensin I/ml per h, 4 rats in each group) were measured 1 h
after the second of 2 injections of captopril following the same
dosage schedule as used in the behavioural part of the experiment
except that rats were not allowed to drink or eat. The mean renin
concentration of adrenalectomized rats was 16.5 ±1.3 compared ˙with

4.4 $\pm$ 2.5 in normal rats, a significant (p<0.01) difference. Renin levels were further increased by low (59.0 $\pm$ 6.0, p<0.001) and high (88.0 $\pm$ 29.2, p<0.05) dosage schedules of captopril, compared with untreated adrenalectomized rats. On the other hand, even after the high dosage schedule of captopril plasma renin was undetectable in 7 DOC-treated rats.

These results could be interpreted as meaning that the additional stimulus provided by adrenalectomy sensitises the mechanisms responsible for sodium appetite so that increases in angiotensin II generated in the brain by moderate dosage of captopril produce an earlier increase in NaCl intake instead of the more gradually developing intake seen in unoperated animals. That renin is partly responsible for the increase in sodium appetite following adrenalectomy is suggested by the reduction in NaCl intake of adrenalectomized rats after the high dosage schedule of captopril. Against this interpretation is the finding that in the first 3 h after pharmacological activation of the renin-angiotensin system with isoprenaline or phentolamine, the adrenalectomized rat increased its water intake but not its NaCl intake[6]. Further experiments are needed to establish whether this is an effect of dose, timing or different pharmacological properties of the various drugs, or whether it represents a more fundamental objection to the hypothesis that angiotensin is a physiological stimulus to sodium appetite.

GOLDBLATT HYPERTENSION

Occlusion of the abdominal aorta between the renal arteries leads to severe hypertension, polyuria and polydipsia[17]. The left kidney distal to the occlusion secretes excessive amounts of renin, the plasma concentration reaching a maximum by about the 5th day after operation and returning to near normal by 30 days. Rats went into fluid and electrolyte deficit after partial aortic occlusion and they increased their intake of 2.7 % NaCl and water when both fluids were offered (Fig.4)(Costales, Fitzsimons & Vijande, this workshop)[18]. Arterial blood pressure rose immediately after occlusion, before the onset of increased drinking, presumably in response to increased renin secretion. Up to 3 weeks after operation the incidence and severity of the hypertension did not appear to depend on the spontaneous changes in intake of water or hypertonic NaCl, though when changes in intake are imposed, blood pressure may change[19].

The increase in intake of 2.7 % NaCl preceded the increase in water intake. Drinking during the day increased greatly, which together with the drop in food intake that also occurred and the rise in 24 h fluid intake, meant that the normal association between feeding and drinking was lost. The increased intakes of

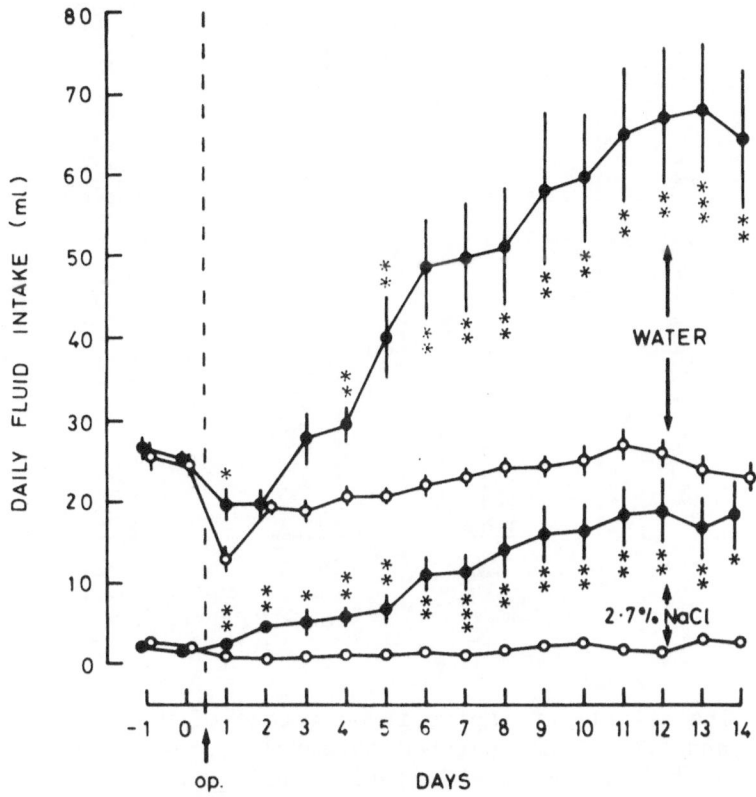

Fig. 4. Mean 24 h intakes (± SEM) of water and 2.7 % NaCl of
female rats subjected to partial aortic occlusion (n = 15)
or sham operation (n = 10) and offered both fluids. *
p<0.05, ** p<0.01, *** p<0.001, compared with the intake
by sham-operated rats on the same day by unpaired t test.
From reference 18.

2.7 % NaCl and water were in response to fluid losses from both the
extracellular and cellular compartments. Increased sodium appetite
and polydipsia were abolished by removal of the ischaemic, renin-
producing kidney. High dosage of captopril (50 mg/kg s.c., twice
in 24 h) (Fig. 5), or low dosage (0.5 mg/kg s.c., twice) combined
with simultaneous intracranial injection of captopril in μg
quantities, also reduced the intakes of both fluids to near normal.
These results suggest that renal renin contributes to both
behaviours. However, the contribution is likely to be a variable
one in view of the lack of correlation between changes in blood
pressure, which initially are entirely renin-dependent, and the
changes in drinking. An additional reason for concluding that
renin must interact with other stimuli in both behaviours is the
dissociation between the increases in intakes of the two fluids.
Increase in intake of 2.7 % NaCl preceded the polydipsia.

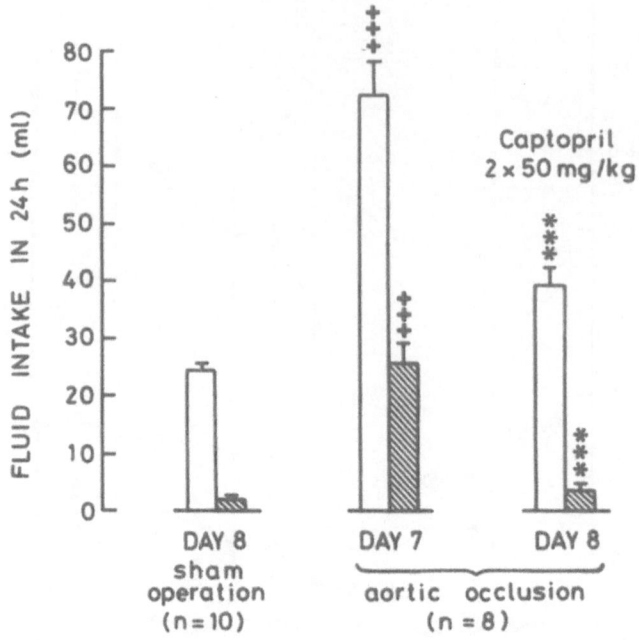

Fig. 5. The effects of captopril on intake of water (open columns) and 2.7 % NaCl (hatched columns) on day 8 after partial aortic occlusion. Comparison before and after captopril in the 8 rats subjected to aortic occlusion was by paired t test, *** p<0.001. Comparison with 10 rats subjected to a sham operation 8 days previously was by unpaired t test, +++ p<0.001. From reference 18.

CONCLUSIONS

Increased secretion of renin by the kidney, leading to increased generation of angiotensin II in the circulation, may be as important a stimulus to sodium appetite as it is to hypovolaemic thirst. The short-term effect of renin is to stimulate water intake, seen here especially in captopril-treated rats with ligated ureters. Since renin is released in cardiovascular emergencies, it is fitting that it should, first of all, contribute to increased water intake, water being the physiological volume expander.

In the longer-term, however, sodium (with appropriate anion) is also needed in hypovolaemia. Therefore it is equally fitting that renin should stimulate sodium appetite, but the response, unlike that for water, does not have to be immediate. From the experiments described here, it seems that for endogenously generated angiotensin II to stimulate sodium appetite, prolonged exposure of the brain to angiotensin, or the presence of an additional signal to sodium appetite is required. Such signals are understimulation of volume receptors as occurs in hypovolaemia, reduced sodium concentration in the cerebrospinal fluid, increased secretion of mineralocorticoids and increased secretion of the hormones of pregnancy and lactation[19]. But there are other possibilities still to be explored. Normally all of these signals occur when, for one reason or another, the body is depleted of sodium, the condition par excellence in which sodium appetite is increased. All are associated with prolonged rises in circulating angiotensin II.

REFERENCES

1. D. B. Avrith and J. T. Fitzsimons, Renin-induced sodium appetite: effects on sodium balance and mediation by angiotensin in the rat, J. Physiol. 337: 479-496 (1983).
2. Emma Chiaraviglio, Effect of renin-angiotensin system on sodium intake, J. Physiol. 255: 57-66 (1976).
3. S. J. Fluharty and A. N. Epstein, Sodium appetite elicited by intracerebroventricular infusion of angiotensin II in the rat: II synergistic interaction with systemic mineralocorticoids, Behav. Neurosci. 97: 746-758 (1983).
4. J. T. Fitzsimons, "The Physiology of Thirst and Sodium Appetite," Monographs of the Physiological Society, No. 35, Cambridge University Press, Cambridge (1979).
5. D. B. Avrith, J. T. Fitzsimons and S. Nicolaidis, The effects of long-term intravenous infusions of angiotensin II on thirst, sodium appetite and water and sodium balance in the rat, 7th IUPS Internat. Conf. Physiol. Food and Fluid Intake, Warsaw (1980).
6. J. T. Fitzsimons and J. B. Wirth, The renin-angiotensin system and sodium appetite, J. Physiol. 274: 63-80 (1978).
7. R. W. Bryant, A. N. Epstein, J. T. Fitzsimons and S. J. Fluarty, Arousal of a specific and persistent sodium appetite in the rat with continuous intracerebroventricular infusion of angiotensin II, J. Physiol. 301: 365-382 (1980).
8. T. Kent Keeton and W. B. Campbell, The pharmacologic alteration of renin release, Pharmacol. Rev. 32: 81-227 (1980).
9. J. T. Fitzsimons, The role of a renal thirst factor in drinking induced by extracellular stimuli, J. Physiol. 201: 349-368 (1969).

10. R. M. Elfont and J. T. Fitzsimons, Renin-dependence of captopril-induced drinking after ureteric ligation in the rat, J. Physiol. 343: 17-30 (1983).

11. C. C. Barney, M. J. Katovich and M. J. Fregly, The effect of acute administration of an angiotensin converting enzyme inhibitor, captopril (SQ 14,225), on experimentally induced thirsts in rats, J. Pharmacol. Exp. Ther. 212: 53-57 (1980).

12. M. J. Katovich, C. C. Barney, M. Fregly and R. McCaa, Effect of an angiotensin converting enzyme inhibitor (SQ 14,225) on β-adrenergic and angiotensin-induced thirsts, Europ. J. Pharmacol. 56: 123-130 (1979).

13. M. J. Fregly, Effect of the angiotensin converting enzyme inhibitor, captopril, on NaCl appetite of rats, J. Pharmacol. Expt. Ther. 215: 407-412 (1980).

14. R. M. Elfont, A. N. Epstein and J. T. Fitzsimons, Involvement of the renin-angiotensin system in captopril-induced sodium appetite in the rat, J. Physiol. 354: 11-27 (1984).

15. K. K. Rice and C. P. Richter, Increased sodium chloride and water intake of normal rats treated with desoxycorticosterone acetate, Endocrinol. 33: 106-115 (1943).

16. R. M. Elfont and J. T. Fitzsimons, The role of angiotensin in increased sodium appetite after adrenalectomy, J. Physiol. 320: 70P (1981).

17. J. M. Rojo-Ortega and J. Genest, A method for production of experimental hypertension in rats, Canad. J. Physiol. Pharmacol. 46: 883-885 (1968).

18. Marina Costales, J. T. Fitzsimons and M. Vijande, Increased sodium appetite and polydipsia induced by partial aortic occlusion in the rat, J. Physiol. 352: 467-481 (1984).

19. D. Denton, "The Hunger for Salt," Springer-Verlag, Berlin, Heidelberg, New York (1983).

HORMONAL SYNERGY AS THE CAUSE OF SALT APPETITE

Alan N. Epstein

Department of Biology
and Institute of Neurological Sciences
University of Pennsylania, Philadelphia, PA 19104

It is well known that angiotensin and aldosterone, together, permit the kidney to conserve body sodium[1]. Renin which generates angiotensin from its substrate is released by hyponatremia, angiotensin then releases aldosterone which acts on the kidney to promote tubular reabsorption of sodium.

I have proposed[2] that the same hormones, acting in synergy, are the cause of the appetite for salt which is the behavioral complement of renal tubular reabsorption and which is the only means by which sodium is introduced into the body as a pure commodity. The proposal unites the behavioral and renal contributions to sodium homeostasis in the same endocrine network. It is schematized in Figure 1 in which natural sodium depletions (diets poor in salt, gestation and lactation, loss of large volumes of sodium containing fluid as the result of grooming or sweating during thermolysis, and pathological states that deplete extracellular fluid) set in motion

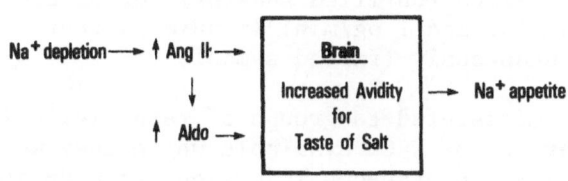

Figure 1. The synergy hypothesis for the arousal of sodium appetite. Sodium depletion increases angiotensin II which releases aldosterone, and both hormones acting at physiological levels then increase the brain's avidity for salt. Each hormone can individually do so but only when it is in physiological excess and only after long periods of treatment. From: Zhang et al. (1984).

the endocrine changes that lead to concurrent elevations of angiotensin and aldosterone which, when raised to levels within their
normal physiological ranges, act together on the brain to increase
its avidity for the taste of salty substances which the animal then
seeks and ingests.

This concept of the mechanism by which sodium appetite is aroused solves several of the paradoxes that have emerged from investigation of the phenomenon since it was first produced experimentally by
Richter in 1936[3]. It explains why the animal's contemporary sodium
content does not control its salt drinking behavior[4,5]. According
to the concept decreases in blood sodium are essential for the initiation of the hormonal events that arouse the behavior but once they
are initiated it is these endocrine consequences of sodium deficiency,
not the deficiency itself, that are sufficient to produce the behavior
even if at the time of salt drinking sodium levels are normal or even
in excess. The synergy hypothesis also explains why there are long
latencies between the onset of Na+ depletion and the expression of
the appetite[5]. Endocrine events, especially those involving steroids,
are slow often requiring several hours for their behavioral effects.
And lastly, the hypothesis explains why the hormones of sodium conservation can each produce the appetite when acting alone[6,7]. The
concept proposes that the hormones are the message which apprises the
brain of the body's need for sodium and it acknowledges that each of
them can do so in the absence of the other, but under such abnormal
conditions each hormone will generate the appetite only when it is
raised to pathological or pharmacological concentrations and only
when it is at those levels for long periods. The hypothesis predicts
that, when acting together, as they do during episodes of natural
sodium depletion, angiotensin and aldosterone synergize and produce
sodium appetite at concentrations of both hormones that are within
the endogenous range of each and, furthermore, it predicts that they
do so rapidly.

These predictions have been confirmed by Fluharty and myself[8].
We have produced an appetite for 3% NaCl solution in excess of 15 ml/
day in sodium replete rats by treating them concurrently with systemic DOCA and continuous intracerebroventricular (cICV) angiotensin
II. The data on the left in Figure 2 shows the effects on salt
drinking of either cICV isosaline (unfilled squares), or of cICV
Ang II at three doses (100, 10, and 1 pg/min) in animals that received only isosaline subcutaneously (filled symbols).

The treatments were administered to groups of rats living in
their home cages with water and 3% NaCl available while they were infused cICV for 24 hours. The subcutaneous injection was made just
before the infusion began. Salt and water intake and urine output
were measured half hourly. Food was witheld in order not to complicate the measurement of urinary sodium. The control groups which
were treated with cICV isosaline (shaded areas in Figure 3 below)

396

fell into a negative sodium balance of approximately 1 mEq by the
end of the 24 hour infusion. These are the standards of continuous
sodium balance for comparison with the other groups in each experiment.

As you can see on the left in Figure 2 there were very little
effects of any of these treatments. Only the animals given the
highest dose of intraventricular Ang II drank salt. The others,
also given only cICV Ang II, drank no more than untreated rats.

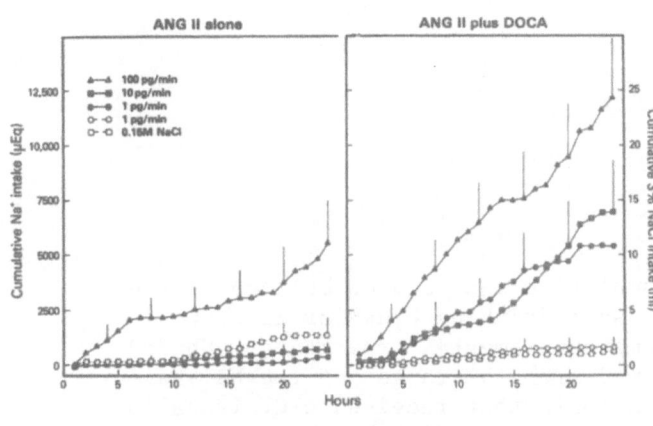

Figure 2. Cumulative
mean intake of 3% NaCl
during 24 hr cICV in-
fusion of various doses
of Ang II alone (left
panel) or in animals
additionally treated
with subcutaneous DOCA
(right panel). Abbre-
viations: Ang II =
angiotensin II, DOCA =
desoxycorticosterone,
cICV=continuous intra-
cerebroventricular).
From: Fluharty &
Epstein (1983).

But as shown on the right, when DOCA was added the outcomes for
the same doses of cICV Ang II were clearly different. First, note
that DOCA by itself (unfilled squares) did not raise an appetite.
But all doses of Ang II were now effective, even the lowest (only
1 pg/min) at which 3 (filled circles) of the 6 rats infused drank
salt, suggesting that it is close to or at threshold. Higher doses
(10 pg/min) became effective or were more effective (100 pg/min) in
all rats studied. The 1 pg/min of cICV Ang II is an exceptionally
low dose. It typically does not increase water intake when given
into the ventricles[9] and it did not do so in these experiments even
when combined with systemic DOCA. This suggests the interesting idea
that the synergy treatment is selective for sodium appetite especially when angiotensin is acting close to the lower limits of its
physiological range.

The continuous sodium balance (intake minus urinary output, calculated half hourly) of the animals in the same experiment is shown
in Figure 3 which is also divided into a panel (on the left) of data
collected from the animals receiving subcutaneous DOCA (the shaded
area which is the variance of the group) and cICV isosaline or from
those receiving cICV Ang II (filled symbols) and subcutaneous isosaline, and a panel (on the right) of data collected from animals
receiving the synergy treatment.

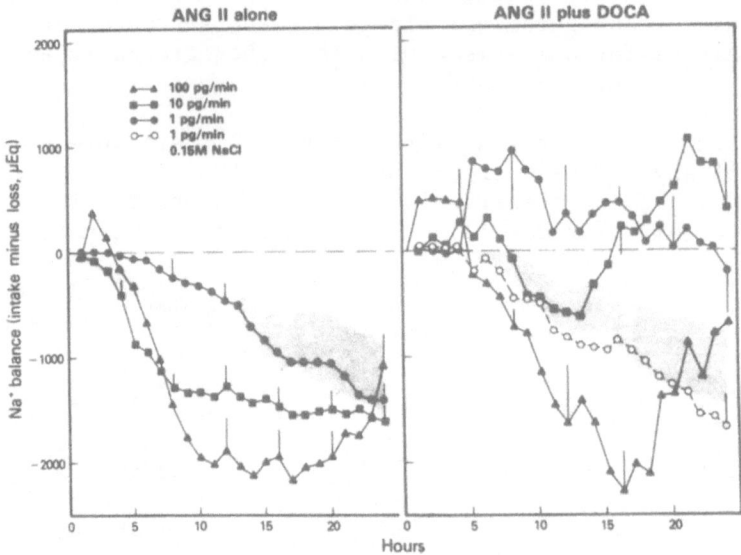

Figure 3. Cumulative mean sodium balance of the animals whose salt drinking behavior is shown in Figure 2. Same experimental treatments. Shaded areas are formed from the ranges of the Na+ balances of the animals that received cICV isosaline From: Fluharty & Epstein (1983).

Note that the addition of systemic mineralocorticoid to continuous intracranial angiotensin produces less negative sodium balance at the same time as it produces more salt intake. All of the balance curves are shifted upward. The animal's receiving the lower doses of Ang II spend the entire 24 hours of treatment in positive sodium balance (compare their curves with the shaded areas which is the variance of the sodium balance of the rats given DOCA and cICV isosaline only). Again the 3 rats infused at 1 pg/min that drank salt (filled circles) are shown separately from those that did not (open circles). Even the animals given the highest dose, although they are still driven into negative sodium balance by this large amount of angiotensin, spend less time in negative balance.

Overall, the outcome of these experiments is that the addition of a mineralocorticoid, at a dose which does not raise a salt appetite, to intracranial angiotensin, at doses that also do not raise the appetite, produces more positive sodium balance and more salt intake in rats that are sodium replete. In other words, animals that do not need sodium drink it.

We have produced the same result with exceptional rapidity by using a second procedure in which rats are pretreated for several days with subcutaneous DOCA (more recently with aldosterone, see free communication by R.R. Sakai, in this voluem) and then receive a pulse intracerebroventricular (pICV) injection of angiotensin II[8].

The results of Fluharty's use of this procedure are shown in Figure 4 which gives, for each animal, the latency to drink 3% NaCl (and water) after the pICV angiotensin as well as the volume of the salt solution drunk in the 30 minutes after the injection (columns of figures to the far right). Three groups of animals were studied. The synergy group, above (five days of SC DOCA at 250 µg/day, 60 ng of Ang II pICV on day 5) and two control groups one of which (middle) received only the pICV Ang II (SC isosaline for 5 days), and the

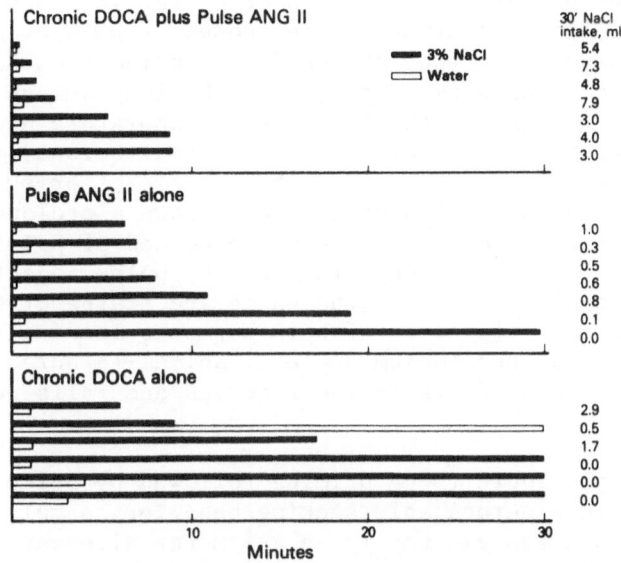

Figure 4. Latencies to drink 3% NaCl (solid horizontal bars) and water (open bars) of rats treated with subcutaneous DOCA for 1 week and with pICV Ang II at time 0 (above), or (middle) with only pICV Ang II (and SC isosaline),or (below) with only Sc DOCA (and pICV isosaline). Volumes of 3% NaCl drunk at end of first 30 min after pICV injection are at far right. From: Fluharty & Epstein (1983).

other (below) which received only the DOCA (SC DOCA for 5 days, pICV isosaline on Day 5). Three percent NaCl was available to the animals continuously from Day 1 in addition to water and a commercial solid food with its usual high salt content. As you can see, the animals injected with Ang drank water (open horizontal bars) very quickly, but only the animals treated with both the mineralocorticoid and angiotensin drank salt with short latencies (4.7 min average for the group). Note that they also drank it in considerable volume (5.8 ml average in first have hour after pICV Ang). Some of the other animals made contact with the salt and therefore had latencies of less than 10 or 20 minutes but they only sampled it and drank very little. Again, there was a clear indication of the selectivity of the synergy treatment for enhancement of salt intake. The water intakes of the groups treated with pICV angiotensin were the same (in the 3 hours after the injection) whether or not they received DOCA pretreatment.

The mobilization of a salt appetite with such rapidity is impressive when it is recalled that it takes hours, even days, for it to be expressed by rats subjected to more drastic procedures such as sodium depletion[5], pharmacological doses of DOCA or aldosterone[6], or adrenalectomy[3] while they are eating a sodium rich commercial food. Our result suggests, first, that the long delays that are typical of the arousal of the appetite by laboratory procedures may be artifacts of its unnatural production, and, second, that the mechanism for its natural evocation may be capable of brisk activation especially when a surge of angiotensin follows a prolonged period of elevated aldosterone production. The method of priming with mineralocorticoid for several days followed by pulse injection of angiotensin may be a close mimic of the endocrine status of the sodium deficient (moderate elevation of both hormones leading to maintenance of eunatremia) and then sodium depleted animal (a surge of angiotensin as it loses the last of its sodium reserves and falls into hyponatremia).

The same procedure (DOCA priming followed by pulse angiotensin) has been used to generate salt seeking behavior as well as its consumption[10]. Rats ran vigorously in a 1 meter alleyway and for as long as 30 to 40 minutes for rewards of a few drops of concentrated salt solution when treated with this combination of hormones, thus, expressing a true appetite for salty substances. The animals were sodium replete when tested (they were endocrinologically intact, had been receiving exogenous DOCA, 3% NaCl was continuously available to them in their home cages and they were eating a salt rich food). Nevertheless, as shown in Figure 5, salt seeking behavior was aroused by the synergy treatment. The animals were run for salt (data on the right) or water (on the left) on different days. Pulse injection of angiotensin elicited high-speed running for water both with and without DOCA priming (open and filled circles, on the left), but the only treatment that elicited running for salt was the combination of the two hormones (filled circles, on the right). Note that, again, when the hormones were given individually (open circles and filled triangles, on the right) the animals showed very little inclination to search for salt.

All of the above demonstrates the power of the synergy treatment to generate a salt appetite in animals that do not need sodium. In more recent work we have been altering the expression of the appetite in sodium depleted rats with pharmacological blockade of the production of endogenous angiotensin II. If the action of angiotensin is necessary for arousal of the appetite then it should be suppressed or abolished by drugs that prevent conversion of Ang I

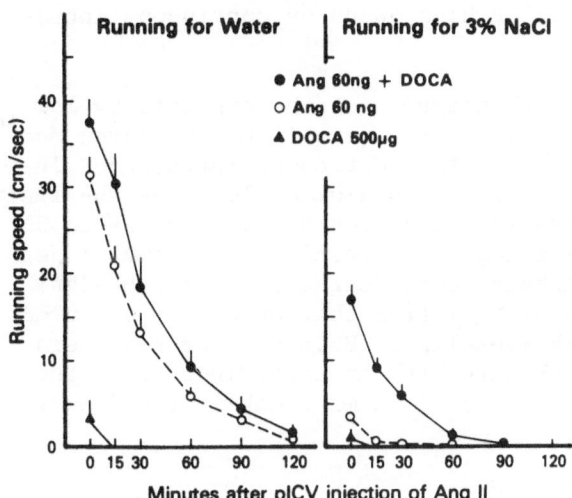

Figure 5. Running speed for two drops of water (left) or for 3% NaCl (right) of rats treated with only SC DOCA (filled triangles), only pICV Ang II (open circles), or with both SC DOCA and pICV Ang II (filled circles). From: Zhang et al (1984).

to Ang II. Captopril (Squibb), a blocker of angiotensin converting enzyme[11], has allowed us to do this in animals expressing the appetite as the result both of sodium depletion (a large dose of the natriuretic drug lasix followed by one night of sodium deficient diet) or of adrenalectomy. In both instances we find that the appetite is markedly suppressed by doses of the drug that are high enough to cross the blood-brain barrier[12] and therefore to block angiotensin II production in both the periphery and in the brain.

In the adrenalectomized rat suppression of the appetite is achieved with lower doses of captopril than are necessary for comparable reduction in the sodium depleted endocrinologically intact rat. The greater sensitivity of the adrenalectomized rat's sodium appetite to blockade by captopril suggests that it is more dependent on endogenous angiotensin II for its expression. The appetite after adrenalectomy may therefore be an instance in which angiotensin has been driven to its physiological ceiling by sodium deficiency and has remained there long enough for it to induce an avidity for salt even in the absence of aldosterone.

Lower doses of systemic captopril given either PO, IV or SC, that block conversion only outside the brain, enhance the appetite both in the sodium depleted intact rat and in the adrenalectomized rat. They also produce the appetite in the sodium replete rat[13]. This enhancement disappears when the drug is admitted to the brain either by trauma into the blood-brain barrier or by cICV infusion of it at doses (1 or 2 μg/hr) that are sufficient to block conversion to the cerebral ventricles. It appears therefore to be due to overflow of unconverted angiotensin I into some compartment within the brain into which captopril can not penetrate and within which the excess angiotensin I is converted to equally excessive angiotensin II[14] which then generates the appetite just as it does when it is introduced into the brain at high doses by continuous intra-cerebroventricular infusion[15].

Both of these phenomena (enhancement of the appetite by low doses, blockade by high) are shown by the experiments of Karen Moe[16]. Figure 6 is her dose-response curve for continuous intravenous infusion of captopril into rats that were sodium depleted by the lasix plus sodium deficient diet treatment and then allowed access to 3% NaCl for two hours the next morning. The results are shown as percentage change (increase or enhancement, decrease or suppression) in the animals' appetites for 3% NaCl from that which was measured after a night of infusion with isosaline. Points at or near zero mean that the treatment did not alter the appetite induced by the depletion which we find to be from 8 to 12 ml drunk in the 2 hour

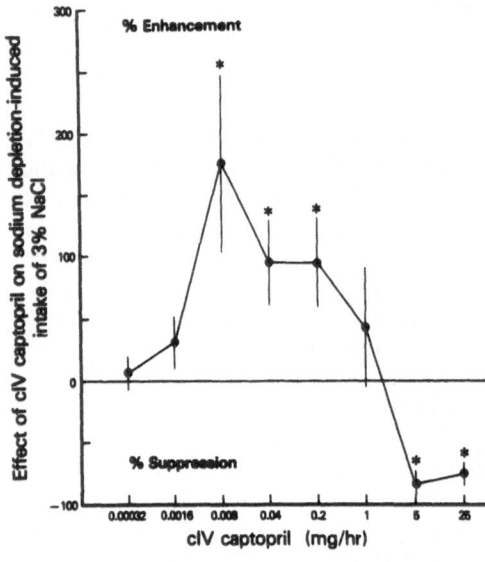

Figure 6. Effects of overnight intravenous infusion of captopril on sodium-depletion-induced intake of 3% NaCl. Animals had access to salt during the 2 hrs at the end of the period of captopril infusion. Intakes are represented as average % change from those produced by the depletion alone.
From: Moe et al (1984).

access period, usually in the first half hour. Very low doses of intravenous captopril that do not block angiotensin I conversion in the periphery have no effect, the enhancement occurs with doses that prevent conversion in the periphery, and doses of 5 mg/hr or more produce a nearly complete suppression of sodium appetite. These same doses (~5mg/hr) do not interfere with salt drinking aroused by pharmacological doses of DOCA, and the 5 mg/hr dose does not reduce the urinary sodium loss produced by the lasix treatment. Note that doses in the range of 1 mg/hr have no effect on the endocrinologi-cally intact rat (a balance seems to be reached at which captopril penetration into the brain is not yet sufficient to block conversion there). A similar dose (1 mg/ml) given in the drinking water to adrenalectomized animals abolished their appetite. The greater sensitivity of the adrenalectomized rat to suppression of its appetite by systemic captopril is discussed above where it is suggested that it may be more dependent than the intact rat on endogenous angiotensin.

It is difficult to foresee the species generality of the mechanism proposed above because sodium appetite has been studied in so few animals. All of the work summarized here was done with only the domestic rat. In the other well studied species[17] sodium appetite appears to be controlled by brain sodium content (sheep) or it can be evoked by treatment with several of the hormones of pregnancy (wild rabbit). Nevertheless, it can already be suggested that animals equipped by their evolution with a hormonal mechanism for arousal of sodium appetite would have an advantage over those that have to seek salt only after they have become sodium depleted. The elevations of angiotensin and aldosterone that would be produced by small declines in blood sodium would have two complimentary effects. They would achieve sodium homeostasis by renal conservation and they would give early warning of imminent sodium depletion. That is, the hormones would simultaneously allow the kidneys to conserve sodium, and would act on the brain to begin the process of arousal of a sodium appetite. Several such corrected episodes of mild sodium deficiency may be all that is required for frank expression of salt seeking behavior. Having acted on the brain in synergy, angiotensin and aldosterone will have aroused an appetite for salt and will have led the animal to ingest it before frank sodium depletion occurs. The animal could therefore express the behavior without the disabilities of sodium depletion. By this scheme the appetite has evolved, not as an emergency solution to the problem of an existing sodium depletion, but as a means for avoiding it.

REFERENCES

1. I.H. Page & F.M. Bumpus, "Angiotensin", Springer-Verlag, N.Y (1974).

2. A.N. Epstein, Mineralocorticoids and cerebral angiotensin may act together to produce sodium appetite. Peptides, 3: 493-494, (1982).

3. C.P. Richter, Increased salt appetite in adrenalectomized rats. Am. J. Physiol. 115: 155-161 (1936).

4. M. Nachman & D.A. Valentino, Roles of taste and postingestional factors in the satiation of sodium appetite in rats. J. Comp. Physiol. Psychol. 62: 280-283 (1966)

5. J.E. Jalowiec & E.M. Stricker, Sodium appetite in rats after apparent recovery from acute sodium deficiency. J. Comp. Physiol. Psychol. 73: 238-244 (1970).

6. K.K. Rice & C.P. Richter, Increased sodium chloride and water intake of normal rats treated with desoxycorticosterone acetate. Endocrinol. 33: 106-115 (1943).

7. A.L.R. Findlay & A.N. Epstein, Increased sodium intake is somehow induced in rats by intravenous angiotensin II. Hormones & Behav. 14: 86-92 (1980).

8. S.J. Fluharty & A.N. Epstein, Sodium appetite elicited by intra-cerebroventricular infusion of angiotensin II in the rat: Synergistic interaction with mineralocorticoid. Behav. Neurosci. 97: 746-758 (1983).

9. J.B. Simpson, A.N. Epstein & J.S. Camardo, The localization of receptors for the dipsogenic action of angiotensin II in the subfornical organ. J. Comp. Physiol. Psychol. 92: 581-608 (1978).

10. D.M. Zhang, E. Stellar & A.N. Epstein, Together intracranial angiotensin and systemic mineralocorticoid produce avidity for salt in the rat. Physiol. Behav. 32: 677-681 (1984).

11. Z.P. Horovitz, "Angiotensin Converting Enzyme Inhibitors", Urban and Schwarzenberg, Baltimore (1981).

12. M.L. Cohen & K.D. Kurz, Angiotensin converting enzyme inhibition in tissues from spontaneously hypertensive rats after treatment with captopril or MK-421. J. Pharm. Exptl. Therap. 220: 63-69 (1982).

13. M.J. Fregly, Effect of the angiotensin converting enzyme inhibitor, captopril, on NaCl appetite in rats. J. Pharm. Exptl. Therap. 215: 407-412 (1980).

14. R.M. Elfont, A.N. Epstein & J.T. Fitzsimons, Involvement of the renin-angiotensin system in captopril-induced sodium appetite. J. Physiol. (London) in press.

15. R.W. Bryant, A.N. Epstein, J.T. Fitzsimons and S.J. Fluharty, Arousal of a specific and persistent sodium appetite in the rat with continuous intracerebroventricular infusion of angiotensin II. J. Physiol. (London) 301: 365-382 (1980).

16. K.E. Moe, M.L. Weiss & A.N. Epstein, Sodium appetite during captopril blockade of endogenous angiotensin II formation. Am. J. Physiol. in press.

17. D. Denton, "The Hunger for Salt", Springer-Verlag, N.Y. (1982).

Footnote

The original research reviewed here was supported by NS 03469 from the USPHS.

SODIUM APPETITE INDUCED BY SODIUM DEPLETION IS SUPPRESSED BY INTRACEREBROVENTRICULAR CAPTOPRIL

Mark L. Weiss

Department of Biology
and Institute of Neurological Sciences
University of Pennsylvania, Philadelphia, PA   19104

INTRODUCTION

We have previously shown[1] that systemic captopril (CAP) has effects that implicate angiotensin II in the expression of sodium appetite induced by sodium depletion in the rat. In addition, that report suggested that it is angiotensin II (Ang II) in the brain that was important.

The work described here takes the next experimental step by treating sodium deficient rats with CAP infused directly into the cerebral ventricles.

METHODS

Continuous intracerebroventricular infusion (cICV)

Adult male Sprague-Dawley rats were fitted with chronic intracerebroventricular (ICV) cannulae opening into the anterior cerebral ventricles. After recovery from surgery they were adapted to continuous ICV (cICV) infusion by attachment to a watertight swivel and spring-guarded PE tubing through which a vehicle solution (isosaline to which a prophylactic dose of 1.5% gentamycin had been added) was driven for several days by a Harvard infusion pump at a rate of 5 μl/hr. This system allows complete control and flexibility of the parameters of the infusate. Rats are given ad lib access to distilled HOH, a 3% NaCl solution, and Purina chow pellets (approx. 1% NaCl by weight). All liquids are offered at room temperature.

Sodium Depletion

Sodium appetite was elicited by depletion produced by subcutaneous injection of furosemide (Lasix), a potent natriuretic and diuretic drug (2 injections of 5 mg/0.5 ml/rat, separated by 2 hrs). With the first injection, chow pellets were replaced by a sodium deficient powdered food (ICN #902903) offered in glass cups, the 3% NaCl was removed, and the cages were wiped to remove adherent salt. Twenty-two to twenty-four hours later 3% NaCl was returned to the rats, and their consumption of it and of HOH was recorded at 15, 30, 60 and 120 min. The latency to drink each solution was also measured. Appetite was defined as intake elicited by the depletion treatment in the two hour drinking test minus the spontaneous intake that the non-depleted rat demonstrated in a 2 hr access to 3% NaCl a few days before. Rats whose depletion-aroused saline intake was less than 3 ml in the two hour test were allowed at least 2 days of ad lib access to 3% NaCl and Purina pellets to replete themselves prior to another depletion treatment.

Experiment 1

Captopril Administration

To examine the role of angiotensin II in the brain, the angiotensin converting enzyme inhibitor, captopril (CAP) was chronically infused into the cerebral ventricles at doses of 1.2 μg, 120 ng, 12 ng and 1.2 ng/hr, beginning one hour before the start of the sodium depletion treatment, and continuing through the end of the two hour drinking test (total infusion time: 25–28 hrs). Immediately following the drinking test, the effectiveness of the blockade of conversion of Ang I to Ang II by CAP was behaviorally assessed by giving each rat pulse ICV (pICV) injection of 6 ng/1 μl of Ang I. Ang I is dipsogenic when given to rats by pICV, but only after conversion to Ang II by endogenous angiotensin converting enzyme (ACE)[2]. Failure of the rat to drink to Ang I was thus taken as behavioral evidence that angiotensin conversion was blocked. The dipsogenic response to Ang II, which should remain intact following CAP treatment, was then measured to demonstrate the behavioral competence of the animals and their capability to respond to intraventricular Ang II in the presence of a blocking dose of CAP. Data was accepted only from animals that drank to pICV injection of Ang II with a latency of less than 120 s and that consumed more than 3 ml of HOH in 15 min. To check for nonspecific effects of cICV CAP, overnight HOH and food intakes were measured when the rats received the three highest doses of CAP.

RESULTS

Experiment 1

        The first figure shows that the three highest doses of capto-
pril suppressed the expression of the 3% NaCl appetite in the two
hour test, the lowest dose had no effect on 3% NaCl intake. Note
also that the two hour HOH intake was also suppressed. However, it
is important to recall that: 1) these rats are water replete at the
start of the two hour drinking test; 2) rats drink little or no HOH
during the daylight cycle; 3) the HOH intake which occurs in the two
hour test is secondary to the 3% NaCl intake and will be suppressed
when 3% NaCl intake is suppressed. Note also that centrally administered CAP never caused an enhancement of sodium appetite, as is produced by low doses of the drug given systemically[1].

        CAP's effects on Ang I drinking are shown in the next figure.
Note that the increasing doses of CAP that suppressed the expression
of sodium appetite also blocked the drinking response to Ang I in a
consistently graded fashion beginning with the lowest dose which had
no effect on either behavior and culminating in the highest dose which
markedly suppressed both.

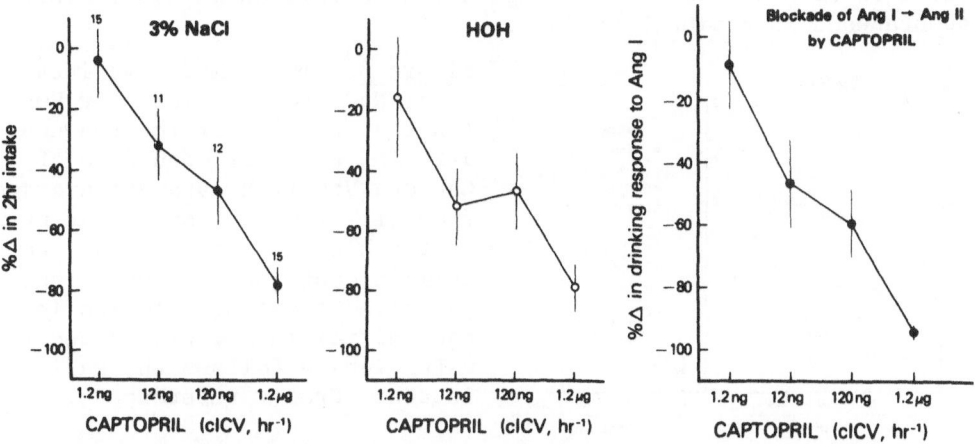

Figure 1. Inhibition of sodium depletion induced sodium appetite
(3% NaCl on the left, HOH on the right) by chronic intracerebroventricular infusion of CAP. Percentage of inhibition was calculated
from the appetite expressed when the animal received cICV VEH infusion minus the appetite expressed when the animals received cICV
CAP divided by the appetite expressed after cICV VEH. Significant
inhibition of the 3% NaCl intake was produced by the three highest
doses of CAP ($p < 0.075$, using the Bonferoni manipulation). From:
ref. 4.

Figure 2. Blockade of converting enzyme by cICV CAP, seen as inhibition of the dipsogenic response to Ang I (6 ng in 1 µl) in a
15 minute test. This test immediately followed the two hour drinking

test (after 24-28 hrs of CAP infusion). Percentage of inhibition was calculated by subtracting the HOH intake induced by a pICV Ang I injection (6   ng: 1 μl) after cICV VEH infusion from that evoked after cICV CAP infusion, then dividing by the pICV Ang I induced HOH intake following VEH infusion. Significant blockade of converting enzyme was produced by the three highest doses of CAP (p$<$0.015 using the Bonferoni manipulation). From: reference 4.

Thus, taken together these data show that blockade of Ang II synthesis in the brain suppresses the expression of depletion-induced sodium appetite.

This suppression of sodium appetite was not the result of non-specific behavioral suppression or malaise for several reasons. First, all animals receiving the higher doses of CAP drink some 3% NaCl at the beginning of the two hour access period, see figure 3. And in fact, their latency to drink saline was not altered. This means that they are avid for the salt, competent to drink it, but terminate their intake prematurely. In addition, overnight food and HOH intake of rats receiving doses of CAP that suppressed sodium appetite were not affected. Moreover, the rats responded to Ang II at the end of the drinking test. Thus the animals were behaviorally competent and able to respond to angiotensin II at the end of 26 hr of CAP infusion. These facts show that converting enzyme inhibition does not disturb spontaneous ingestion behaviors.

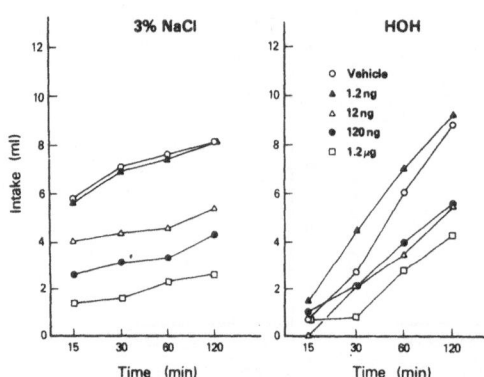

Figure 3. The cumulative intake of 3% NaCl (on the left) or HOH (on the right) over the two hour drinking test for each dose of CAP and VEH (VEH data are averaged from all groups). Note that even at the highest doses of CAP, the animals continue to consume the aversive 3% saline solution in the initial 15 min period and that water intake follows the saline intake. From: reference 4.

As an additional test of the specificity of the effects of cICV CAP, it was given to animals which were expressing a sodium appetite induced by daily injections of DOCA rather than by deple-tion. The DOCA appetite is known to be independent of the renin-angiotensin system as DOCA treatment suppresses the release of renin[3]. Therefore, this sodium appetite should not be affected by CAP. Animals received 2 mg/day of DOCA for seven days, subcutaneously, and after receiving CAP at 1.2 μg/hr overnight were given two hour access to 3% NaCl and tests of the blockade of conversion (pICV Ang I) and of their competence to respond to Ang II (pICV Ang I) as described above. Animals were excluded if they failed to drink to pICV Ang II, as described above.

As shown in table 1, DOCA-induced sodium appetite was not at all affected by the highest dose of CAP. Again indicating that CAP's effects on sodium appetite are not the result of malaise of behavioral incompetence.

Table 1

cICV CAP does not effect the salt intake aroused by DOCA
From: reference 4

	cICV Vehicle	cICV Captopril (1.2μg/hr)
Chronic DOCA (2mg/d, sc, 1wk) N=8	9.3±1.2	10.8±1.7

ns, p>0.1

Finally, the highest dose of CAP, 1.2 μg/hr, had no effect on the total sodium excretion induced by the sodium depletion treatment, because the cumulative sodium loss of rats receiving cICV CAP was not different from those receiving VEH.

Experiment 2

Sar[1]-Ile[8]-angiotensin II (SAR), like Saralasin, is an analogue of angiotensin II that acts as an antagonist at its receptor site. This makes it an ideal drug to confirm that captopril's suppression of the sodium appetite is the result of interference with Ang II's actions and not due to other pharmacological effects of CAP. Thus, animals with chronic cannulae were sodium depleted, as described above. 30 minutes before and then again just before they were given access to 3% NaCl, they received a 1 μl pICV injection of SAR (5 μg in 1 μl of SAR) or vehicle. The drinking test was abbreviated to thirty min in this experiment. Confirmation of blockade of Ang II receptors followed immediately by examination of the drinking response to a pICV injection of Ang II (6 ng in 1 μl). The competence of the animals to drink in response to pICV chemical injection was then demonstrated by evoking a dipsogenic response to a pICV injection of carbachol (40 ng/μl: 1 μl).

Result Experiment 2

Eight of the nine rats tested showed a marked suppression of their appetites expressed in the thirty min test (10.7 ± 1.6 ml of 3% NaCl VEH vs. 6.35 ± 1.4 ml with SAR; t(7) = 4.27, p < 0.01).

These rats also demonstrated blockade of the drinking response to Ang II (9.6 $\pm$ 1.5 ml HOH after VEH; 2.3 $\pm$ 0.7 ml SAR). The blockade of the receptors did not affect water intake in the thirty min drinking test (0.5 $\pm$ 0.3 ml VEH vs. 0.7 $\pm$ 0.4 ml SAR)nor did it impair their drinking response to pICV injection of carbachol (10.6 $\pm$ 1.6 ml HOH in 15 minutes). The remaining animal in this group was not blocked by the SAR injections. Rather, his response to pICV injection of Ang II was enhanced from 2.2 ml HOH when he received VEH to 8.7 ml after the SAR injections. His sodium appetite was also enhanced to 10.6 ml of 3% after SAR treatment from 5.0 ml after VEH.

All rats appeared aroused by the injection of the antagonist, seen as increased grooming, rearing and sniffing the spout inlets. There was no evidence of agonistic actions caused by the first SAR pICV injection; e.g., no dipsogenic effect the first 30 min. Again, the suppression of the appetite did not affect the latency to drink saline, indeed there is no difference between the groups at the five min reading (4.7 $\pm$ 0.5 ml 3% NaCl after VEH vs. 4.8 $\pm$ 1.0 ml after SAR, five min reading).

In Summary

We have shown that the sodium depleted rat can not fully express a sodium appetite when 1) conversion of Ang I is prevented in the brain or 2) when Ang II receptors in the brain are blocked. Our doses of CAP are below that which have been shown to leak into the periphery[3] and it is unlikely that the Ang II analogue as used here was active in the serum. Our results are therefore best understood as the consequence of action of the agents within the brain itself. Similar results have been reported by Buggy and Jonklaas[5].

Our results confirm the importance of cerebral Ang II in the expression and maintanence of a naturally occurring sodium appetite, that of sodium depletion.

ACKNOWLEDGEMENTS

This work was supported by NS 03469 to Alan Epstein and by MH 15902 to Mark Weiss. The author thanks E. Carl Jameson, Barbara Stamoutsos, Karen Linderman, Dr. Karen Moe and Dr. Alan Epstein for advice, constructive criticism and technical assistance.

REFERENCES

1. K. Moe, M.L. Weiss and A.N. Epstein, Sodium appetite during captopril blockade of endogenous angiotensin II formation, Am. J. Physiol., in press (1984).

2. J.T. Fitzsimons, A.N. Epstein and A.K. Johnson, <u>Brain Res</u>. <u>153</u>: 319-331 (1978).

3. W.A. Pettinger, M. Marchelle and L. Augusto, <u>Am. J. Physiol</u>. <u>221</u>: 1071-1074 (1971).

4. M.L. Weiss, K. Moe and A.N. Epstein, Central blockade of angiotensin II's actions suppresses the expression of sodium appetite induced by sodium depletion. <u>Am J. Physiol</u>., (1985).

5. J. Buggy and J. Jonklaas, Sodium decreased by central angiotensin blockade, <u>Physiol. Behav</u>., <u>32</u>: (5)- 737-742.

INCREASED SODIUM APPETITE AND POLYDIPSIA IN GOLDBLATT HYPERTENSION

Manuel Vijande, Marina Costales and James T. Fitzsimons

Department of Physiology, University of Oviedo, Spain
and The Physiological Laboratory, Downing Street
Cambridge CB2 3EG, England

INTRODUCTION

Two kidney Goldblatt hypertension can be induced in the rat by complete ligation of the aorta between the renal arteries[1]. The ischaemic (left) kidney secretes excessive amounts of renin, resulting in plasma renin concentrations which peak during the first week after operation and return to near normal after about a month[2]. Occlusion of the aorta leads to polydipsia and polyuria[3], with reduced preference for NaCl in the chronic stages of hypertension 4-6 weeks postoperatively[4]. Partial aortic occlusion, which is less lethal than complete occlusion but which produces a similar sequence of changes, was introduced in order to examine the polydipsia and changes in Na appetite that occur in renal hypertension[5]. The relation between blood pressure and spontaneous fluid intake was also investigated.

METHODS

Male or female Wistar or Sprague-Dawley rats (250-350g) were individually housed and maintained on standard laboratory diet (Na content measured by flame photometry, 0.22 mmol/g) and tap water alone, or tap water and 2.7 % NaCl. The abdominal aorta was approached through a midline ventral incision and it was partly occluded with a fine silk thread (no. 4/0) placed just below the superior mesenteric artery and above the left renal artery. A stylus (o.d. 0.5 mm) was included within the ligature and was removed after the tie had been made. Sham-operated animals were subjected to similar manipulations but the aorta was not occluded. Left nephrectomy was performed through a mid-line dorsal incision.

413

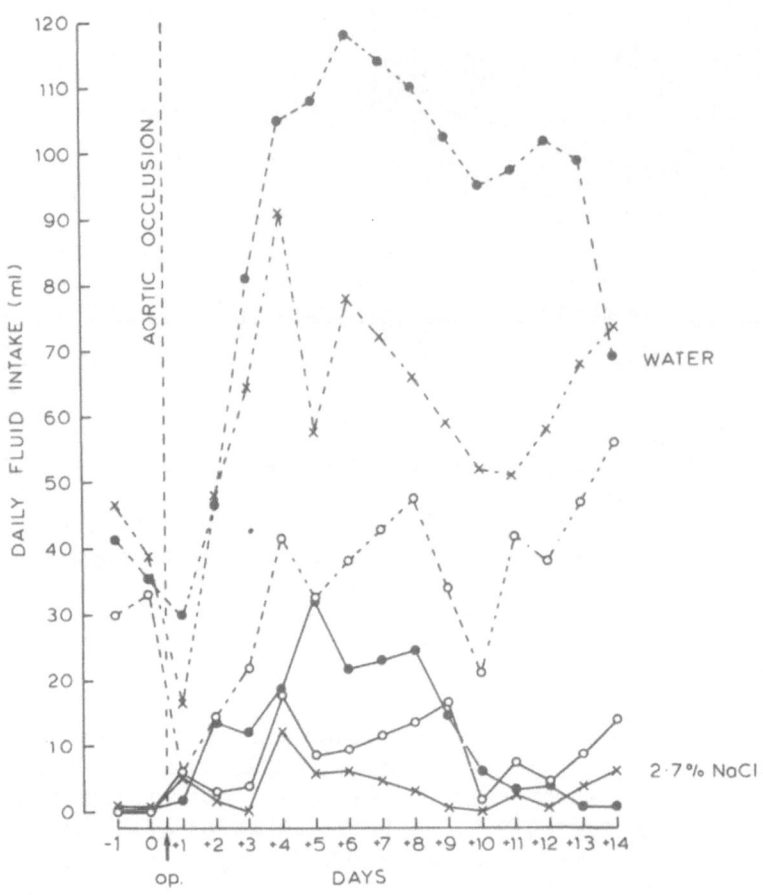

Fig. 1. Daily intakes of water (dashed lines) and 2.7 % NaCl
(continuous lines) by 3 male rats after partial aortic
occlusion and offered both fluids. Each symbol represents
one rat.

Arterial blood pressure was measured in conscious rats through a
catheter implanted in the common carotid artery and exteriorized to
the back of the neck. Daily fluid and electrolyte balances were
determined for 3 days before and 7 days after operation in 11 male
rats, of which 8 were subjected to partial aortic occlusion and 3
to a sham operation. The rats were housed in individual metabolism
cages. Urinary electrolytes were measured by flame photometry.
Faecal sodium was also measured by flame photometry after heating
the dried faeces with 100 volumes H_2O_2 and taking up the desiccated
residue in distilled water. Captopril (Squibb) was administered in
2 s.c. doses of 50 mg/kg body weight separated by 12 h, with in
addition 1 mg/ml dissolved in the drinking water.

POLYDIPSIA AND INCREASED SODIUM APPETITE IN GOLDBLATT HYPERTENSION

After partial aortic occlusion, rats offered food and water but no NaCl, developed polydipsia which reached a peak on the 13th postoperative day. When 2.7% NaCl and water were available, intake of NaCl also increased; individual results for 3 male rats are given in Fig. 1. The increase in mean intake of 2.7 % NaCl reached significance even earlier than the increase in mean water intake (Fig. 2; see also Fig. 4 in "Endogenous Angiotensin and Sodium Appetite", Fitzsimons, this workshop). Results for male and female rats were similar. Food intake decreased and body weight fell during the first 7 days after aortic occlusion despite the increased intakes of water and 2.7% NaCl. The amounts of fluid drunk during the day increased greatly, which together with the fall in food intake and the overall increase in fluid intake, meant that drinking was less associated with feeding than it is in the normal rat.

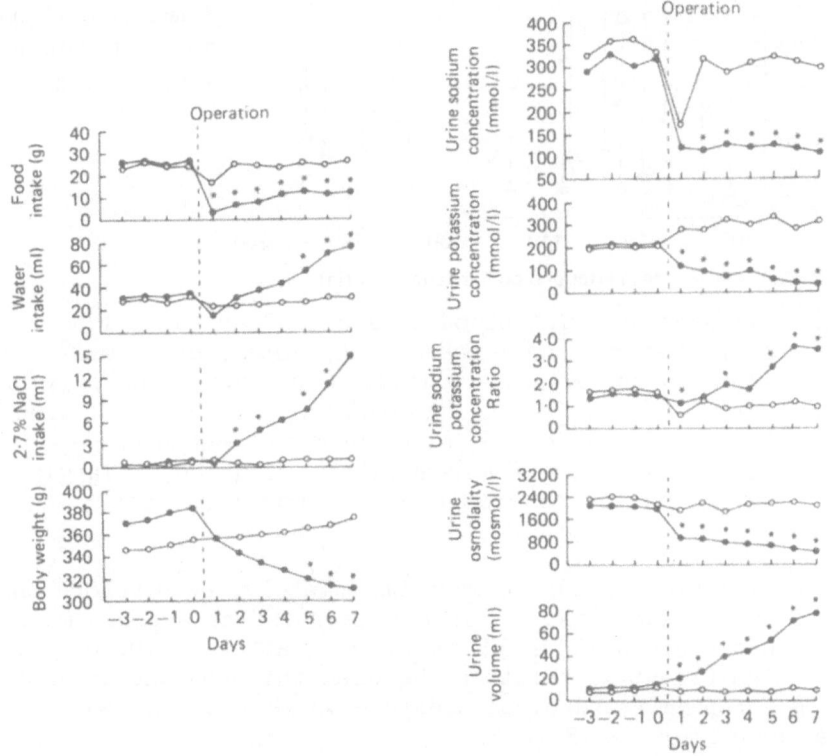

Fig. 2. Daily intakes of food and fluid, body weights and urine values of 8 male rats subjected to partial aortic occlusion (filled circles) and 3 to a sham operation (open circles). * p<0.05 compared with sham-operated rats on the same day by Wilcoxon rank sum test. From reference 5.

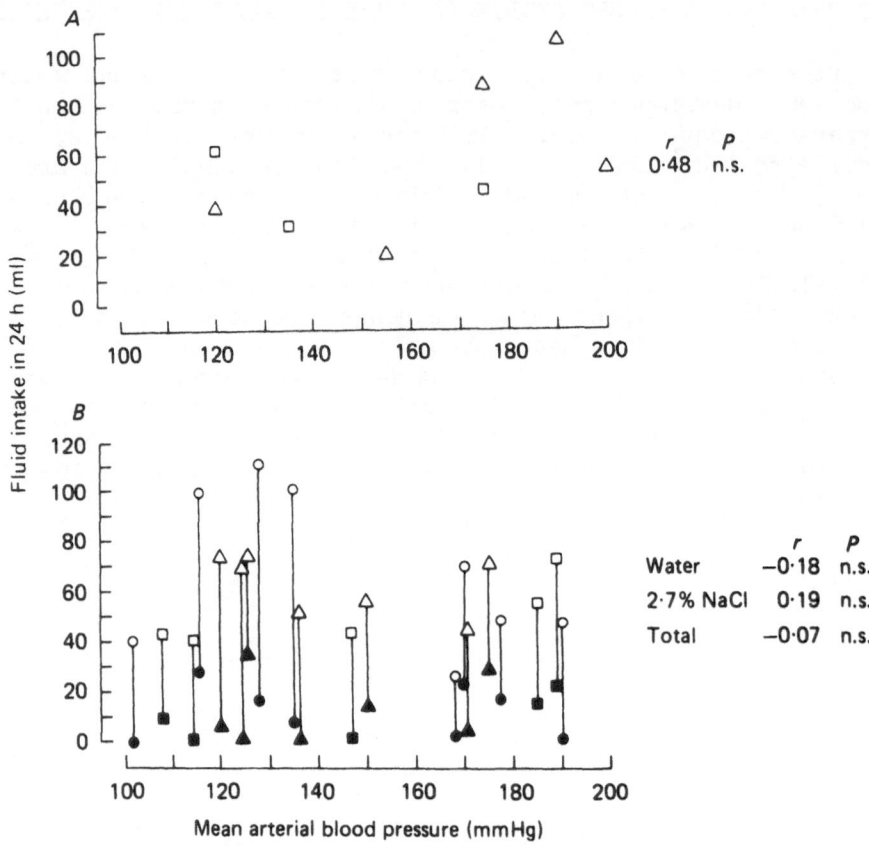

Fig. 3. The mean arterial blood pressure after partial aortic
occlusion and the intakes of A, water, or B, water (open
symbols) and 2.7 % NaCl (filled symbols), the pair of
values for one rat being joined by the vertical lines, on
the day preceding the blood pressure measurements.
Circles are values 7 days after occlusion, triangles 13-16
days, and squares 21-24 days. From reference 5.

Arterial blood pressure rose immediately, even before the
onset of increased drinking. In 6 male rats the average mean
pressure had reached 150 + 9.5 mmHg, 24 h after occlusion. There
were no significant correlations between the spontaneous intakes of
water or 2.7 % NaCl and blood pressure in 28 rats, measured 7-24
days after occlusion (Fig. 3).

Renal function was unimpaired 7 days after occlusion. Urinary
Na and K concentrations and osmolality fell but urinary volume
rose. The rats developed deficits of Na and K owing to a decrease
in dietary intake and continuing excretion of electrolytes. The
cumulative Na balance was -1.74 mmol by day 3. From day 4 the
daily balances were positive as the rats started drinking more 2.7 %

416

NaCl but the cumulative Na balance only became positive again on day 7 (+0.48 mmol). The cumulative K balance was still negative at this time (-1.24 mmol). Plasma osmolality increased from 291 $\pm$ 1.9 mosmol/l (n = 27) pre-operatively, to 294 $\pm$ 3.1 (n = 12) on day 7 and 299 $\pm$ 2.4 on day 14 (n = 14, $p<0.05$). Plasma Na concentration did not change significantly, being 152 $\pm$ 2.2 mmol/l (n = 27) preoperatively, 149 $\pm$ 3.7 (n = 11) on day 7 and 150 $\pm$ 2.5 (n = 19) on day 14. Plasma K concentration increased from 3.8 $\pm$ 0.1 mmol/l (n = 27) pre-operatively, to 4.3 $\pm$ 0.2 (n = 11, $p<0.05$) on day 7 and 5.1 $\pm$ 0.2 (n = 19, $p<0.001$) on day 14.

Left nephrectomy 7 or 14 days after partial aortic occlusion, caused an immediate fall in intakes of water and 2.7 % NaCl to the levels of sham-operated rats, and a gain in weight. Captopril in large dosage had the same effect as removing the ischaemic kidney; polydipsia and excessive NaCl intake were attenuated (see Fig. 5 in "Endogenous Angiotensin and Sodium Appetite", Fitzsimons, this workshop).

CONCLUSIONS

Partial occlusion of the abdominal aorta between the renal arteries caused rats to increase their intake of 2.7% NaCl and water. Even before this, blood pressure had risen causing a pressure diuresis, the increased urine flow being maintained by the increased fluid intake. Renal function was not impaired at this stage. Increased intake of 2.7% NaCl (an aversive concentration) within 24 h of the onset of a two kidney Goldblatt hypertension has not been described before, though rats offered 0.9% NaCl and water 10 days after constriction of one renal artery increased their intake of this palatable solution and retained Na[6]. Others have found that rats go into negative Na balance after unilateral renal artery constriction[7], which agrees with the present results with aortic occlusion. The severity of the hypertension was not correlated with the changes in spontaneous fluid intake.

The Na and K deficits following aortic occlusion, indicate that body water losses were shared between the two main fluid spaces. These losses would have accounted for increased thirst and Na appetite[8]. Angiotensin II is a known stimulus to thirst[8] and Na appetite[9,10], so that it is reasonable to expect that increased renin secretion after aortic occlusion would contribute to the behaviours. Left nephrectomy resulted in an immediate reversal of symptoms. However, since removal of the ischaemic kidney abolishes hypertension and therefore urinary fluid loss, the more conclusive evidence for a role for renin is the reduction in intakes of water and 2.7 % NaCl after high dosage of captopril. Captopril is diuretic and natriuretic in normotensive and hypertensive rats[11]. The reduction in drinking is therefore most reasonably explained by the action of captopril in preventing formation of angiotensin II.

Our results on Na appetite contrast with those of Forman and Falk[4] who found that 4-6 weeks after aortic occlusion there was a relative rejection of 0.9-1.5% NaCl. Differences in NaCl intake in the acute and chronic stages of renal hypertension may be related to differences in renin secretion. Increased Na appetite after aortic occlusion may require a a concomitant neural signal arising from the fluid deficit in addition to hyper-reninaemia.

(We are grateful for financial support from Grant FIS 83/0923).

REFERENCES

1. J. M. Rojo-Ortega and J. Genest, A method for production of experimental hypertension in rats, Can. J. of Physiol. Pharmacol. 46: 883-885 (1968).
2. M. Fernandes, G. Onesti, A. Weder, R. Dykyj, A. B. Gould, K. E. Kim and C. Swartz, Experimental model of severe renal hypertension, J. Lab. Clin. Med. 87: 561-567 (1976).
3. J.M. Rojo-Ortega, F. P. Queiroz and J. Genest, Effects of sodium chloride on early and chronic phases of malignant hypertension in rats, Am. J. Physiol. 236: H665-H671 (1979).
4. S. Forman and J. L. Falk, NaCl solution ingestion in genetic (SHR) and aortic-ligation hypertension, Physiol. & Behav. 22: 371-377 (1979).
5. Marina Costales, J. T. Fitzsimons and M. Vijande, Increased sodium appetite and polydipsia induced by partial aortic occlusion in the rat, J. Physiol. 352: 467-481 (1984).
6. J. Mohring, Maria Petri and Barbel Mohring, Salt appetite during the early phase of renal hypertension in rats, Pflugers Arch. 356: 153-158 (1975).
7. J. D. Swales, H. Thurston, F. P. Queiroz and A. Medina, Sodium balance during the development of experimental hypertension, J. Lab. Clin. Med. 80: 539-547 (1972).
8. J.T. Fitzsimons, "The Physiology of Thirst and Sodium Appetite," Monographs of the Physiological Society, No. 35, Cambridge University Press, Cambridge (1979).
9. D. B. Avrith and J. T. Fitzsimons, Increased sodium appetite in the rat induced by intracranial administration of components of the renin-angiotensin system, J. Physiol. 301: 349-364 (1980).
10. R. M. Elfont and J. T. Fitzsimons, The role of angiotensin in increased sodium appetite after adrenalectomy, J. Physiol. 320: 70P (1981).
11. R. G. Bengis, T. G. Coleman, D. B. Young and R. E. McCaa, Long-term blockade of angiotensin formation in various normotensive and hypertensive rat models using converting enzyme inhibitor (SQ 14,225), Circ. Res. 43: suppl. I, I45-I53 (1978).

INFLUENCE OF SODIUM LOAD ON ANGIOTENSIN-INDUCED SODIUM APPETITE

Louise M. Fuller and J.T. Fitzsimons

The Physiological Laboratory, Downing
Street, Cambridge CB2 3EG, England

INTRODUCTION

Whether or not angiotensin II (Ang II) plays a role in Na appetite is an open question[1]. Intracerebroventricular (I.C.V.) administration of Ang II or renin causes rats to drink hypertonic NaCl as well as water when both solutions are available[2,3,4,5]. But because I.C.V. Ang II also causes natriuresis[6], it has been suggested that NaCl intake is secondary to Na loss in the urine. According to Fluharty and Manaker[7], I.C.V. infusion of Ang II causes a "small and brief" early phase of Na ingestion which is not the result of natriuresis, and "larger and more sustained bouts of Na ingestion" which are secondary to Na loss in the urine. On the other hand, rats given a single I.C.V injection of 50 mu renin drank sufficient hypertonic NaCl to go into positive Na balance within the first hour of injection and remain in positive balance for at least 24 h afterwards[5].

In experiments discussed here the primary stimulating effect of Ang II on Na appetite was assessed in the presence of pre-existing increases in Na appetite and differing initial conditions of Na loading[8]. The effect was compared with that of carbachol (Carb) which also increases Na excretion[9]. Carb increases water intake but it inhibits NaCl intake in the normal[2,3] and adrenalectomized[10] rat.

Rats, individually housed in metabolism cages, were divided into five groups of which all except group I had continuous access to 2.7 % NaCl, as well as to water and food (Na content, 0.16 mmol/g), between experiments: group I, adrenalectomized rats deprived of 2.7 % NaCl solution for 24 h immediately prior to testing; group II, adrenalectomized rats; group III, normal rats; group IV, normal rats given 30 ml 0.9 % NaCl I.P. daily for 3 days up to and on the test day; and group V, normal rats given 1 mg deoxycorticosterone (DOC) S.C. daily for 5 days up to and on the test day.

During the test, the rat was taken from its metabolism cage and the urinary bladder was emptied by suprapubic pressure. The test solution (1.0 µl of 0.9 % NaCl, alone or containing 100 pmol Ang II or 300 ng Carb) was injected over a period of 15 sec into the preoptic region through a cannula implanted at least a week previously. The rat was weighed to the nearest 0.1 g and returned to its metabolism cage. The latency to the onset of drinking was timed from the start of the injection and cumulative intakes of water and 2.7 % NaCl were measured over 6 h. Food was not available during this time. Spontaneously voided urine was collected at 1 h. Further collections of urine were made at 3 and 6 h, the bladder being emptied on these occasions by suprapubic pressure. The rat was reweighed at 3 and 6 h, after the bladder had been emptied. Rats were tested once a week.

DRINKING BEHAVIOUR

Ang II caused increased intakes of water and 2.7 % NaCl whereas Carb stimulated water intake only (Fig. 1). All groups except group I started drinking water before NaCl. There were differences between groups in the patterns of cumulative intakes of the two fluids after Ang II. The largest intakes of NaCl were by groups I (NaCl-deprived adrenalectomized) and V (DOC-treated), in which initial Na balances were at opposite extremes. However, the NaCl intake of group I after Ang II was not significantly different from that of the corresponding 0.9 % NaCl controls. The water intakes of the adrenalectomized groups (groups I and II) after Ang II and, to a lesser extent, after Carb, were appreciably less than the intakes of the other groups.

The effects of Ang II and Carb in the NaCl-deprived adrenalectomized rats (group I) merit further consideration. In contrast to all other groups, the 0.9 % NaCl-injected animals of group I drank large amounts of NaCl when it was returned to them. Ang II had no effect on this intake but Carb caused a significant depression in NaCl intake and it stimulated water intake.

The relative effect of Ang II or Carb on intakes of NaCl and water in the different groups was assessed by expressing the volume of 2.7 % NaCl drunk as a percentage of total fluid intake. Except for group I, these percentages were much less after Carb than after Ang II. After Ang II, the adrenalectomized rats of group II and DOC-treated animals (group V) drank a greater percentage of their

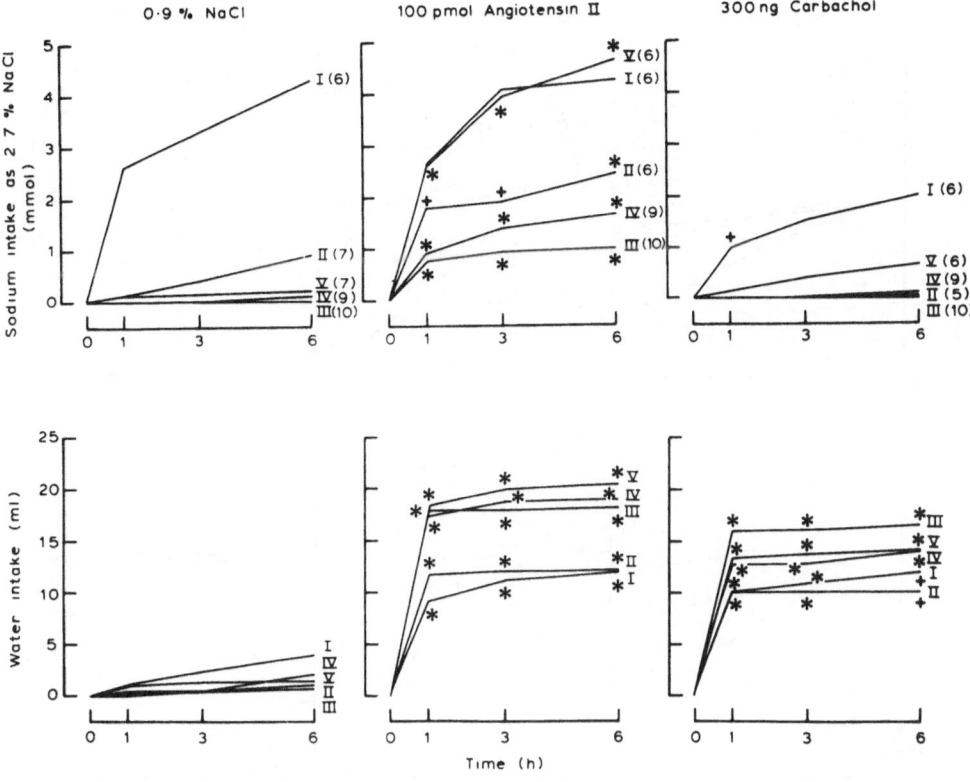

Fig. 1. Mean cumulative intakes of Na and water in response to
I.C.V. 0.9 % NaCl, 100 pmol Ang II or 300 ng Carb. In
this and Fig. 2, roman numerals are the groups as
described in the text, with number of rats in parentheses.
+ p<0.05, * p<0.01, compared with the corresponding 0.9 %
NaCl group by unpaired Wilcoxon.

fluid intake as NaCl than did normal rats (group III). At 1 h the
percentages were 7.9 + 2.2 for group III, 20.7 + 4.9 for group II
and 24.1 + 4.3 for group V. Enhancement, compared with group III,
of the percentage NaCl intake both in group II rats (p<0.05) which
lack mineralocorticoid, and in group V rats (p<0.05) which were
treated with DOC, suggests that Na appetite is multifactorial.

EFFECT ON SODIUM BALANCE

After Ang II, urine flow was least in adrenalectomized rats
and greatest in NaCl-loaded normal rats. After Carb, urine flow
was low in adrenalectomized and in DOC-treated rats, and was
greatest in NaCl-loaded rats. After Ang II, Na excretion at 1 h
was significantly more than in controls in non-adrenalectomized
groups, whereas after Carb, Na excretion at 1 h was significantly
greater than in controls in all groups despite no (group II) or

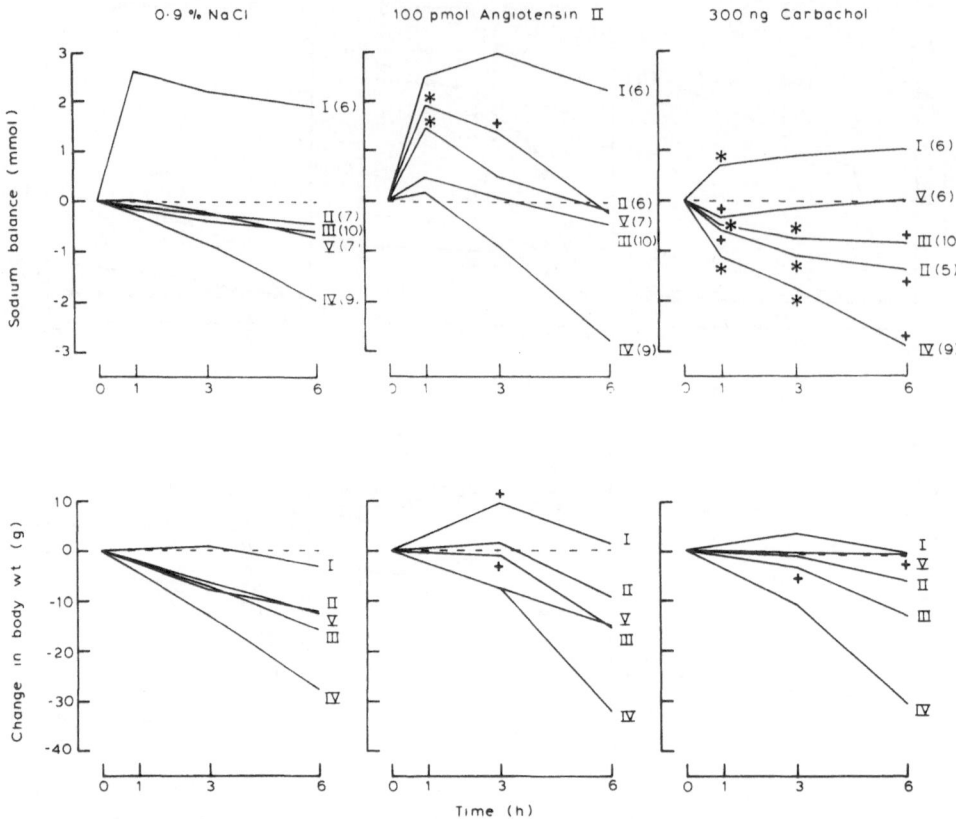

Fig. 2. Mean cumulative Na balance and change in body weight of
the groups of rats whose intakes are given in Fig. 1. For
further explanation, see the legend of Fig. 1.

depressed (group I) intake of NaCl. The initial natriuresis after
Carb was not sustained, whereas Na excretion after Ang II increased
as rats drank NaCl and went into positive Na balance.

From the cumulative intake and excretion patterns of Na for
all five groups, it was apparent that Na intake preceded Na
excretion after injection of Ang II, whereas excretion preceded
whatever small intake of Na there was after Carb. The absent or
depressed intakes of NaCl in all groups after Carb despite the
greater initial natriuresis compared with after Ang II was a
striking feature of this comparison. Carb-induced natriuresis was
also associated with a kaliuresis. In general the patterns of
potassium excretion did not vary very much between groups,
including, interestingly, group V, the DOC-treated group.

Because of different patterns of Na intake and excretion after
Ang II, groups II and V were in significantly greater positive Na
balance at 1 h than the corresponding controls (Fig. 2). Though
the other Ang II-injected groups were generally in positive balance
at 1 and 3 h, they were not in significantly greater positive

422

Na balance than the controls. In contrast, at 1 h all Carb-injected animals were in more negative Na balance than the controls, and groups II, III and IV remained so throughout the experiment. Even though the NaCl-deprived adrenalectomized rats drank some NaCl after Carb and went into positive Na balance, they drank less than the controls and were in significantly less positive Na balance at 1 h. Changes in body weight, measured at 3 and 6 h only, showed the expected changes, but the differences between experimental and control groups were not great.

CONCLUSIONS

In a number of different initial conditions, Ang II-stimulated rats were more likely to go into positive Na balance than controls, whereas Carb-stimulated rats were more likely to go into negative balance. In adrenalectomized, DOC-treated and normal rats, I.C.V Ang II caused increased intake of hypertonic NaCl and water whereas I.C.V. Carb caused increased water intake only. Ang II caused significant increases in Na excretion in normal, 0.9 % NaCl-loaded and DOC-treated rats, but not in adrenalectomized rats in which there were also Ang II-induced intakes of NaCl. On the other hand, Carb caused significant increases in Na excretion at 1 h in all groups despite the absence of a Na appetite. The striking result after Carb was the inhibition of NaCl intake in the NaCl-deprived adrenalectomized rats, though these animals drank water vigorously. Since Carb also caused significant natriuresis, this group of animals ended in more negative Na balance than the controls.

Although the large Ang II-induced NaCl intake in DOC-treated rats is compatible with Na appetite being aroused by synergy between Ang II and mineralocorticoid[11], the smallness of the intake in normal rats, does not support the idea that the synergy occurs with normal mineralocorticoid levels. Only the normal animals showed a significant kaliuresis 1 h after Ang II, suggesting endogenous mineralocorticoid release[12]. Therefore, higher levels of, or longer exposure to, mineralocorticoid is required before a synergistic action with Ang II on Na appetite becomes apparent.

These results suggest that a combination of hypovolaemia, mineralocorticoid secretion and increased endogenous Ang II may interact with exogenously injected Ang II to increase Na appetite. The idea that Ang II is just one of a number of factors that stimulate Na appetite accords well with our view of its role in hypovolaemic thirst, where it is thought to make a contribution but not an essential one, to increased drinking[13]. It seems fitting that a function as important for survival as Na appetite should result from the action of several factors, whose relative contribution depends on the particular circumstances associated with the increased appetite.

REFERENCES

1. D. A. Denton, "The Hunger for Salt. An Anthropological, Physiological and Medical Analysis," Springer Verlag, Berlin, Heidelberg, New York (1982).
2. J. Buggy and A. E. Fisher, Evidence for a dual central role for angiotensin in water and sodium intake, Nature, 250: 733-735 (1974).
3. D. B. Avrith and J. T. Fitzsimons, Increased sodium appetite in the rat induced by intracranial administration of components of the renin-angiotensin system, J. Physiol. 301: 349-364 (1980).
4. R. W. Bryant, A. N. Epstein, J. T. Fitzsimons and S. J. Fluharty, Arousal of a specific and persistent sodium appetite in the rat with continuous intracerebroventricular infusion of angiotensin II, J. Physiol. 301: 365-382 (1980).
5. D. B. Avrith and J. T. Fitzsimons, Renin-induced sodium appetite: effects on sodium balance and mediation by angiotensin in the rat, J. Physiol. 337: 479-496 (1983).
6. W. B. Severs, Anne Daniels-Severs, Joan Summy-Long and G. J. Radio, Effects of centrally administered angiotensin II on salt and water excretion, Pharmacol. 6: 242-252 (1971).
7. S. J. Fluharty and S. Manaker, Sodium appetite elicited by intracerebroventricular infusion of angiotensin II in the rat: I relation to urinary sodium excretion, Behav. Neurosci. 97: 738-745 (1983).
8. J. T. Fitzsimons and Louise M. Fuller, Influence of sodium load on angiotensin-induced sodium appetite in the rat, J. Physiol. (in Press).
9. S. E. A. Perez, C. R. Silva-Netto, W. A. Saad, A. A. Camargo and J. Antunes-Rodriguez, Interaction between cholinergic and osmolar stimulation of the lateral hypothalamic area (LHA) on sodium and potassium excretion, Physiol. & Behav. 32: 191-194 (1984).
10. J. T. Fitzsimons and J. B. Wirth, The renin-angiotensin system and sodium appetite, J. Physiol. 274: 63-80 (1978).
11. S. J. Fluharty and A. N. Epstein, Sodium appetite elicited by intracerebroventricular infusion of angiotensin II in the rat: II. Synergistic interaction with systemic mineralocorticoids, Behav. Neurosci. 97: 746-758 (1983).
12. W. F. Ganong, The brain renin-angiotensin system. Ann. Rev. Physiol. 46: 17-31 (1984).
13. J. T. Fitzsimons, "The Physiology of Thirst and Sodium Appetite," Monographs of the Physiological Society, No. 35. Cambridge University Press, Cambridge (1979).

THE HORMONES OF RENAL SODIUM CONSERVATION ACT SYNERGISTICALLY TO AROUSE A SODIUM APPETITE IN THE RAT

Randall R. Sakai

Department of Biology, University of Pennsylvania

Philadelphia, PA   19104

Central or systemic administration of angiotensin[1,2,4] as well as peripheral administration of either desoxycorticosterone acetate (DOCA) or aldosterone[8,9] will arouse a sodium appetite in the rat but only when large doses of the hormone are used or when negative sodium balance is induced[6,7]. Recently, Epstein[3] has suggested that these same hormones may act in a synergistic manner at physiological concentrations, to produce a sodium appetite and that their concurrent action may underly the sodium appetite which occurs naturally in states of sodium depletion. Fluharty and Epstein[5] have since supported this hypothesis by arousing a rapid and robust sodium appetite upon central administration of angiotensin II (Ang II) to sodium replete rats given subcutaneous injections of DOCA. The doses of both hormones were insufficient to arouse a sodium appetite when given alone, and the appetite occured in animals that were in positive sodium balance. We have now extended this line of investigation to the naturally occuring mineralocorticoid, aldosterone, enquiring whether it is the synergistic partner with angiotensin II in generating the appetite.

METHODS

Subjects

Male Sprague-Dawley rats (280-325 gms at time of surgery) were fitted with intracerebroventricular (ICV) cannula which terminated in the anteroventral portion of the third ventricle. Animals were individually housed in a temperature controlled room with water, 3% NaCl and Purina pellets ad lib, except as described below. Intake of fluid and body weight were recorded daily.

Hormone Treatment

Aldosterone-acetate (Sigma), corticosterone-acetate (Sigma) or the sesame oil vehicle in a volume of 0.2 ml was injected subcutaneously twice daily at approximately 900 and 1700 hours. Thus each rat received 40 μg aldo/day, 500 μg cort/day, or sesame oil.

PROCEDURE

Experiment 1: Elicitation of Sodium Appetite

The experiment was run as a 2X2 design, eight rats per group, with hormone or oil vehicle and Ang II or isosaline vehicle. Each rat received the two daily hormone injections for five days. On the fifth day each rat also received a pulse intracerebroventricular (pICV) injection of Ang II (0.6, 6, 60ng) or the isosaline vehicle. Following pICV injection, the latency to drink water and 3% NaCl was recorded as were water and 3% NaCl intake at 0.5, 3, and 24 hours.

Experiment 2: Sodium Balance

Rats were individually housed in stainless steel metabolism cages with water, 3% NaCl and Purina pellets ad lib. There were 4-5 rats in each group. Rats received hormone treatment as in Experiment 1. All food was removed at 1100 hours on the fifth day of hormone treatment at which time each rat received a pICV injection of 6ng of Ang II. Water and 3% NaCl intake and urine volume were recorded at 0.5, 3, and 24 hours following pICV injection. A portion of the urine was reserved for determination of sodium concentration by flame photometry. In a separate group of rats treated for five days with either aldosterone or sesame oil, trunk blood was collected for plasma sodium and haematocrit measurements of animals prior to administration of Ang II.

Experiment 3: The Effects of pICV Carbachol or SC Corticosterone

As in experiment 1, rats received aldosterone or sesame oil treatment for five days. On day five all rats received a pICV injection of the dipsogen, carbachol (Sigma, 40ng), an acetylcholine mimetic. The animals' water and 3% NaCl intake were recorded thereafter. Another group of rats received either two daily injections of corticosterone (250 μg/injection), the glucocorticoid of the rat, or the sesame oil vehicle for five days. On day five all rats received a pICV injection of 6ng Ang II. Latencies and fluid intakes were recorded as in experiment I. There were 4-5 animals in each group with no animals receiving more than one treatment.

STATISTICAL ANALYSIS

Appropriate ANOVA tests were performed on all data. Statistically significant effects were further evaluated by post hoc comparisons.

RESULTS

Experiment 1: Elicitation of Sodium Appetite

Animals treated with both hormones drank more 3% NaCl in the 30 minutes after a single pICV injection of 0.6, 6 or 60ng of Ang II ($p < .05$) when compared to those that received only Ang II (fig. 1.). The latency to drink saline was shorter (4 mins vs 15 mins) and the amount consumed in the first 30 minutes post pICV Ang II increased with increasing doses of Ang II.   The rats in both groups responded

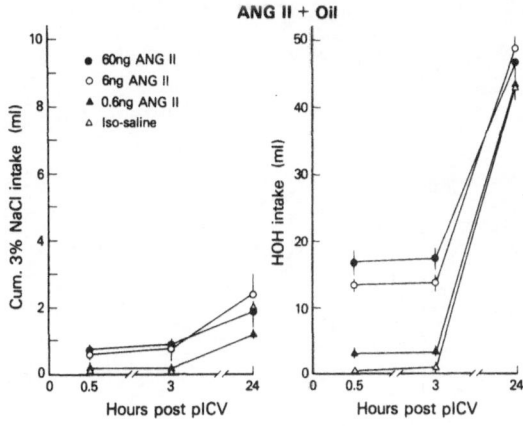

Figure 1A. Dose response analysis of cummulative 3% NaCl (on the left) and water (right) intakes (mean $\pm$ SE) of rats following pICV injection of Ang II alone.

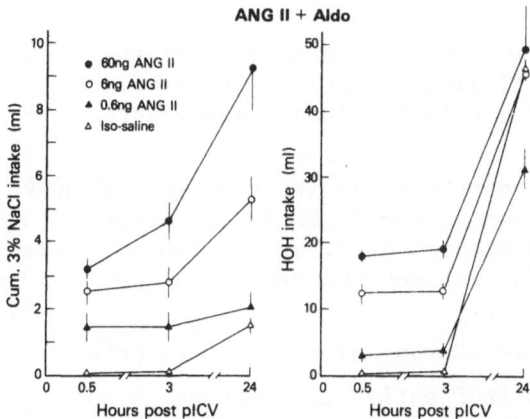

Figure 1B.   Dose response analysis of cummulative 3% NaCl (on the left) and water (right) intakes (mean $\pm$ SE) of rats following pICV injection of Ang II in rats also treated with aldosterone (40 µg/day for five days).

immediately by drinking water. There was no difference between groups in latency to drink or volume of water consumed at any time during the test. Angiotensin alone was ineffective in arousing a sodium appetite during the test period at all doses. During the five days of aldosterone treatment prior to the pICV Ang II there was no increase of daily 3% NaCl intake.

Experiment 2:   Sodium Balance

The arousal of the sodium appetite in rats treated with both
hormones occured while they were in normal sodium balance (fig. 2).
After five days of aldosterone treatment plasma sodium concentration
(150.5 ± 1.5 mEq/l) and haematocrit (47.7 ± 0.8) were not different
from control (oil treated) rats 153.1 ± 0.7 mEq/l and 46.2 ± 0.5)
prior to pICV Ang II. Moreover, at the end of 30 minutes post pICV
Ang II, aldosterone treated rats were in positive sodium balance.

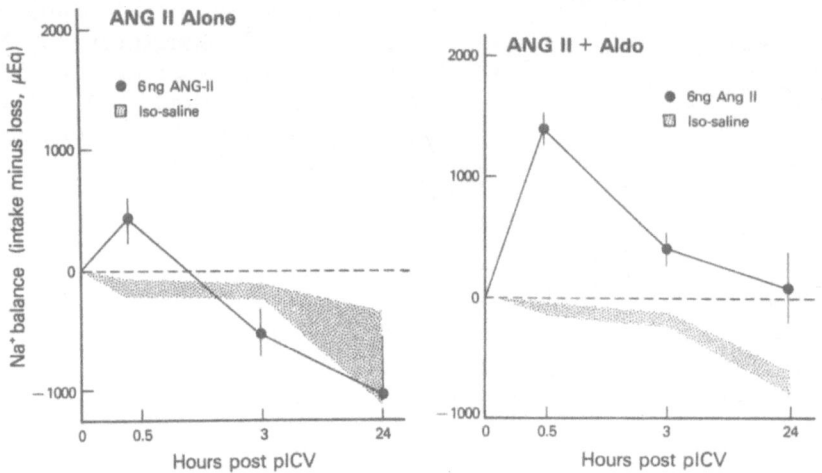

Figure 2. Cummulative mean sodium balance following
pICV injection of Ang II alone (left) or in animals
treated with aldosterone (right). Shaded area represents control range.

The combined treatment kept these rats in positive sodium balance
largely as the result of their own sodium ingestion. Except for a
small amount of sodium ingested within the first 30 minutes, animals
receiving Ang II alone had sodium intakes and excretions similar to
control 24 hour food deprived rats.

Experiment 3:   The Effects of pICV Carbachol or SC Corticosterone

The dipsogen carbachol was ineffective in arousing 3% NaCl drinking when given as a pICV injection to rats pretreated with aldosterone, but the animals did drink a volume of water in 30 minutes (9.6
± 3.4 ml) that was comparable to that drunk by animals receiving
6ng of Ang II (12.2 ± 1.4 ml) in experiment 1. Animals receiving
corticosterone and Ang II did not drink more 3% NaCl than animals
receiving Ang II alone after 30 minutes post pICV Ang II (1.2 ± 0.5ml
vs. 1.1 ± 0.5 ml). Both groups of animals drank comparable amounts
of water 30 minutes after the Ang II injection (9.3 ± 1.0 ml vs.
11 ± 1.7ml).

428

DISCUSSION

Present experiments show that the natural mineralocorticoid, aldosterone is a synergist with angiotensin in generating a sodium appetite. The dose of aldosterone chosen was that reported by Fregly and Waters[7] to be optimum for suppressing the sodium appetite of the adrenalectomized rat. In experiment 1, aldosterone was administered daily for five days with no increase in spontaneous 24 hour 3% NaCl intake or change in plasma sodium concentration. When these rats received a pulse injection of angiotensin, they rapidly ingested both water and 3% NaCl. Aldosterone was an effective synergist in arousing a sodium appetite with angiotensin at doses as low as 0.6ng of angiotensin. With increasing doses of angiotensin, the animals 3% NaCl consumption increased in a dose dependent fashion. In addition, the combined treatment with exogenous angiotensin and aldosterone elicited an increase in 3% NaCl consumption with no change in the dipsogenic action of the pICV angiotensin.

Experiment 2 demonstrated that the arousal of the sodium appetite in the animals treated with both hormones occured in the absence of negative sodium balance. In fact, the animals that were given both hormones were in positive sodium balance throughout the entire 24 hour period of observation, due mainly to their own ingestion of saline.

In experiment 3, the specificity of the hormones used to arouse a sodium appetite was demonstrated. Neither treatment with corticosterone and pICV Ang II nor treatment with aldosterone and pICV carbachol evoked a sodium appetite in the rat.

The work described here demonstrates that angiotensin and aldosterone are effective synergists in the arousal of sodium ingestion and suggests that the hormones of renal sodium conservation are also the hormones of sodium appetite.

ACKNOWLEDGEMENTS

The author would like to thank Alan N. Epstein for the support and encouragement, Barbara A. Stamoutsos for technical assistance and Keely I. Byford for help in preparation of this manuscript.

REFERENCES

1. D.B. Avrith & J.T. Fitzsimons, Increased sodium appetite in the rat induced by intracranial administration of components of the renin-angiotensin system. J. of Physiol. 301: 349-364 (1980).

2. R.W. Bryant, A.N. Epstein, J.T. Fitzsimons & S.J. Fluharty, Arousal of a specific and persistent sodium appetite in the rat with continuous intracerebroventricular infusion of angiotensin II. J. of Physiol. 301: 365-382 (1980).

3. A.N. Epstein, Mineralocorticoid and cerebral angiotensin may act together to produce a sodium appetite. Peptides, 103: 60-65 (1982).

4. A.R.L. Findaly & A.N. Epstein, Increased sodium intake is somehow induced in rats by intravenous angiotensin II. Hormones & Behav. 14: 86-92.

5. S.J. Fluharty & A.N. Epstein, Sodium appetite elicited by intracerebroventricular infusion of angiotensin II in the rat: II. Synergistic interaction with systemic mineralocorticoids. Behav. Neurosci. 97: 746-758 (1983).

6. S.J. Fluharty & S. Manaker, Sodium appetite elicited by intracerebroventricular infusion of angiotensin II in the rat: I. Relationship to urinary sodium excretion. Behav. Neurosci. 97: 738-745.

7. M.J. Fregly & I.W. Waters, Effect of mineralocorticoids on spontaneous sodium chloride appetite of adrenalectomized rats. Physiol. & Behav. 1: 65-74 (1966).

8. K.K. Rice & C.P. Richter, Increased sodium chloride and water intake of normal rats treated with desoxycorticosterone acetate. Endocrinol. 33: 106-115 (1943).

9. G. Wolf & P.J. Handal, Aldosterone-induced sodium appetite: Dose response and specificity. Endocrinol. 78: 1120-1124 (1966).

Footnote: This research was supported by grant NS 03469 (NIH) to Alan N. Epstein.

SODIUM APPETITE DURING CAPTOPRIL BLOCKADE OF ENDOGENOUS

ANGIOTENSIN FORMATION

Karen Moe[1] and Alan N. Epstein

Department of Biology
and Institute of Neurological Sciences
University of Pennsylvania, Philadelphia, PA 19104

Activation of the renin-angiotensin-aldosterone system during sodium deficiency is directly or indirectly responsible for many compensatory physiological responses which act to conserve both sodium and water[1]. Moreover, there is a large and increasing body of evidence for a role of angiotensin (ANG) in salt appetite, the behavioral partner to these conservation mechanisms. For example, it is well-established that administration of ANG can arouse a salt appetite in sodium-replete rats[2,3]. However, in natural situations, ANG probably acts in conjunction with aldosterone. Epstein and colleagues have recently shown that when the two hormones are administered together at very low doses they act synergistically to produce the appetite[4]. Thus, concurrent elevation of the two hormones, as in sodium deficiency, may be an important mechanism by which the appetite is aroused.

If ANG does play an important role in the appetite, then blockade of its biosynthesis should reduce the appetite. Use of captopril (a competitive inhibitor of angiotensin-converting enzyme) for this purpose has produced conflicting results[2,5,6]. In the following experiments captopril (CAP) was administered to rats by several routes and over a wide range of doses to resolve these apparent contradictions.

METHODS

Male Sprague-Dawley rats were housed with continuous access to Purina rat chow pellets, distilled water and a 3% NaCl solution. Some of the rats were implanted with intravenous (IV) catheters

[1] Present address: Department of Psychology, Washington State University, Pullman, WA 99164-4830.

and after recovery from surgery were adapted to chronic IV infusion of isotonic saline (0.4 ml/hr). Another group of rats were stereotaxically implanted with a cannula guide shaft whose tip opened into the anterior ventricles. After at least 3 days of recovery, their dipsogenic response to a pulse intracerebroventricular (ICV) injection of ANG II (6 ng/μl/rat) was assessed. Only rats that drank at least 4 ml water within 15 min in response to this injection were used for subsequent tests. Before those tests, they were first adapted to continuous ICV infusion (5 μl/hr).

For most rats, salt appetite was elicited by sodium depletion. This was produced by subcutaneous (SC) injection of 10 mg furosemide (2 injections of 5 mg/0.5 ml, separated by 2 hr), followed by immediate removal of saline from the cages and substitution of a powdered sodium-deficient food for the chow pellets. The following morning, after an 18-hr depletion period, saline was returned to the rats and their consumption of it and of water was measured for 2 hr. Overnight water, and in some cases food, intake were also recorded.

The salt appetite of each rat was assessed twice in this way: once while treated with CAP (D-3-mercapto-2-methylpropanoyl-L-proline, Squibb) and once while treated with its vehicle solution, in random order and with 3-6 days between the 2 tests. For rats with an IV catheter, CAP was infused intravenously over a wide range of doses (0.00032 - 25 mg/hr). The infusion began 30 min before the first furosemide injection and continued throughout the entire 18-hr sodium depletion period and subsequent 2-hr appetite test. Other rats received CAP orally (PO); 2 hr before the furosemide injection, their drinking water was replaced with captopril-adulterated water (0.1 or 1 mg/ml) which was the only source of fluid during the overnight depletion. A third group of rats was treated SC as well as orally with CAP. In addition to adulterated drinking water, they received 50 mg/kg CAP SC 15 min before each furosemide injection, and then again 2 hr before the salt appetite test and 30 min before the test.

Finally, another group of rats received CAP both orally and in the brain. First, the disruption and healing of the blood-brain barrier that follows cannula implantation[7] was assessed by examining the dipsogenic response to a pulse ICV injection of ANG 1 (6 ng/μl/ rat) after overnight treatment with oral CAP (0.1 mg/ml drinking water). Intracranial ANG I is dipsogenic only after its conversion to ANG II[8], and that conversion is prevented by intracranial administration of CAP[9]. CAP given at this low dose (0.1 mg/ml) in the drinking water does not cross the blood-brain barrier of neurologically intact rats[10] and therefore should not interfere with the dipsogenic response to ANG I[9]. But the stab wound produced by cannula implantation temporarily compromises the integrity of the blood-brain barrier[7], and, until the barrier is healed, the orally administered CAP should have access to the brain and will therefore reduce the dipsogenic response to intracranial ANG I.

432

While the blood-brain barrier was still open to CAP (as determined above), depletion-induced salt appetite was measured both with and without the oral CAP treatment described above. These two tests were repeated later after the orally ingested CAP no longer had access to the brain. Finally, the depletion-induced appetite following oral CAP treatment combined with infusion of CAP directly into the brain (1.2 µg/hr) was compared to the appetite following oral CAP and ICV infusion of vehicle. The effectiveness of the central CAP at blocking conversion centrally was tested by assessment of the rats' dipsogenic response to intracranial injection of ANG I and ANG II.

In another experiment, salt appetite was aroused by daily SC injection of deoxycorticosterone acetate (DOCA; 2 mg/day in 1-2 ml sesame oil). Rats in this experiment had continuous access to Purina pellets and water. In addition, 3% NaCl was offered daily for the hour immediately preceding the DOCA injection, and ingestion of both saline and water was recorded. Once all rats were demonstrating a reliable salt appetite, they received the SC plus PO CAP treatment described above. That is, they received two SC injections of CAP just before a daily DOCA injection plus an injection 2 hours before and 30 minutes before the next day's salt access, and CAP in their drinking water overnight.

Finally, sodium excretion was measured in one group of rats: once during sodium depletion plus IV infusion of 5 mg/hr CAP and once during depletion plus IV vehicle infusion. Rats had no access to food during urine collection, to prevent contamination of samples. Sodium concentration of samples was determined by flame photometry.

All data are presented as means ± standard error. Paired t tests were used to evaluated the data. Statistical significance was set at $p < .05$, except for the intravenous CAP experiment. For that experiment, multiple Bonferroni t tests were performed, one for each of the eight pairwise comparisons. Statistical significance for each of these pairwise comparisons was set at $p < .01$, making the overall significance level for the entire experiment $p < .08$.

RESULTS

Low doses of IV CAP enhanced and high doses suppressed depletion-induced salt appetite (Fig. 1; Table 1). An intermediate dose had no effect.

Qualitatively similar effects were produced by PO CAP at low doses or SC CAP at high doses. Orally administered CAP enhanced salt appetite at both the 1 mg/ml dose (5.8 ±1.5 ml saline consumed following depletion when only water was available overnight compared to 10.0±1.8 ml saline with CAP overnight) and the 0.1 mg/ml dose (5.30 ±1.1 ml with water during depletion compared to 12.8±1.4ml with CAP). Subcutaneous administration of high doses of CAP suppressed the depletion-induced appetite, from 10.4±1.0 ml with vehicle injection to 4.1 ± 0.8 ml with CAP.

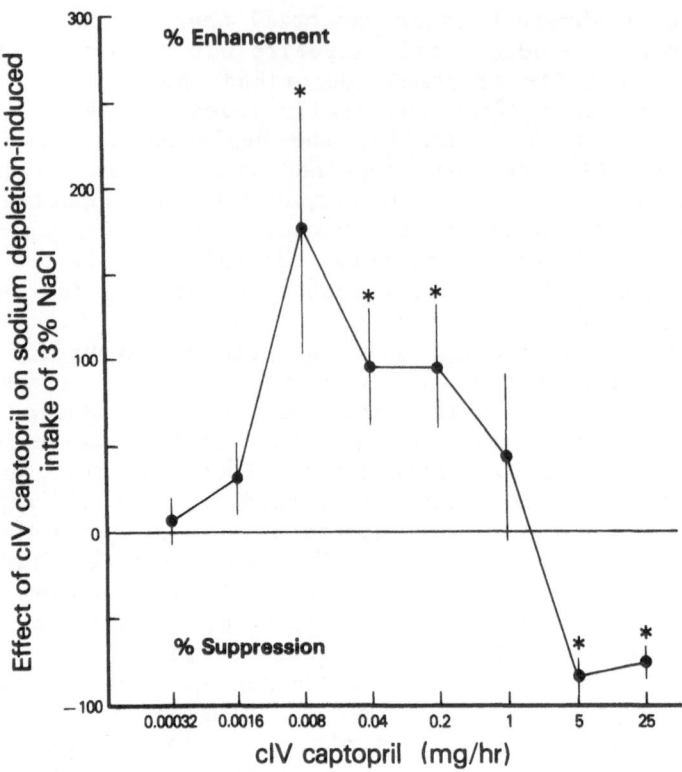

Fig. 1.   Effect of overnight IV infusion of captopril on sodium depletion-induced intake of 3% NaCl during subsequent 2-hr appetite test.   Intake after captopril infusion is reported as average % change (± SEM) from intake produced by depletion alone.

Table 1.   Mean consumption (±SE) of 3% NaCl in 2-hr test following overnight sodium depletion and concurrent infusion of captopril or its vehicle.

IV captopril dose/hr	2-hr intake of 3% NaCl (ml)	
	Depletion + VEH	Depletion + CAP
0.32 μg	7.6±0.6	8.2±1.3
1.6 μg	7.6±0.6	9.4±0.8
8 μg	4.9±1.9	9.9±2.2
40 μg	6.3±0.8	11.8±1.5
200 μg	6.4±0.9	12.2±0.2
1 mg	7.1±1.2	8.6±1.7
5 mg	5.4±1.0	1.1±0.7
25 mg	5.2±0.8	1.2±0.5

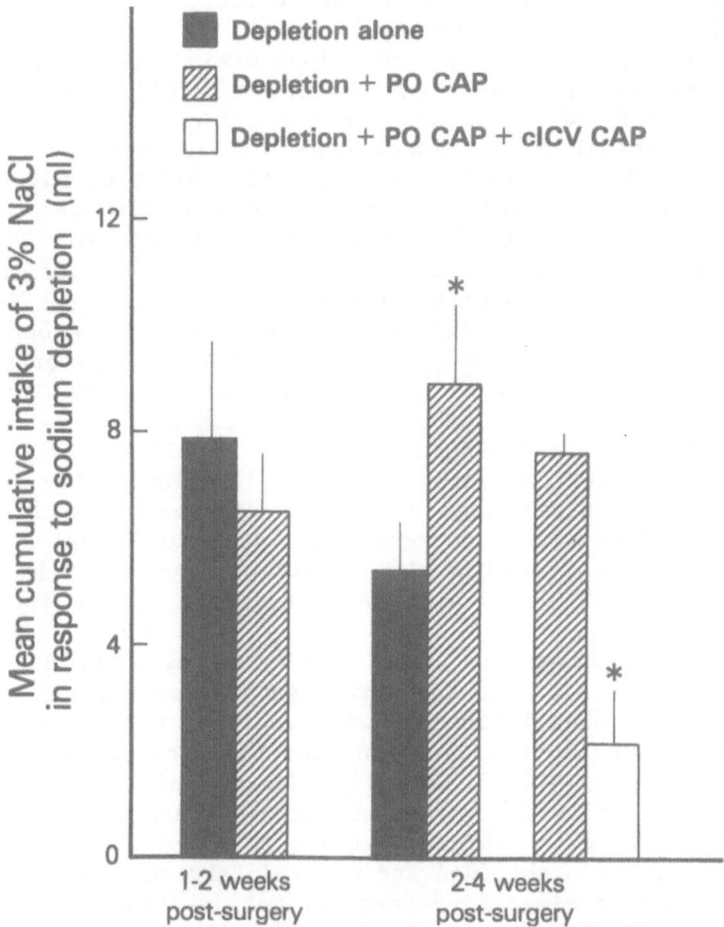

Fig. 2. Two-hour intake of 3% NaCl by sodium-depleted rats following overnight treatment with (1) no captopril, (2) CAP-
adulterated water (0.1 mg/ml), or (3) CAP-adulterated water
and ICV infusion of CAP (1.2 μg/hr). Rats were tested 1-2
and 2-4 weeks after surgical implantation of intracranial
cannula.

The enhancement of salt appetite by low doses of CAP was
abolished by the addition of CAP to the brain (Figure 2).
The first demonstration of this occurred one to two weeks after
surgery. At this time, the blood-brain barrier was still damaged
from cannula implantation and consequently CAP had access to the
brain (as indicated by a reduced dipsogenic response to ICV pulse
injection of ANG 1: 5.4±1.9 ml water consumed following overnight
oral CAP versus 9.4±2.3 ml following control treatment). In
response to overnight sodium depletion, the rats drank 7.9±1.8 ml
saline. When PO CAP was combined with the depletion (a procedure

435

that produced enhanced intake in intact rats; see above), they drank only 6.5±1.1 ml saline.

Two to four weeks later, the blood-brain barrier was restored, as rats treated with oral CAP now drank 12.6±2.2 ml water in response to ICV ANG 1 compared to 9.9±2.3 ml with no CAP. In addition, oral CAP now enhanced depletion-induced salt appetite (8.9±1.5 ml saline with CAP compared with 5.4±0.9 ml with vehicle). However, this enhancement was abolished when CAP was infused directly into the brain, in combination with the oral CAP. Only 2.2±1.0 ml saline was consumed following sodium depletion + oral CAP + ICV CAP, compared with 7.6±0.4 ml following depletion + oral CAP + ICV vehicle. The infusion was effective at blocking the central conversion of ANG I to ANG II, as the rats drank only 2.1±0.7 ml water to ICV ANG I when treated with ICV CAP. When infused with ICV vehicle, they drank 8.4±2.6 ml. However, ICV CAP had no effect on the dipsogenic response to ICV ANG II (7.0±2.2 ml water following ICV CAP versus 6.4±0.6 ml following ICV vehicle).

DOCA-elicited salt appetite was unaffected by the high dose SC CAP injections which suppressed the depletion-induced appetite. For the two days preceding CAP injection, average 1-hr saline intake was 11.4±0.9 ml. With CAP treatment, saline intake was 11.3±1.6 ml.

Sodium excretion was also unaffected by high doses of peripherally administered CAP which suppressed depletion-induced salt appetite. When infused IV with vehicle, rats lost a total of 3.44±0.08 meq during the depletion period. Following IV CAP at 5 mg/hr, they lost 3.13±0.27 meq. Though not statistically significant, total urine volume was somewhat decreased by CAP, from 49.8±3.5 ml with vehicle infusion to 40.7±2.4 ml with CAP.

Finally, overnight food and water intake remained normal, even when rats received high doses of CAP in their drinking water. Rats whose salt appetite was suppressed by SC CAP drank 41.2±2.3 ml water overnight compared with 37.5±3.5 ml when they were not given CAP. When sodium-depleted rats were given 0.1 mg/ml CAP for their overnight drinking fluid, they consumed 50.3±5.4 ml during the night compared to 34.2±3.8 when they had unadulterated water.

DISCUSSION

Peripheral administration of high doses of captopril suppresses depletion-induced salt appetite, whereas lower doses can enhance it. These results support the hypothesis that ANG II is important for the expression of depletion-induced salt appetite in the rat.

However, at first glance the enhancement of the appetite by low doses of peripheral CAP (Figure 1) is paradoxical. If ANG II is important for the expression of salt appetite, then why does blockade of its synthesis under these circumstances enhance rather than suppress salt intake? Lehr, Goldman and Casner[11] found a similar enhancement of angiotensin-induced water intake with peripheral administration of a related converting enzyme inhibitor (teprotide). To explain their results, they proposed a "spill-over" hypothesis, which relies on two findings. First, low doses of peripherally administered CAP block converting enzyme activity and subsequent ANG II synthesis only in the periphery; i.e., CAP is excluded from the brain at these doses[9,10]. Second, the resulting decrease of peripheral ANG II levels removes negative feedback inhibition of ANG II on renin release by the kidney and consequently leads to increased plasma levels of renin and ANG I[12]. This is added to the already high levels in the circulation that are produced by a thirst challenge or by sodium depletion. Lehr et al hypothesized that at such high concentrations, peripherally generated ANG I gains access or spills over into some compartment of the brain which has not been penetrated by effective amounts of CAP. Brain converting enzyme then converts ANG I to ANG II locally, producing even higher levels of ANG II and therefore an exaggerated appetite. Thus, peripheral administration of low doses of CAP may indirectly increase the formation of ANG II in the brain.

This hypothesis predicts that addition of CAP to the brain under these circumstances would eliminate the enhanced ingestion of water or salt. Lehr et al showed that intracranial injection of teprotide did reduce the enhanced drinking produced by peripheral administration of the drug. Similar findings have been reported by other investigators[9]. In the experiments reported here, the enhanced salt appetite that accompanies peripheral administration of low doses of CAP was no longer demonstrated when the CAP was present in the brain. This was shown in two ways (Figure 2). First, rats with a damaged blood-brain barrier which allowed leakage of CAP into the brain did not exhibit the enhancement. Second, direct application of CAP to the brain in combination with the low, enhancing dose of CAP also prevented the enhancement, and in fact suppressed salt intake.

Of the doses of peripheral CAP used here, those which were high enough to block conversion in the brain[9,10] as well as the periphery suppressed salt appetite. Lower doses which enhanced the appetite are reported to block conversion only in the periphery[9,10]. The two lowest doses of IV CAP which have no effect on salt intake appear to be ineffective at blocking conversion anywhere[13]. (See ref. 14 for a more complete discussion of this issue.)

Captopril has been reported to have many actions in addition to its effect on angiotensin biosynthesis. However, it is unlikely that these other actions explain the results reported here. For example, CAP-induced natriuresis appears only with very high doses or with chronic administration of low doses[15]. The low, enhancing doses used here apparently have no effect on natriuresis over the time course when they were used. In addition, as reported above, a high dose of CAP which greatly suppresses salt appetite has no effect on natriuresis.

Other reported actions of CAP also do not offer satisfactory explanation of our results (see ref. 14). Some of these actions simply do not occur under the conditions of these experiments; many of them occur only at the very highest doses used here.

Further evidence for the specificity of CAP in these experiments is its lack of effect on DOCA-elicited salt appetite. A high dose which suppresses depletion-induced salt intake had no effect on the intake elicited by daily DOCA treatment. Moreover, there was no sign that CAP-treated animals were behaviorally incompetent or suffering from malaise. Even at the highest CAP doses used, overnight food and water intake were normal, as was the dipsogenic response to intracranial injection of ANG II.

A rise in ANG II levels is one of the consequences of sodium deficiency and it is an important mediator of the physiological responses that lead to the restoration of sodium balance. It may also be important for the expression of the behavioral partner to these responses, i.e., sodium appetite. These results offer strong support for this hypothesis. They also explain why the use of captopril has produced contradictory reports about the role of ANG in salt appetite. When increased levels of ANG were prevented in the brain of sodium-depleted rats by central administration of CAP or peripheral administration of high doses, salt appetite was suppressed. Low peripheral doses of CAP which may actually increase ANG II production in the brain enhanced the appetite.

ACKNOWLEDGMENTS

The excellent technical assistance of J. Butler, C. Jameson, B. Stamoutsos, M.A. Stevenson and M.L. Weiss are gratefully acknowledged. This research was supported by NIH grant NS-03469 to A.N.E and NIH grant NS-07304 to K.E.M. We thank the Squibb Institute for Medical Research for the generous donation of captopril.

REFERENCES

1. W.S. Spielman and J.O. Davis, The renin-angiotensin system and aldosterone secretion during sodium depletion in the rat. Circ. Res. 35:615 (1974).
2. D.B. Avrith and J.T. Fitzsimons, Increased sodium appetite in the rat induced by intracranial administration of components of the renin-angiotensin system. J. Physiol. 301:349 (1980).
3. R.W. Bryant, A.N. Epstein, J.T. Fitzsimons and S.J. Fluharty, Arousal of a specific and persistent sodium appetite in the rat with continuous intracerebroventricular infusion of angiotensin II. J. Physiol. 301:365 (1980).
4. S.J. Fluharty and A.N. Epstein, Sodium appetite elicited by intracerebroventricular infusion of angiotensin II in the rat. II. Synergistic interaction with systemic mineralocor-

ticoids. Behav. Neurosci. 97:746 (1983).

5. R.M. Elfont and J.T. Fitzsimons, The role of angiotensin in increased sodium appetite after adrenalectomy. J. Physiol. 320:70P (1981).

6. M.J. Fregly, Effects of the angiotensin converting enzyme inhibitor, captopril, on NaCl appetite of rats. J. Pharmacol. Exp. Ther. 215:407 (1980).

7. L. Persson and H.A. Hansson, Reversible blood-brain barrier dysfunction to peroxidase after a small stab wound in the rat cerebral cortex. Acta Neuropathol. 35:333 (1976).

8. W.B. Severs, J. Summy-Long and A.E. Daniels-Severs, Effects of a converting enzyme inhibitor (SQ 20,881) on angiotensin-induced drinking. Proc. Soc. Exp. Biol. Med. 142:203 (1973).

9. M.D. Evered, M.M. Robinson and M.A. Richardson, Captopril given intracerebroventricularly, subcutaneously or by gavage inhibits angiotensin-converting enzyme activity in the rat brain. Eur. J. Pharmacol. 68:443 (1980).

10. M.L. Cohen and K.D. Kurz, Angiotensin converting enzyme inhibition in tissues from spontaneously hypertensive rats after treatment with captopril or MK-421. J. Pharmacol. Exp. Ther. 220:63 (1982).

11. D. Lehr, H.W. Goldman and P. Casner, Renin-angiotensin role in thirst: paradoxical enhancement of drinking by angiotensin converting enzyme inhibition. Science 182: 1031 (1973).

12. E.L. Schiffrin, J. Gutkowska and J. Genest, Mechanisms of captopril-induced renin release in conscious rats. Proc. Soc. Exp. Biol. Med. 167:327 (1981).

13. B. Rubin, R.J. Laffan, D.G. Kotler, E.H. O'Keefe, D.A. Demaio and M.E. Goldberg, SQ 14,225 (D-3-mercapto-2-methyl propanoyl-L-proline), a novel orally active inhibitor of angiotensin I-converting enzyme. J. Pharmacol. Exp. Ther. 204:271 (1978).

14. K.E. Moe, M.L. Weiss and A.N. Epstein, Sodium appetite during captopril blockade of endogenous angiotensin II formation. Am. J. Physiol. 247:R356 (1984).

15. M.D. Evered and M.M. Robinson, The effects of captopril on salt appetite in sodium-replete rats and rats treated with desoxycorticosterone acetate (DOCA). J. Pharmacol. Exp. Ther. 225:416 (1983).

THE RENIN-ANGIOTENSIN-ALDOSTERONE SYSTEM AND SODIUM APPETITE IN RATS[1]

Melvin J. Fregly and Neil E. Rowland

Department of Physiology and Psychology
University of Florida, Colleges of Medicine and Liberal
Arts and Sciences, Gainesville, FL 32610

Richter (1) was the first to show that bilateral adrenalectomy induced a specific appetite for sodium chloride (NaCl) solutions in rats. This appetite was normalized by adrenal cortical transplants, implicating an hormonal factor which we now recognize as aldosterone. Accordingly, administration of graded doses of aldosterone or of desoxycorticosterone acetate (DOCA) to adrenalectomized rats produced a graded reduction in salt appetite (Figure 1, top panel). However, doses of DOCA in excess of about 1 mg/kg/day increased NaCl intake, resulting in a U-shaped dose-response relationship. In these experiments 0.15 M NaCl solution was given in simultaneous choice with distilled water, the intake of which was the mirror image of the NaCl curve (Figure 1, lower panel). High doses of aldosterone or DOCA also induce sodium appetite in intact rats (2,3).

Increases in production of aldosterone are known to inhibit renin secretion from the kidneys, and hence decrease circulating levels of angiotensin II (AII) (Figure 2). The possible ways in which these peripheral changes may influence brain levels of AII (but not necessarily endogenous brain AII) are also indicated in Figure 2. Studies from the laboratories of Fitzsimons (4) and Epstein (5) have found that chronic intracerebroventricular (IVT) infusions of AII induce sodium appetite in rats. The inference that brain AII levels may modulate salt appetite is also indicated in Figure 2. The ways in which aldosterone may interact with brain AII to influence sodium appetite are currently speculative, but several areas of the brain contain high affinity mineralocorticoid receptors (6) suggesting the possibility of such interactions.

Table 1 is a summary of selected experimental procedures which have been shown in this and other laboratories to induce NaCl appetite in rats. Also shown are the corresponding changes in circulating

441

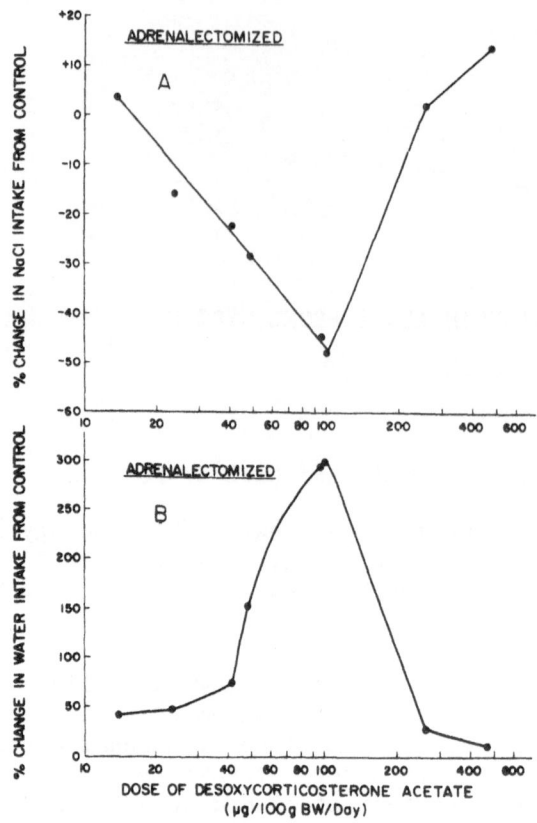

Figure 1 The effect of graded doses of desoxycorticosterone acetate on the percentage change in intakes of 0.15 M NaCl solution (A) and water (B) from adrenalectomized control rats is shown. Data from (2).

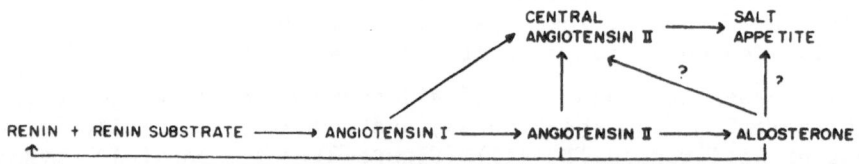

Figure 2 This schema illustrates potential interactions between the central and peripheral renin-angiotensin systems. Peripherally formed AI may enter the central nervous system (CNS) where it is converted to AII. Peripherally formed AII may also enter the CNS. It is unknown, however, whether central levels of AII reflect peripheral levels. Aldosterone may induce a salt appetite by affecting the regulation of central AII receptors. It may also interact at central neural levels mediating the salt appetite. The question marks indicate that the relationship is uncertain.

components of the renin-angiotensin-aldosterone system. While aldosterone is certainly not necessary for induction of salt appetite (as in adrenalectomy), it has been suggested that it may normally have a

442

Table 1. The renin-aldosterone system and induction of salt appetite in rats

Experimental Paradigm	Plasma Concentrations of:				Brain Levels of:		Salt Appetite
	Renin*	AI	AII	Aldo.	Aldo.[+]	AII[‡]	
Low Na+ Diet	↑	↑	↑	↑	↑?	↑?	↑
Adrenalectomy	↑	↑	↑	↓	↓?	↑?	↑
Adrenalectomy + DOCA	↓	↓	↓	↓	↑DOCA?	↓?	↓Low ↑High
Intact + DOCA	↓	↓	↓	↓	↑DOCA?	↓?	↑
Captopril	NC	↑	↓	↓	↓?	↑?	↑
Captopril + DOCA	↓	↓	↓	↓	↑DOCA?	↓?	↓Low ↑High
AII IVT	?	?	?	↓	↓?	↑	↑
Thyroidectomy	↓	↓	↓	↓	↓?	↓?	↑

Abbreviations: Aldo. = aldosterone; AI, AII = angiotensin I and II; IVT = intracerebroventricular.
* Renin substrate levels are assumed not to contribute; plasma renin concentration is assumed to be rate limiting in the formation of AII.
+ It is assumed that brain levels of mineralocorticoid hormone will reflect plasma levels of Aldo. or exogenous DOCA.
‡ Indicates the AII reaching brain from periphery; no data are discussed relevant to the brain renin-angiotensin system. ↑ = increase; ↓ = decrease; NC = no change from normal levels; ? = unknown or speculation; Low = low dose range; high = high dose range.

synergistic action with brain AII (7). With a low sodium diet, both AII and aldosterone are increased, and a brisk salt appetite is observed, consistent with the synergy hypothesis. However, when high doses of DOCA are administered to adrenalectomized rats, renin secretion is decreased and a salt appetite occurs which may be driven by mineralocorticoid alone. It is of note that the daily dose of aldosterone which produces a maximal reduction in salt intake of adrenalectomized rats (c.f. the apex of the U-shaped curve in Figure 1) is 10 µg/100 g/day, which is very close to the observed 24 hr production of aldosterone as measured in adrenal vein effluent (8). Thus, under normal maintenance conditions rats may be at the apex of the U-shaped curve and regulate their sodium intake at a minimum. The administration of DOCA to intact rats will thus move the rat from this minimum and induce a sodium appetite despite decreased AII levels.

Captopril blocks the conversion of AI to AII and thus decreases aldosterone production. However, the increased blood levels of AI may be sufficient to cross the blood-brain barrier in substantial amounts and be cleaved to AII in the CNS (low doses of captopril apparently do not produce a complete blockade of central converting enzyme activity (9). Thus, from the hormonal profile, captopril-induced salt appetite resembles that induced by adrenalectomy. Accordingly, one might anticipate that graded doses of DOCA might produce graded reductions in captopril-induced NaCl intake, and this is precisely the result we have observed (10). Finally, thyroidectomy, induced surgically or by antithyroid drugs, is accompanied by a salt appetite (13). Aldosterone secretion is also significantly reduced (8).

Preference Thresholds For NaCl Following Induction Of Sodium Appetite

Table 2. Mean bottle sodium chloride solution preference thresholds (mEq/l) in rats following diverse treatments that induce a sodium appetite

| | Groups: | | |
Treatments	Control	Experimental	Reference
Adrenalectomy	9.6	0.5	11
Hypothyroid[a]	23	4	13
Salivariectomy	50	7.5	14
Oral Contraceptive[b]	30	15	15
Diuretic[c]	29	10	16
Captopril (0.7 g/kg food)	30	0.6	17

[a] Propylthiouracil (1 g/kg food).
[b] Enovid (7.5 mg/kg food).
[c] Hydrochlorothiazide (0.6 g/kg food).

Richter (11) noted not only that adrenalectomized rats drank more NaCl than controls, but also that they exhibited a much lower preference threshold for salt over water. Subsequent studies, notably those of Nachman (12) established that this occurs too rapidly after adrenalectomy to be accounted for by postingestive events, suggesting that taste is a major factor in the salt appetite of these animals. In Table 2 we summarize our own data and those of others regarding the effect of various experimental procedures on the preference threshold concentration relative to water; that is, the concentration at and above which more fluid is ingested from the salt bottle than from the water bottle offered simultaneously. In addition to the effects of adrenalectomy on preference threshold, similar downward shifts in preference threshold and increased sodium appetite are seen with hypothyroidism (13), salivariectomy (14), and treatment with oral contraceptives (15), diuretics (16), and captopril (17). Insofar as we are aware, there has been no experimental dissociation between the preference threshold shifts and the induction of salt appetite.

Electrophysiological studies have found no change in the whole chorda tympani response to salt stimuli in adrenalectomized rats (18), and more recent studies have suggested a decreased responsiveness of single "specialist" sodium fibers (19). Thus, no immediate functional explanation for the behavioral preference changes has been forthcoming. However, it is possible that gustatory sensory input interacts with angiotensin and/or mineralocorticoid elements in the brain to induce the behavioral changes. Gustatory afferent input to the nucleus tractus solitarius is modulated by lateral hypothalamic activity (20), and this suggests that our speculation may have a physiological basis.

REFERENCES

1. Richter, C.P., Increased salt appetite in adrenalectomized rats, Am. J. Physiol. 115:155 (1936).
2. Fregly, M.J. and I.W. Waters, Effect of mineralocorticoids on

spontaneous sodium chloride appetite of adrenalectomized rats, Physiol. Behav. 1:65 (1966).

3. Wolf, G., Effect of desoxycorticosterone on sodium appetite of intact and adrenalectomized rats, Am. J. Physiol. 208:1281 (1965).

4. Avrith, D.B. and J.T. Fitzsimons, Increased sodium appetite in the rat induced by intracranial administration of components of the renin-angiotensin system, J. Physiol. (London) 301:349 (1980).

5. Bryant, R.W., A.N. Epstein, J.T. Fitzsimons and S.J. Fluharty, Arousal of specific and persistent sodium appetite in the rat with continuous intracerebroventricular infusion of angiotensin II, J. Physiol. (London) 301:365 (1980).

6. Beaumont, K. and D.D. Fanestil, Characterization of rat brain aldosterone receptors reveals high affinity for corticosterone, Endocrinology 113:2043 (1983).

7. Fluharty, S.J. and A.N. Epstein, Sodium appetite elicited by intracerebroventricular infusion of angiotensin II in the rat II: Synergistic interaction with systemic mineralocorticoids, Behav. Neurosci. 97:746 (1983).

8. Fregly, M.J., J.R. Cade, I.W. Waters, J.A. Straw and R.E. Taylor, Jr., Secretion of aldosterone by adrenal glands of propylthiour-acil-treated rats, Endocrinology 77:777 (1965).

9. Evered, M.D., M.M. Robinson and M.A. Richardson, Captopril given intracerebroventricularly, subcutaneously or by gavage inhibits angiotensin-converting enzyme activity in the rat brain, Europ. J. Pharmacol. 68:443 (1980).

10. Fregly, M.J., N.E. Rowland and W.G. Luttge, Effect of chronic administration of desoxycorticosterone acetate (DOCA) on salt appetite of captopril-treated rats, Fed. Proc. 43:717 (1984).

11. Richter, D.P., Salt taste thresholds for normal and adrenalect-omized rats, Endocrinology 24:367 (1939).

12. Nachman, M., Taste preference for sodium salts by adrenalect-omized rats, J. Comp. Physiol. Psychol. 55:1124 (1962).

13. Fregly, M.J. and I.W. Waters, Effect of propylthiouracil on the preference thresholds of rats for NaCl solution, Proc. Soc. Exp. Biol. Med. 120:637 (1965).

14. Thrasher, T.N. and M.J. Fregly, Factors affecting salivary sodium concentration, NaCl intake, and preference threshold and their interrelationship, in Biological And Behavioral Aspects Of Salt Intake, eds. M.R. Kare, M.J. Fregly and R.A. Bernard, Pergamon Press, NY, p. 145 (1980).

15. Fregly, M.J., Effect of an oral contraceptive on NaCl appetite and preference threshold in rats, Pharmac. Biochem. Behav. 1: 61 (1973).

16. Fregly, M.J., Effect of hydrochlorothiazide on preference thresh-old of rats for NaCl solutions, Proc. Soc. Exper. Biol. Med. 125: 1079 (1967).

17. Fregly, M.J., Effect of the angiotensin converting enzyme inhib-itor, captopril, on NaCl appetite of rats, J. Pharm. Exp. Therap. 215:407 (1980).

18. Pfaffman, C. and J.K. Bare, Gustatory nerve discharges in normal

and adrenalectomized rats, J. Comp. Physiol. Psychol. 43:320 (1950).

19. Contreras, R.J., T. Kosman and M.E. Frank, Activity in salt taste fibers: peripheral mechanism for mediating changes in salt intake, Chemical Senses 8:275 (1984).

20. Matuso, R., N. Shimizu and K. Kosano, Lateral hypothalamic modulation of oral sensory afferent activity in nucleus tractus solitarius neurons of rats, J. Neurosci. 4:1201 (1984).

---

[1]Supported by grant AM31837-02 from the National Institutes of Health, Bethesda, MD.

SODIUM APPETITE AFTER ADRENALECTOMY OR DEOXYCORTICOSTERONE-TREATMENT IN DIABETES INSIPIDUS (BRATTLEBORO) RATS

Louise M. Fuller

The Physiological Laboratory, Downing
Street, Cambridge CB2 3EG, England

INTRODUCTION

Brattleboro (BB) rats with hereditary hypothalamic diabetes insipidus have undetectable levels of vasopressin in either the brain or plasma[1,2]. These animals are polyuric owing to vasopressin deficiency[3] and consequently they are dehydrated[4]. Hypovolaemia, increased plasma renin activity[5] and increased plasma osmolality[6] are factors that could contribute to the characteristic polydipsia[3] of the BB rat. Surprisingly, plasma aldosterone is reduced in BB rats, either as a result of decreased production[7], or of increased synthesis combined with high turnover[8].

In normal animals many stimuli which increase water intake, such as haemorrhage, hyperoncotic dialysis and caval ligation, also cause delayed increases in sodium appetite[9]. However, despite the impressive thirst in diabetes insipidus rats, they do not normally drink hypertonic NaCl if it is offered with water. Indeed, rats with surgically induced diabetes insipidus are less tolerant of NaCl than normal rats[10]. In these animals the NaCl acceptance-rejection function is altered so that hypertonic NaCl is rejected at concentrations highly palatable to normal rats[11]. In this study the effects of adrenalectomy or mineralocorticoid-treatment, both established stimuli to sodium appetite in normal animals, are examined in BB rats.

ADRENALECTOMY-INDUCED SODIUM APPETITE

Bilateral adrenalectomy in normal rats results in renal sodium loss, leading to negative sodium balance, but these animals will survive if they are allowed to drink NaCl[12]. In BB rats extracellular fluid volume is reduced and intracellular fluid volume expanded[13]. These fluid shifts are similar to those occuring in adrenalectomized (ADX) animals, suggesting that the BB rat may normally be adrenal insufficient. It has also been reported that ADX male BB rats do not go into negative sodium balance[14], a surprising finding since adrenalectomy in rats with surgically induced diabetes insipidus results in gradual increases in 1% NaCl intake[15].

   To examine the effect of adrenalectomy on sodium intake in diabetes insipidus animals, BB rats and rats of the parent Long-Evans (LE) strain, both with continuous access to tap water and 2.7% NaCl, were either ADX or sham-operated. Intakes of 2.7% NaCl and water were measured for 2 days before, and 8 days after operation (Fig. 1). Over the 5 days following adrenalectomy both strains gradually increased their intakes of 2.7% NaCl, a solution

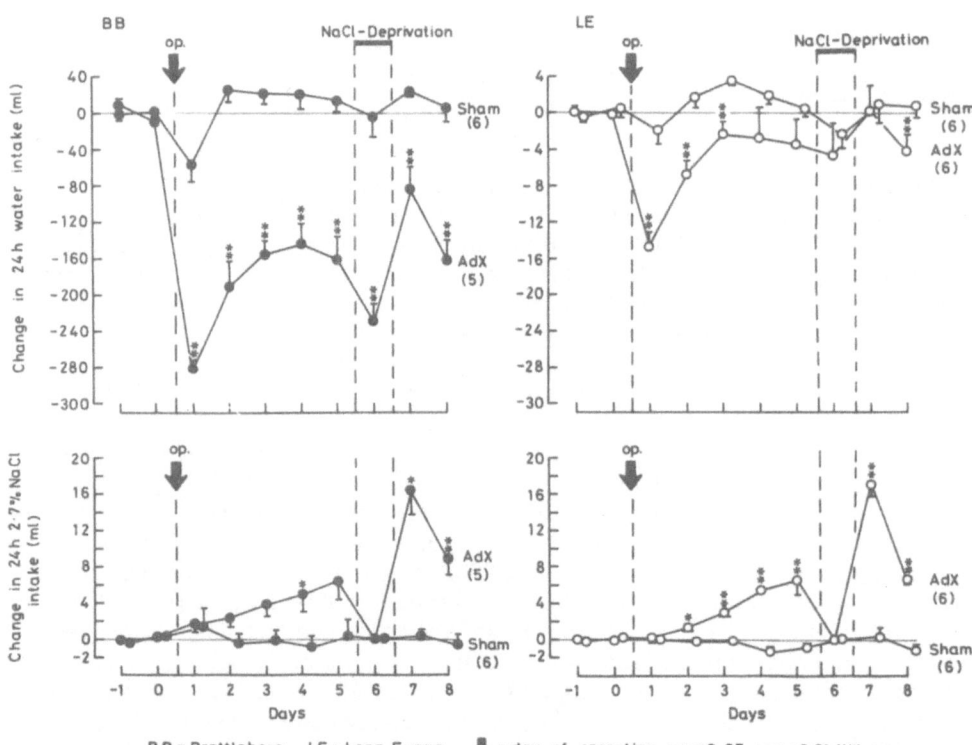

BB = Brattleboro, LE = Long Evans, ⬇ = day of operation, *p < 0·05, **p < 0·01, Wilcoxon.

Fig. 1   Change in 24 h water (upper panels) and 2.7% NaCl (lower panels) intakes from baseline in ADX or sham-operated BB (●) rats and LE (o) rats. All comparisons between sham-operated and ADX animals.

which both strains normally find aversive. Sham-operated rats of
either strain did not develop a sodium appetite. On day 6 after
operation, all rats were deprived of 2.7% NaCl for 24 h. When NaCl
was returned on day 7 all tne ADX animals immediately began
drinking it. There were no significant differences between the 24
h 2.7% NaCl intakes of BB and LE rats at any time after
adrenalectomy.

    The spontaneous polydipsia of BB rats is reduced by bilateral
adrenalectomy[16] (Fig. 1). Possible reasons for this are, decreases
in urine volume, free water clearance, and glomerular filtration
rate, changes in urinary electrolyte excretion, or a fall in food
intake. In contrast LE rats showed only slight reductions in water
intake.

    To discover whether or not the sodium appetite of these two
groups of animals is due to a developing sodium deficit, the sodium
balance of ADX BB and LE rats was examined before, during and after
a 48 hour period of NaCl deprivation (Fig.2). At the end of the
period of NaCl deprivation BB and LE rats had gone into negative
sodium balance compared to the pre-deprivation period. Therefore
adrenalectomy causes sodium depletion-induced sodium appetite in
both BB and LE rats, despite reduced plasma mineralocorticoid and
increased plasma renin activity in Brattleboros.

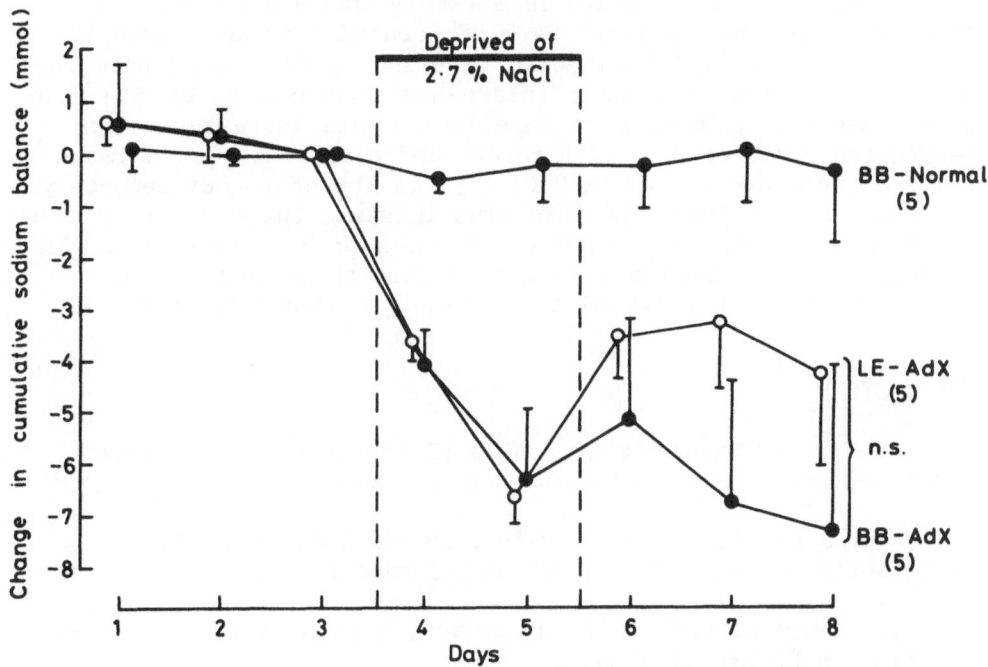

Fig. 2. Change in cumulative sodium balance (mmol) from baseline
         in ADX-BB (●), ADX-LE (o) and unoperated BB (●) rats.
         n.s. = not significant by unpaired Wilcoxon.

MINERALOCORTICOID-INDUCED SODIUM APPETITE

Mineralocorticoid-treated normal rats retain sodium and lose potassium because of the action of the mineralocorticoid on the distal nephron. This results in development of positive sodium and fluid balance with consequent suppression of renal renin release. Despite sodium retention chronic mineralocorticoid-treatment causes a paradoxical increase in sodium appetite[18]. In a similar way to normal rats, DOCA-treated BB rats go into positive sodium balance[22], with renal renin release being suppressed in both DOCA/salt Brattleboros[23] and in those on a normal sodium diet[24].

To examine the effect of chronic DOCA-treatment on sodium appetite in diabetes insipidus, a group of BB and LE rats, with access to both 2.7% NaCl and tap water, were treated either with DOCA (25 mg/kg/day s.c.) or Arachis oil (s.c.) for 6 days. DOCA-treated BB and LE rats gradually increased their hypertonic NaCl intakes subsequent to day 2 of treatment. The increase in NaCl intake was not significantly different between the two rat strains on any treatment day. On day 6 of DOCA-treatment the 24 hour intakes of 2.7% NaCl were 10.9 ± 2.5 ml in BB rats (n=9) and 10.0 ± 2.0 ml in the LE (n=9). Arachis oil-treatment had no effect on NaCl intake in either strain.

In BB rats water intake is markedly increased by DOCA-treatment. During the first 24 h of treatment water intake is increased by 49.0 ± 8.0 ml above baseline (p <0.01 from baseline), and, on day 6 the 24 h water intake has increased to 211.5 ± 12.6 ml above baseline (p<0.001 from baseline). This increase occurs before the onset of 2.7% NaCl intake and also occurs in animals which do not have access to NaCl. It is therefore not secondary to increased NaCl intake. LE rats also increase their water intakes during DOCA-treatment but not until day 2 or 3 of treatment. The volumes of water drunk are much less than those in BB rats, increasing to 13.1 ± 1.3 ml above baseline (p<0.001) on day 6.

CONCLUSIONS

1. After adrenalectomy BB and LE rats develop a sodium appetite, which is secondary to sodium loss.

2. The polydipsia of BB rats is markedly reduced by adrenalectomy, but water intake of LE rats is unaffected.

3. Treatment with DOCA causes very similar increases in sodium appetite in BB and LE rats.

4. The polydipsia of BB rats is augmented by DOCA-treatment. LE rats exhibit smaller, more delayed, increases in water intake.

5. The increase in sodium appetite of BB rats after either adrenalectomy or DOCA-treatment is independent of changes in water intake.

6. The presence of vasopressin is not necessary for the generation of a sodium appetite.

REFERENCES

1. H. Valtin, W. H. Sawyer, and H. W. Sokol, Neurohypophyial principles in rats homozygous and heterozygous for hypothalamic diabetes insipidus (Brattleboro strain). Endocrinol. 77:701 (1965).
2. B. Mohring and J. Mohring, Plasma ADH in normal Long-Evans and in Long-Evans rats heterozygous and homozygous for hypothalamic diabetes insipidus. Life Sci. 17:1307 (1975).
3. H. Valtin and H. A. Schroeder, Familial hypothalamic diabetes insipidus in rats (Brattleboro strain). Am. J. Physiol. 206:425 (1964).
4. L. B. Kinter, Water balance in the brattleboro rat: Considerations for hormone replacement therapy. N.Y. Acad. Sci. 394:448 (1982).
5. L. B. Kinter, D. N. Shier, A. C. Barger, W. Flanmenbaum and R. Beeuwkes III Comparison of plasma renin activity in conscious and anaesthetised Brattleboro strain rats. Fed. Proc. 38(3 part 1):892 (1979).
6. B. Mohring, J. Mohring, G. Dauda, D. Haack, Potassium deficiency in rats with hereditary diabetes insipidus. Am. J. Physiol. 227:916 (1974).
7. H. W. Sokol and J. Mohring, Morphological correlates of renin and aldosterone secretion in the Brattleboro rat. Ann N.Y. Acad. Sci. 394:291 (1982).
8. L. B. Kinter, J. B. Melby, D. N. Shier, T. Wilson, and R. Beeukes III, Aldosterone in conscious Brattleboro strain rats. Kidney Int. 16:873 (1979).
9. J. T. Fitzsimons. in: "The physiology of thirst and sodium appetite," Monographs of the Physiological Society, No 35. Cambridge University Press, Cambridge (1979).
10. H. G. Swann and B. J. Penner, The effect of salts on the diabetes insipidus following posthypophysectomy in the rat. Endocrinol. 24:253 (1939).
11. L. F. Titlebaum, J. L. Falk, and J. Mayer, Altered accceptance and rejection of NaCl in rats with diabetes insipidus. Am. J. Physiol. 199:22 (1960).
12. C. P. Richter, Increased salt appetite in adrenalectomised rats. Am. J. Physiol. 115:155 (1936).
13. S. M. Freidman and C. L. Friedman, Salt and water distribution in hereditary and induced hypothalamic diabetes insipidus in the rat. Can. J. Physiol Pharmacol. 43:699 (1965).

14. R. J. Balment, I. W. Henderson, I. Chester Jones, and W. Mosley, Water and electrolyte balance in adrenalectomized rats with Diabetes Insipidus (Brattleboro Strain) given antidiuretic hormone. Gen. and Comp. Endocrinol. 33:428 (1977).

15. G. M. A. Palmieri and S. Talesnik, Intake of NaCl solution in rats with diabetes insipidus. J. Comp. Physiol. Psychol. 68:38 (1969).

16. R. J. Balment, I. Chester Jones, I. W. Henderson and J. Ann Oliver, Effects of adrenalectomy and hypophysectomy on water and electrolyte metabolism in male and female rats with inherited hypothalamic Diabetes Insipidus (Brattleboro strain). J. Endocrinol. 71:193 (1976).

17. K. K. Rice, and C. P. Richter, Increased sodium chloride and water intake of normal rats treated with desoxycorticosterone acetate. Endocrinol. 33:106 (1943).

18. A. Schomig, H. Schomig-Breckner, B. Mohring, and J. Mohring, Effects of ADH and DOCA on sodium and potassium balances in rats with hereditary hypothalamic Diabetes Insipidus. Naunyn Schm. Arch. Pharmak. 274 (Suppl.) R100, (1972).

19. W. B. Campbell and W. A. Pettinger, Sodium chloride suppression of renin release in the unanaesthetised rat. Endocrinol. 97:1394 (1975).

20. R. D. Murray, K. W. Cho, and R. L. Malvin, Deoxycorticosterone Acetate-induced renin supression in the absence of Antidiuretic hormone. Proc. Soc. Exp. Biol. Med. 165:137 (1980).

DOPAMINERGIC MODULATION OF CHOICE IN SALT PREFERENCE TESTS

Steven J. Cooper and David B. Gilbert

Department of Psychology
University of Birmingham
Birmingham B15 2TT, U.K.

INTRODUCTION

Pharmacological blockade of dopamine receptors by some neuro-leptic drugs reduces fluid consumption in rats. For example, subcutaneous injections of haloperidol reduced water intake in rats which were induced to drink by water deprivation, or by subcutaneous injections of hypertonic saline, polyethylene glycol, or isoproterenol[1]. Three butyrophenones, droperidol, haloperidol and spiperone, each produced dose-dependent attenuation of deprivation-induced drinking[2]. Intracerebroventricular administration of several neuroleptics (haloperidol, fluphenazine, cis-flupenthixol, sulpiride, pimozide, chlorpromazine), in each case, caused inhibition of angiotensin-induced drinking in male rats[2]. Thus, there may be some dopaminergic involvement in the control of drinking responses.

An important, recent development is the evidence that central dopamine projections to the amygdala may interact with gustatory projections to the same region to modulate responses to salt taste[4]. In a two-bottle preference test, water-deprived rats had a choice between water and a non-preferred 1.5% NaCl solution. Injection of 0.2 µg spiperone (a dopamine receptor antagonist) into the central nucleus of the amygdala produced a significant increase in saline intake, without affecting water intake. Dopamine receptor blockade, therefore, may under certain circumstances affect sodium appetite and preference.

DOPAMINE ANTAGONISTS AND SALT PREFERENCE

In the absence of other data, we began a series of experiments

453

to investigate this question in detail. A number of important independent variables have been taken into account. Firstly, preference tests were conducted using a range of salt concentrations. Two-bottle and single-stimulus tests have clearly demonstrated that the preference and acceptance measures vary as a function of NaCl concentration[5,6,7]. Dilute salt solutions are typically preferred to water in rats, whereas more concentrated solutions may be aversive, and are rejected. Effects of drug treatments may depend on the concentration of NaCl used in the test. Secondly, a variety of compounds have been investigated. In addition to familiar neuroleptics (pimozide, clozapine), drugs which appear to block specific dopamine receptor sub-types[8] have been used. Thus, the substituted benzamides (sulpiride, sultopride) are considered to be specific dopamine D2 receptor antagonists[9,10]. In contrast, the novel compound SCH 23390 is thought to act specifically as a D1 receptor antagonist[11]. Thirdly, extensive dose-effect data were collected for each compound.

Methods

The general procedures have been described in detail elsewhere[12]. Briefly, adult rats (hooded General strain), bred in our laboratory, were adapted to a 22 h water-deprivation schedule, and were familiarised with a wooden test box, in which they had access to two drinking spouts. For a particular salt concentration, a separate group of animals was trained in the choice between the salt solution and water. Within each of these groups, each animal was tested over a number of occasions with a series of doses of the compound under investigation. All drugs were administered intraperitoneally, with the exception of SCH 23390 which was given subcutaneously. Control injections matched the drug vehicles, and injection-test intervals were chosen on the basis of the published literature. The test lasted 15 min, and salt and water intakes were separately measured. The results were described in terms of the total fluid intake (ml), and in terms of saline preference, i.e. (saline intake/total intake) x 100%. The data were analysed using the analysis of variance, followed by Dunnett's t test. In the Figures, data are shown as group means (N=8 in each case), with standard errors represented by vertical lines. Significant differences between individual drug dose conditions and corresponding control scores are denoted as follows: *P<0.05; **P<0.01.

Results

In male rats, pimozide (0.1-3.0 mg/kg) and clozapine (0.3-10 mg/kg) produced dose-dependent reductions in the total fluid intake (ml), in tests of preference between 0.125% NaCl, 0.6% NaCl and 1.7% NaCl, respectively (Fig. 1). Sulpiride (1.0-3.0 mg/kg), on

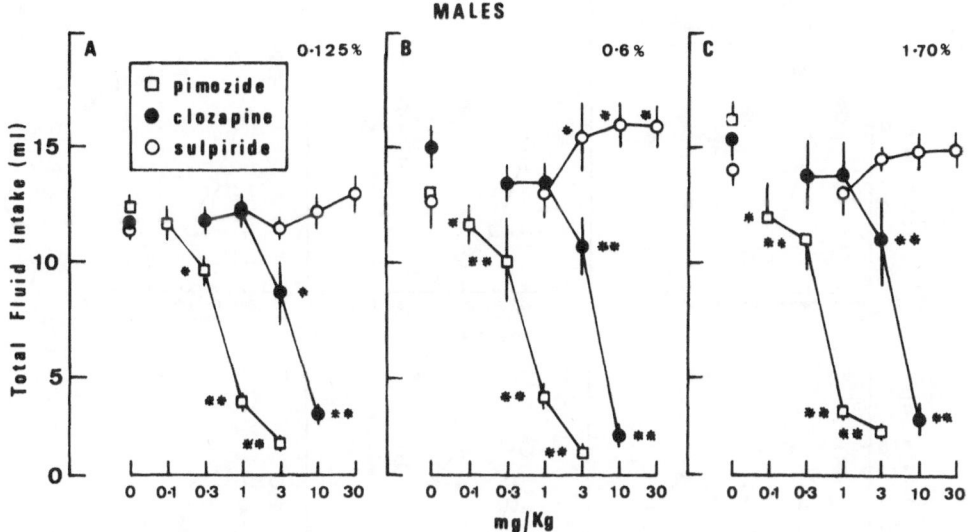

Fig. 1 Effects of neuroleptics on total fluid intake in salt
preference tests in male rats

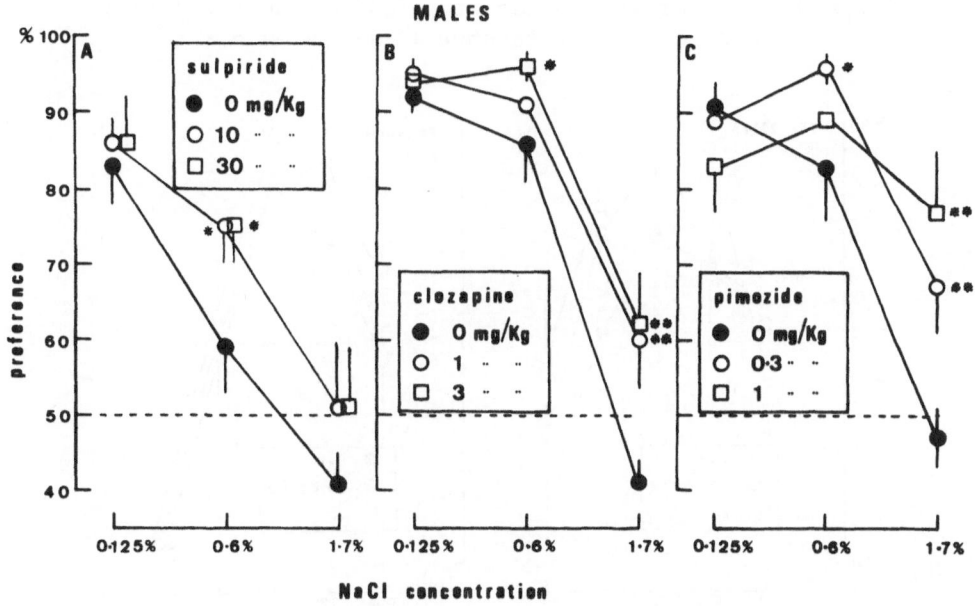

Fig.2 Effects of neuroleptics on salt preferences in male
rats

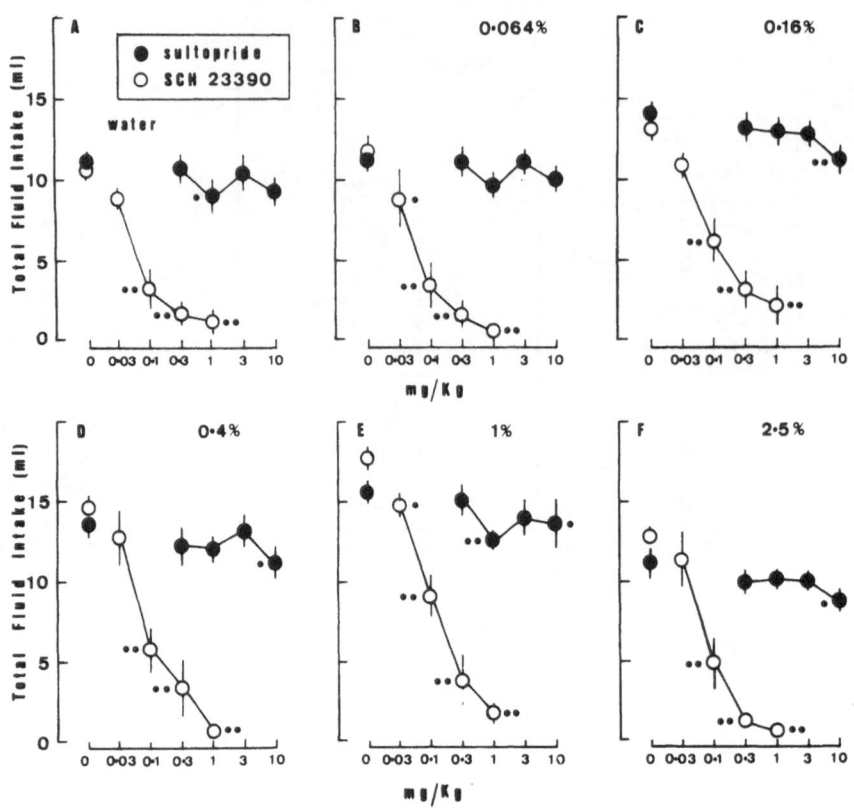

Fig.3   Effects of sultopride and SCH23390 on total fluid
intake in salt preference tests in male rats.

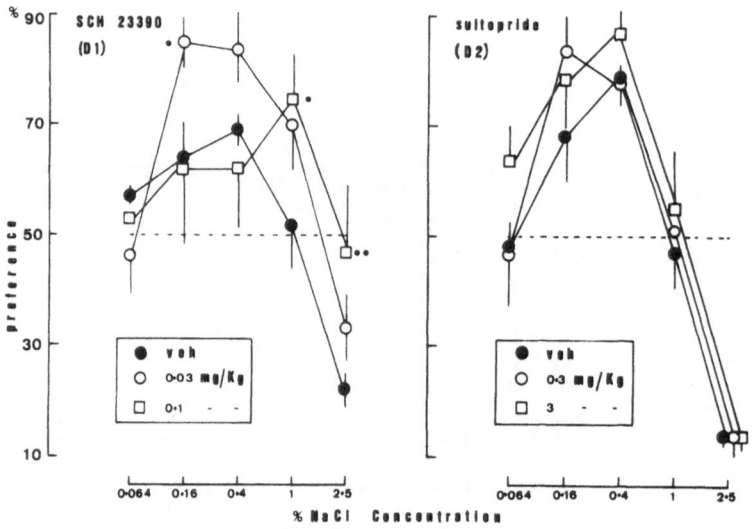

Fig.4   Effects of sultopride and SCH23390 on salt
preference in male rats.

the other hand, did not reduce fluid intake. Indeed, in the 0.6% NaCl condition, fluid intake was significantly enhanced (Fig. 1). The preference measure results are shown in Fig. 2. Significant increases were detected, and there were no cases in which saline preference was diminished. Sulpiride (Fig. 2A) significantly enhanced the preference for the palatable hypotonic 0.6% NaCl solution. This effect was also seen with clozapine (Fig. 2B) and pimozide (Fig. 2C), in doses which did not produce a major suppression of the total fluid intake. Clozapine and pimozide, but not sulpiride, also significantly enhanced the preference for a 1.7% NaCl solution.

In a separate series of experiments, the specific D1 receptor antagonist, SCH 23390, produced dose-dependent reductions in the total fluid intake (ml), in tests of preference between 0.064% NaCl, 0.16% NaCl, 0.4% NaCl, 1.0% NaCl, and 2.5% NaCl, respectively (Fig. 3). In contrast, the substituted benzamide, sultopride (a specific D2 receptor antagonist), given in doses 0.3-10 mg/kg, did not consistently affect the total fluid consumption except for a modest reduction at the highest dose (Fig. 3). Results for the saline preference measure are shown in Fig. 4. Following vehicle (veh) injections, typical inverted-U shaped preference-aversion curves were obtained, as a function of NaCl concentration. A small dose of SCH 23390 (0.03 mg/kg), which did not significantly reduce overall fluid consumption (see Fig. 3), produced an elevation in salt preference over the range of NaCl concentrations, from 0.16% to 2.5% NaCl. On the other hand, sultopride raised the preference measure on the ascending limb of the curve, but failed to affect preference at the two highest salt concentrations (Fig. 4).

DISCUSSION AND CONCLUSIONS

Our data extend and qualify earlier reports that dopamine receptor antagonists potently reduce fluid consumption. Following pimozide, clozapine and haloperidol (data not shown) treatments, total fluid consumption in the preference tests was reduced dose-dependently. This antidipsogenic effect was shared by the specific dopamine D1 receptor antagonist SCH 23390. In contrast, two substituted benzamines, sulpiride and sultopride, thought to be specific D2 receptor antagonists, did not reduce fluid consumption. Dopamine D1 receptor blockade appears to be sufficient to reduce fluid intake, whereas D2 receptor blockade alone may not be sufficient.

The neuroleptic-induced changes in the saline preference measure were almost always increases. This is not an inevitable consequence of drug treatments, as our work with naloxone shows[13]. Treatments with the benzamines produced selective effects, enhancing the preference for initially preferred solutions, but have little effect on the choice of non-preferred solutions. In contrast, pimozide and clozapine significantly enhanced the preference for both hypotonic and hypertonic solutions. The selective D1 receptor antagonist, SCH 23390, acted similarly to pimozide and clozapine.

It appears, therefore, that the pharmacological distinction between D1 and D2 receptor blockade is a significant one, so far as salt preference is concerned. Our data suggest that there are important distinctions to be drawn between responses to preferred and non-preferred salt solutions, in terms of the details of dopamine receptor involvement.

REFERENCES

1. M. L. Block and A. E. Fisher, Cholinergic and dopaminergic blocking agents modulate water intake elicited by deprivation, hypovolemia, hypertonicity and isoproterenol, Pharmacol. Biochem. Behav. 3:251 (1975).
2. N. Rowland and D. J. Engle, Feeding and drinking interactions after acute butyrophenone administration, Pharmacol. Biochem. Behav. 7:295 (1977).
3. C. Summer, G. N. Woodruff, J. A. Poat and K. A. Munday, The effect of neuroleptic drugs on drinking induced by central administration of angiotensin or carbachol, Psychopharmacol. 60:291(1979).
4. G. J. Mogenson and M. Wu, Electrophysiological and behavioral evidence of interaction of dopaminergic and gustatory afferents in the amygdala, Brain Res. Bull. 8:685 (1982).
5. J. K. Bare, The specific hunger for sodium chloride in normal and adrenalectomized white rats, J. comp. physiol. Psychol. 42:242 (1949).
6. K. T. Borer, Disappearance of preferences and aversions for sapid solutions in rats ingesting untasted fluids, J. comp. physiol. Psychol. 65:213 (1968).
7. I. H. Weiner and E. Stellar, Salt preference of the rat determined by a single-stimulus method, J. comp. physiol. Psychol. 44:394 (1951).
8. J. W. Kebabian and D. B. Calne, Multiple receptors for dopamine, Nature. 277:93 (1979).
9. J. W. Kebabian, K. Tsuruta, T. E. Cote and G. W. Grewe, The activity of substituted benzamides in biochemical models of dopamine receptors, in: "The Benzamides: Pharmacology, Neurobiology and Clinical Aspects", J. Rotrosen and M. Stanley, eds., Raven Press, New York (1982).
10. P. Jenner, A. Theodorou and C. D. Marsden, Specific receptors for substituted benzamide drugs in brain, in: "The Benzamides" ibid.
11. J. Hyttel, SCH 23390 - the first selective dopamine D-1 antagonist Europ. J. Pharmacol. 91:153 (1983).
12. S. J. Cooper and D. B. Gilbert, Naloxone suppresses fluid consumption in tests of choice between sodium chloride solutions and water in male and female water-deprived rats, Psychopharmacology. (1985) in press.

CONTOL MECHANISMS OF SALT APPETITE

Susan Kaufman

Department of Medicine
University of Alberta
Edmonton, Alberta, Canada, T6G 2G3

INTRODUCTION

There is considerable evidence that intravascular volume is
monitored by the low pressure stretch receptors found in the atria
and great veins near the heart.[1] Thus it was gratifying, but not
altogether surprising, to confirm that stimulation of these receptors attenuated drinking to extracellular fluid deficits but was
without effect on drinking to intracellular fluid deficits.[2]
However, maintenance of extracellular fluid volume depends upon
the ability to regulate not just water but also salt balance. A
study was therefore made of the effect on salt appetite of stimulating the right atrial receptors. Since there is evidence that
the renin-angiotensin system is involved in regulating salt intake
[3], two very different models of salt appetite were investigated.
The first, namely sodium depletion by removal of isotonic fluid,
results in activation of the renin-angiotensin system[4,5]. In the
second experiment, salt appetite was induced by injection of
deoxycorticosterone acetate (DOCA). Under these circumstances the
renin-angiotensin system is known to be suppressed[6].

METHODS

Animals

Male Wistar rats weighing between 300 and 400g were housed in
temperature and humidity controlled rooms with the lights on from
0700 hr. to 1900 hr. They had free access to water and to a
sodium controlled diet (0.1% Na, Bioserv) except where noted
otherwise. At least one week after delivery they were prepared

with venous cannulae, atrial balloons and, for the first experiment, peritoneal dialysis sacs. The rats were always put in the metabolism cages for a sham experiment several days before the definitive experiment so as to familiarize them with the procedure.

Surgery

Surgery was performed at least one week after delivery. Under pentobarbital anaesthesia the right jugular vein was exposed and a 3 mm incision was made. Through this, a small inflatable balloon was advanced down the right superior vena cava so that its tip lay at the vein/atrial junction. This position was stabilized by ligating the cannula to the right clavicle (See Kaufman[1] for details). This balloon does not interfere with venous return from the head since, in the rat, there is a left superior vena cava which joins the inferior vena cava.

An indwelling cannula was also implanted non-occlusively into the inferior vena cava according to the method of Kaufman.[7]

Sacs were fashioned from a 12 cm. length of Spectrapor II dialysis tubing tied around a fenestrated cannula so as to allow material to be injected and withdrawn. There dialysis sacs were implanted into the peritoneal cavity at the time of venous cannulation. The three cannulae from the balloon, the inferior vena cava and the dialysis sac were led subcutaneously to the nape of the neck where they were attached to stainless-steel tubing which protruded through the skin. These tubes could be capped to prevent fluid escape. The animals were allowed at least one week to recover from surgery before experiments were started.

Chemicals

Polyethylene glycol (mol. wt. 20,000 Sigma) was dissolved with stirring in warm distilled water (25% w/w) about 1 hr before use.

Statistics

Results are expressed as mean $\pm$ SE of mean. In graphs, the SE of mean is delimited by the vertical bars. The significance of the difference between means was determined by Student's t-test.

RESULTS

Sodium depleted rats

For at least two weeks before the experiments started, the rats had been maintained on a sodium controlled diet (0.1% Na, Bioserv). Polyethylene glycol (15 ml/kg body wt) was then injected into the dialysis sacs and the rats were left overnight (17 hr)

with distilled water to drink but no food. The next morning the
dialysis sacs were drained and the rats were given access to 2.7%
saline, at which time the balloons were inflated in the experimental group. After two hours the balloons were deflated. Salt
intake was measured for a total of 6 hr, after which time the
saline was removed. The rats were again left overnight with only
distilled water to drink and the experiment was repeated the next
day.

During the 2 hr period of balloon inflation the experimental
group drank significantly less than the control rats (Fig. 1).
However, after the balloons were deflated the experimental group
drank more than the control rats so that the total sodium intake
of the two groups was identical. The same phenomenon was observed
on the second day of testing.

The mean net volume withdrawn from the dialysis sacs was
19.9 ± 1.6 ml (n=11) for the experimental group and 19.8 ± 1.0 ml
(n=13) for the control group. This resulted in sodium deficits of
1.006 ± 0.089 meq. and 1.086 ± 0.046 meq. respectively.

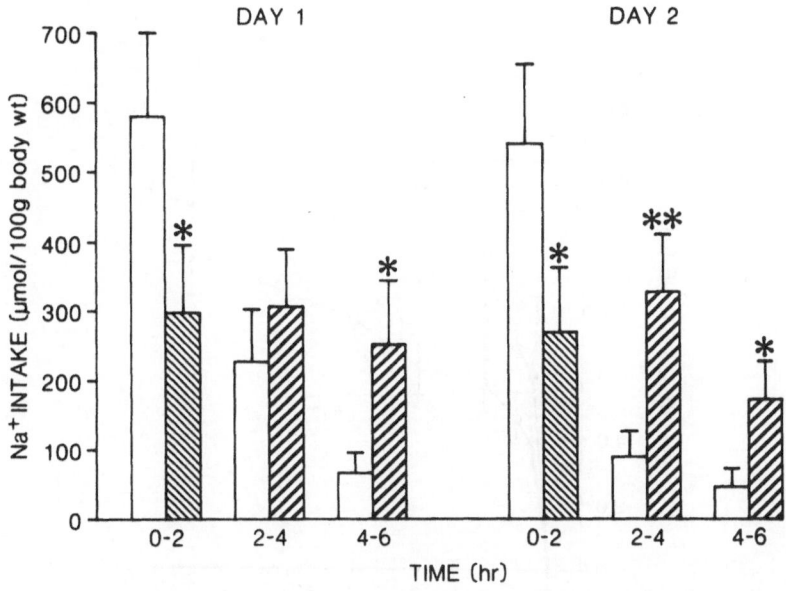

Fig. 1. Sodium intake (2.7% saline) of rats which have extracellular fluid deficits produced by peritoneal dialysis with hyperoncotic colloid. Right atrial balloons were inflated in the experimental group (hatched bars) for the first 2 hr after offering
saline.

DOCA treated rats

The rats were maintained on the sodium controlled diet for at
least two weeks before the experiments started. Twenty rats were
prepared with right atrial balloons. They all received daily DOCA
injections (10mg/kg body wt. dissolved in oil). On day 8 they were
given access to 3% saline to drink for 6 hr. This served to
familiarize the rats with the testing procedure. Previous experiments had shown that saline intakes were stable after 8 days of
DOCA treatment. On day 9, the balloons were inflated in half the
animals immediately before presenting the saline. After 2 hr the
balloons were deflated and saline intakes were measured for a
further 4 hr. The next day the experiment was repeated except that
the balloons were inflated in the other rats ie. each animal served
as its own control. The rats were then sacrificed and the posi-
tioning of the balloon at the right atrium was checked.

It can be clearly seen in Fig. 2 that the intake of 3% saline
was significantly depressed during the 2 hr period of balloon
inflation, but not significantly thereafter.

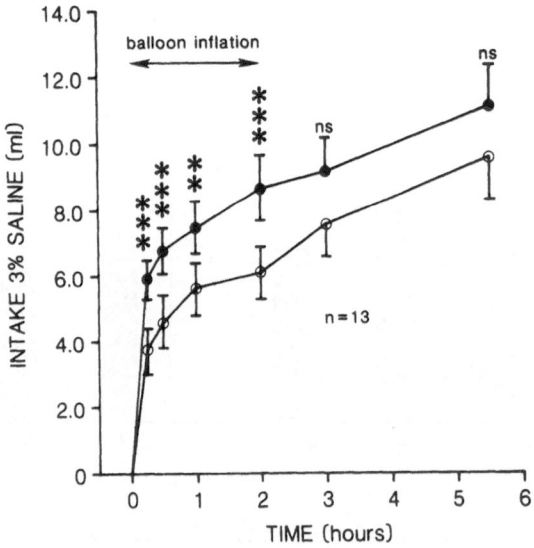

Fig. 2. Sodium intake (3% saline) of rats injected daily with
DOCA (10 mg/kg body wt). The experiment took place after 9 days
of injection. Right atrial balloons were inflated in the experiment group (O) for the first 2 hr after offering saline.

462

DISCUSSION

Angiotensin II is widely held to be a mediator of salt appetite[3]. Certainly, the extracellular fluid deficits obtained in this first experiment (about 20%) were more than adequate to activate the renin-angiotensin system. Previous experiments using, a similar model, have shown that a deficit of only 14% causes PRA to increase from 1.5 ng/ml/hr to 7.4 ng/ml/hr[4]. In these last-mentioned experiments there was some indication that an extended period of right atrial stretch might reduce PRA. Since salt depleted rats seem to drink saline with less urgency, ie. less quickly, than a thirsty rat drinks water, we could not eliminate the possibility that the attenuation of salt-appetite, seen when the right-atrial balloons were inflated in the volume depleted rats, was secondary to suppression of the renin-angiotensin system.

The second experiment addressed itself to this question. When DOCA is administered daily to rats they initially retain sodium and then after a few days, depending on their daily salt intake, they return towards normal sodium balance. This phenomenon is known as mineralocorticoid escape. If rats are given access to hypertonic saline during this period, they will increase their intake each day until it stabilizes after 7 days of DOCA injection at about 9 ml/rat/6 hr. At this time the rats are still in slightly positive sodium balance (unpublished observation) and the renin-angiotensin system is suppressed.[6] Under these conditions inflation of the right atrial balloon could not reflexly inhibit renin secretion and hence the attenuation of salt appetite that was observed in the DOCA treated rats could not have been secondary to a reduction in PRA. Therefore, pathways must exist, independent of the renin-angiotensin system, whereby information obtained from the cardiac volume receptors regarding the state of filling of the vasculature may regulate salt intake.

REFERENCES

1. R.J. Linden & C.T. Kappagoda. Atrial Receptors. Publ. Cam. U. Press. (1982).
2. S. Kaufman. Role of right atrial receptors in the control of drinking in the rat. J. Physiol. 349: 389 (1984).
3. J.T. Fitzsimons. The Physiology of Thirst and Sodium Appetite. Publ. Cam. U. Press. (1979).
4. S. Kaufman. Relationship between right atrial stretch and plasma renin activity. This publication.
5. F.H.H. Leenen, E.M. Stricker, R.H. McDonald & W. DeJong. Relationship between increase in plasma renin activity and drinking following different types of dipsogenic stimuli in: Control Mechanisms of Drinking, ed. G. Peters, J.T. Fitzsimons & Lisette Peters - Haefeli p. 84. Publ. Springer-Verlag (1975)
6. T.K. Keeton & W.B. Campbell. The pharmacological alteration of renin release. Pharmacol. Rev. 32: 81 (1980).

THE NATURE OF THE SALT APPETITE OF ADRENALECTOMIZED RATS

Christian Mueli and Georges Peters

Institut de pharmacologie de l'Université
CH-1011 Lausanne (Switzerland) *)

*) (Supported by Swiss National Science Foundation,
   Grant Nr. 3.220-082)

INTRODUCTION

An increased salt appetite is usually considered to be present when animals, choosing between water and hypertonic NaCl solution (1.8%, 2.7 or 4.5%) drink more of the apparently aversive salt solution than normal animals (1,5,7). Overcoming the aversive taste of the concentrated NaCl solution is taken as an expression of motivation for salt intake. We investigated the validity of this assumption. If it is correct, animals with an increased salt appetite should also drink increased amounts of a 0.9% NaCl solution made aversive by adding bitter substances and given as a choice to water, provided that the taste of the NaCl is not masked by the bitter substance. If salt-appetite is regulatory, more bitter 0.9% NaCl than 1.8 or 2.7 NaCl should be chosen by salt-hungry rats.

Bilaterally, adrenalectomized rats show an increased salt-appetite even if they are salt-satiated (1,2,3,4,5,6). With a choice between 0.9% NaCl and water, they may (e.g. present experiments) or may not (4) drink more saline than normal controls depending on the saline intake of the latter(4). Choosing between hypertonic NaCl solution and water, they drink more saline than normal controls (3,6,7), and thus show motivated salt appetite. We re-investigated this motivated salt intake by offering adrenalectomized rats a choice between bitter 0.9% saline and water.- In all choice situations between water and saline, however, adrenalectomized animals also drink less water than normal controls : they show an aversion against water (4).

In the choice situation between water and a standard concentration of hypertonic NaCl, the intensity (strength) of the motivation cannot be evaluated. A stronger motivation should result in an appreciable intake of more concentrated NaCl solutions than tolerated by normal control animals, in the choice situation against water. We attempted to measure the strength of this motivation.

MATERIAL AND METHODS

Experimental animals were adult male rats of the Wistar type (Madörin strain, Füllinsdorf, Basle, Switzerland) kept in individual cages at constant temperature ($22-25^o$ C) and light from 6.00 a.m. to 6.00 p.m. . The animals had permanent free access to rat pellets containing 104 mEq of Na^+ per kg. and two different drinking fluids from two bottles previously tested for equal rates of delivery. The position of the two bottles was changed every day. Animals which showed a constant preference for one bottle, or for one position were excluded from the experiments. Fluid intake from both bottles was measured over 24 hour intervals (9.00 a.m. to 9.00 a.m.) for periods of usually 10 days. The data shown under "Results" are those of the 4th day of any given choice situation.

Bitter 0.9% NaCl solutions were prepared by adding either quinine hydrochloride (purity Ph. H. VI) or dry extract of gentian, Ph. H. VI (obtained from Siegfried, Switzerland). "Equi-aversivity" of bitter solutions of 0.9% NaCl with concentrated, (1.8% or 2.7%) NaCl solutions was determined in normal rats by selecting a concentration of quinine or of gentian extract which, given as a choice to the hypertonic NaCl solution, was consumed in equal amounts."Equi-aversivity" was confirmed by giving normal rats a choice between the bitter 0.9% NaCl solution and water, and another group of rats a choice between the hypertonic saline solution and water : equi-aversivity was considered as established if the intakes of saline solution in the two groups were approximately equal. A 0.9% NaCl solution containing 50 mg/l (0.12 mM) quinine hydrochloride or 5 g/l dry extract of gentian was found to be equi-aversive to 1.8% NaCl. A 0.9% NaCl solution containing 200 mg/l (0.48 mM) quinine hydrochloride or 20 g/l dry extract of gentian) was found to be equi-aversive to 2.7% NaCl.

Numerical results are given as means $\pm$ SEM for groups of 8-10 rats. Whenever more than one out of 10 rats died during the 10 day observation period in a given choice situation, this fact is shown in the graphs by a black cross next to the columns.- Adrenalectomy was carried out, under ether anesthesia, one week before experiments from a dorsal incision. Adrenalectomized animals and their controls were maintained on 0.9% NaCl and water from operation to experiments.

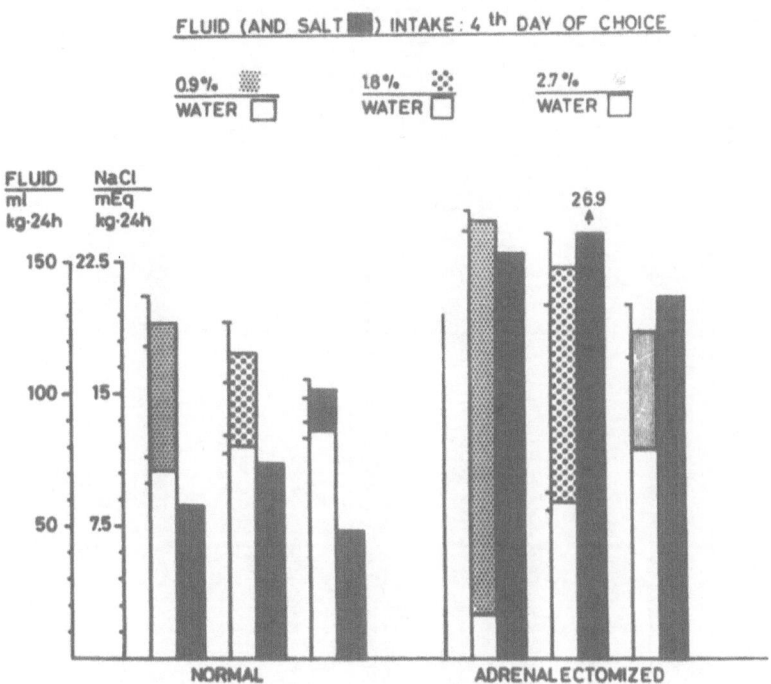

FIG. 1

RESULTS

Normal rats given a choice between 0.9%, 1.8% or 2.7% NaCl and water drank more water than saline, the relative saline intake decreasing with increasing NaCl concentration. The (superfluous) salt intake secured in drinking fluid varied between 7-10 mEq/kg · 24 h. (black columns). Adrenalectomized rats drank less water than normal controls in the same choice situation and always drank much more saline, the relative saline consumption decreasing equally with increasing NaCl concentrations. They secured a drinking fluid salt intake of 20-27 mEq/kg · 24 h. (black columns).

When given a choice between two different salt solutions (1.8%/0.9% ; 2.7%/0.9% ; 2.7%/1.8% ; 4.5%/2.7%), normal rats always drank larger amounts of the less concentrated solutions (Fig. 2). A concentrated salt solution (e.g. 1.8 NaCl) which was drunk in small amounts when offered as a choice to 0.9% NaCl was drunk in large amounts when offered as a choice to 2.7% NaCl. With increasing salt concentrations, the total saline fluid intake in normal rats increased up to 2.7%/1.8% : much smaller amounts of fluid (and salt) were drunk with a choice 4.5%/2.7%. None of the normal animals appeared to suffer or died within 10 days.

467

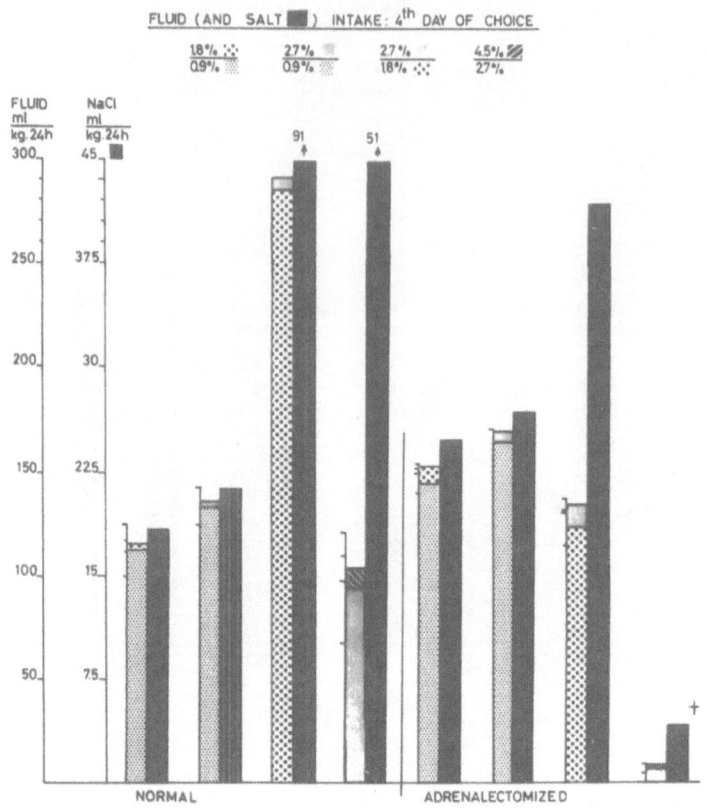

FLUID (AND SALT ■) INTAKE: 4ᵗʰ DAY OF CHOICE

FIG. 2

Adrenalectomized rats also consistently drank larger volumes of the less concentrated saline solution. With 1.8%/0.9% or 2.7%/ 0.9%, they drank larger amounts of both solutions than normal controls. However, choosing between 2.7% and 1.8% NaCl, the adrenalectomized rats drank less of both solutions than normal controls. Their motivation for salt intake, as measured by the taste aversivity which they were ready to accept, thus, appeared to be weaker than in the normal animals. In a choice between 4.5% and 2.7% NaCl, the adrenalectomized animals drank extremely small amounts of both solutions and, thus, committed suicide. The total intake of fluid and of salt became too small to permit survival : 7 out of 10 animals died within 10 days, while the survivors (not shown) slightly increased their intakes.

When quinine was added to both water and 0.9% or a 1.8% NaCl solution and the rats were given a choice between quinine containing water and a quinine containing saline, normal animals consistently drank more bitter water than bitter saline while adrenalectomized animals consistently drank much more bitter saline than bitter water. (Fig. 3). Similar results (not shown) were obtained with water and 0.9% NaCl solution made bitter by adding gentian.

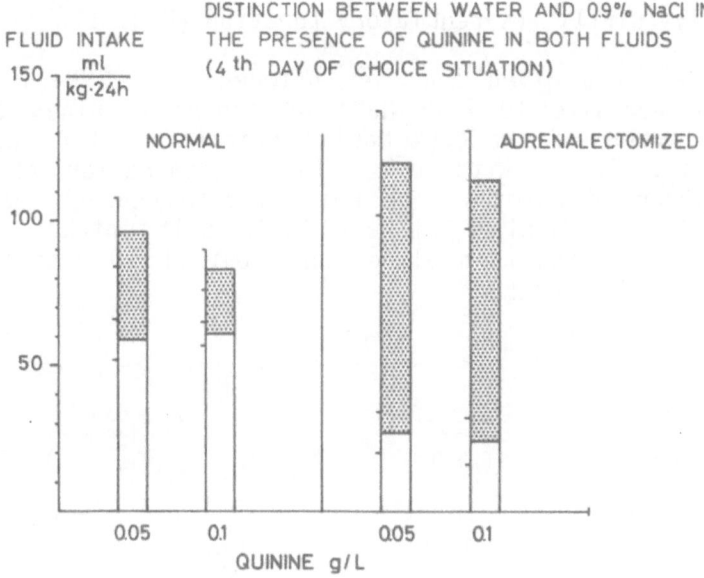

FIG. 3

Given a choice between 0.9% saline made equi-aversive to 1.8% solution by (50 mg/l) quinine and water, normal rats drank much more water than bitter saline (Fig. 4).

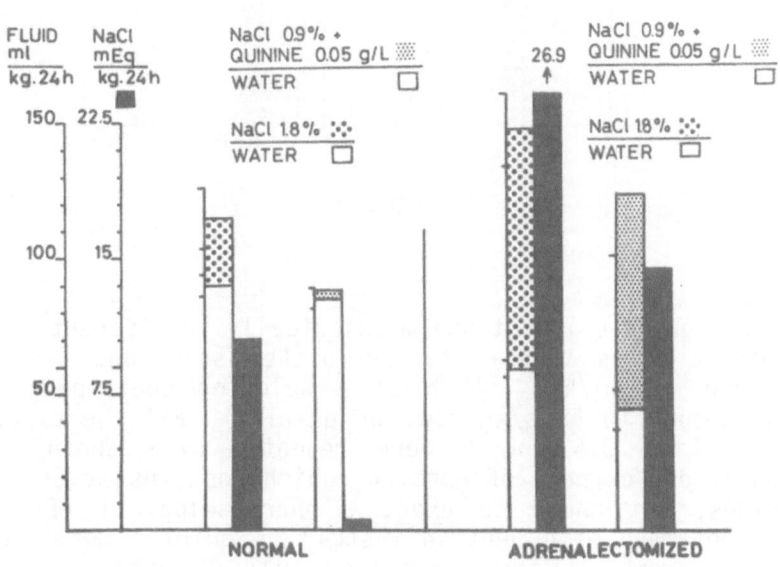

FIG. 4

Adrenalectomized rats increased their intake of NaCl- quinine solution as compared to normal controls but did not increase
it sufficiently to secure a salt intake as great as with 1.8%
NaCl solution and water. Their salt appetite,thus, failed to
completely fulfill its regulatory function (Fig. 1).

Normal rats given a choice between a 0.9% NaCl solution,
made equi-aversive to 2.7% NaCl by adding quinine, and water
drank as little of the 0.9% NaCl + quinine solution as of 2.7%
NaCl. (Fig. 5). - With this high concentration of quinine,
adrenalectomized animals were not motivated enough to increase
their saline + quinine intake : their salt intake was insufficient for keeping them alive and 5 out of 10 rats "committed
suicide" within 10 days.

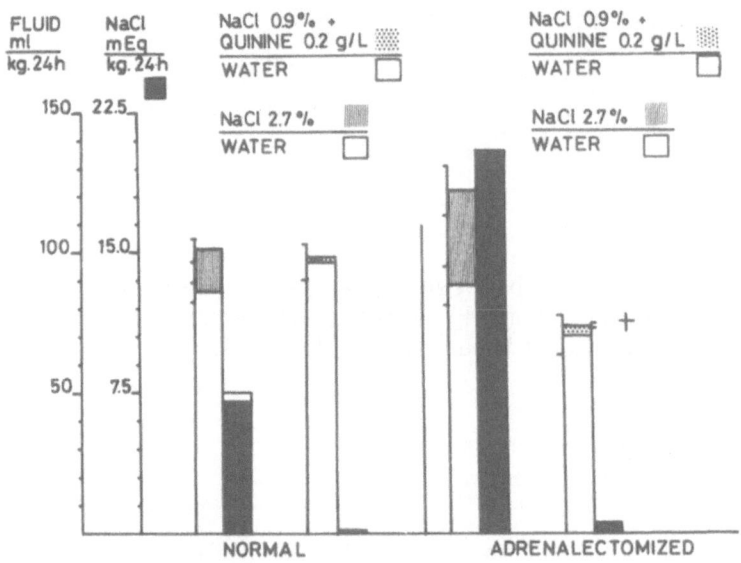

FIG. 5

Since quinine is not pharmacologically indifferent and since
the quinine doses taken with the bitter solutions were fairly
large (e.g. 50 ml/kg · 24 h of a solution containing 0.2 g/l
quinine amount to 5 mg/kg b.w. of quinine), all the experiments
shown on Fig. 3.4 and 5 were repeated with another bitter
substance, dry extract of gentian, which contains several bitter
glycosides, not known to exert a pharmacological effect. The
results obtained with gentian instead of quinine were, however,
nearly the same as those obtained with quinine (not shown).

DISCUSSION

When given a choice between water and moderately aversive concentrated NaCl solutions, or between two salt solutions, or between saline made aversive by adding quinine or gentian, adrenalectomized animals are <u>motivated</u> to increase their saline intake to such an extent that they secure a much larger salt intake than normal rats. Their motivation to ingest salt, however, is weaker than that of normal animals since most adrenalectomized rats are not prepared to drink sufficient amounts of highly aversive concentrated saline or bitter saline solutions to stay alive. With a choice between two highly aversive concentrated NaCl solutions, normal rats overcome their aversion against their taste to a sufficient extent to stay alive. The motivated salt appetite of adrenalectomized rats could also be demonstrated when they were given a choice between water and isotonic saline solutions made equi-aversive to hypertonic NaCl solutions which adrenalectomized animals chose in larger amounts than normal aninmals ; the salt intake secured, however, was smaller than with a choice between water and the corresponding concentrated NaCl solution. The salt appetite, thus, did not actually fulfill its hypothetical regulatory function.

ABSTRACT

In choosing between water and 0.9%-2.7% NaCl solutions, adrenalectomized animals drank increased amounts of aversive concentrated saline and thus secured a greater salt intake. Choosing between 0.9/1.8% NaCl, or 0.9%/2.7%, or 1.8%/2.7%, or 2.7%/4.5% NaCl, normal and adrenalectomized rats drank more of the less concentrated salt solution, and drank more of a given aversive NaCl solution when it was offered as a choice against a more concentrated than when it was given as a choice to a less concentrated NaCl solution. Normal rats tremendously increased their intake of saline solutions up to 1.8%/2.7% NaCl, but also drank fairly great amounts of 2.7%/4.5% NaCl. Adrenalectomized animals drank less than normal rats with 1.8%/2.7% NaCl. Most of them completely refused 2.7%/4.5% NaCl and, thus, committed suicide. - 0.9% NaCl solution made equi-aversive to 1.8% NaCl by adding quinine or extract of gentian, and offered as a choice against water, was ingested at increased amounts by adrenalectomized rats securing an increased salt-intake which, however, was smaller than with 1.8% NaCl. A 0.9% NaCl solution made equi-aversive to 2.7% NaCl by adding the bitters, however, was not drunk by most adrenalectomized animals, a number of which committed suicide by this refusal. - The increased saline intake of adrenalectomized rats was motivated but not strictly regulatory. The strength of the motivation for salt intake in the adrenalectomized animals, however, was smaller than in normal controls.

REFERENCES

1. J. K. Bare, The specific hunger for sodium chloride in normal and adenalectomized white rats. J.Comp.Physiol. Psychol., 42, 242-249 (1949).
2. W. G. Clark, D.F. Clausen, Dietary "self-selection" and appetites of untreated and treated adrenalectomized rats. Am.J.Physiol., 139, 70-79 (1943)
3. A. N. Epstein, E. Stellar, The control of self-preference in the adrenalectomized rat. J.Comp.Physiol.Psychol., 48, 167-181, (1955).
4. G. Peters, Aufnahme und renale Ausscheidung von Wasser und Salzen bei freiem Nahrungs- und Trinkflüssigkeits-angebot : ihre Beeinflussung durch Adrenalektomie und Behandlung mit Nebennierenrindenhormonen bei der Ratte.- Arch. exp. Path. u. Pharmakol., 235, 205-229, (1959).
5. G. Peters, Nebennierenrinden-Inkretion und Wasser-Elek-trolythaushalt. Leipzig, VEB Georg Thieme-Verlag, (1960). Pg. 122-127 for references.
6. C. P. Richter, J.F. Eckert, Mineral metabolism of adren-alectomized rats studied by the appetite method. Endocrinology, 22, 214-225, (1938)
7. G. Wolf, J. F. McGovern, L.V. Di Cara, Sodium appetite : some conceptual and methodologic aspects of a model drive system. Behavioral Biol., 10, 27-35, (1974).

THE EFFECT OF LOCAL CHANGE IN CSF [Na] IN THE ANTERIOR THIRD VENTRICLE ON SALT APPETITE

E. Tarjan, P. Cox, D.A. Denton, M.J. McKinley and
and R.S. Weisinger

Howard Florey Institute of Experimental Physiology and
Medicine, University of Melbourne
Parkville, Victoria 3052, Australia

The physiological changes inducing salt appetite or the satiation of appetite after ingestion have been investigated in our laboratory for many years. It has been reported that this complex behaviour may be significantly altered by experimental manipulations of the cerebrospinal fluid (CSF) [Na] in sheep. When the CSF [Na] was increased by infusion of hyperosmotic artifical CSF containing 0.5 M NaCl into the lateral cerebral ventricle the salt intake of sodium deficient sheep decreased by about 70 per cent. The CSF [Na] was found to be 162 mM after 1 hour of infusion. Infusion of 0.7 M mannitol-CSF into the lateral cerebral ventricle decreased the CSF [Na] to 134 mM. Sodium intake of salt deficient sheep increased about twofold during the infusion of mannitol[1]. It was postulated that sensors in the brain which are accessible from the CSF and some distance from the ventricular wall monitor the [Na] of the brain extracellular fluid (ECF) and modify in turn the salt intake of the sodium deficient animal.

The aim of the present studies was to investigate whether localized changes in the CSF [Na] would lead to differential changes in the salt intake of sodium deficient sheep. We attempted to localize an area in the anterior third ventricle of sodium deficient sheep which when irrigated with solutions of different [Na] would induce a change in the sodium intake of sodium deficient sheep more effectively than other areas.

The experiments were performed in ovariectomized ewes with a parotid fistula. The animals were surgically prepared with guide tubes above the anterior third ventricle. They lost daily 1.5-3 l of sodium rich saliva. The animals were trained to press two pedals in their cages: one delivered 50 ml of water into a drinking cup at each press, the other pedal delivered 15 ml of 0.6 M $NaHCO_3$ into another cup. Water was continuously available, while sodium was available for two hours daily. All deliveries were consumed.

On the day of the experiment a blocker was removed from the guide tube and a push-pull cannula was inserted into the anterior third ventricle. The push-pull cannula was made so that the internal needle was 1 mm longer than the external one. Solutions were infused through the inner, thin needle and CSF-solution mixture was withdrawn via the shorter outer needle, both at a rate of 1.2 ml/h. The withdrawal of infused solution was not complete: ca 20-30 per cent of the infusate leaked into the CSF and mixed into the third ventricular CSF. There was, however, a volume of infusate around the tip of the infusion needle. Localized perfusion of normal, 150 mM Na-CSF, hyperosmotic 200 mM Na-CSF and hyposmotic 100 mM Na-CSF was carried out in two sites in the anterior third ventricle: dorsal (AD3V), at or above the level of anterior commissure and ventral (AV3V), 4-7 mm below the dorsal perfusion site, at about the level of the optic recess. Perfusion started one hour before the access to $NaHCO_3$ and stopped after two hours of perfusion. The animals thus had two hours access to sodium and in the second hour there was no perfusion.

Statistical analysis of cumulative numbers of deliveries between two baseline (no perfusion) days and experimental days was performed by two-way analysis of variance (repeated measures design) with subsequent t-tests.

Localized perfusion of 150 mM Na-CSF did not alter the salt intake of sodium deficient sheep neither at AD3V nor AV3V perfusion sites. Perfusion with 200 mM Na-CSF in the AV3V reduced the sodium intake of sodium deficient sheep during the first hour of sodium access relative to control, and intake increased during the second hour of access so that final intake was close to baseline levels. Perfusion with 200 mM Na-CSF in the AV3V halved the sodium intake during the first hour of access and total intake remained lower than during the previous baseline days (Fig.1). Perfusion with 100 mM Na-CSF in the AD3V caused an increase in salt intake during the second hour of bar pressing, while the same concentration perfused in the AV3V failed to alter sodium intake (Fig.2).

Infusion of both hyperosmotic 0.7 M mannitol-CSF and 0.27-0.3 M mannitol-CSF solutions into the lateral ventricle of sodium deficient sheep has been shown to increase the salt intake[1]. For localized perfusion we used 0.3 M mannitol solution in artificial CSF, the [Na] of which was 22 mM. Perfusion of 0.3 M mannitol-CSF in the AD3V tended to cause a small increase in sodium intake during the first hour of sodium access. Perfusion of 0.3 M mannitol-CSF in the AV3V resulted in an increase in the salt intake during the first hour (Fig. 3).

474

In a few animals the effect of localized perfusion with 10^{-6} M ouabain and 10^{-5} M monensin was tested. Both drugs reduced the salt intake of sodium deficient sheep at AD3V and AV3V perfusions (Fig.4)

The results of these experiments support the hypothesis that there are sensors located around the anterior third ventricle in the brain of the sheep which are accessible from the CSF and which will modify the salt intake of sodium deficient sheep. The effect of ouabain, a Na-K-ATP-ase blocker, and monensin, a monovalent ionophore, suggest that the parameter to which the sensors reacted was the intracellular [Na], and increase of intracellular [Na] acted to decrease the sodium intake of sodium deficient sheep.

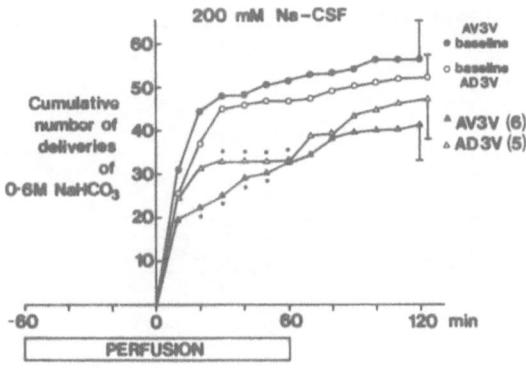

Fig. 1.   Cumulative number of deliveries of 0.6 M NaHCO$_3$ drunk during and after perfusion with 200 mM Na-CSF. Cumulative mean ± SE of deliveries for the baseline and the experimental days are shown (number of animals).

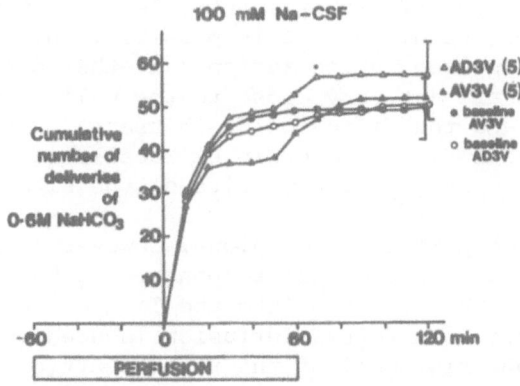

Fig. 2.   Cumulative number of deliveries of 0.6 M NaHCO$_3$ drunk during and after perfusion with 100 mM Na-CSF.

475

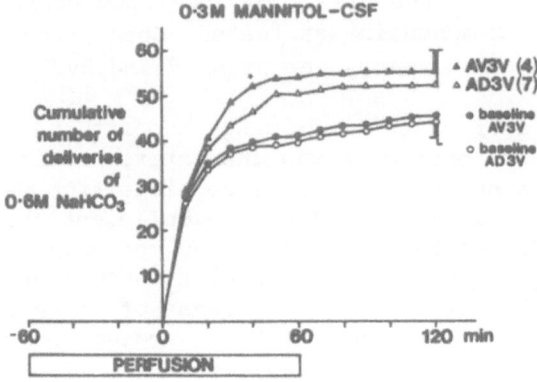

Fig. 3. Cumulative number of deliveries of 0.6 M NaHCO₃ drunk during and after perfusion with 0.3 M mannitol-CSF.

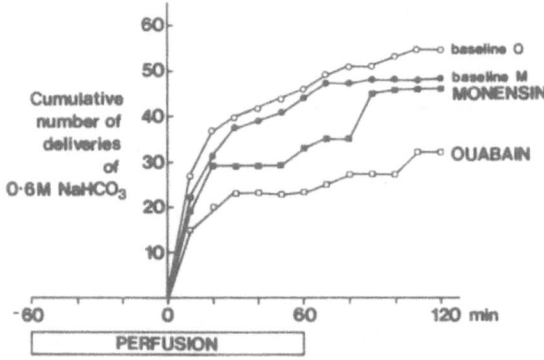

Fig. 4. Cumulative number of deliveries of 0.6 M NaHCO₃ drunk during and after AD3V perfusion with 10^{-6} M ouabain, and 10^{-5} M monensin in one animal.

Localization of the sensors, however, to a discrete area in the region of the anterior third ventricle was not possible with this experimental technique. It is possible, that these sensors are diffusely dispersed in the region. Another possibility is that the sensors are not very close to the wall of the anterior third ventricle and the diffusion of infusate, or the changes caused by the infusate in the brain extracellular fluid, are equal at the site of the sensor irrespective of the site of perfusion.

The localized perfusion technique, however, was suitable for the localization of a biological response. It has been reported that hyperosmotic NaCl infused into the lateral cerebral ventricle, or used in ventriculocisternal perfusion induced natriuresis[2,3]. Lesions of the anterior wall of the third ventricle impaired the natriuretic response to intraventricular administration of hyperosmotic NaCl[4,5]. On the basis of these data in another set of experiments in sodium replete, non-fistulated sheep, the changes in water and sodium excretion were investigated before, during and

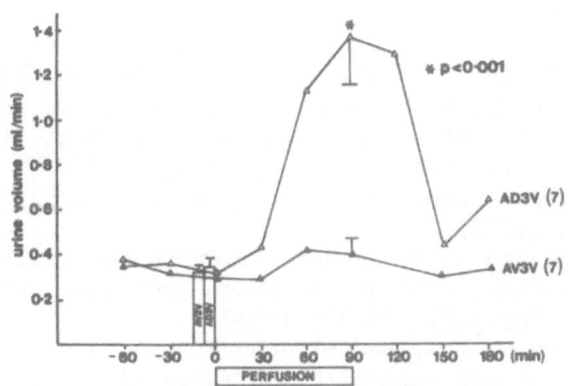

Fig. 5. The effect of localized perfusion of 200 mM Na-CSF in the
AD3V and AV3V on urine volume (number of animals).
Statistical analysis: two-way analysis of variance with
subsequent t-test.

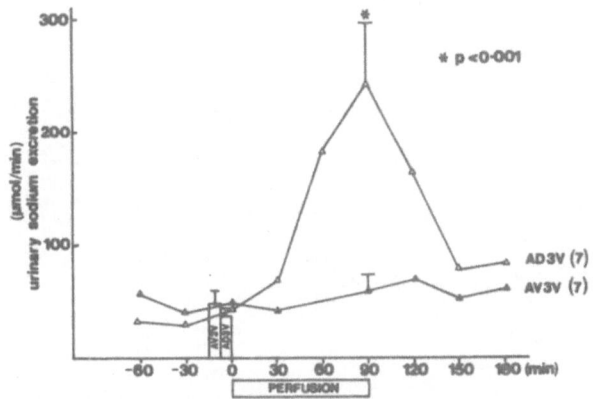

Fig. 6. The effect of localized perfusion of 200 mM Na-CSF in the
AD3V and AV3V on urinary sodium excretion (number of
animals).

after localized perfusions in the AD3V, AV3V, and the posterior
dorsal and ventral third ventricle with 200 and 150 mM Na-CSF.
Perfusion with 200 mM Na-CSF in the AD3V resulted in a striking
increase in urine volume, while perfusion in the AV3V did not
change the diuresis (Fig.5). The increase in urine volume during
perfusion with 200 mM Na-CSF in the AD3V coincided with an increase
in urinary sodium excretion. Perfusion with 200 mM Na-CSF in the
AV3V did not influence urinary sodium excretion (Fig.6). Perfusions
in the posterior III ventricle with 200 mM Na-CSF did not alter the
urine flow or the urinary sodium excretion. Perfusion of isomotic
150 mM Na-CSF did not influence urinary water and sodium excretion
at either perfusion sites.

REFERENCES

1. R. S. Weisinger, P. Considine, D. A. Denton, L. Leksell, M. J. McKinley, D. R. Mouw, A. F. Muller and E. Tarjan, Role of sodium concentration of the cerebrospinal fluid in the salt appetite of sheep, Am. J. Physiol. 242:R51 (1982).
2. B. Andersson, Regulation of body fluids, Annu. Rev. Physiol. 39:185 (1977).
3. D. R. Mouw and A. J. Vander, Evidence for brain Na receptors controlling renal Na excretion and plasma renin activity, Am. J. Physiol. 219:822 (1970).
4. B. Andersson, L. G. Leksell and F. Lishajko, Perturbations in fluid balance induced by medially placed forebrain lesions, Brain Res. 99:261 (1975).
5. S. L. Bealer, Impaired natriuresis following intraventricular administration of hypertonic saline solutions in rats with lesions of the anterioventral third ventricle (AV3V) region, Federation Proc. 39:1086 (1980).

PERIPHERAL GUSTATORY MECHANISMS OF SALT INTAKE IN THE RAT

Robert J. Contreras and Therese Kosten

Yale University
Department of Psychology
Box 11A Yale Station
New Haven, CT 06520

The focus of this research is to determine the nature of the contribution of taste sensitivity in controlling salt intake in the rat. We know that salt taste sensitivity is important to the salt intake of the rat for several reasons. First, the normal preferences for salt are reduced after severing the taste nerves[1] or after gustatory thalamic lesions[2]; moreover, denervation of taste afferents in sodium deficient adrenalectomized rats causes them to not increase their salt intake which ultimately leads to their death[3]. Second, rats select sodium solutions the first time they become sodium deficient, and they do so immediately upon tasting the salt[4]. Third, this sodium appetite is triggered by sodium ions as rats chose sodium salts over nonsodium salt and other taste compounds; lithium salts are the only exception as sodium deficient rats can be fooled into consuming the toxic LiCl simply because it tastes similar to NaCl[4]. Fourth, the rat has taste receptors and fibers that are keenly sensitive to sodium (and lithium) salts; these receptors and fibers evolved presumably to protect the animal from sodium deficiency[5].

Salt taste sensitivity is important for the initiation of salt intake by providing a recognition system for the identification of the stimulus as tasting 'salty'. This particular sensory input from the taste receptors is believed to be inextricably bound to a brain motivational system which automatically adds a hedonic quality once the stimulus' representation enters the brain. The salty stimulus at moderate concentrations is innately pleasant and is therefore a motivating force in consumption. Thus, taste provides the animal with a mechanism for locating the nutrient and for encouraging consumption.

As summarized previously[6], there are at least three reasons why our studies are focused on the peripheral taste afferent neurons. First, we are able to assess the contribution of taste alone without the influences of converging inputs from trigeminal and visceral sensory neurons and from both local and distal influences from within the brain as found even in the first central relay nucleus for taste. Second, we assume that sensory receptors in general are not passive receptacles for incoming information, but rather are dynamic and can be influenced by the changing physiological state of the organism. Third, the effects of one synaptic event and convergence are already present by the time we assess the functional properties of taste afferents. This complexity allows for a greater potential of plasticity due to changes in the physiological state of the organism.

In our studies of peripheral mechanisms mediating changes in salt intake, our preparations are examined after being fed a sodium deficient diet for 10 days or after removal of the adrenal glands; these conditions enhance the salt intake of the behaving rat. Our neurophysiological studies are of the chorda tympani nerve, which innervates the taste receptors on the tongue. Of the three cranial nerves that innervate receptors, the chorda tympani (a branch of the facial nerve) has the greatest salt sensitivity. We have obtained both whole nerve recordings from the entire chorda tympani nerve as well as from single axons of that nerve in normal sodium replete and sodium deprived rats.

In our earlier work[7], we showed that the whole nerve responses of the chorda tympani to NaCl stimulation of the tongue are reduced as a result of sodium deprivation. This reduction in responsivity occurred only to suprathreshold NaCl concentrations; neural thresholds for NaCl stimulation were the same for sodium deprived and sodium replete animals. Furthermore, the whole nerve responses to sucrose, quinine hydrochloride, and hydrochloric acid were similar for sodium deprived and control animals.

These changes in neural responsivity of the chorda tympani suggest that suprathreshold NaCl solutions taste weaker to sodium deprived rats compared to the taste intensity sensed by normal sodium replete rats. Reductions in suprathreshold taste intensity for NaCl would have obvious consequences on the animals' salt intake because NaCl intake varies with concentration. We have therefore determined what the control animals' salt intake would be if they had a reduced neural signal as well as the lowered salivary sodium level similar to that seen following sodium deprivation[6]. These determined values agree fairly well with the actual intakes of sodium deprived animals.

480

The whole nerve response of the chorda tympani is composed of the firing frequencies of two major types of single units[8]. The H-best unit is a generalist, responding most to HCl, but also responding to NaCl and to quinine hydrochloride. The N-best unit is a specialist responding specifically to NaCl and little, if at all, to the other stimuli representing the three remaining basic tastes. We have suggested that N-best units may be mainly responsible for transmitting the salty quality elicited by NaCl solutions from the receptors to the brain[8]. Therefore, the responses of N-best units, but not H-best units, should be influenced by sodium deprivation.

We have shown that N-best unit responses, like those of the whole nerve, are reduced to NaCl stimulation in sodium deprived rats[6,7]. But the responses of H-best units to NaCl stimulation are unaffected by sodium deprivation. Although both N-best and H-best units respond to NaCl, these data suggest that the effect of sodium deprivation on whole nerve responses to NaCl is mediated only through N-best unit responses. Two possible implications from these findings are that the receptors innervated by N-best units have different structural and functional properties than the receptors innervated by H-best units.

We have used a Beidler[6] plot analysis to show that the stimulus-receptor interactions contacted by N-best units seem to differ in number from those contacted by H-best units. A Beidler plot is useful in examining the interaction between the stimulus (sodium) molecules and the receptor membrane. Linear regression analysis yields straight line plots of NaCl responses in N-best and H-best units when the ratio of concentration over response is plotted as a function of concentration. As a result of this analysis, the data indicate that under normal conditions stimulus-receptor interactions of N-best units differ from those of H-best units. Application of the same analysis to the responses of units from sodium deprived rats reveals that receptor mechanisms of N-best units are altered, but not those of H-best units. On the basis of Beidler's model, the effect of sodium deprivation is to either reduce the effectiveness of the interaction between NaCl and its receptor sites or the number of receptor sites for NaCl.

The apparent differential effects of sodium deprivation on N-best units encouraged us to examine further the contribution of receptor mechanisms. Our interest is to determine whether the salt taste receptor is a specific sodium receptor or a general salt receptor which is sensitive to both sodium and nonsodium salts. If the salt taste receptor along with the taste fibers are important in mediating changes in salt intake, then we might expect the receptor to be a specific sodium receptor because sodium deficient rats select sodium over nonsodium salts. And in fact the little data that we have obtained so far have confirmed this expectation, although more research is needed. We have recorded the responses from three N-best units and three H-best units to show that LiCl activates N-best and H-best units to the degree as does NaCl, while KCl activates only H-best units[8].

Another way of addressing this same question is to apply the specific sodium transport blocker, amiloride, to the taste receptor membrane and examine amiloride's effects on the rats' neural taste responses and salt intake. So far, we have examined only the effects of amiloride on the whole nerve responses of the chorda tympani. We have confirmed the findings of two recent reports showing that NaCl and LiCl responses are reduced after amiloride application, but KCl responses are unaltered[9,10]. Thus, it would appear that the salt taste receptor is a specific sodium (LiCl) receptor. We intend to examine in the near future the effects of amiloride on N-best and H-best unit responses. Amiloride might be expected to have greater effects on NaCl responses of N-best units than of H-best units. We also intend to examine the effects of amiloride on the salt intake of the rat. If the salt taste receptor is indeed important in mediating changes in intake, which varies with NaCl concentration (intensity), then we should be able to mimic the effects of salt deprivation with amiloride application on the tongue prior to an intake test.

ACKNOWLEDGEMENTS

We thank Nancy Moses Cobb for her adept technical assistance. This research was supported by the National Institute of Health Grant HL-28952.

REFERENCES

1. C. Pfaffmann, Taste preference and aversion following lingual denervation, J. Comp. Physiol. Psychol. 45:393 (1952).
2. B. Oakley and C. Pfaffmann, Electrophysiologically monitored lesions in the gustatory thalamic relay in the albino rat, J. Comp. Physiol. Psychol. 55:155 (1962).
3. C. P. Richter, Salt appetite in mammals: its dependence on instinct and metabolism, in: "L'instinct dans le comportement des animaux et de l'homme," Masson et Cie., Paris (1956).
4. M. Nachman and L. P. Cole, Role of taste in specific hungers, in: "Handbook of Sensory Phusiology," L. Beidler, ed., Springer-Verlag, New York (1971).
5. D. A. Denton, "The Hunger for Salt. An Anthropological, Physiological and Medical Analysis," Springer-Verlag, Berlin (1982).
6. R. J. Contreras, T. Kosten, and M. E. Frank, Activity in salt taste fibers: peripheral mechanism for mediating changes in

salt intake, Chem. Senses, 8:275 (1984).

7. R. J. Contreras and M. E. Frank, Sodium deprivation alters neural responses to gustatory stimuli, J. Gen. Physiol. 73:569 (1979).

8. M. E. Frank, R. J. Contreras, and T. P. Hettinger, Nerve fibers sensitive to ionic taste stimuli in chorda tympani of the rat, J. Neurophysiol. 50:941 (1983).

9. G. L. Heck, S. Mierson, and J. A. DeSimone, Salt taste transduction occurs through an amiloride-sensitive sodium transport pathway, Science, 223:403 (1984).

10. J. G. Brand, J. H. Teeter, and W. L. Silver, Effect of amiloride concentration on reduction of chorda tympani responses to alkali chlorides, Assoc. Chemoreception Sci. Abst. 6:27 (1984).

CEREBRAL Na SENSORS AND Na APPETITE OF SHEEP

R.S. Weisinger, D.A. Denton, M.J. McKinley, J.B. Simpson and E. Tarjan

Howard Florey Institute of Experimental Physiology and Medicine, University of Melbourne, Parkville Victoria, 3052, Australia

Previous research[4] has shown that the Na intake of sheep can be altered by lateral ventricular (IVT) infusions which alter the cerebrospinal fluid (CSF) Na concentration. For example, infusion of 0.7 M mannitol-CSF decreased CSF [Na] by 15-20 mM and increased the Na intake of both Na-deplete and Na-replete sheep. IVT infusion of 0.5 M NaCl-CSF increased the CSF [Na] by 15-20 mM and decreased Na intake of Na-deplete sheep.

The Na appetite of Na-deplete sheep is decreased by intracarotid (IC) infusion of 4 M NaCl. This infusion increases plasma and CSF [Na] by 15-20 mM and replaces 40-65% of Na deficit. IVT infusion of 0.7 M mannitol-CSF prevented the increase in CSF [Na] and the decrease in Na intake observed during IC infusion of 4 M NaCl[3]. Thus, the inhibition or 'turn-off' of Na appetite of Na-deplete sheep by systemic infusion of hypertonic NaCl seems to be mediated, at least in part, by cerebral Na sensors.

The present experiments investigated the role of cerebral Na sensors in the hedonic Na intake of Na-replete sheep and in sheep where Na intake is stimulated by IVT infusion of angiotensin II (AII). In these experiments, the effect on Na intake of IVT or IC infusion of hypertonic NaCl was evaluated. In addition, the effect on Na intake of increasing CSF osmolality without changing CSF [Na] was evaluated in both Na-deplete and Na-replete sheep. In these experiments, IVT infusion of 0.34 M mannitol-0.33 M NaCl was used.

METHODS

Animals   Thirty-one crossbred Merino ewes, 30-40 kg body weight were used. Prior to experimentation, all animals were surgically prepared with a guide tube (17 g stainless steel needle 42 mm long) implanted 6-10 mm above each lateral brain ventricle. The sheep were ovariectomized and had each carotid artery exteriorized in a skin loop. In addition, 20 of the sheep had a unilateral parotid fistula so that NaHCO3-rich saliva was continually lost (daily Na loss = 300-500 mmol). The sheep were maintained in metabolism cages which contained two pedals. The animals were trained to press one pedal in order to obtain delivery of Na to a drinking cup (sheep with a parotid fistula: Na delivery=15 ml of 0.6 M NaHCO3=9 mmol; sheep without a parotid fistula: Na delivery=25ml of 0.5 M NaCl=12.5 mmol) and a second pedal in order to obtain delivery of water (50 ml/ delivery) to a second drinking cup. The animals had 2-h daily access to Na (1200-1400 h) and continuous access to water. All deliveries were consumed. The number of deliveries was counted and recorded continuously by computer (IBM Series 1).

Infusions. IVT infusion (1 ml/h) of 0.5 M NaCl-CSF or 0.34 mannitol-0.33 M NaCl-CSF began 1-h prior to Na access and continued for 3 h, i.e., until the end of the Na access period. The IVT infusion of angiotensin II (AII) at 3.8 µg/h began at the end of a Na access period and continued for 24 or 48 h (i.e., until the end of the next one or two Na access periods). When IVT infusion of 0.5 M NaCl-CSF was given with IVT infusion of AII, the infusion of 0.5 M NaCl was into the opposite lateral ventricle to that being infused with AII and was started 21 h after the start of the infusion of AII. IC infusion (1.6 ml/min) of 4 M NaCl began 10 minutes prior to Na access and continued for 30 minutes. The contralateral carotid artery was occluded. When IC infusion of 4 M NaCl was given with IVT infusion of AII, the infusion of 4 M NaCl was started 21 h and 50 minutes after the start of the infusion of AII.

Data Analysis. Two way analysis of variance (repeated measures design) and subsequent t-tests were used to compare baseline and experimental values.

RESULTS

Experiment 1   Na-depletion induced Na intake

In Na-deplete sheep, water intake was increased while Na intake was not changed during IVT infusion of 0.34 M mannitol-0.33 M NaCl-CSF (Figure 1). CSF [Na] was unchanged but CSF osmolality was increased from 298 ± 1 to 343 ± 9 mosmol/kg ($P<0.001$) by this infusion.

Experiment 2 Hedonic Na Intake

In sheep with high baseline Na intake, Na intake was greatly decreased by IVT infusion of 0.5 M NaCl-CSF or IC infusion of 4 M NaCl (Figure 2) and, similar to Na-deplete sheep, was not changed by IVT infusion of 0.34 M mannitol-0.33 M NaCl CSF (Figure 3).

486

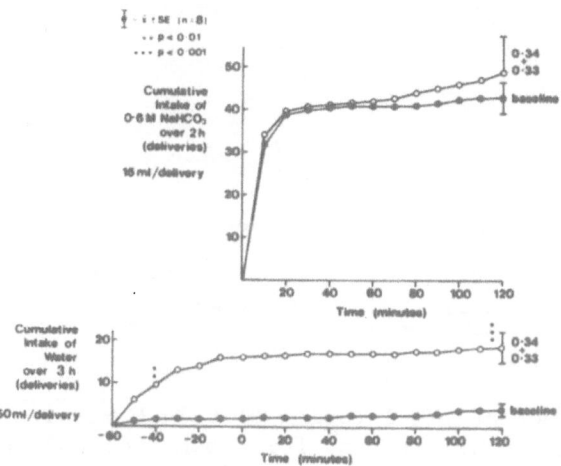

Figure 1. Effect of IVT infusion of 0.34 M mannitol-0.33 M NaCl-CSF
on bar pressing for 0.6 M NaHCO3 and water. The sheep
were Na-deplete as a result of 22 h parotid salivary
loss. The 3-h infusion started 1-h prior to 2-h access
to Na bar press. Cumulative mean (and SE) deliveries for
0.6 M NaHCO3 and water on the baseline (i.e. no infusion
day preceding the experimental day) and the IVT infusion
days are shown. First interval at which a significant
difference occurs and significance at last interval are
shown.

Water intake was increased during IVT infusion of 0.5 NaCl-CSF, IC
infusion of 4 M NaCl (Figure 2) or IVT infusion of 0.34 M mannitol-
0.33 M NaCl-CSF (Figure 3).

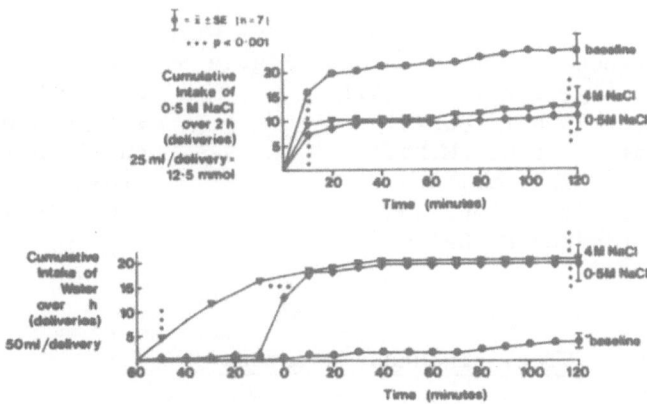

Figure 2. Effect of IVT infusion of 0.5 M NaCl-CSF or IC infusion
of 4 M NaCl on bar pressing for 0.5 M NaCl and water.
The sheep were selected because of their high, need free
intake of Na.

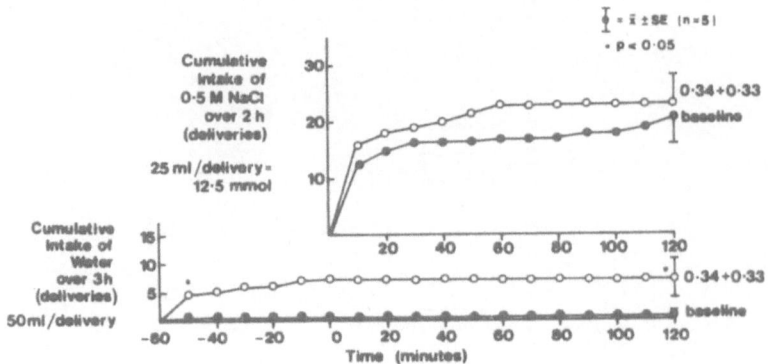

Figure 3. Effect of IVT infusion of 0.34 M mannitol-0.33 M NaCl-
CSF on bar pressing for 0.5 M NaCl and water in Na-replete
sheep.

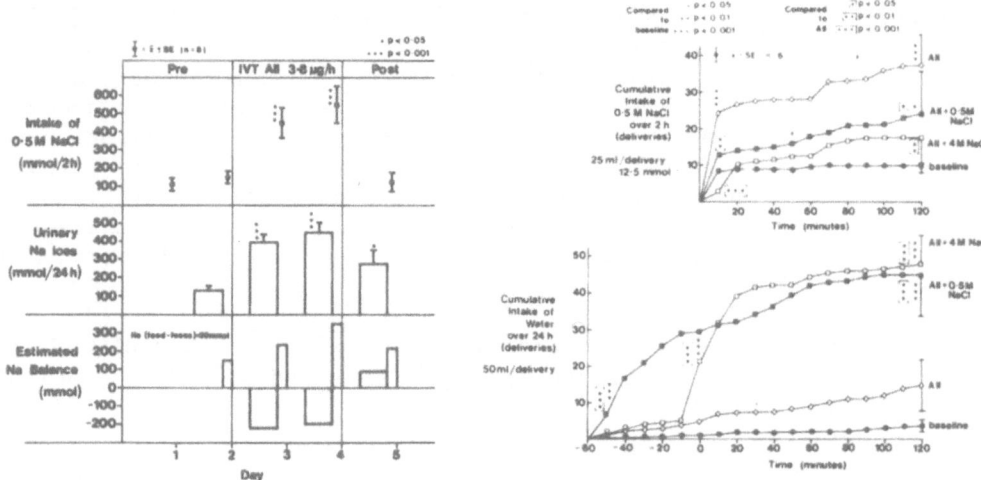

Figure 4. Effect of IVT infusion
of angiotensin II (AII)
over 48 h (3.8µg/h) on
intake of 0.5 M NaCl
(mmol/2h), urinary Na
loss (mmol/24h) and
estimated Na balance
(mmol)

Figure 5. Effect of IVT infusion
of 0.5 M NaCl or IC
infusion of 4 M NaCl
on IVT AII-induced bar
pressing for 0.5 M
NaCl and water.

Experiment 3   IVT AII induced Na intake

Figure 4 shows that IVT AII (3.8 µg/h) caused a large increase
in Na loss prior to a large increase in Na intake. That is, a large
Na deficit (e.g., 200-220 mmol) preceded each Na access period
during infusion. Na intake, however, greatly exceeded the prior Na
loss and at the end of each Na access period during infusion a

488

positive Na balance (e.g., 220–250 mmol) was observed. Figure 5 shows that IVT AII (3.8μg/h) induced Na intake was decreased by IVT or IC infusion of hypertonic NaCl.

IVT AII increased water intake within the first 10 minutes of infusion. Mean 2-hrly and cumulative water intake remained elevated over baseline during the entire period of IVT AII infusion. Water intake was elevated further by IVT or IC infusion of hypertonic NaCl.

DISCUSSION

Previous observations[4,5], utilizing, in the main, hypersomotic infusates (i.e., osmolality=1000–1050 mosmol/kg), demonstrated that the Na intake of Na-deplete sheep is increased by IVT infusions which decrease CSF and, presumably, brain ECF [Na]. The Na intake of Na-deplete sheep is decreased by IVT infusions which increase CSF and brain ECF [Na][4]. In the present experiments, Na intake of Na-deplete sheep was not changed by IVT infusion of hypertonic mannitol-hypertonic NaCl-CSF which increased osmolality without altering [Na] of CSF and brain ECF. Thus, altered [Na] and not osmolality of CSF and brain ECF is involved in the regulation of Na appetite of Na-deplete sheep. Na intake of Na-deplete sheep is decreased by IC infusion of 4 M NaCl and the effect of this infusion on Na intake seems to be mediated, in part, via its effects on brain ECF [Na][3]. For both hedonically-determined and IVT AII-stimulated Na intake, IVT 0.5 M NaCl-CSF and IC 4 M NaCl were used to evaluate the role of cerebral Na sensors.

Similar to the results obtained in Na-deplete sheep, the Na intake of Na-replete sheep with a high, need-free intake of Na was decreased by IVT or IC infusion of hypertonic NaCl but not by IVT hypertonic mannitol-hypertonic NaCl-CSF. Previously, it has been shown that the Na intake of Na-replete sheep is increased by IVT infusion of hypertonic or isotonic mannitol solutions which decrease CSF and brain ECF [Na][4]. Thus, it would appear that both taste factors and changes in brain ECF [Na] are involved in determining the Na intake of Na-replete sheep.

IVT infusion of AII caused a large increase in Na intake. The increase in intake was preceded by increased Na loss. Given that the increased Na intake exceeded the increased Na loss, it would appear that AII-induced Na intake may be in part, but not entirely due to Na loss. The preliminary evidence obtained showed that AII-induced Na intake is decreased by IVT or IC infusion of hypertonic NaCl. Thus, changes in brain ECF [Na] may be involved in the Na appetite induced by IVT AII.

In regard to water intake, IVT infusions which increase CSF and brain ECF osmolality with decreased[4,5] or normal [Na][1,] present experiments, caused increased water intake. However, the increased water intake produced under these circumstances is not as consistent nor as large as that produced by IVT infusions which increase both [Na] and osmolality. Thus, in contrast to Na appetite, both osmoreceptors and cerebral Na sensors seem to be involved in thirst as previously reported[2].

In conclusion, at present the evidence suggests that cerebral Na sensors are involved in the Na appetite induced by Na-depletion, IVT infusion of AII and that observed in Na-replete sheep in the absence of any need.

ACKNOWLEDGEMENTS

This work was supported by the National Health and Medical Research Council of Australia. I would like to thank Brett Purcell, Don Reilly and Susan Eason for their skilled technical assistance.

REFERENCES

1. M. J. McKinley, P. Considine, D. A. Denton, L. G. Leksell, E. Tarjan and R. S. Weisinger, Cerebral sodium sensors, thirst and sodium balance in sheep, Adv. Physiol. Sci. 11: 637 (1981).
2. M. J. McKinley, D. A. Denton and R. S. Weisinger, Sensors for antidiuresis and thirst-osmoreceptors or CSF sodium detectors Brain Res. 141: 89 (1978)
3. A. Muller, D. A. Denton, M. J. McKinley, E. Tarjan and R. S. Weisinger, Lowered cerebrospinal fluid sodium antagonizes effect of raised blood sodium on salt appetite, Am. J. Physiol. 244: R810 (1983).
4. R. S. Weisinger, P. Considine, D. A. Denton, L. G. Leksell, M. J. McKinley, D. Mouw, A. Muller and E. Tarjan, Role of sodium concentration of the cerebrospinal fluid in the salt appetite of sheep, Am. J. Physiol. 242: R51 (1982).
5. R. S. Weisinger, D. A. Denton, M. J. McKinley, A. Muller and E. Tarjan, Cerebrospinal fluid sodium concentration and sodium appetite, Appetite, 4: 239 (1983).

THE EVOLUTION AND EXPRESSION OF SALT APPETITE

Jay Schulkin

Institute of Neurological Sciences
University of Pennsylvania
Philadelphia, Pennsylvania 19104

Bunge some time ago suggested that herbivores, rather than carnivores, are under selective pressure to show an appetite for salt; he thought the appetite was for potassium and not sodium salts[1]. More recently Denton has substantively elaborated on the notion that there are selective pressures on herbivores for special behavioral salt homeostatic mechanisms; but the appetite is geared towards sodium and not potassium. In elegant ecological studies it has been shown that rabbits and kangaroos emerge from areas of depleted sodium resources to ingest salt from implanted salt licks elsewhere. Moreover, many kinds of herbivorous mammals are known to migrate great distances to ingest from salt licks, or other sources of sodium, when sodium sources in their grazing pastures are depleted (see 2 for review).

But while there may be selective pressure for behavioral mechanisms for salt appetite in herbivores, it is nonetheless present in carnivores e.g. the dog[3,4]. Interestingly the carnivorous cat shows only a slight preference for low concentrations of saline, while the herbivorous rabbit shows a larger one when sodium replete[5]. Neither looks anything like the omnivorous rat who is a salt glutton in two senses: both in a need free condition and following the elicitation of a salt appetite: rabbits (or sheep) do not ingest more salt than they need following the elicitation of a salt appetite[2]. This salt gluttony in the omnivorous rat may be a consequence of the broad feeding capacities of the animal.

Other omnivores are known to show an appetite for salt. There is some evidence, though not compelling, that primates have a salt appetite (see 2 for review). There is also clinical as well as laboratory evidence that humans display an appetite for salt when sodium deficient[6],[7]. But the evidence of salt appetite resulting from sodium deficiency in higher organisms is not strong, although salt gluttony in sodium replete humans is well documented[2]. Moreover, a recent study failed to observe a salt appetite in the Rhesus monkey[8]. The natrorexigenic manipulation that was used in the study (9 days on a sodium deficient diet), however, may not have been strong enough to unmask the phenomenon; the authors acknowledged this possibility.

We decided to take another look[9]. What we found was that the 5 Rhesus monkeys tested displayed an increase in their saline intake over baseline conditions. For our natrorexigenic manipulation we placed them on a sodium deficient diet for 6 days in addition to giving them two sc injections of 10 mg furosemide to promote sodium loss on the second and the fourth day of the diet. The behavioral results from this study are revealed in Table 1. Table 1 provides the individual scores of 5 monkeys and some biographical notes. Note that the appetite was specific for the sodium salt. You should also note that the sodium concentration that we were forced to use was 9% NaCl. This occurred because we discovered that these monkeys would ingest up to 80 ml a day of either 3 or 6% sodium. They were salt gluttons. The 9% concentration reduced their intakes, though it still remained high. The potassium solution was 3% KCl. All 5 monkeys expressed an increase in their sodium ingestion when salt hungry.

Table 1.

Individual Subject Comsumption Data (in ml) for 24 hrs.

Monkey	Sex	Weight(in Kg)	Pretreatment/Intake KCl	Pretreatment/Intake NaCl	Posttreatment/Intake KCl	Posttreatment/Intake NaCl
114	M	10	35	27	37	36
125	M	11	21	10	36	68
120	M	8	20	19	13	39
106	F	6	33	62	41	86
93	F	5	63	27	30	35

We have also gathered evidence that the phenomenon of salt appetite may not be exclusively a mammalian phenomenon. The herbivorous pigeon is known to be responsive to thirst challenges[10], and there is reason to think that pigeons would also display an appetite for salt following natrorexigenic treatment. We have a little bit of evidence to share with you that supports this claim[11]. The three pigeons tested displayed an increase in salt appetite following a sc 15 mg injection of furosemide. They ingested 12 ml over baseline conditions of 2% saline. Further research is now under way to determine the specificity of the appetite.

Behaviorally, the salty taste appears to be one very important gustatory characteristic which triggers the ingestion of salty sources in animals deficient in sodium[12]. This same ingestion of salty sources may also be triggered when animals are deficient in other minerals beside sodium. There is evidence, that for example, potassium and calcium deficient rats prefer sodium over potassium or calcium salts[13,14]. In both cases the rats will eventually ingest the salt they need. But why should the potassium or calcium deficient animal ingest the sodium salt? One possibility is that the salty taste of sodium (and NaCl is the prototype) is attractive not only during sodium deficiency but also when deficient in potassium or calcium; in the natural wild environment these minerals would usually be compounded. The salty taste is a basic gustatory category and a behavioral strategy built into the mineral deficient animal may be to search for salty sources, e.g. salt licks, containing several salts.

Table 2.

Mean Cumulative Fluid Intakes (ml)

Solution	Dep. Group			Hours			
3% NaCl	NA	$\bar{X}$	11.2	11.2	20.6	25.0	36.4
		SD	3.5	3.5	11.6	12.5	16.2
	K	$\bar{X}$	4.4	5.3	15.1	19.7	33.2
		SD	2.0	2.3	8.5	10.8	19.0
	CA	$\bar{X}$	1.7	3.4	15.4	17.7	28.5
		SD	2.2	3.4	10.0	11.8	15.9
	THI	$\bar{X}$	0.6	1.1	6.5	10.2	11.3
		SD	0.7	1.0	4.5	6.1	6.1
	Cont	$\bar{X}$	0.0	0.0	5.5	5.5	5.9
		SD	0.0	0.0	4.5	4.5	4.7
1.5% KCl	NA	$\bar{X}$	0.2	0.2	2.4	2.4	4.5
		SD	0.5	0.5	2.3	2.3	2.7
	K	$\bar{X}$	2.2	2.8	15.8	20.5	32.1
		SD	1.7	2.6	5.6	10.3	15.0
	CA	$\bar{X}$	0.5	1.1	10.0	17.5	26.6
		SD	0.6	1.8	9.1	15.1	19.3
	THI	$\bar{X}$	0.0	0.3	1.6	2.3	3.2
		SD	0.0	0.5	2.3	2.6	2.9
	Cont	$\bar{X}$	0.6	0.6	2.3	5.6	9.3
		SD	1.3	1.3	2.5	4.0	4.5
0.9% CaCl$_2$	NA	$\bar{X}$	0.2	1.0	4.0	4.5	4.9
		SD	0.5	0.8	2.0	2.6	2.9
	K	$\bar{X}$	0.6	0.8	4.2	5.1	8.7
		SD	0.7	0.7	5.6	6.5	7.9
	CA	$\bar{X}$	0.8	1.1	12.0	16.9	32.9
		SD	1.7	2.2	10.2	14.5	21.7
	THI	$\bar{X}$	0.0	0.5	0.5	1.4	3.7
		SD	0.0	0.7	0.7	0.8	2.3
	Cont	$\bar{X}$	0.0	0.0	5.6	5.6	5.6
		SD	0.0	0.0	3.8	3.8	3.8

Table 2 shows that young rats (45 days) who were rendered sodium, potassium or calcium deficient (by sodium -NA, potassium -K, or calcium -CA deficient diets until their body weights were 20% lower than control animals) all ingest the NaCl solution[15]. While both the potassium or calcium deficient rats eventually ingested the potassium or calcium respectively, they continued to ingest the NaCl throughout the self selection period. The calcium deficient rats also continued to ingest the KCl-which tastes somewhat salty. We ran a thiamine (THI) deficient control group and we were encouraged by our initial results where we saw only a mild increase in their NaCl consumption, although a follow up study revealed NaCl intakes that were larger than our initial observations. The results remain suggestive, as well as the hypothesis, that the search and ingestion of the salty taste may be an important factor in the behavioral organization of mineral deficient animals.

It wasn't until saline was unavailable that brain mechanisms became necessary in organizing the appetite for salt, when the evolutionary chain moved out of the sea. Possibly the origins of salt appetite are to be found in fresh water fish, but evolving into a full appetite in reptiles e.g. turtles. The neural mechanisms which organize the initiation of the appetite may require the participation of the upper brainstem and forebrain[2,16]. We have recently shown that the chronic supracollicular decerebrated rat (a preparation which disconnects hypothalamic, thalamic and forebrain mechanisms from the caudal brainstem) does not show the consummatory ingestive responses to intraorally delivered saline when made hungry for salt as shown in Figure 1,[17]. These rats also do not show the hedonic shift in the perception of the saline, as measured by their facial displays as intact rats do when rendered salt hungry[18,19]. They also do not show osmotic thirst; but they do increase their sucrose ingestion when rendered hungry[20]. The evolution of salt appetite and body fluid homeostatic behavioral mechanisms more generally, seems to have evolved later than that for feeding[21]; this is reflected in more rostral organization of behavioral control within the neural axis[22].

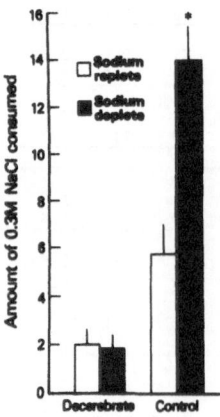

Figure 1.

Finally, despite the fact that many kinds of vertebrate animals may express the appetite, the physiological mechanisms which generate the behavior may be somewhat different, in fact recent evidence in the sheep[2], the rat[10,23,24], and the dog[4] suggest just that. For example, aldosterone and angiotensin II play an important role in the salt seeking behavior of the omnivorous rat[10,23,24,25,26,27,28]; it does not play a prominent role in the herbivorous sheep[2] or carnivorous dog[4]. While dimunition of brain sodium causes sheep to ingest salt[29], it does not have this effect on the rat[24]. Nature seems to have produced a variety of physiological events for the same behavioral function. Analogously, in the design of artificial systems it is known that a variety of physical systems can produce the same behavioral function. The same idea may hold for biological systems.

REFERENCES

1. H. Kaunitz. Causes and consequences of salt consumption. Nature, 178: 1141-1145, (1956).
2. D.A. Denton. The hunger for salt. New York: Springer-Verlag, (1982).
3. J.T. Fitzsimons & M.J. Moore-Gillon . Drinking and antidiuresis in response to reductions in venous return in the dog: Neural & endocrine mechanisms, J. Physiol., 308: 403-416, (1980).
4. D.J. Ramsay & I.A. Reid. Salt appetite in dogs. Soc. Neurosci. Abstrs. (1981).
5. J.A. Carpenter. Species differences in taste preferences. J. Comp. & Physiol. Psych., 49: (2) 139-144, (1956).
6. R.A. McCance. Experimental sodium chloride deficiency in man. Proc. of the Royal Soc. Series A. 119: 245-263 (1936).
7. L. Wilkins & C.P. Richter. A great craving for salt by a child with cortico-adrenal insufficiency. JAMA. 114: 866-868, (1940).
8. T.M. McMurray & C.T. Snowdon. Sodium preferences and responses to sodium deficiency in Rhesus monkeys. Physiol. Psych. 5: 477-482, (1977).
9. J. Schulkin, D. Leibman, R.N. Ehrman, N.S. Norton and J.W. Ternes. Salt hunger in the Rhesus monkey. Behav. Neurosci., 98: (4), 753-756 (1984).
10. J.T. Fitzsimons. The physiology of thirst & sodium appetite. Cambridge, England (1979).
11. R. Allan & J. Schulkin. Unpublished Observations.
12. J. Schulkin. Behavior of sodium deficient rats: The search for a salty taste. J. Comp. Physiol. Psych. 96: 4, 623-624, (1982).
13. W.R. Adams & J.K. Dewborn. Effect of potassium depletion on mineral appetite in the rat. J. Comp. & Physiol. Psych, 78: (1) 54-58, (1972).
14. M. Lewis. Discrimination between drives for sodium chloride and calcium. J. Comp. & Physiol. Psych. 65: (2), 208-212 (1965).
15. J. Schulkin. Unpublished Observations.
16. G. Wolf & J. Schulkin. Brain lesions and sodium appetite: an approach to the neurological analysis of homeostatic behavior. In: M. Kare, eds. Biological and Behavioral Aspects of Salt Intake, New York, Academic Press, 331-339, (1980).

17. H.J. Grill & J. Schulkin. Sodium homeostasis in decerebrate rats. Manuscript in press. (1985).

18. K.C. Berridge, F.W. Flynn. J. Schulkin & H.J. Grill. Sodium depletion enhances salt palatability in rats. Behav. Neurosci., 98: 652-660, (1984).

19. F.W. Flynn, K.C. Berridge, J. Schulkin & H.J. Grill. Unpublished Observations.

20. R. Norgren & H.J. Grill. Brain-Stem Control of Ingestive Behavior, pp 99-131. In: The Physiology of Motivation. (ed. Pfaff) Springer-Verlag N.Y. (1982).

21. C.J. Herrick. The Brain of the Tiger Salamander, Ambystoma tigrinum. Chicago: Univ. of Chicago, Press, (1948).

22. J. Taylor. Selected Writing of John Hughlings Jackson. Vols. 1 & 2, ed. Staple Press, London, (1958).

23. S.J. Fluharty & A.N. Epstein. Sodium appetite elicited by intracerebroventricular infusion of angiotensin II in the rat: II. Synergistic interaction with systemic mineralocorticoids. Behav. Neurosci., 97: 746-758, (1983).

24. A.N. Epstein, D.M. Zhang, J. Schultz, M. Rosenberg, P. Kupsha & E. Stellar. The failure of ventricular sodium to control sodium appetite in the rat. Physiol & Behav., 32: 683-686, (1984).

25. K.K. Rice & C.P. Richter. Increased sodium chloride and water intake of normal rats treated with desoxycorticosterone acetate. Endocrinology., 33: 106-115, (1943).

26. G. Wolf. Sodium appetite elicited by aldosterone. Psychonomic Science. 1: 211-212, (1964).

27. M.J. Fregly & I.W. Waters. Effects of mineralocorticoids on spontaneous sodium chloride appetite of adrenalectomized rats. Physiol. & Behav. 1: 65-74, (1966).

28. E.M. Stricker. Thirst and sodium appetite after colloid treatment in rats: Role of the renin-angiotensin-aldosterone system. Behav. Neurosci. 97: 725-737, (1983).

29. R.S. Weisinger, P. Considine, D. Denton., D. Leksell, L. McKinley, D. Mouw, A. Muller, & E. Tarjan. Role of sodium concentration of the cerebrospinal fluid in the salt appetite of sheep. American Journal of Physiology. Regulatory Integrative Comparative Physiology 242: R51-R53, (1982).

BEHAVIORAL DYNAMICS IN THE APPETITE FOR SALT IN RATS

Jay Schulkin

Institute of Neurological Sciences
University of Pennsylvania
Philadelphia, Pennsylvania   19104

I am going to briefly present some laboratory findings. Hope-
fully, they will provide you with an idea of the behavioral facts
in sodium homeostasis.

The appetite for salt that results from sodium deficiency is
innate, as is well known; from the very first time rats are made
sodium deficient, they display a salt appetite by immediately ingesting salty solutions when they are presented[1,2], and see Denton's
analysis for an analogous phenomenon in sheep[3]. Rats can also recall how saline was obtained[4] and where the saline was located[5] even
though they encountered the saline while they were not hungry for
salt. The very first time they are driven to ingest salt they express the knowledge of how the saline can be obtained and where it
was located. These findings are further substantiated by the following recent results.

Lesions of the central gustatory system at the level of the thalamic taste relay impairs the expression of salt appetite; however,
if the rats simply taste saline preoperatively (with or without the
drive for salt) they are protected from the usual lesion-induced deficits in salt appetite[6]. Are these rats protected (despite impairments in their taste sensibility) because they recall postoperatively
where the saline was located? The answer seems to be yes[7]. This is
because only those rats who had the saline located in the same place
both pre and postoperatively were protected from these lesion-induced
deficits in salt appetite. They ingested 9.5 ml of saline if the saline remained in the same place; if it was not, they ingested 1.0 ml
following the elicitation of a salt appetite. Intact rats ingested
9.0 ml of the saline. In other words, they learned where the saline
was located when they did not need it and returned there when they did.

Because the appetite for salt that results from sodium deficiency is innate, one would expect that just a brief exposure to the relevant gustatory stimulus would be sufficient to trigger the knowledge of how it can be obtained and where it can be located. And there is evidence that shows just that.

First, a brief exposure to saline (over a 2 minute period) is sufficient to trigger the knowledge of how saline was obtained (pressing a bar in a Skinner box) while not under the drive state of salt appetite; when subsequently made hungry for salt for the first time the rats demonstrated that they had learned how saline could be obtained[8,9]. Second, we also found that a brief preoperative exposure to saline protects rats from the deficits in salt appetite, that results from central gustatory damage at the level of the thalamus[10]. A 30 second exposure to saline (even though there is no hunger for it) before a lesion of this part of the central nervous system is sufficient to trigger the protective effect. This effect was specifically related to tasting the saline since the preoperative exposure to sucrose was not protective. The group of rats preoperatively exposed to the saline ingested 10.5 ml of the saline postoperatively following the elicitation of a salt appetite, while the preoperative sucrose group ingested 0.5 ml. Intact rats ingested 9.0 ml. Note that the protective effects from tasting the saline were completely implemented within the brief preoperative exposure. When taken together these results show that the rat is prepared to learn in seconds about how or where saline is found.

The rat also seems to be prepared to associate the saline when it is mixed with other sources that have different taste properties. In fact, one would expect that sources of saline in the wild would be a mixture of a variety of tastes. In the laboratory, rats who have been given a mixture of saline with quinine or banana when not under the drive state of salt appetite will subsequently ingest the quinine or the banana when it is given alone the very first time they are salt hungry. They do so because they associate the banana or quinine with the saline[11].

But is it the salty taste that the rat is searching for? The results of several studies[12,13] and a close reading of the literature suggests that sodium hungry rats will ingest non-sodium salts, like lithium and potassium. Why should that be? Possibly because they both taste salty. We looked at the salt choices of salt hungry rats offered a choice of sodium salts, or non-sodium salts, and of a non-sodium and a sodium salt. In each case one salt was judged by humans to be more salty. If the salt hungry rat is looking for a salty taste it ought to ingest the salt which tastes more salty. Sodium chloride (NaCl), it should be noted, is the prototypical salty taste: all other salts are either less salty or have some other taste quality in addition to saltiness. This would explain why it is preferred over other salts by the salt hungry rat; and we have provided evidence that shows that sodium deficient rats in contrast to when they are thirsty or hungry will ingest the saltier of two salt solutions[14]. For example they ingest .12 molar NaCl over Na acetate (84% vs. 47%); they also preferred the more salty tasting potassium chloride over calcium chloride (76% vs. 58%) and lithium chloride over Na bicarbonate (64% vs. 40%) when hungry for salt vs when thirsty in this case. In each case the more preferred, when salt hungry, was the saltier substance. Nature may have insured that sodium sources would be ingested by sodium deficient animals because they are the more salty tasting.

Does the animal know what it is looking for when it is salt hungry? We know that they recognize salty tastes immediately and will return to the place where they have sampled them. But are they guided in their search for salt by an active perceptural representation of salt? It may be the case that when the animal is salt hungry it generates a central representation of salt which serves to guide the ingestive behavior. This is a difficult question to answer. But we have attempted to do so in the following manner.

Rats can learn a taste aversion to salt; and it is easily overcome the first time they are made salt hungry[15]. But will they show less of an aversion to salt if they are first made salt hungry (without access to salt), then given time to recover from the natrorexigenic treatment, and then given access to saline (1.2%) when thristy and made sick by its ingestion (by lithium). If the drive for salt generated the representation of the salt, then this should produce less of an aversion to the salt; it would be as if they had tasted the salt itself. We found, in fact, that the pre-treated salt hungry rats displayed a reduced aversion to the saline[16]. The aversion to the saline was reduced by only 2 ml in the salt hungry pretreated group; for the control group it was by 7 ml. We used a low dose of lithium to produce the aversion so as not to mask what we predicted would be a small effect. The effect was indeed small, but present. Note that the initial ingestion of the saline (4 days following the last DOCA salt hunger treatment) was about the same for both groups, (23 ml DOCA pre-treated, 25 ml for controls); so that they didn't appear to have been salt hungry at the time of testing; we are running further controls at this time. One tentatively suggests from these results, and others; e.g. anticipatory running (or returning) to the location of salt when sodium hungry, that the drive for salt alone may activate the central representation of the salty taste.

This search for the salty taste when hungry should also produce hedonic changes in the motivation to ingest salt and in two ways, one appetitive and one consummatory. In an appetitive test in salt hungry rats we have looked at how rats run for different concentrations of saline and for different salts[17]. Figure 1 just shows how they run for the different concentrations of hypertonic saline and water. They run about the same rate for every concentration of

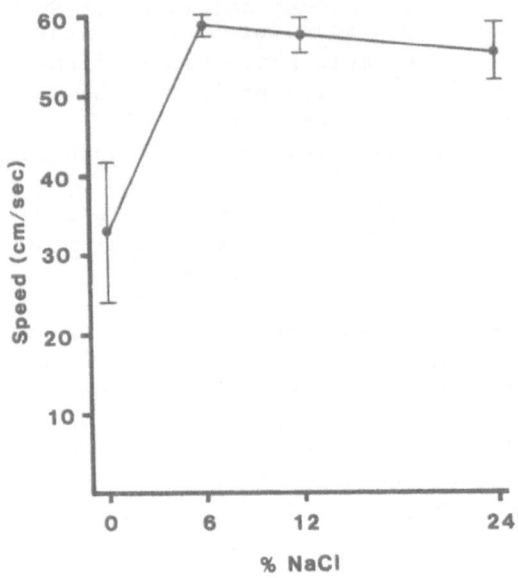

Figure 1.

saline up to 24%. They also run for less concentrated saline. And
they avidly run for the salty tasting LiCl. Different degrees of
salt hunger produce differential running speeds for saline solutions.
Analogously different degrees of salt hunger produce diffential motivation to bar press for saline solutions[18].

		INGESTIVE					AVERSIVE				
		PL	LTP	TP	MM	PD	G	CR	FW	FF	HS	LO
WEEKS I	REPLETE	0.2	0.4	0.4	1.2	0.6	0.8	1.0	2.4	1.2	1.0	0.4
	DEPLETED	0	2.0	4.4	4.6	0.4	0.4	0.2	0.4	0.4	0	0.2
II	REPLETE	0.2	0	0	0.6	2.2	0.4	1.4	5.4	6.6	2.0	0.4
	DEPLETED	2.4	3.8	5.4	4.2	0	0.2	0	0.2	0	0	0

Figure 2. Taste-elicited fixed action patterns of 5 rats to 0.5 M
NaCl when either sodium replete or sodium deplete. (Values represent
the mean number of reponses per rat. Ingestive responses are paw
licking [PL], lateral tongue protrusions [LTP], tongue protrusions
[PL], and mouth movements [MM]. Failure to show any consummatory
response, either ingestive or aversive, results in the passive
drip of fluid from the mouth [PD]. Aversive responses are gapes
[G], chin rubs [CR], face washing [FW], forelimb flailing [FF],
head shakes [HS], and locomotion [LO].)

Finally, we have shown in a consummatory test that the taste of 3% NaCl, while only mildly aversive when intraorally delivered in small amounts to the oral cavity, (invoking a mixture of aversive and ingestive responses) nonetheless becomes hedonically more pleasing (a change in the palatability judgement resulting in mainly ingestive responses)to the salt hungry rat as shown by their video-taped facial responses[19]. These results can be seen in Figure 2. This change in the response to the saline can be found the first time they are rendered salt hungry, and remains constant over repeated trials. We also found that the shift in this palatability judgement is specifically related to the saline.

Let me summarize. The rat not only ingests salt immediately but can also recall later on, when it is made salt hungry for the first time, how or where salt is to be found. The triggering of this knowledge requires just a brief exposure to the salty taste. They are driven to ingest salty tastes; sodium salts are generally the more salty tasting salts, insuring their ingestion. The rats are also prepared to associate the salty taste of the saline with other tastes. The search for the salty taste may be guided by a central representation of the salt which guides the ingestive act. The sodium hungry state motivates an avidity to search and then ingest salty solutions, reflecting an hedonic shift in the perception of salt.

REFERENCES

1. M. Nachman & L.P. Cole. Role of the taste in specific hungers. In: L. Beidler, ed., Handbook of sensory physiology: Vol 1. Pt. 7. Chemcial senses (pp. 337-362), Berlin: Springer (1971).
2. G. Wolf. Innate mechanisms for regulation of sodium intake. In: C. Pfaffmann, ed., Olfaction and taste: Proceedings of the Third International Symposium (pp. 548-553), New York: Rockefeller University Press, (1969).
3. D.A. Denton. The Hunger for Salt. New York, Springer-Verlag, (1982).
4. E.E. Krieckhaus & G. Wolf. Acquisition of sodium by rats: Interaction of innate mechanisms and latent learning. J. Comp. & Physiol. Psych., 65: 197-201, (1968).
5. E.E. Krieckhaus. "Innate recognition" aids rats in sodium regulation. J. Comp. & Physiol. Psych. 3: 117-122, (1970).
6. G.A. Ahern, M.L. Landin & G. Wolf. Escape from deficits in sodium intake after thalamic lesions as a function of preoperative experience. J. Comp. & Physiol. Psych., 92: 544-554, (1978).
7. R.A. Paulus, R. Eng & J. Schulkin. Preoperative latent place learning preserves salt appetite following damage to the central gustatory system. Behav. Neurosci. 98: 68-75, (1984).
8. C.R. Wirsig & H.J. Grill. Contribution of the rat's neocortex to ingestive control: I. Latent learning for the taste of sodium chloride. J. Comp. & Physiol. Psych. 96: 615-627, (1982).
9. R. Bregar, N. Strombakis, R. Allan & J. Schulkin. Brief exposure to a saline stimulus promotes latent learning in the salt hunger system. Soc. Neurosci. Abstrs. (1983).

10. A.K. Hartzell, R.A. Paulus & J. Schulkin. Brief preoperative exposure to saline protects rats against behavioral impairments in salt appetite following central gustatory damage. Manu. in press.

11. R.A. Rescorla. Simultaneous associations. In: P. Harzem & M.D. Zeiler, eds., Predictability, correlation and contiguity. (pp. 47-80) New York: Wiley (1981).

12. J.L. Falk. Limitations to the specificity of NaCl appetite in sodium-depleted rats. J. Comp. & Physiol. Psych., 60: 393-396, (1965).

13. M. Nachman. Taste preferences for lithium chloride by adrenalectomized rats. Am. J. Physiol., 250: 219-221, (1963).

14. J. Schulkin. Behavior of sodium deficient rats: The search for a salty taste. J. Comp. & Physiol. Psych. 96: 628-634, (1982).

15. E.M. Stricker & N.E. Wilson. Salt seeking behavior of rats after acute sodium deficiency. J. Comp. & Physiol. Psych., 72: 416-420, (1970).

16. R. Hashemiyoon, P. Rozin & J. Schulkin. Unpublished Observations.

17. J. Schulkin, P. Arnell & E. Stellar. Running in anticipation of the taste of salt in mineralocorticoid treated rats. Manu. under review.

18. D. Quartermain, N.E. Miller & G. Wolf. Role of experience in relationship between sodium deficiency and rate of bar pressing for salt. J. Comp. & Physiol. Psych., 63: 417-420 (1967).

19. K.C. Berridge, F.W. Flynn, J. Schulkin & H.J. Grill. Sodium depletion enhances salt palatability in rats. Behav. Neurosci., 98: 652-660, (1984).

EFFECT OF CEREBROVENTRICULAR INFUSION OF HYPERTONIC SODIUM

SOLUTIONS ON SODIUM INTAKE IN RATS

Emma Chiaraviglio and María F. Pérez Guaita

*Instituto de Investigación Médica
 Mercedes y Martín Ferreyra
 Casilla de Correo 389
 5000 Córdoba, Argentina

INTRODUCTION

The onset of specific sodium appetite as a result of body sodium loss, is slow. In an early report it has been shown that in rats, acute sodium depletion by peritoneal dialysis (pd) produced a sudden and significant sodium deficit within half an hour, but sodium appetite is made evident usually after 12-14 hours, being the higher intake of sodium salts 20-24h after pd (1).

Sodium depletion by pd was associated with a concomitant drop in blood volume and serum sodium concentration [Na] . However, the [Na] in the extracellular fluid does not appear to have a direct effect on sodium appetite. In the rat, sodium appetite persisted even though serum [Na] and blood volume returned to normal values (1) .

Sodium intake was reduced by increasing sodium concentration in the cerebrospinal fluid (CSF) in sheep depleted of sodium by parotid salivary loss, but it was enhanced by lowering CSF [Na] by intra - cerebro ventricular (icv) infusion of mannitol (2).

The aims of the present work were: first, to study the time-dependent profiles of changes of sodium and potassium concentration in CSF and serum from $\emptyset$ to 24 hours after pd of male rats, and second, to evaluate the effect on sodium appetite of hypertonic CSF infused at discrete periods after peritoneal dialysis.

*Supported by Consejo Nacional de Investigaciones Científicas y Técnicas , Argentina.

METHODS

The experiments were carried out on adult male albino rats.
Animals were depleted of sodium by an intraperitoneal injection of 5%
glucose solution warmed at 37°C, in a volume equivalent to 10% of the
rat body weight (bw). After one hour, the ascitic liquid was recovered.
The amount of NaCl withdrawn by this method was 0.84 ± 0.02 mM/100g
rat (mean ± SE N=18 rats) (1). The dialysed rats were caged individually without food and with distilled water as the only drink. Samples
of CSF and blood were taken at 0-0.5-1-4-14 and 24h after pd in six
groups of 12 animals each. Sodium and potassium concentration was
measured. The animals were anesthetized with tribromoethanol (200mg/
Kg bw). To sampling CSF the cisterna magna was punctured; bloody
samples were discarded. Blood samples were taken from the jugular
vein.

RESULTS

Comparatively CSF [Na] was higher than serum [Na] in non-deprived
rats. Cerebrospinal fluid potassium concentration was lower than serum
potassium concentration. Thirty minutes after pd, CSF [Na] decreased
significantly, but the minimum value was observed 4h after pd. The
profile of changes in CSF [Na] after pd, showed that there was a rapid
decrease, followed by a period of slow recovery (fig. 1).

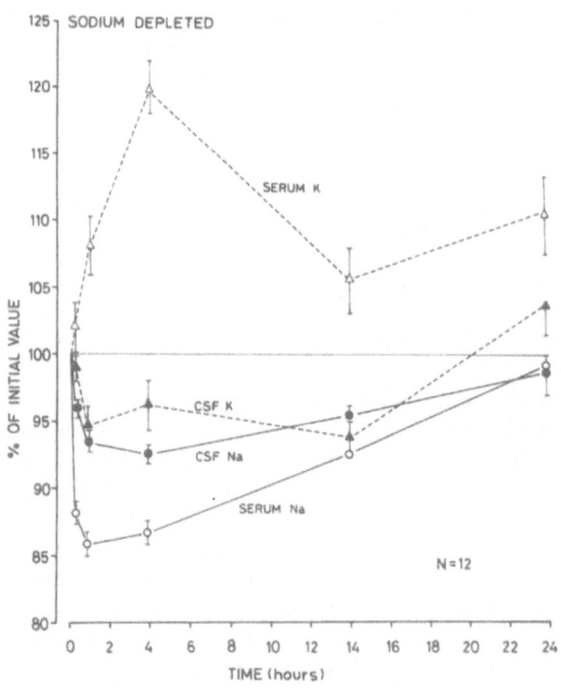

Fig. 1. Changes from initial value in CSF and serum Na and K concentration in six groups of rats at different times after sodium
depletion . Mean ± SE.

Serum [Na] showed similar pattern than CSF. Comparatively,serum [Na] showed more marked decline and faster recovery than CSF [Na] Twenty four hours after pd, CSF [Na] was 1.5% below whereas serum [Na] almost reached initial value. It is worth noting that during this period rats did not have access to sodium salts. Cerebrospinal fluid [K] significantly decreased from Ø to 24h after dialysis, whereas serum [K] increased between Ø and 24h .

Two experiments were designed to investigate whether intracerebro ventricular (icv) reposition of Na at different times after pd could reduce sodium appetite: first, long term infusion from one hour before to 24h after pd, and second short term infusion in which 3 periods post-pd of 8h each were selected.

Under tribromoethanol anesthesia the rats were implanted with a stainless steel cannula 15 mm length and 0.6 mm o.d. into the 3rd brain ventricle according to the Köning and Klippel stereotaxic coordinates (3). Animals were allowed to recover during a week.

Continuous infusion of artificial CSF, during 24th (long term infusion) were made by means of an osmotic minipump (mp) Alzet, delivering 1μl/h . Infusions with artificial CSF made hypertonic with different concentrations of NaCl, during 8 or 16 hours (short term infusion) were accomplished by a continuous infusion pump adapted to a micro syringe to deliver 1μl/h. During infusion, the animals were unrestrained, moving freely into a metabolic cage having distilled water for drinking. In the long term infusion experiments rats were implanted with a filled osmotic minipump under the loose skin of the back. The pump was connected by a polyvinyl catheter to the guide cannula. One hour later the animals were depleted of sodium by pd as described before. After 24th, the infusion was stopped and two bottles test of 1.8% NaCl and distilled water was given. Three groups of rats were infused: (i) with artCSF [Na,150mM] N=5; the total Na given was 3.6μM; (ii) with Na free-CSF [Na,ØmM] N=6 and (iii) with distilled water N=5. Each animal was used as its own control.

In short term infusions three groups of six rats each were used: (i) infused with CSF [Na,450mM] from -1 to 8h post-pd. (ii) with CSF [Na,225mM] , starting 8h, and ending 24h post-pd and (iii) with CSF [Na,450mM] , from 16 to 24h post-pd. Intake test was given 24h after pd . Each rat was infused a second time with CSF [Na,150mM] . Data on sodium and water intake,urine volume and urine Na and K excretion were taken as control values.

The infusion of CSF [Na,150mM] at a rate of 1μl/h,from-1 to 24h after pd,significantly decreased the intake of 1.8% NaCl to 2.6+0.4 ml/100g bw; as compared with the control value (4.2+0.5ml/100g bw,the difference was significant (P <0.001). The groups infused with Na free-CSF or distilled water did not show a significant difference with its control;however they differed with the group which received infusion of CSF [Na,150mM] (Fig.2) .

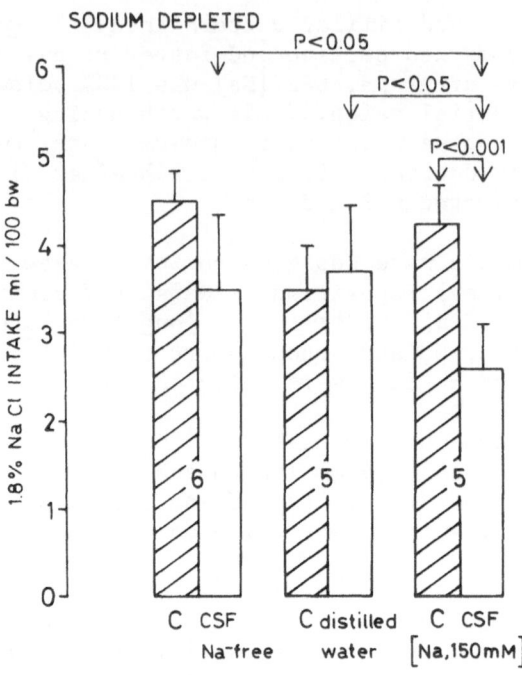

Fig. 2.   Effect of continuous icv infusion of CSF [Na,150mM],
          Na free-CSF and distilled water (1µl/h) on two hours
          intake of 1.8% NaCl solution. Infusion was given -1 to
          24h after sodium depletion. Mean+SE.

    As in the former experiment,rats infused with CSF [Na,225mM]
received during 16h the same total amount of Na. As can be seen in
Fig.3 the intake of 1.8% NaCl decreased by 56%. The volume drunk was
2.8+0.4ml/100g bw. Compared with control values (4.5+0.3ml/100g bw)
the difference was significant. In contrast with rats infused from -1
to 8h post-dialysis, in the present experiment animals drank less
fluid than the urine voided,remaining in a negative fluid balance. The
urinary sodium excretion was less in experimental rats compared with
the same rats infused with CSF [Na,150mM] .

    Rats were infused with CSF [Na,450mM] the same concentration used
in the first of this series of experiments. But, in this case,the infusion was given during the last 8h before test, when serum and CSF
[Na] were returning to its initial level. Sodium intake was reduced in
40% (1.5+0.5ml/100g bw); compared with control value (3.8+0.4ml/100g bw)
the difference was significant.

DISCUSSION

    Long term infusion (24h) of normal CSF [Na,150mM] decreased
intake of sodium solution induced by pd. This is the first evidence

that icv reposition of sodium partially suppresses the specific behaviour in rats. A similar finding has been reported in the sheep (2).
This fact rised up the question whether infusion of hypertonic artCSF
during the critical period of lowest [Na] seen immediately after pd,
could inhibit sodium appetite, assuming that circulating CSF affects
the putative Na sensors located on or near the periventricular walls
(4,5,6). However, the first short term experiment shows that [Na] and
delay in onset of specific sodium intake were not directly related,
since infusion of hypertonic CFS during the period of lowest Na after
pd (8h) did not interfere with Na intake. A similar conclusion was
drawn by studies on the relationship between fall in serum sodium and
sodium intake (1).

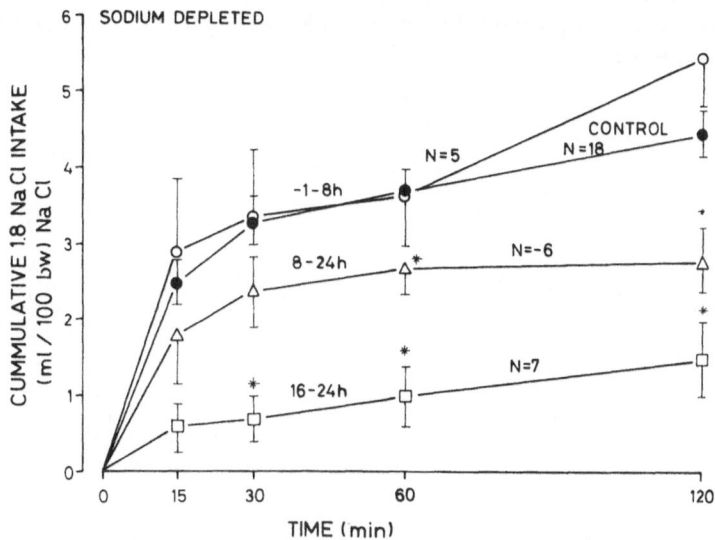

Fig. 3. Effect of continuous infusion of hypertonic CSF at different time after Na depletion, on the intake of 1.8 NaCl.
Mean ± SE. * P < 0.001 as compared with control group.

These experiments indicate that CSF [Na] drops suddenly after
sodium depletion by pd and slowly recovers the initial level during a
period of 24h; icv infusion of hypertonic CSF at the time of maximal
fall in CSF [Na] did not affect sodium intake, while infusion made
during recovery of [Na] significantly decreased sodium intake.

REFERENCES

1. M.C. Ferreyra and E. Chiaraviglio, Changes in volemia and natremia
and onset of sodium appetite in sodium depleted rats, Physiol.
Behav. 19:197 (1977).
2. R.S. Weisinger, P. Considine, D.A. Denton, M.J. McKinley and D. Mouw,
Rapid effect of change in cerebrospinal fluid sodium concentration on salt appetite, Nature, 280:490 (1979).

3. J.F.R. Koning and R.A. Klippel, "The rat brain. Stereotaxic atlas of the forebrain and lower parts of the brain stem". The Williams and Wilkins Company, Baltimore (1963).
4. B. Anderson, The effect of injection of hypertonic NaCl solutions into different parts of the hypothalamus of goats, Acta physiol. scand. 28:188 (1953).
5. E.M. Stricker and G. Wolf, Behavioral control of intraventricular fluid volume. Thirst and sodium appetite, Annals of the New York Academy of Sciences, 157:553 (1969).
6. S. Nicolaïdis, Responses unitaries dans les aires anterieures et medianes de l'hypothalamus associées a des variations de pression arterielle et de volemie, C.R. Acad. Soc.,Paris, 270:839 (1970).

This work was supported by Fundación Alberto J. Roemmers, Buenos Aires, Argentina.

508

ROLE FOR α_2-ADRENOCEPTORS IN EXPERIMENTALLY-INDUCED DRINKING IN RATS[1]

Melvin J. Fregly and Neil E. Rowland

Depts. of Physiology and Psychology, University of

Florida, Gainesville, Florida 32610

INTRODUCTION

The β-adrenergic agonist, isoproterenol, is well known to increase water intake within 1 hr of its administration to rats (1,2). One of the major contributory mechanisms to this drink is via isoproterenol-induced release of renin from the kidneys (3,4) and the subsequent elevation of circulating angiotensin II (AII) levels. (See also references 5-7 for other mechanisms). AII, administered either peripherally or into circumventricular brain regions such as subfornical organ or anteroventral third ventricle, is itself a potent dipsogen (8-11). However, the evidence that normal levels of AII in blood gain access to these regions of the brain is not compelling, and the inference that peripheral AII exerts its dipsogenic action by a direct effect on the brain is thus based on indirect evidence. For example, peripheral administration of the competitive AII inhibitor, saralasin (sar^1,ala^8-AII) inhibits the drinking response to centrally-administered AII (12), and lesions of the circumventricular organs inhibit the drink to peripherally-administered AII (11). Further, the brain has its own intrinsic renin-angiotensin system (13) including high affinity receptors for AII in the circumventricular organs (14), yet the ways, if any, in which the central and peripheral renin-angiotensin systems interact have not yet been elucidated.

Clonidine is an α_2-adrenoceptor agonist currently in clinical use as an antihypertensive agent. Its central α_2-agonistic effects include a reduction in the sympathetic, and an increase in parasympathetic, outflow to the periphery (15). Among the resulting peripheral effects are decreases in blood pressure, heart rate, and renin secretion (15,16). Thus, clonidine at doses approximating

509

30 µg/kg, i.p. inhibited basal renin release within 10 min in the conscious rat, and also inhibited the isoproterenol-stimulated release of renin, both in vivo and in vitro (15,16). These observations prompted our initial study (17) to investigate whether clonidine might inhibit isoproterenol-induced water intake in rats. A robust inhibitory effect was indeed observed. Further studies were then carried out to determine whether clonidine would alter the drinking response to other dipsogens, and to assess the effect of α_2-adrenergic antagonists on AII-related drinking. Finally, we have studied the effects of cerebroventricular administration of clonidine, as well as isomers of octopamine and synephrine, on AII-related drinking.

METHODS

General

Female rats of the Blue Spruce (Sprague-Dawley) strain weighing 180-280 g were used. They were maintained in groups of three to four in stock cages and allowed tap water and Purina Laboratory Chow ad libitum. The colony room and the adjacent testing room were maintained at 25 ± 1° C. and illuminated from 7 AM to 7 PM.

On test days, the rats were weighed, injected with appropriate drug or vehicle (details are given with results) and each rat was placed into an individual stainless steel cage. No food was available during the experiment. Water (25° C.) intakes were measured gravimetrically at 30, 60 and 120 min after administration of the dipsogen. All experiments were set up with groups of at least 6 rats according to a statistical factorial design to assess the effects of treatment and interaction (18).

Experiment 1 Effect of Peripherally-Administered Clonidine On Experimentally-Induced Drinking

The full details of these studies have been published elsewhere as indicated.
A. Isoproterenol (17). Rats were administered various doses of clonidine (Catapres ®, Boehringer Ingelheim), or dipsogen, or both agents. Control groups received equivalent volumes of the isotonic saline vehicle.
B. Angiotensin II (19). Rats were administered either clonidine (6, 12 or 24 µg/kg, i.p.), or angiotensin II (Sigma, 150 µg/kg, i.p.), or both drugs.
C. Hypertonic NaCl (19). Rats were administered either saline or clonidine (6, 12, or 24 µg/kg, i.p.) at the same time as 1 M NaCl (1 ml/100 g body weight, i.p., warmed to 37° C.).
D. Pilocarpine (19). Rats received either distilled water or clonidine (6 or 12 µg/kg) at the same time as the parasympathomimetic agonist, pilocarpine (3 mg/kg, i.p.).
E. Water Deprivation (19). Rats were dehydrated by 24 hr water deprivation with food available. At the end of this time the rats received either saline or clonidine (6 or 12 µg/kg, i.p.) followed

Table 1

Water intakes (ml/kg body weight) of adult female sprague dawley rats
treated with various dipsogens by various doses of clonidine (i.p.)

Dipsogen	Time Intake Measured	Clonidine (µg/kg)				
		0 (vehicle)	6	12	24	Reference
A. Isoproterenol (25 µg/kg, s.c.)	1 hr	16+2*	ND	10+3[†]	4+3[‡]	17
B. Angiotensin II (150 µg/kg, s.c.)	1 hr	19+1	8+3[‡]	5+3[‡]	2+2[‡]	19
C. Hypertonic Saline (10mM/kg)	1 hr	22+2	13+4[†]	11+3[‡]	6+2[‡]	19
D. Pilocarpine (3 mg/kg, i.p.)	2 hr	12+3	4+2[‡]	4+2[‡]	ND	19
E. Dehydration (24 hr)	1 hr	28+2	22+3	20+2	7+2[‡]	19
F. 5-Hydroxytryptamine (2 mg/kg, s.c.)	2 hr	16+1	6+2[‡]	ND	ND	20
G. 5-Hydroxytryophan (25 mg/kg, s.c.)	1 hr	11+2	5+2[†]	2+1[‡]	ND	21

* One standard error of mean.
† Significantly different from vehicle control (P < 0.05).
‡ Significantly different from vehicle control (P < 0.01).
ND = Not done

by the standard drinking test in the absence of food.
F. 5-Hydroxytryptamine (20). Rats were administered either
clonidine (6 µg/kg, i.p.) or 5-hydroxytryptamine (serotonin) (2 mg/
kg, s.c.), or both.
G. 5-Hydroxytryptophan (21). Rats were administered either
clonidine (6 and 12 µg/kg, i.p.) or 5-hydroxytryptophan (25 mg/kg,
s.c.), or both.

RESULTS

The results are summarized in Table 1. Peripheral administration
of clonidine significantly inhibited the drinking responses to all
of the dipsogens in a dose-related manner. In each case, with the
exception of water deprivation, the first statistically significant
reduction occurred at the lowest dose studied (6 µg/kg). Although
not included in our original study, we have subsequently found this
low dose to be effective against isoproterenol. In the case of
water deprivation, a fourfold higher dose was required. This finding has precedent with many other pharmacological blockers of drinking which are less potent against dehydration than other types of
drinking (22). The thirst induced by hypertonic NaCl is due to
cellular dehydration, while those induced by isoproterenol, pilocarpine, 5-hydroxytryptamine and 5-hydroxytryptophan appear to be
renin-dependent (1,19-21).

Experiment 2   Central Versus Peripheral Site Of Action Of Clonidine
And Angiotensin II

Six separate studies were carried out in which two injections were given either simultaneously or with a delay, as specified. The objective of the delay was to allow for the first drug to be absorbed prior to administration of the second. All rats were fitted with indwelling 23 gauge cannulas terminating in the right lateral cerebral ventricle (IVT). They were implanted at least one week prior to the studies using a stereotaxic method. For IVT injections, the rats were hand-held and 10 µl of solution delivered from a micro-syringe over a 10 sec. period.

A. <u>Peripheral AII and IVT Clonidine (23)</u>. Rats received either saline or clonidine (0.5 or 2 µg/rat, or approximately 2 or 8 µg/kg body weight) followed after a 10 min delay with AII (200 µg/kg, s.c.).

B. <u>Peripheral Clonidine And IVT AII (23)</u>. Rats received either saline or clonidine (12 µg/kg, i.p.) followed after a 5 min delay by AII (1 or 5 ng IVT or approximately 4 and 20 ng/kg body weight).

C. <u>Peripheral Clonidine And IVT Carbachol (23)</u>. Rats received either saline or clonidine (12 µg/kg, i.p.) followed after a 5 min delay by the cholinergic agonist, carbachol (300 ng IVT or approximately 1.2 µg/kg).

D. <u>IVT Clonidine And IVT AII (23)</u>. Clonidine (2 µg/rat or 8 µg/kg, IVT), AII (5 ng/rat or 20 ng/kg, IVT), or the combination were given 10 min apart.

RESULTS

For brevity, the results are expressed in Table 2 as the percentage change from the corresponding group that received AII alone (either IVT or s.c.). In each case the AII-treated groups drank 11 - 22 ml/kg during the first hour.

Clonidine at a dose of 2 µg IVT, but not 0.5 µg, attenuated the water intake induced by peripheral angiotensin II (Table 2). In an additional control experiment, IVT clonidine (2 and 3 µg) alone did not stimulate drinking (data not shown). Peripheral administration of clonidine (12 µg/kg, i.p.) inhibited the drinking response to centrally administered AII (1 ng IVT). However, an increase in the

Table 2

Effects of various combinations of cerebroventricular (IVT) and peripheral administrations of clonidine and all on water intake of female rats (EXPT. 2)

Study	Treatment[‡] 1	2	Water Intake (% of AII) 0.5	1.0 hr
A.	Clonidine (0.5 µg IVT)	AII (200 µg/kg s.c.)	72 ± 22*	108 ± 16
A.	Clonidine (2 µg IVT)	AII (200 µg/kg s.c.)	20 ± 6[†]	25 ± 10[†]
B.	Clonidine (12 µg/kg, ip)	AII (1 ng IVT)	27 ± 9[†]	27 ± 9[†]
B.	Clonidine (12 µg/kg, ip)	AII (5 ng IVT)	120 ± 15	105 ± 17
C.	Clonidine (12 µg/kg, ip)	Carbachol (300 ng IVT)	25 ± 12[†]	44 ± 12[†]
D.	Clonidine (2 µg IVT)	AII (5 ng IVT)	50 ± 15[†]	50 ± 20

*One standard error of mean.
[†]Significantly different from control group or treatment with clonidine (P<0.01).
[‡]Treatment 1 was given 5 min prior to angiotensin II except in study D where the period was 15 min.

dose of AII to 5 ng IVT apparently overcame the antidipsogenic effect of this dose of clonidine. Peripherally administered clonidine also inhibited the drinking response to centrally administered carbachol (300 ng IVT). When both clonidine (2 ng) and AII (5 ng) were administered IVT, the dipsogenic effect of AII was inhibited.

Experiment 3   Effect Of Peripherally Administered Yohimbine On AII-Induced Drinking, And The Antidipsogenic Effect Of Clonidine

A. Isoproterenol (24). Rats received either distilled water or yohimbine (150, 300 or 600 µg/kg, s.c.) at the same time as isoproterenol (25 µg/kg, s.c.).
B. Angiotensin II (24). Rats received either distilled water or yohimbine (150 or 300 µg/kg, s.c.) at the same time as AII (200 µg/kg, s.c.).
C. Clonidine Plus Angiotensin II (24). Rats received AII (200 µg/kg, s.c.) as the dipsogen. They also received either vehicle, clonidine (8 µg/rat or 40 µg/kg, IVT), yohimbine (300 µg/kg, i.p.), or clonidine plus yohimbine. In this study yohimbine was given 20 min, and clonidine 10 min, prior to AII.

RESULTS

Peripheral administration of yohimbine significantly attenuated the water intake induced by peripheral administration of angiotensin II or isoproterenol (Sections A and B, Table 3). Study C showed that the inhibition of AII-induced drinking by IVT clonidine could

Table 3

Effect of yohimbine on water intake induced by isoproterenol and angiotensin II

	Water Intake (ml/kg) During:	
Experimenatl Group	1.0	2.0 hr
A. Isoproterenol (25 µg/kg)		
Control	15 + 1*	16 + 2
Yohimbine (250 µg/kg, s.c.)	16 + 3	17 + 3
Yohimbine (500 µg/kg, s.c.)	30 + 2‡	33 + 3‡
B. Angiotensin II (200 µg/kg, s.c.)		
Control	7 + 2	8 + 2
Yohimbine (150 µg/kg, s.c.)	12 + 2	13 + 2
Yohimbine (300 µg/kg, s.c.)	14 + 2†	14 + 2†

* One standard error of mean
† Significantly different from control (P <0.05).
‡ Significantly different from control (P <0.01).

513

be reversed by yohimbine (Table 4). Similar results have been obtained with peripheral clonidine (data not shown).

Experiment 4 Effect of IVT Administration Of m-Octopamine And m-Synephrine On AII-Induced Water Intake

The rats had indwelling IVT cannulas as discussed above. They first received either distilled water or yohimbine (300 µg/kg, s.c.). This was followed by IVT injection of either saline or m-octopamine (10 µg, IVT) or m-synephrine (20 µg, IVT). After another delay, all rats received AII (200 µg/kg, s.c.) and the drinking test was begun.

Table 4

Reversal by yohimbine (300 µg/kg, sc) of the antidipsogenic effect of clonidine, m-octopamine, and m-synephrine

Treatments #		Cumulative Water Intakes (ml/kg)		
1 (IP)	2 (IVT)	0.5	1.0	2.0 hr
Experiment 3 (Clonidine)				
Vehicle	Vehicle	14 ± 2*	14 ± 2	16 ± 2
Vehicle	Clonidine (8 µg)	5 ± 1‡	6 ± 1‡	7 ± 1†
Yohimbine	Vehicle	17 ± 3	19 ± 3	24 ± 2
Yohimbine	Clonidine	24 ± 3†	24 ± 4†	30 ± 4‡
Experiment 4 (m-Octopamine)				
Vehicle	Vehicle	14 ± 2	18 ± 2	18 ± 2
Vehicle	m-Octopamine (10 µg)	6 ± 1‡	9 ± 2‡	9 ± 2‡
Yohimbine	Vehicle	15 ± 2	21 ± 3	22 ± 3
Yohimbine	m-Octopamine	15 ± 3	23 ± 3	25 ± 3
Experiment 4 (m-Synephrine)				
Vehicle	m-Synephrine (20 µg)	5 ± 3	11 ± 3	15 ± 3
Yohimbine	m-Synephrine	19 ± 6¶	28 ± 7¶	40 ± 6¶

* One Standard Error of Mean
# All rats received AII (200 µg/kg) as a third and final treatment. The delays between treatments 1 and 2, and treatments 2 and 3, were 10 min (Experiment 3) or 5 min (Experiment 4) as described in the text.
† Significantly different from the control group receiving AII alone ($P < 0.05$).
‡ Significantly different from the control group receiving AII alone ($P < 0.01$).
¶ Significantly different from vehicle-m-synephrine group ($P < 0.01$).

RESULTS

The IVT administration of m-octopamine and m-synephrine significantly inhibited AII-induced water intake (Table 4). The o- and p-isomers of these agents were largely ineffective (25). Yohimbine completely reversed the antidipsogenic effects of m-octopamine and m-synephrine on AII-induced drinking (Table 4).

DISCUSSION

Acute peripheral administration of clonidine, an α_2-adrenoceptor agonist, to rats inhibited their drinking response to a number of dipsogenic stimuli including isoproterenol, angiotensin II, hypertonic saline, pilocarpine, dehydration, 5-hydroxytryptamine and 5-hydroxytryptophan (Table 1) (17,19,20,21). The dehydration data confirm an earlier report (26).

Administration of either yohimbine or tolazoline, both α_2-adrenoceptor antagonists, significantly enhanced the water intake induced either by isoproterenol or AII (Table 3) but had no effect on basal or unstimulated day time drinking (24). These studies thus suggest that α_2-adrenoceptors play a bidirectional role in modulating water intake induced by isoproterenol and AII.

We have further established a dose-inhibition relationship between the dose of clonidine administered IVT and inhibition of the drinking response to peripherally administered AII (25). The drinking response was reduced to about 50% by a dose of 4 μg/kg IVT. The dipsogenic response to centrally administered AII was attenuated by both peripheral and IVT administration of clonidine. Yohimbine also reversed the antidipsogenic effect of centrally administered clonidine on AII-induced water intake (Table 4). These results support the concept that clonidine exerts its antidipsogenic effect via α_2-adrenoceptors in the brain.

The area of the central nervous system at which clonidine acts to inhibit AII, as well as carbachol-induced drinking is not addressed in these studies. These central dipsogens mediate drinking in quite different ways (1), but may provide convergent inputs to a final common pathway for drinking (27,28). Although such a pathway has not been identified anatomically, clonidine probably acts at or beyond the level of any integrative mechanism for the various types of drinking (29).

The effective IVT antidipsogenic dose of clonidine is similar to, or only slightly less than, an equi-effective peripheral dose. One possible explanation for this finding is that IVT clonidine may not act within the brain, but instead virtually all of it may leak into the periphery. However, the effective doses of clonidine are of the order of those found to have cardiovascular effects when injected directly into the nucleus tractus solitarius (30,31). These and other data reviewed elsewhere (23) constitute indirect evidence against the "leak hypothesis". A second and more interesting possibility is that peripherally-administered clonidine may have better access to the effective site in the brain than after its IVT administration. This proposition rests upon the initial assumption that only a small fraction of a peripheral dose of clonidine will actually be available to the brain.

The effect of IVT administration of m-octopamine and its metabolite, m-synephrine, on AII-induced drinking was also studied. Both compounds attenuated AII-induced drinking. The antidipsogenic effect of these compounds was blocked by prior administration of yohimbine. These results suggest that m-octopamine and m-synephrine also exert their antidipsogenic effect via α_2-adrenoceptors.

Additional studies reviewed elsewhere (22,29) have shown that naloxone, an opioid antagonist, can inhibit the dipsogenic responses to both antiotensin II and isoproterenol. Yohimbine reversed the antidipsogenic effect of naloxone on AII-induced drinking while clonidine potentiated the suppression by this dose of naloxone. These results suggest that the α_2-adrenoceptor mechanisms described above interact with or are modulated by opioid systems.

Both opioid agonists (32,33) and antagonists (22,29) suppress experimentally-induced drinking in rats, although naloxone can reverse the antidipsogenic effects of enkephalins and endorphins (32,33). It has been inferred from these results that the endogenous opioids are involved in both AII and osmoreceptor drinking pathways. We have built upon the models of Oatley (34) and Toates (28) to provide a partial explanation both for the antidipsogenic effect of clonidine and the apparent paradox that both opioid agonists and antagonists suppress drinking in rats. The new model (Figure 1) involves a simplified diagram of two receptors (P_1 and P_2), which represent the two drinking pathways (osmoreceptor and AII) and their convergence to form a final common pathway (F_1 and F_2). Since clonidine and naloxone inhibit responses to stimulation of both drinking pathways (P_1 and P_2), then their actions are likely to be at some point on the final common pathway (I and II). Whether the sensitive sites are at the same terminal, or in series (as shown in III), cannot be determined at this time. Although experimental evidence that m-octopamine and m-synephrine can inhibit hypertonic saline- and carbachol-induced drinking is not yet available, it seems likely that these compounds act at similar sites.

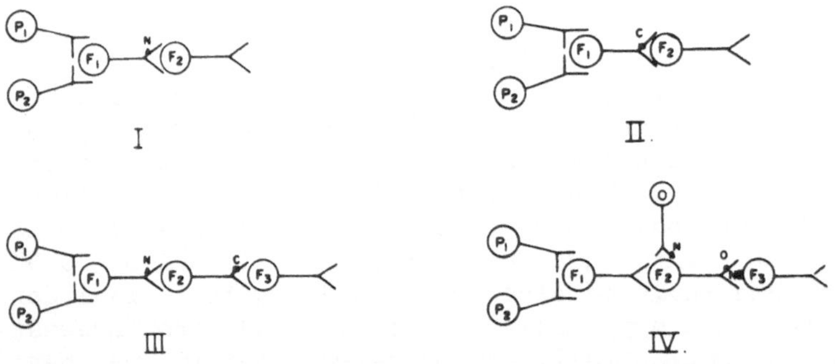

Figure 1. A model illustrating the possible sites of naloxone (N), clonidine (C), and endogenous opioids (O) along the final pathway involved in drinking responses. P_1 and P_2 represent the AII and osmoreceptor pathways, respectively, which converge into a final pathway (F_1, F_2 and F_3). NE = Norepinephrine. (From 29).

The fourth panel (IV) considers the suppressive effects of the endogenous opioids (O) on drinking induced by stimulation of osmoreceptor and AII pathways. Analogous to the actions of naloxone and clonidine, enkephalins and β-endorphin may also exert their effect along the final common pathway. The possibility exists that α_2-adrenoceptors may be involved in opioid inhibition of experimentally-induced drinking. In addition, the antidipsogenic effect of naloxone in the absence of opioid influence may not be opioid-related, but rather a consequence of α-adrenergic mediation. Thus, the inhibition of experimentally-induced drinking by naloxone, as well as endogenous opioids, may be manifested by inhibition of norepinephrine (NE) release from pre-synaptic α-adrenergic terminals. This suggestion must, however, remain speculative in the absence of any other evidence that the effects of opioid peptides or naloxone involve direct mediation by α-adrenoceptors.

REFERENCES

1. Fitzsimons, J.T., The Physiology Of Thirst And Salt Appetite, Cambridge Univ. Press, Cambridge (1979).
2. Lehr, D., J. Mallow and K. Krukowski, Copious drinking and simultaneous inhibition of urine flow elicited by beta-adrenergic stimulation and contrary effect of alpha-stimulation, J. Pharm. Exp. Therap. 158:150 (1967).
3. Peskar, B., D.K. Meyer, U. Tauschman and G. Herting, Influence of isoproterenol, hydralazine, and phentolamine on the renin activity of plasma and renal cortex of rats, Europ. J. Pharmacol. 9:394 (1970).
4. Weinberger, M.H., W. Aoi and D.P. Henry, Direct effect of beta-adrenergic stimulation on renin release by the rat kidney slice in vitro, Circ. Res. 37:318 (1975).
5. Abraham, S.F., R.N. Baker, E.H. Blaine, D.A. Denton and M.J. McKinley, Water drinking induced in sheep by angiotensin- a physiological or pharmacological effect? J. Comp. Physiol. Psychol. 88:503 (1975).
6. Stricker, E.M. and W.G. Bradshaw, The renin-angiotensin system in thirst: a reevaluation, Science 194:1169 (1976).
7. Tang, M. and J.L. Falk, Sar[1]-ala[8]-angiotensin II blocks renin angiotensin but not beta adrenergic thirst, Pharmac. Biochem. Behav. 2:401 (1974).
8. Buggy, J. and A.K. Johnson, Angiotensin-induced thirst: effects of third ventricle obstruction and periventricular ablation, Brain Res. 149:117 (1978).
9. Epstein, A.N., J.T. Fitzsimons and B. Rolls, Drinking induced by angiotensin in the brain of the rat, J. Physiol. (London) 210: 457 (1970).
10. Mogenson, G.J. and J. Kucharcyk, Evidence that the lateral hypothalamus and mid-brain participate in drinking elicited by intra-cranial angiotensin, in Control Mechanisms of Drinking, eds. G. Peters, J.T. Fitzsimons and L. Peters-Haefeli, Springer-Verlag, New York, pp. 127 (1975).
11. Simpson, J.B. and A. Routtenberg, Subfornical organ: site of drinking elicitation by angiotensin II, Science 181:1172 (1973).
12. Hoffman, W.E. and M.I. Phillips, Evidence for sar[1]-ala[8]-angio-tensin crossing the blood cerebrospinal fluid barrier to ant-

gonize central effects of angiotensin II, Brain Res. 109:541 (1976).

13. Hirose, S., H. Yokosawa and T. Inagami, Immunochemical ident-
    ification of the renin in rat brain and distinction from acid
    proteases, Nature 274:392 (1978).

14. Sirett, N.E., A.S. McLean, J.J. Bray and J.I. Hubbard, Distribut-
    ion of angiotensin II receptors in rat brain, Brain Res. 122:
    299 (1977).

15. Laubie, M. and M. Schmitt, Sites of action of clonidine: Central-
    ly mediated increase in vagal tone, centrally mediated hypoten-
    sive and sympatho-inhibitory effects, Prog. Brain Res. 47:337
    (1977).

16. Pettinger, W.A., T.K. Keeton, W.B. Campbell and D.C. Harper,
    Evidence for a renal α-adrenergic receptor inhibiting renin
    release, Circ. Res. 38:338 (1976).

17. Fregly, M.J. and D.L. Kelleher, Antidipsogenic effect of clonid-
    ine on isoproterenol-induced water intake, Appetite 1:279 (1980).

18. Goldstein, A., Biostatistics, An Introductory Text, MacMillan,
    New York (1964).

19. Fregly, M.J., D.L. Kelleher and J.E. Greenleaf, Antidipsogenic
    effect of clonidine on angiotensin II-, hypertonic saline-, pilo-
    carpine-, and dehydration-induced water intakes, Brain Res. Bull.
    7:661 (1981).

20. Kikta, D.C., C.C. Barney, R.M. Threatte, M.J. Fregly, N.E. Rowland
    and J.E. Greenleaf, On the mechanism of serotonin-induced dipso-
    genesis in the rat, Pharmac. Biochem. Behav. 19:519 (1983).

21. Threatte, R.M., M.J. Fregly, T.M. Connor and D.C. Kikta, L-5-
    Hydroxytryptophan-induced drinking in rats: Possible mechanisms
    for induction, Pharmac. Biochem. Behav. 14:385 (1981).

22. Rowland, N., Comparison of the suppression by naloxone of water
    intake induced in rats by hyperosmolality, hypovolemia and angio-
    tensin, Pharmac. Biochem. Behav. 16:87 (1982).

23. Fregly, M.J., N.E. Rowland and J.E. Greenleaf, Clonidine antag-
    onism of angiotensin-related drinking: a central site of action,
    Brain Res. 298:321 (1984).

24. Fregly, M.J., N.E. Rowland and J.E. Greenleaf, Effects of yohim-
    bine and tolazoline on isoproterenol- and angiotensin II-induced
    water intake in rats, Brain Res. Bull. 10:121 (1983).

25. Fregly, M.J., N.E. Rowland and J.E. Greenleaf, A role for pre-
    synaptic α2-adrenoceptors in angiotensin II-induced drinking in
    rats, Brain Res. Bull. (in press).

26. LeDouarec, J.C., H. Schmitt and B. Lucet, Influence de la clon-
    idine et des substances α-sympathomimetique sur la prise d'eau
    chez le rat assoiffe, J. Pharmac. (Paris) 2:435 (1971).

27. Greenleaf, J.E. and M.J. Fregly, Dehydration-induced drinking:
    peripheral and central aspects, Fed. Proc. 41:2507 (1982).

28. Toates, F.M., Homeostasis and drinking, Behav. Brain Sci. 2:
    95 (1979).

29. Wilson, K.M., N. Rowland and M.J. Fregly, Drinking: A final
    common pathway? Appetite (in press).

30. Rockhold, R.W. and R.W. Caldwell, Effect of lesions of the
    nucleus tractus solitarii on the cardiovascular actions of
    clonidine in conscious rats, Neuropharmacology 18:347 (1979).
31. Rockhold, R.W. and R.W. Caldwell, Cardiovascular effects follow-
    ing clonidine microinjection into the nucleus tractus solitarii
    of the rat, Neuropharmacology 19:919 (1980).
32. Summy-Long, J.Y., L.C. Keil, K. Deen, L. Rosella and W.B. Severs,
    Endogenous opioid peptide inhibition of the central actions of
    angiotensin, J. Pharm. Exp. Therap. 217:619 (1981).
33. Summy-Long, J.Y., L.M. Rosella and L.C. Keil, Effects of central-
    ly administered endogenous opioid peptides on drinking behavior,
    increased plasma vasopressin concentration and pressor response
    to hypertonic sodium chloride, Brain Res. 221:343 (1981).
34. Oatley, K., Stimulation and theory of thirst, in The Neuro-
    psychology Of Thirst: New Findings And Advances In Concepts,
    eds. A.N. Epstein, H.R. Kissileff and E. Stellar, Winston,
    Washington, D.C., p. 199 (1973).

_____

[1]Supported by contract NCA2-OR204-101 from NASA, Moffett Field, CA.

HUMAN THIRST: THE CONTROLS OF WATER INTAKE IN HEALTHY MEN

Barbara J. Rolls*, P.A. Phillips, J.G.G. Ledingham
M.L. Forsling, J.J. Morton and M.J. Crowe

*Department of Psychiatry, Johns Hopkins University
School of Medicine, Baltimore, Maryland
Department of Experimental Psychology, University
of Oxford, Oxford OX1 3UD, England

The purpose of these experiments was to apply the knowledge gained from experiments in animals to understanding the controls of water intake in healthy men. The unique advantage of studying thirst in humans is that the measurement of physiological parameters can be combined with the simultaneous assessment of subjective sensations associated with thirst and fluid intake. In the series of experiments to be described, factors involved in the initiation and termination of drinking are considered as well as changes in thirst which are associated with aging.

INITIATION OF DRINKING

Following 24 hr water deprivation, plasma osmolality, sodium, and protein, and thirst sensations assessed by visual analog rating scales, were increased significantly,[1] but hematocrit and plasma angiotensin II concentrations were not altered significantly.[1,2] To determine which stimuli may be involved in the initiation of drinking, hypertonic saline and angiotensin II were infused IV in fluid replete subjects.

Hypertonic Saline

The effects of double-blind IV infusions of hypertonic (0.45M) and isotonic (0.15M) saline were compared in seven healthy young men. Only the hypertonic saline significantly increased plasma sodium concentration, plasma osmolality, plasma vasopressin concentration, or visual analog ratings of thirst sensations (fig. 1). Throughout a 60 min drinking period, following the infusions, water intake was significantly greater after the hypertonic saline than after the isotonic saline (fig. 1). The main subjective sensation of thirst was a dry, unpleasant tasting mouth. The experiment clearly demonstrates that hypertonic saline stimulates thirst and water intake in man at plasma sodium and osmolality levels within the physiological range and indicates that cellular dehydration is likely to be an important thirst stimulus in man.[3,4]

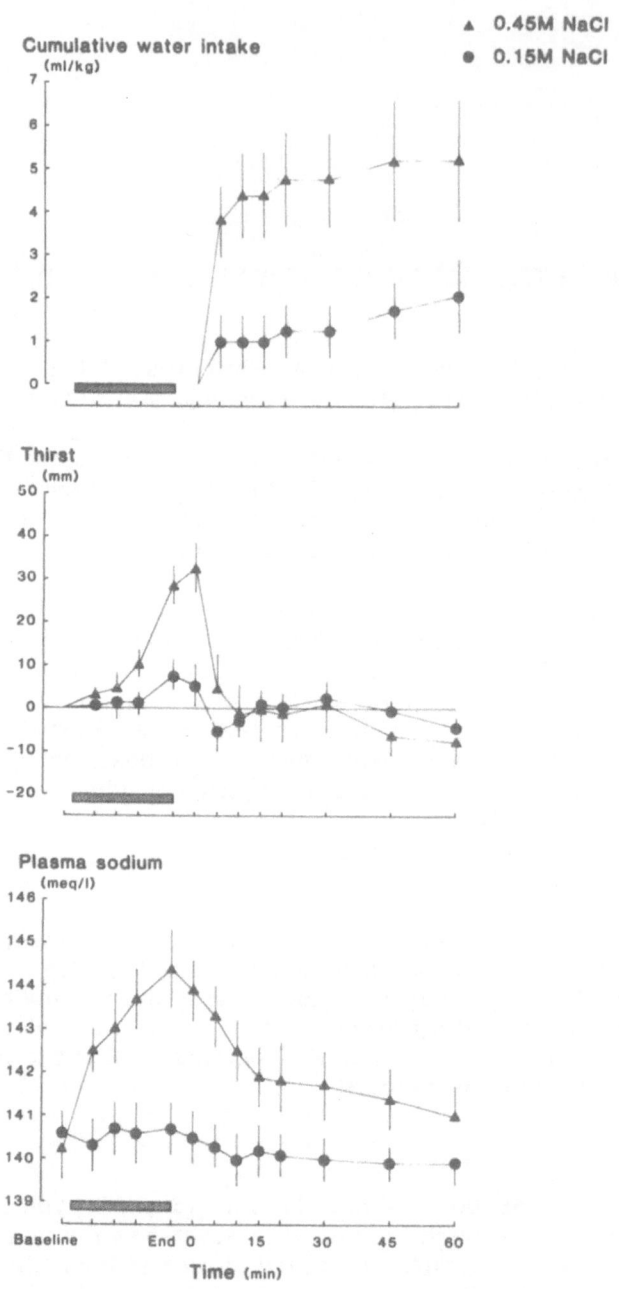

Figure 1: The effect of double-blind IV infusions of hypertonic saline (0.45 M NaCl) and isotonic saline (0.15 M NaCl) on mean cumulative water intake, changes in subjective ratings of thirst, and plasma sodium concentration in fluid replete men. The hatched bar indicates the infusion period.

522

Angiotensin II

Although angiotensin has been implicated in the thirst of malignant and renovascular hypertension in man, the effects of IV infusions of angiotensin on thirst in man have not previously been investigated. Therefore, angiotensin II-amide (Hypertensin, CIBA) was infused IV into 10 fluid replete young men at mildly pressor doses (2-16 ng/kg/min for 60 min). It was found that four of the 10 subjects responded to the angiotensin infusion with significantly greater water intakes compared to the single blind isotonic saline control condition (fig. 2). The responders also showed the greatest increases in thirst, mouth dryness and plasma vasopressin levels (fig.2). There were no differences between responders and nonresponders in plasma AII levels, other plasma variables, or in the elevation of blood pressure. Thus, angiotensin can stimulate thirst and water intake in some men, but the plasma levels of AII in this study (19.6 ± 1.7 to 442.1 ± 83.4 pg/ml) were far in excess of those seen under physiological conditions. This finding coupled with the observation that there was no change in plasma AII after 24 hr fluid deprivation, suggests that angiotensin may not play a significant role in the normal physiological regulation of human thirst. However, more studies are required to determine under what conditions angiotensin may be an important stimulus to thirst in man, e.g. some pathological states. The general importance of hypovolemia for thirst in man is still poorly understood. [2,3]

Ad Libitum Thirst and Drinking

To investigate whether human thirst and drinking during ad libitum access to water occur in response to body fluid deficits, blood samples and visual analog scale thirst ratings were obtained from five young men at hourly intervals and when they were thirsty during a normal working day. Although there were significant increases in ratings of thirst, pleasantness of drinking water, mouth dryness and unpleasantness of the taste in the mouth when subjects were thirsty enough to drink compared with intervening intervals, there were no concomitant changes in body fluid variables (hematocrit, plasma osmolality, sodium, potassium, protein, and angiotensin II concentrations). Subjects drank mainly in association with eating and were not overhydrated as indicated by constantly hypertonic urine and significant tubular reabsorption of free water over the experimental period. The results indicate that during free access to water, humans become thirsty and drink before body fluid deficits develop, perhaps in response to subtle oropharyngeal cues, and so provide evidence for anticipatory thirst and drinking in man. [5]

DRINKING TERMINATION

Rehydration following 24 hr fluid deprivation was investigated to determine what factors could be involved in the termination of drinking in man. The dehydration caused significant cellular and extracellular fluid depletions and was associated with a dry, unpleasant tasting mouth. During rehydration, subjects drank 65% of their total intake within 2.5 min. The marked decrease in drinking rate thereafter, and the alleviation of thirst, occurred before plasma dilution had become significant. This attenuation of drinking was attributed to stomach fullness. Presystemic factors may therefore be important for drinking termination in humans. [1]

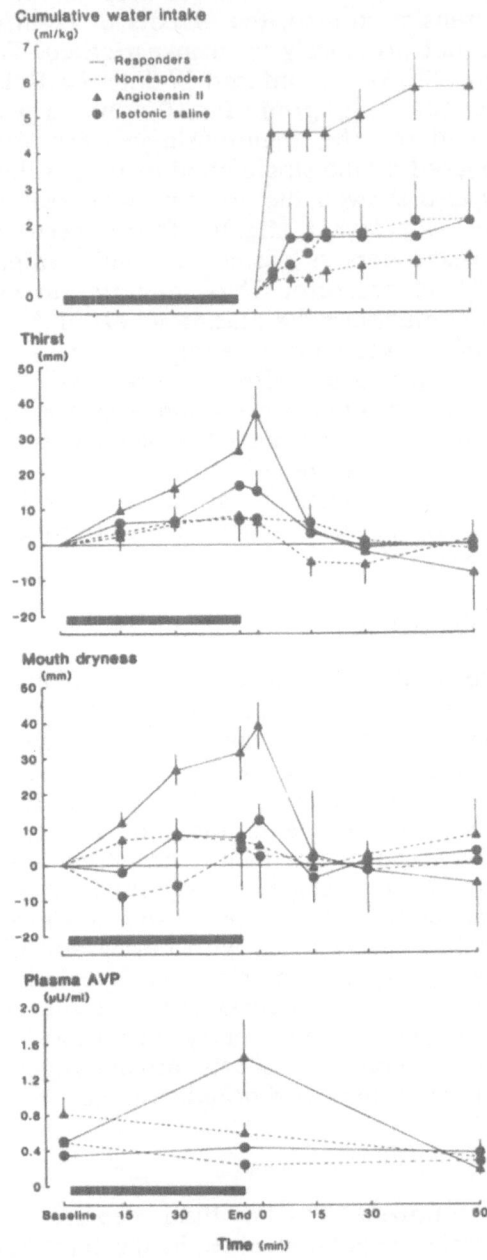

Figure 2: The effect of single-blind IV infusions of angiotensin II or isotonic saline on the mean cumulative water intake, changes in subjective ratings of thirst and mouth dryness, and plasma vasopressin concentrations for subjects that drank in response to angiotensin (responders, n=4) and subjects that did not drink (nonresponders, n=6). The hatched bar indicates the infusion period.

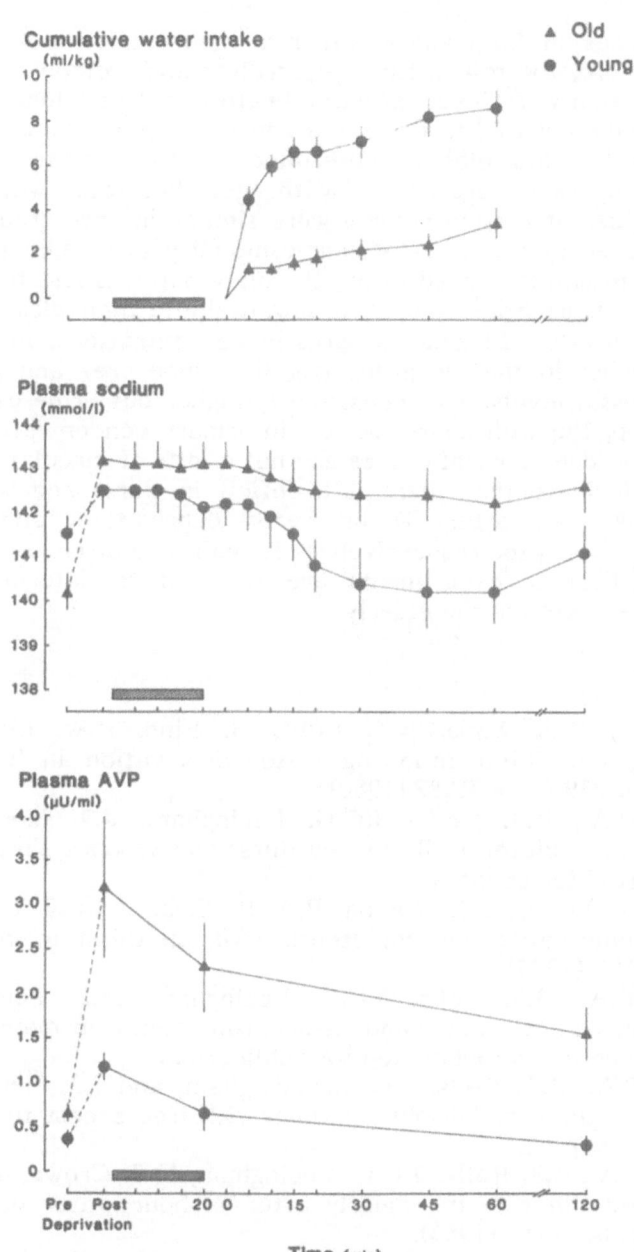

Figure 3: Mean cumulative water intakes, plasma sodium concentrations and plasma vasopressin levels for old and young subjects before and after 24 hr fluid deprivation, during a 20 min sham infusion (hatched bar), and during rehydration (0-120 min). Despite greater increases in plasma sodium, the elderly subjects experienced little thirst and drank insufficient water to rehydrate the body fluids.

REDUCED THIRST IN THE ELDERLY

Disturbances in fluid and electrolyte homeostasis are common in the elderly. Therefore, the thirst, fluid, electrolyte and hormonal responses to 24 hr fluid deprivation were investigated in healthy, active, elderly men (n=7, 67-75 years old) and compared to those in a young group matched for body weight loss (n=7, 20-31 years old) to determine whether thirst and vasopressin responses to dehydration are altered with age. Following water deprivation, changes in indices of plasma volume were similar in both groups, but the old group had greater increases in plasma osmolality and sodium concentration (fig. 3). Despite similar dehydration, the old group was less thirsty and drank much less after deprivation (fig. 3), failing to dilute their plasma and urine to predeprivation levels. Plasma vasopressin concentrations, in contrast, were consistently higher in the old group (fig.3). Since pre- and postdeprivation plasma vasopressin levels were consistently higher but urine osmolality lower in the old group, the well known deficit in urinary concentrating ability with age seems to be due to renal causes and not a lack of circulating vasopressin. These results indicate that there is a deficit in thirst and water intake in healthy elderly men after 24 hr water deprivation, and suggest that vasopressin osmoreceptor responsiveness is maintained, or even increased, in the elderly.[6,7] Further experiments are required to determine why thirst sensitivity is decreased in the elderly.

REFERENCES

1. Rolls, B.J., R.J. Wood, E.T. Rolls, H. Lind, R.W. Lind, and J.G.G. Ledingham, Thirst following water deprivation in humans, Am. J. Physiol. 239:R476-R482 (1980).
2. Phillips, P.A., B.J. Rolls, J.G.G. Ledingham, J.J. Morton, and M.L. Forsling, Angiotensin II-induced thirst and vasopressin release in man. Submitted for publication.
3. Ledingham, J.G.G., J.J. Morton, P.A. Phillips, and B.J. Rolls, Effects of hypertonic saline and angiotensin (AII) on thirst in man, J. Physiol. 345: 114P (1983).
4. Phillips, P.A., B.J. Rolls, J.G.G. Ledingham, M.L. Forsling, and J.J. Morton, Osmotic thirst and vasopressin release in man: a double-blind crossover study. Submitted for publication.
5. Phillips, P.A., B.J. Rolls, J.G.G. Ledingham, and J.J. Morton, Body fluid changes, thirst and drinking in man with free access to water, Physiol. Behav. In press.
6. Phillips, P.A., B.J. Rolls, J.G.G. Ledingham, M.J. Crowe, and L. Wollner, Reduced thirst in the elderly after 24-hour water deprivation Clin. Sci. 64: 61P-62P (1983).
7. Phillips, P.A., B.J. Rolls, J.G.G. Ledingham, M.L. Forsling, J.J. Morton, M.J. Crowe, and L. Wollner, Reduced thirst following water deprivation in healthy elderly men. Submitted for publication.

EFFECTS OF NARCOTIC ANALGESICS ON WATER AND FOOD INTAKE IN NORMAL RATS

Yerner Arslan, Ruth Burckhardt, Kovvuri Jawaharlal,
Kurt Ornstein and Georges Peters

Institue de Pharmacologie de l'Université
CH-1011 Lausanne (Switzerland) *)
*) (Supported by Swiss National Science Foundation
Grant Nr. #.220-082)

INTRODUCTION

S.c. morphine (0.5 to 8.0 mg/kg) in normal satiated rats induces water and food intake (1,3) beginning 1-2 hours and terminated 6 hours after injection, when given during the day (light) but not during the night (dark) period (4,5). Water and food ingestion is depressed between 6 and 24 hours after injection (3). The drinking and eating responses appear to be independent of each other. A second morphine injection 6 hours after the first induces drinking, but no eating.   Bilateral nephrectomy suppresses the drinking, but not the eating response (4). With daily repetition of the injections, no tolerance develops within 17 days for the drinking and eating responses to 0.5 to 16 mg/kg of morphine (2,3).

The present experiments were undertaken in order to determine whether other analgesic agents produce similar responses, i.e. whether the drinking and eating response is related to the analgesic activity. Furthermore, the effect of morphine receptor blocking agents was investigated.

MATERIAL AND METHODS

The experimental animals were Wistar type rats of a strain bred by Tierzucht Institut of the University of Zurich, Switzerland. During experiments they were kept in individual cages with permanent free access to water (drinking bottle) and food (rat pellets containing approximately 100 mEq/kg pellet Na$^+$),   available continuously. They were kept in a temperature controlled (20-25° C) room with alternate 12 hour (6.00 a.m. to 6.00 p.m.

(light) and dark periods. Injections were given s.c. between 8.00 and 10.00 a.m. Water and food intake was then measured, hourly for 6 hours and subsequently for the period from the 7th to the end of the 24th hour. Water and food intake was expressed per kg b.w. and indicated in the figures for the periods from the end of the 2nd to the end of the 6th hour, and from the end of the 6th to the end of the 24th hour after injection; in one graph, food and water intake is also indicated for the period from injection to the end of the 2nd hour. All numerical data are given as means ± SEM. The analgesic agents morphine, levorphanol, methadone, pethidine, tilidine and pentazocine were given in dose ranges approximately equi-analgesic with morphine. The antagonist used was naltrexone injected 10 min. before the analgesic agents. Each group of rats comprized 6-10 animals. When animals were used repeatedly 3-week intervals were observed.

RESULTS

The drinking response to morphine was usually preceded by an equally dose-dependent depression of water-intake during the first 2 hours after injection (Fig. 1).

Morphine induced its known dose-dependent eating and drinking response (Fig. 2).

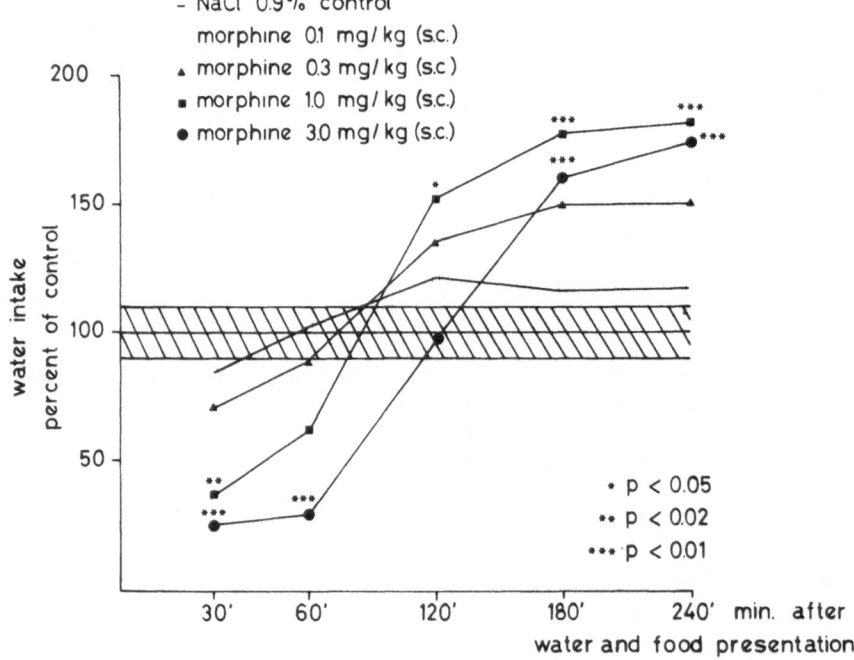

Fig. 1.

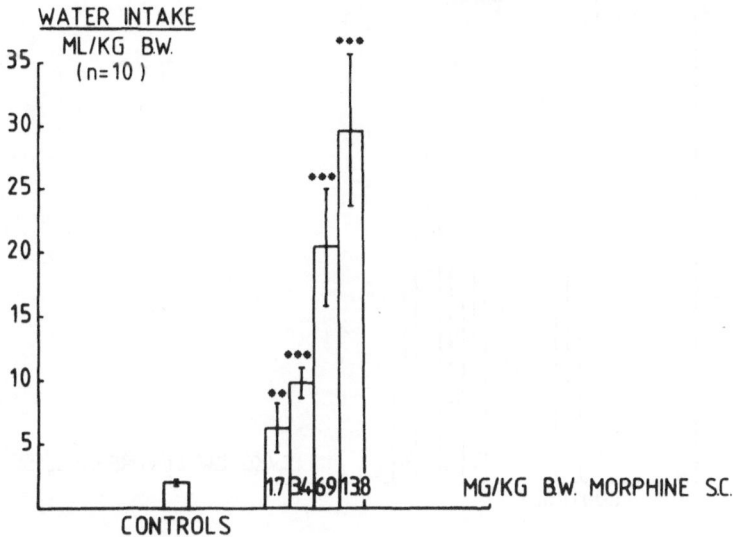

Fig. 2

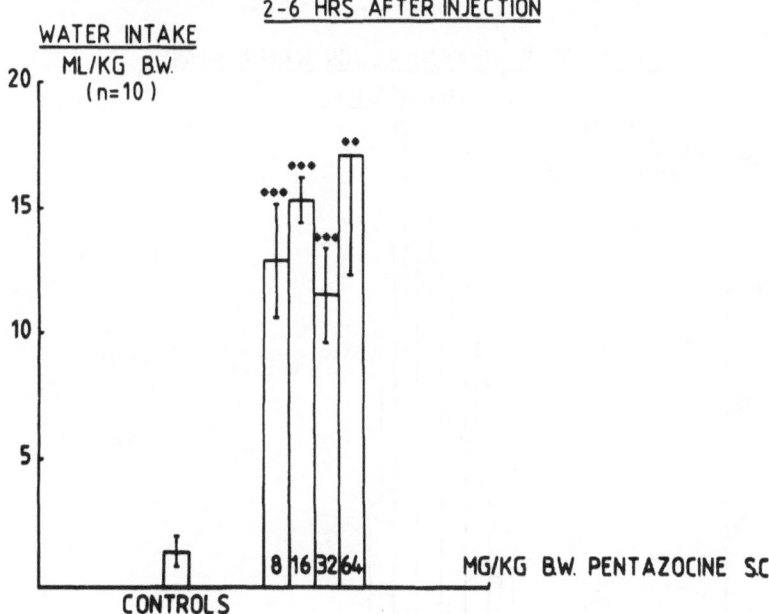

Fig. 3

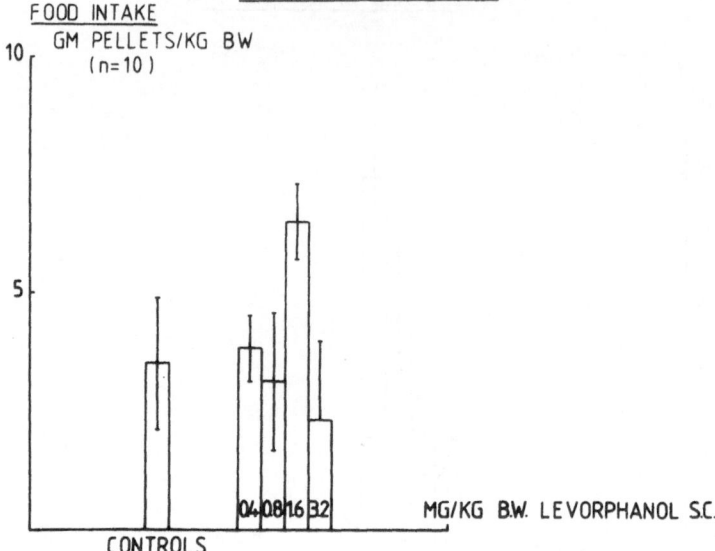

INFLUENCE OF S.C. LEVORPHANOL ON FOOD INTAKE
2-6 HRS AFTER INJECTION

FOOD INTAKE
GM PELLETS/KG BW
(n=10)

04 08 16 32   MG/KG B.W. LEVORPHANOL S.C.

CONTROLS

Fig. 4

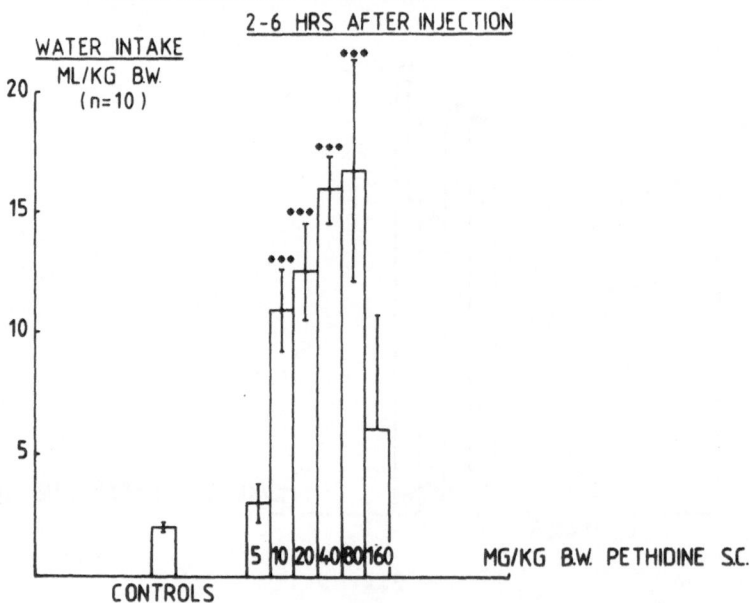

INFLUENCE OF S.C. PETHIDINE ON WATER INTAKE
2-6 HRS AFTER INJECTION

WATER INTAKE
ML/KG B.W.
(n=10)

5 10 20 40 80 160   MG/KG B.W. PETHIDINE S.C.

CONTROLS

Fig. 5

The ingestive responses were followed by a dose-dependent depression of water and food intake during the 6th-24th hour period (not shown).

Pentazocine (8-64 mg/kg) (Fig. 3) and levorphanol (0.4-3.2 mg/kg) produced a similar drinking, but a less conspicuous eating response between 2-6 hours after injection (Fig. 3), equally followed by a depression of ingestion 6-24 hours after injection.

Pethidine (5-80 mg/kg) produced a dose-dependent very marked drinking response (Fig. 5).

Pethidine, however, did not increase but rather depressed eating during the 2-6 hour periods (Fig. 6).

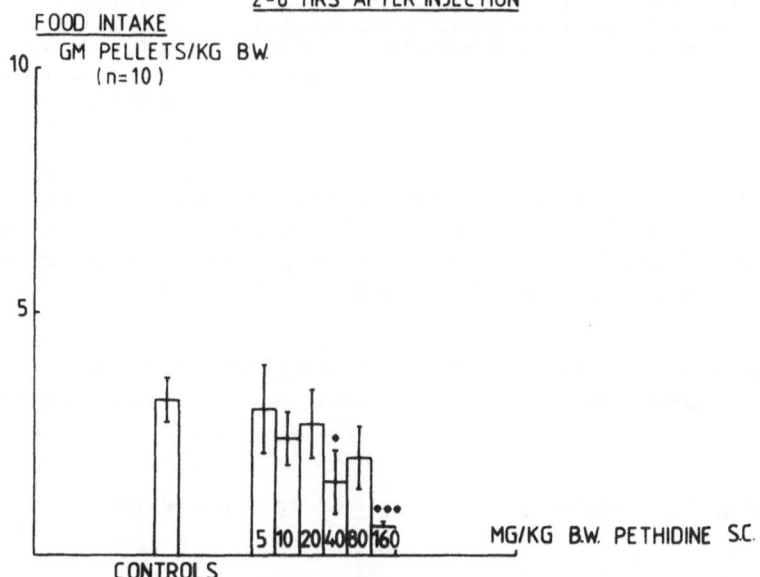

Fig. 6

Like pethidine, methadone (2-8 mg/kg) induced drinking, but no eating.

Tilidine (40-80mg/kg) induced a strong acute drinking response 2-6 hours after injection but a depression rather than an increase of food intake during the same period.

In spite of this fact, both food and water intake were dose dependently depressed during the 6-24 hour period (Fig. 5).

The drinking responses to morphine, (Fig. 7) levorphanol, pethidine and methadone were competitively inhibited by 0.3 mg/kg naltrexone.

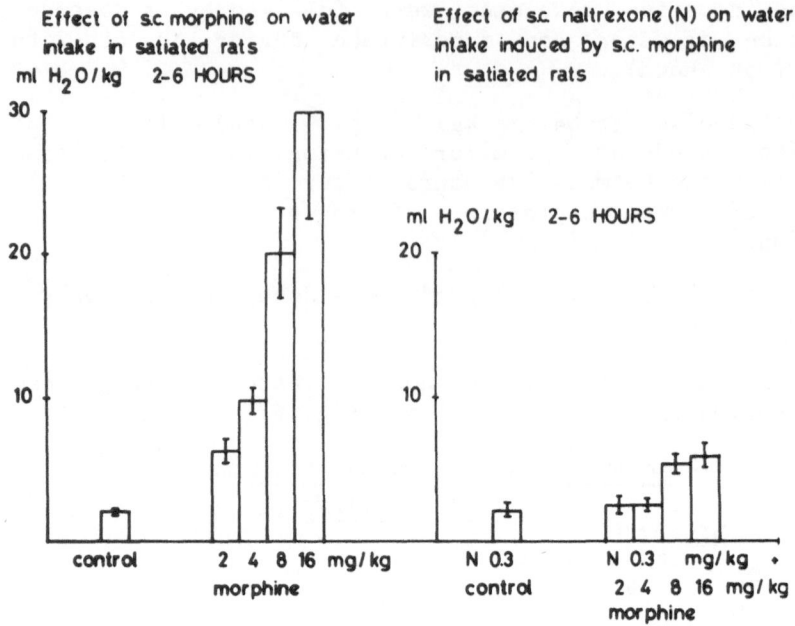

Fig. 7

Naltrexone did not inhibit eating responses. The drinking response to tilidine was equally depressed by naltrexone but the inhibition of food intake produced by pethidine was re-enforced by naltrexone pre-treatment (Fig. 8).

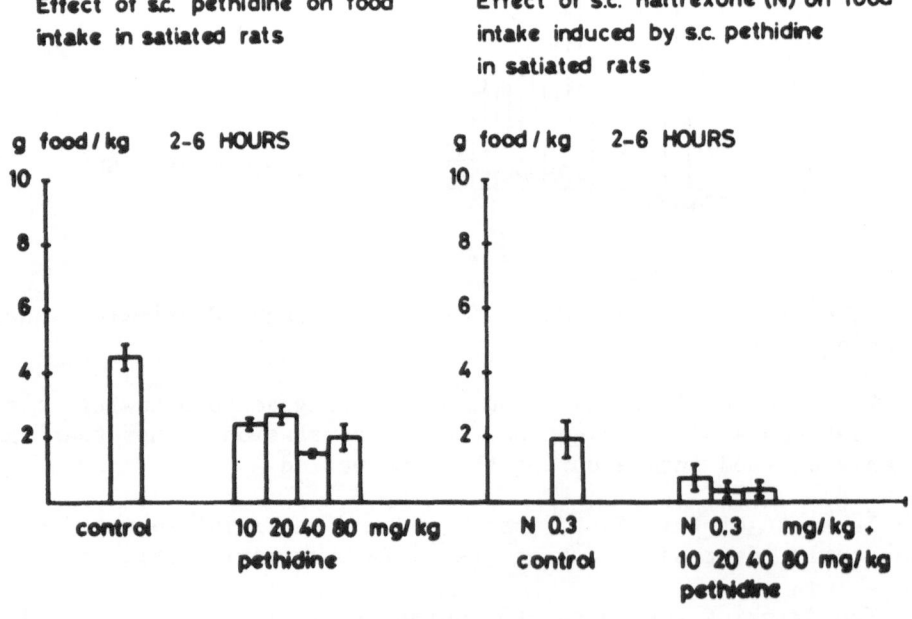

Fig. 8

DISCUSSION

All analgesic agents tested induced a dose-dependent increase of water intake between 2-6 hours, irregularly preceded by a depression of water intake during the first 2 hours, and followed by a depression of water intake from 6-24 hours after injection. This drinking response, thus, appears to be connected with the analgesic effect. The drinking responses were competitively antagonized by naltrexone.

Morphine, levorphanol and pentazocine also induced a bout of eating 2-6 hours after injection followed by a depression during the 6-24 hour period. Pethidine, methadone or tilidine did not induce eating between 2-6 hours after injection. Tilidine actually depressed it. In spite of this fact, the 3 agents entailed a depression of food intake 6-24 hours after injection. The eating response, thus, appears to be confined to morphine, drugs chemically ressembling morphine and pentazocine, like levorphanol, but not to be related to the typical analgesic effect.

Naltrexone blocked drinking responses but had no consistent effect on eating responses.

ABSTRACT

We investigated, during the light period, the influence of s.c. morphine, methadone, levorphanol, pethidine, tilidine and pentazocine on water and food intake, 2-6 hours, and subsequently 6-24 hours after injection. All the analgesic agents caused a usually dose-dependent increase in water intake in the 2-6 hour period. For morphine, levorphanol and pentazocine, this increase was accompained by a bout of eating. Methadone and pethidine, on the other hand, failed to increase food intake while tilidine actually depressed it. 6-24 hours after injection, (i.e. after their elimination) all the agents markedly depressed water and food intake independently of the presence or the absence of a preceding increase in the 2-6 hour period. Naltrexone (0.3 mg s.c., 10 min before analgesics) antagonized the drinking but not the eating response.

REFERENCES

1. R. Burckhardt, G. Peters, Morphine stimulation of food and water intake in the rat. Abstracts, 7th International Congress of Physiology, Paris, Abstr. 453, pg. 177, (1978).
2. R. Burckhardt, G. Peters, Morphine-induced drinking and eating in the rat. Abstracts, 7th International Conference on the physiology of food and fluid intake, Warsaw, (1980).

3. R. Burckhardt-Stuker, Experimentelle Analgetika-Abhängigkeit bei der Rattle (Ph. D. thesis). Berne, Frankfurt, New York, Peter Lang (1983), Pg. 218-248.

4. K. Jawaharlal, J. Atkinson, Role of the renin-angiotensin system in morphine-induced drinking (Abstract), Experientia 38:755 (1982).

5. G. Peters, Mécanismes de réglage de l'ingestion d'eau. J. Physiol. (Paris) 76:295 (1980).

6. G. Peters, K. P. Besseghir, H.P. Kaesermann, L. Peters-Haefeli, Effects of drugs on ingestive behavior. Pharmacol. Ther. 5:485 (1979).

BIOCHEMICAL, PHYSIOLOGICAL AND PATHOLOGICAL PROPERTIES OF ENDOGENOUS DIGITALIS-LIKE COMPOUNDS

Jean-François Cloix, Gilbert Deray, Marie-Aude Devynck, Marie-Gabrielle Pernollet, Maryse Crabos, Geneviève Henning, Max Rieux*, Irving W. Wainer** and Philippe Meyer

U7 INSERM, Dept of Pharmacology, Hôpital Necker, 161 rue de Sèvres, 75015 Paris, France
*Centre de Recherches en Endocrinologie, Hôpital Cochin, 27 rue du Fbg St. Jacques, 75014 Paris, France
**Division of Drug Chemistry, US Food and Drug Administration, Washington DC 20204, now resident in U7 INSERM

Sodium and water homeostasis are regulated by intrarenal physical and hormonal factors and by extrarenal hormones. The latter include factors retaining Na^+ and water, and factors eliminating Na^+ while the sodium balance is positive. There is a relationship between arterial blood pressure, Na^+ homeostasis and body fluid volumes (1), and an increase in blood pressure may result from the incapacity of the kidney to eliminate a Na^+ and water loading (2). This renal defect in the excretion of Na^+ leads to an extracellular volume expansion which in turn, increases the activity or the concentration of a natriuretic hormone through hypothalamic stimulation (2).

Based on Na^+,K^+-ATPase inhibition by natriuretic hormone, Blaustein and Hamlyn (3) suggested two hypotheses to explain the participation of this compound in the etiology of hypertension. First, increasing intracellular Na^+ content, it increases intracellular Ca^{2+} concentration by inhibition of Na^+-Ca^{2+} exchange. Second, the alteration of transmembrane Na^+ gradient stimulates the release and decreases the reuptake of catecholamine leading to an hyperactivity of the alpha-adrenoceptor system. These changes might elevate the peripheral resistance and thus blood pressure.

Plasma or plasma extracts from hypertensive patients were

reported to inhibit the ouabain-sensitive Na^+ efflux from leucocytes (4) and from erythrocytes (5), to increase the intra-lymphocytic Na^+ content (6) and, to inhibit dog kidney Na^+, K^+-ATPase activity (7, 8).

This study was undertaken in order to further assess the possible role of this factor in Na^+ homeostasis by its relevance in various diseases having extracellular volume expansion. A partial purification of these substances from hypertensive urine was also performed and some biological properties of such partially purified Na^+,K^+-ATPase inhibitors were studied.

PATIENTS AND METHODS

The ability of plasma, deproteinized by boiling, to compete with ouabain for binding to the digitalis sites of erythrocytes was studied as previously reported (9) and their inhibitory potency quantified by the ratio:

$$\frac{K_{D'} - K_D}{K_D}$$

where $K_{D'}$ and K_D are the apparent affinity constants of ^3H-ouabain binding in the presence and absence of plasma extracts, respectively.

One hundred and eight Caucasian untreated patients coming from Pharmacology Department and Nephrology Unit of Necker Hospital were investigated. They were distributed into 4 groups whose characteristics are given in Table I.

A 24 h-urine was collected from each hypertensive patient and normotensive subjects whose plasma inhibited ouabain binding to human erythrocytes (9). A 24 h-urine was also collected from normotensive non-inhibiting subjects and used as control urine. The samples were centrifuged and the supernatants (1-2 l) were chromatographed on a "flash" chromatography column (52.7 cm x 33 mm ID) packed with a 40 μm octadecyl packing material (J.T. Baker, Deventer, NL). The elution was developed by a water wash followed by a step wise acetonitrile gradient (15, 20, 30, 35, 50 and 100 %, 500 ml each), under a 2-3 bar pressure of nitrogen. Fractions of 30 ml were collected at a flow rate of 15 ml/min , freeze-dried, and assessed for their capacity to inhibit dog kidney Na^+,K^+-ATPase activity (4), to cross-react with anti-digoxin antibodies (10), and to inhibit ouabain binding (9). The fractions inhibiting the three biological tests were saved. The biological characteristics of the active fractions were studied on dog kidney and pig brain Na^+,K^+-ATPase, two commercial sources of Na^+,K^+-ATPase commonly used as target enzyme during purification procedures.

Table 1. Characteristics of Subjects and Patients and Ability of their Plasma Extracts to Compete with ^{3}H-Ouabain for Binding to Digitalis Sites

	Groups	n	Age (years)	♂/♀	Mean BP (mm Hg)	Hypertensive heredity	$\dfrac{K_{D'} - K_D}{K_D}$
1	**Controls**	21	32.1 ± 2.0	13/8	90.8 ± 1.9	-	0.21 ± 0.04
2	**Essential hypertension**	54	44.0 ± 2.0	29/25	122.7 ± 1.7	+(-)	0.83 ± 0.21*
3	**Chronic renal failure**						
	- moderate	8	57.6 ± 4.7		92.0 ± 4.5	-	0.13 ± 0.03
	- end-stage	8	47.7 ± 4.4	4/4	96.7 ± 2.6	-	0.84 ± 0.21*
	- anephrics	6	40.3 ± 4.8	2/4	95.8 ± 6.2	-(4)+(2)	0.75 ± 0.22*
	- after dialysis	14	44.6 ± 3.3	6/8	91.8 ± 2.7	-(12)+(2)	0.27 ± 0.05[+]
4	**Acromegaly**						
	- active	6	43.3 ± 3.3	3/3	< 113	-	2.97 ± 0.41*
	- cured	7	45.0 ± 2.6	2/5	< 113		1.14 ± 0.07*

Mean + SEM
*p < 0.01 when compared to control group
[+]p < 0.01 when compared to the same patients before dialysis
*p < 0.001 when compared to active acromegaly

RESULTS

Scatchard analysis of binding data indicate that high ouabain-like activity was present in plasma extracts from 18 out of 54 untreated patients with moderate essential hypertension, 10 out of 14 patients with end-stage renal failure, and those with active acromegaly. Low ouabain-like activity was found in plasma extracts from control subjects and patients with moderate renal failure. In patients with end-stage renal failure, dialysis decreased its level down to the control values. In acromegalic patients, successful surgery and radiotherapy also brought down the ouabain-like activity to basal levels. The affinity changes induced by plasma extracts are given in Table I.

The flash chromatography resolved the 24 h-urine samples into four peaks inhibiting dog kidney Na^+,K^+-ATPase activity. Two of them cross-reacted with anti-digoxin antibodies and inhibited ouabain binding. These two peaks, named U1 and U4, were eluted off the C18 column by 15 % and 30-35 % acetonitrile, respectively. The fraction U4 was found essentially in urine from hypertensive patients and from normotensive subjects having family history of hypertension.

The dose-dependent curves of these two fractions on Na^+,K^+-ATPase from dog kidney and pig brain showed that there was no difference in the inhibitory capacity of U1 as for ouabain and digoxin. The EC50 value, expressed in term of concentration factor by comparison between the incubation volume and the equivalent volume of starting urine, is 2.5 and 6.0 (n = 3) for dog kidney and pig brain, respectively. Nevertheless, the fraction U4 showed a more potent inhibitory capacity on dog kidney Na^+,K^+-ATPase (EC50 = 8, n = 3) as compared to pig brain enzyme (EC50 = 75, n = 3).

DISCUSSION

Plasma extracts from some hypertensives and normotensives with hypertensive heredity were previously reported to inhibit the Na^+,K^+-ATPase, to behave as apparent competitors of ouabain for binding to the digitalis sites of the Na^+-pump, to partially inhibit 5-HT uptake by blood platelets and to induce a transient rise in blood pressure when injected intracerebroventricularly (11). These observations suggest the presence in plasma of digitalis-like compounds able to inhibit the Na^+-pump.

In the present study, the investigation of the ouabain-like behaviour of circulating compounds was extended to two other diseases, chronic renal failure and active acromegaly. In order to recognize which specific characteristics of these diseases or of high blood pressure are related to high digitalis-like levels, these patients (except anephric patients) were selected as having normal blood pressure and no family history of hypertension. The highest plasma K^+ concentration was 5.4 mM which, at the 20-fold dilution used here, cannot account for the observed inhibition of ouabain binding. Elevated digitalis-like activities were observed in the patients with end-stage renal failure and active acromegaly. The altered physiological characteristics common to these three pathological states are the positive Na^+ balance and the volemic expansion (12).

The proposal that high levels of this circulating compound are related to hypervolemia and positive Na^+ balance is reinforced by the observation that correction of these expanded states, by dialysis in renal insufficiency and surgery and/or cobalt therapy of acromegaly, brings the levels of this compound back to normal values.

The presence of the inhibitor in some anephric patients before dialysis also suggest that kidneys are not a necessary requirement for synthesis or activation of this compound.

The "flash" chromatography resolved the urine samples in, at least, four different Na^+,K^+-ATPase inhibitors. The ions present in urine were not responsible for these effects, since they were eluted off the column during the water wash. Among these four compounds only two have digitalis-like properties as they inhibited ouabain binding and they cross-reacted with anti-digoxin antibodies. These data are consistent with those published by Gruber et al. (13) on purification of these compounds from plasma of volume-expanded dogs.

The dose-response curves of Na^+,K^+-ATPase inhibition show a possible tissue specificity of one of them (U4). This latter appears to be specific for hypertensive urine, and the target tissue appears to be the kidney rather than the brain or the other organs. However, the brain may also be a target tissue if in vivo concentration of this inhibitor is greater in the brain than in the kidney. The main effect of this substance could be a natriuretic effect at low concentration, and side effects, according to Blaustein and Hamlyn (3) and De Wardener and Mac Gregor (2), may occur at a highest level.

REFERENCES

1. A.C. Guyton, T.G. Coleman, A.W. Cowley, R.D. Mannig, R.A. Norman, and J.D. Ferguson, A system analysis approach to understanding long range arterial pressure control and hypertension, Circ.Res. 35: 159 (1974).
2. H.E. De Wardener, and G.A. Mac Gregor, Dahl's hypothesis that a saluretic substance may be responsible for a sustained rise in arterial pressure: its possible role in essential hypertension, Kidney Int. 18: 1 (1980).
3. M.P. Blaustein, and J.M. Hamlyn, Role of a natriuretic factor in essential hypertension: a hypothesis, Ann.Intern.Med. 98: 785 (1983).

4. L. Poston, R.B. Sewell, S.P. Wilkinson, P.J. Richardson, R. Williams, E.M. Clarkson, G.A. Mac Gregor, and H.E. De Wardener, Evidence for a circulating sodium transport inhibitor in essential hypertension, Br.Med.J. 282: 847 (1981).

5. J.F. Cloix, G. Dagher, M. Crabos, M.G. Pernollet, and P. Meyer, Purification from human plasma of endogenous sodium transport inhibitor(s), Experientia 1984 (in press).

6. F.V. Costa, L. Montebugnoli, M.F. Giordani, L. Vasconi, and E. Ambrosioni, Evidence for a plasma factor affecting subjects and in borderline hypertension subjects, Clin.Sci. 63: 530 (1982).

7. J.M. Hamlyn, R. Ringel, J. Schaeffer, P.D. Levinson, B.P. Hamilton, A.A. Kowarski, and M.P. Blaustein, A circulating inhibitor of Na^+,K^+-ATPase associated with essential hypertension, Nature 300: 650 (1982).

8. H. De Thé, M.A. Devynck, J.B. Rosenfeld, M.G. Pernollet, J.L. Elghozi, and P. Meyer, Plasma sodium-pump inhibitor in essential hypertension and normotensive subjects with hypertensive heredity, J.Cardiovasc.Pharmacol. 1984 (in press).

9. M.A. Devynck, M.G. Pernollet, J.B. Rosenfeld, and P. Meyer, Measurement of digitalis-like compound in plasma: application in studies of essential hypertension, Br.Med.J. 287: 631 (1983).

10. T.W. Smith, V.P. Butler, and E. Haber, Determination of therapeutic and toxic serym digoxin concentrations by radio-immunoassay, N.Engl.J.Med. 281: 1212 (1969).

11. J.F. Cloix, M.A. Devynck, J.L. Elghozi, L.A. Kamal, L.C. Lacerda-Jacomlini, P. Meyer, M.G. Pernollet, J.B. Rosenfeld, and H. De Thé, Plasma endogenous sodium pump inhibitor in essential hypertension, J.Hypertension 1(suppl.2): 11 (1983).

12. G. Straugh, A. Lego, and H. Bricaire, Reversible plasma and red blood cell columns increases in acromegaly, Acta Endocrinol. 85: 465 (1977).

13. K.A. Gruber, J.D. Whitaker, and V.M. Buckalew, Endogenous digitalis-like substance in plasma of volume expanded dogs, Nature 287: 743 (1980).

DISORDERLY DRINKING DURING DEVELOPMENT IN SPONTANEOUSLY

HYPERTENSIVE RATS

F. Scott Kraly

Department of Psychology
Colgate University
Hamilton, NY 13346

Adult spontaneously hypertensive rats (SHR) drink more water than do Wistar-Kyoto (WKY) rats when they have free access to solid food and water, and adult SHR exhibit hyperdipsia when eating a meal of solid food.[1] In addition, SHR exhibit a disproportionate (relative to WKY) acute increase in mean blood pressure when drinking water.[2] To examine whether hyperdipsia occurs in young SHR and whether it contributes to the development of hypertension, experiments were done (a) to describe ingestion during development of male SHR and WKY, (b) to examine for a mechanism for hyperdipsia in SHR, and (c) to evaluate whether restriction of water intake can slow or prevent the development of hypertension in SHR.

INGESTION IN DEVELOPING SHR

Daily 24-h Food and Water Intakes

The SHR ate the same amounts of pelleted rat chow as did WKY throughout weeks 7 to 17 of life ($p > .20$). The SHR drank more water throughout development ($p < .001$). They drank more as early as wk 9 of life ($p < .01$) and showed a higher water to food ratio (W:F) as early as wk 8 ($p < .001$). The SHR spilled more uneaten food than did WKY throughout development ($p < .001$) and did so as early as wk 7 ($p < .01$). The SHR exhibited an inverse correlation ($r = -.63$) between 24-h water intake and dry food spilled, in contrast to the positive correlation ($r = +.49$) observed for WKY.

Meal-Related Drinking

When eating in the night phase after 12-h food deprivation, the SHR ate more pelleted chow than did WKY in 1-h tests throughout development (p < .001). The SHR initiated drinking after eating with shorter latency as early as wk 6 (p < .05), they drank more water in the 1-h tests throughout development (p < .001) and showed a higher W:F as early as wk 6 (p < .01). Moreover, the pattern of ingestion in SHR was aberrant: As early as wk 7 and throughout development (all ps < .05 *vs*. WKY) SHR frequently interrupted eating to drink water. In contrast, when rats ate liquid food at 11 wk of age, SHR ate and drank similar amounts (ps > .10) and the frequency of interrupting eating to drink was zero for SHR and WKY.

MECHANISM FOR FOOD-RELATED HYPERDIPSIA IN SHR

Since adult SHR are hyperdipsic in response to s.c. histamine, s.c. angiotensin II (AII) and cellular dehydration produced by i.p. hypertonic NaCl,[1] all putative components of food-related drinking in the rat,[3] we first assessed whether any of these mechanisms could account for food-related hyperdipsia in adult SHR.

Increasing concentrations (0, .45, .9, 1.8, 3.6%) of NaCl in liquid food decreased 24-h food intakes (p < .001) and increased 24-h water intakes (p < .025) for SHR and WKY. The SHR ate more (p < .05) salted food and drank more water (p < .025) than did WKY, such that the W:F did not differ (p > .10) for SHR and WKY. Thus, SHR appeared to appropriately adjust water intake in response to changes in the osmotic property of their food.

The combined antagonism of peripheral H_1 and H_2 receptors for histamine, using 1 mg/kg dexbrompheniramine i.p. plus 16 mg/kg cimetidine i.p., failed to prevent (p > .20) the elevated W:F evident in adult SHR eating a meal of solid chow at the midpoint of the night phase after 24-h food deprivation. In addition, the inhibition of the converting enzyme for AII, using 50 mg/kg SQ14,225 (Captopril) i.g., failed to prevent (p > .20) the elevated W:F in adult SHR eating in these same conditions. Finally, the combination of these histamine and AII antagonists also failed to prevent (p > .20) the elevated W:F. Thus, food-related hyperdipsia in adult SHR is not prevented by systemic antagonism of histaminergic and/or AII mechanisms of putative importance for food-related drinking.[3]

In contrast, infusion of water through a cheek fistula decreased water intake (p < .001), W:F (p < .001) and the frequency of interrupting eating to drink (p < .01) in adult SHR eating solid chow in the night phase after food deprivation. In addition, such treatment improved (p < .02) the efficiency of eating (as measured by g of food spilled per g of food handled) in SHR. Since water in the mouth

appeared to attenuate food-related hyperdipsia, these results suggest
that deficient salivary function may cause hyperdipsia in SHR. Consistent with this notion is the finding that 3.25 mg/kg pilocarpine
nitrate i.p., at a dose which has an apparently maximal secretory
effect in SHR and WKY, elicited 32 percent greater volume of saliva-
tion in WKY than in SHR ($p < .05$).

Finally, hypothesizing that the typically retarded growth of
SHR[4] could result in part from rearing solely on dry chow, SHR and
WKY were maintained from 4-16 wk with access to liquid and solid
foods. The SHR with access to both liquid and solid foods gained
more weight ($p < .01$) during this period of development than did SHR
with access to solid food alone. In contrast, access to liquid food
did not facilitate ($p > .10$) weight gain for comparably treated groups
of WKY. In addition, such treatment did not alter the developmental
course of blood pressure (measured by tail-cuff to determine mean
systolic pressure) for SHR or WKY ($ps > .20$).

ROLE OF HYPERDIPSIA FOR DEVELOPMENT OF HYPERTENSION IN SHR

When SHR were prevented throughout development (4-14 wk of life)
from drinking more water than WKY drank daily, the development of
hypertension (measured by tail-cuff) was retarded ($p < .001$) and body
weight gain was slowed ($p < .001$). Other SHR with unlimited access
to water but restricted access to food, sufficient to enforce a slow-
ing of body weight gain equivalent to that of SHR with restricted
access to water, also showed retarded development of hypertension
($p < .005$). While restricted access to water or food retarded development of hypertension in SHR, the developmental course of blood
pressure for WKY remained considerably lower ($ps < .001$) than that
of both of these groups of SHR. Finally, resumption of unlimited
access to water or food produced within 1 wk mean systolic blood
pressures ($ps > .10$) and body weights ($ps > .05$) not different from
those of SHR having had unlimited access to water and food throughout
development.

CONCLUSION

Spontaneously hypertensive rats are hyperdipsic throughout
development. Moreover, SHR excessively spill dry food, they spill
food in inverse relation to the amount of water they drink, and they
frequently interrupt eating in order to drink. Each of these findings resemble those reported for rats with surgically-produced salivary deficits.[5,6,7] Should a dry mouth be the cause for the behavioral characteristics of hyperdipsia in SHR, minimizing the dryness
of the mouth should ameliorate the hyperdipsia. Consistent with
this notion are the findings that (a) SHR were not hyperdipsic and
did not frequently interrupt eating to drink when eating liquid food,

and (b) oral infusion of water decreased the hyperdipsia and the frequency of interrupting eating to drink in SHR eating solid food. This composite of behavioral evidence indicating salivary deficiency in SHR is corroborated by the finding that pilocarpine was less effective in SHR than in WKY for stimulation of salivary secretion. Taken together these results show that SHR maintained on dry food and water are hyperdipsic in order to facilitate eating dry food with a mouth deficient in saliva.

Hyperdipsia occurs early enough in the lives of SHR to play a role in the development of their hypertension. Restriction of water intake did retard the development of hypertension, but the blood pressure of SHR with restricted access to water was never reduced to that of WKY rats, and within one wk of the return of unlimited access to water, the blood pressure of previously restricted rats was as high as it was for those rats which had unlimited access to water. Thus, restricted access to water throughout development only slowed the development of hypertension in SHR. It also slowed the normal rate of weight gain in developing SHR. Moreover, restricted access to food, sufficient to retard growth to the same extent as did restricted access to water, also retarded the development of hypertension as reported elsewhere.[8,9] These findings together suggest that it was retarded growth, rather than decreased drinking *per se*, which delayed the development of hypertension in the SHR. This finding has a striking parallel in work assessing the effect of restriction of dietary sodium in SHR: Sodium restriction appears to retard the development of hypertension in SHR only when restriction is sufficient to retard growth.[10,11]

In summary, hyperdipsia, apparently caused by deficient salivary function, is not necessary for the development of hypertension in SHR.

REFERENCES

1. F.S. Kraly, A.F. Moore, L.A. Miller, and A. Drexler, Nocturnal food-related hyperdipsia in the adult spontaneously hypertensive rat. Physiol. Behav. 28: 885 (1982)
2. J.E. LeDoux, A. Sakaguchi, and D.J. Reis, Behaviorally selective cardiovascular hyperreactivity in spontaneously hypertensive rats. Hypertension 4: 853 (1982)
3. F.S. Kraly, The physiology of drinking elicited by eating. Psych. Rev. 91: in press (1984)
4. N.C. Trippodo and E.D. Frohlich, Similarities of genetic (spontaneous) hypertension: Man and rat. Circ. Res. 48: 309 (1981)
5. A.N. Epstein, D. Spector, A. Samman, and C. Goldblum, Exaggerated prandial drinking in the rat without salivary glands. Nature 201: 1342 (1964)
6. H.R. Kissileff, Oropharyngeal control of prandial drinking. J. Comp. Physiol. Psych. 67: 309 (1969)

7. E.M. Stricker, Influence of saliva on feeding behavior in the rat. J. Comp. Physiol. Psych. 70: 103 (1970)

8. G.L. Wright, J.P. McMurty, and B.C. Wexler, Food restriction reduces blood pressure of the spontaneously hypertensive rat. Life Sci. 28: 1253 (1981)

9. J.B. Young, D. Mullen, and L. Landsberg, Caloric restriction lowers blood pressure in the spontaneously hypertensive rat. Metabolism 27: 1711 (1978)

10. W.H.Beierwaltes, W.J. Arendshorst, and P.J. Klemmer, Electrolyte and water balance in young spontaneously hypertensive rats. Hypertension 4: 908 (1982)

11. C.B. Toal and F.H.H. Leenen, Dietary sodium restriction and development of hypertension in spontaneously hypertensive rats. Am. J. Physiol. 245: H1081 (1983)

PARTICIPANTS

CANTALAMESSA, F., Camerino, Italy
CHIARAVIGLIO, E., Cordova, Argentina
CLOIX, J.F., Paris, France
CONTRERAS, R.J., New Haven, USA
COOPER, S.J., Birmingham, UK
COSTALES, M., Oviedo, Spain
DE CAMPOS, R.M.M., Setubal, Portugal
DE CARO, G., Camerino, Italy
DENTON, D.A., Parkville, Australia
DI BELLA, L., Modena, Italy
EPSTEIN, A.N., Philadelphia, USA
FELIX, D., Bern, Switzerland
FITZSIMONS, J.T., Cambridge, UK
FRANKMANN, S.P., Philadelphia, USA
FREGLY, M.J., Gainesville, USA
FULLER, L.M., Cambridge, UK
GIBBS, J., White Plains, USA
GONZALES, M.A., Lisbon, Portugal
GROSSMAN, S.P., Chicago, USA
HARDING, J.W., Pullman, USA
IOVINO, M., Angri, Italy
JEULIN, A.C., Paris, France
JOHNSON, A.K., Iowa City, USA
KAUFMAN, S., Edmonton, Canada
KEHOE, P., Baltimore, USA
KENNEY, N.J., Seattle, USA
KRALY, F.S., Hamilton, USA
MANN, J.F.E., Heidelberg, W. Germany
MASSI, M., Camerino, Italy
MAZZARELLA, L., Bari, Italy
MCKINLEY, M.J., Parkville, Australia
MICOSSI, L.G., Camerino, Italy
MISELIS, R.R., Philadelphia, USA
MOE, K., Philadelphia, USA
MOGENSON, G.J., London, Canada
MORALES, A., GRANADA, Spain
NICOLAIDIS, S., Paris, France

PEPEU, G., Florence, Italy
PERFUMI, M., Camerino, Italy
PETERS, G., Lausanne, Switzerland
PUERTO, A., Granada, Spain
RAMSAY, D.J., San Francisco, USA
ROBINSON, M.M., London, Canada
ROLLS, B.J., Baltimore, USA
ROSSI, M.T., Modena, Italy
SAKAI, R.R., Philadelphia, USA
SCHULKIN, J., Philadelphia, USA
SEVERS, W.B., Hershey, USA
SIMPSON, J.B., Seattle, USA
SKADHAUGE, E., Copenhagen, Denmark
SMITH, G.P., White Plains, USA
SZCZEPAŃSKA SADOWSKA, E., Warsaw, Poland
TARJAN, E., Parkville, Australia
THORNTON, S.N., Cambridge, UK
THRASHER, T.N., San Francisco, USA
TÜRKER, R.K., Ankara, Turkey
UNGER, T., Heidelberg, W. Germany
VENTURI, F., Camerino, Italy
VIJANDE, M., Oviedo, Spain
WEISINGER, R.S., Parkville, Australia
WEISS, M.L., Philadelphia, USA
WRIGHT, J.W., Pullman, USA

AUTHOR INDEX

ABHOLD, R.H. 141,333
ARAD, Z. 71
ARSLAN, Y. 527
BIRD, E. 15
BLASS, E.M. 19,25
BOX, B. 361
BRIME, J.I. 181
BURCKHARDT, R. 527
CAMARA, C.G. 141,333
CANTALAMESSA, F. 37
CHIARAVIGLIO, E. 339, 503
CLOIX, J.F. 535
CONTRERAS, R.J. 15,479
COOPER, S.J. 239,453
COSTALES, M. 181,413
COX, P. 473
CRABOS, M. 535
CROWE, M.J. 521
DE CARO, G. 37,213,245,251,257
DENTON, D.A. 1,321,473,485
DERAY, G. 535
DEVYNCK, M.A. 535
DI BELLA, L. 59,65
DORSA, D.M. 115
DUNDORE, R.L. 149
EISELE, S. 199
EPSTEIN, A.N. 13,37,251,395,431
ERICKSON, J.B. ·141,333
EVERED, M.D. 193
FELIX, D. 135,375

FELIX, H. 375
FITZSIMONS, J.T. 383,413,419
FORSLING, M.L. 521
FRANKMANN, S.P. 115
FREGLY, M.J. 441, 509
FULLER, L.M. 419,447
GAMBINO, M.C. 135
GANTEN, D. 123,199
GIBBS, J. 287
GILBERT, D.B. 453
GROSSMAN, S.P. 355
HARDING, J.W. 141.,187,333
HENNING, G. 535
HENRY, R.T. 149
HYDE, T.M. 279
IOVINO, M. 367
JAWAHARLAL, K. 527
JEULIN, A.C. 43
JOHNSON, A.K. 161, 199
KAUFMAN, S. 109,459
KEIL, L.C. 149
KENNEY, N.J. 227
KEHOE, P. 25
KOSTEN, T. 479
KOZŁOWSKI, S. 97
KRALY, F.S. 295,541
LANG, R.E. 123
LEDINGHAM, J.G.G. 521
LEVENTER, M. 321
LOPEZ-SELA, P. 181

LOZADA, C. 339
LUDWIG, G. 123
MANN, J.F.E. 161,199
MARÍN, B. 181
MASSI, M. 251,257
MAZZARELLA, L. 251
McKINLEY, M.J. 1,321,473,485
MEULI, C. 465
MEYER, P. 535
MICOSSI, L.G. 245
MISELIS, R.R. 279,321,345
MOE, K. 31,431
MOGENSON, G.J. 361
MORALES, A. 77
MORSETH, S.L. 187
MORTON, J.J. 521
NICOLAÏDIS, S. 43
ORNSTEIN, K. 527
OSBORNE, P. 1
PARK, R.G. 321
PÉREZ GUAITA, M.F. 503
PERFUMI, M. 37,257
PERNOLLET, M.G. 535
PETERS, G. 465,527
PETERSEN, E.P. 141
PHILLIPS, M.I. 155
PHILLIPS, P.A. 521
PUERTO, A. 77
RAMSAY, D.J. 83,301,327
RETTIG, R. 199
RICE, G.E. 71
RICHARDS-SUMNERS, E. 155
RIEUX, M. 535

RITZ, E. 199
ROBINSON, M.M. 161,193
ROLLS, B.J. 287,521
ROLLS, E.T. 287
ROSSI, M.T. 59,65
ROWLAND, N.E. 441,509
SAKAI, R.R. 115,425
SCHELLING, P. 135
SCHULKIN, J. 491,497
SEVERS, W.B. 149
SHAPIRO, R.E. 279
SIMPSON, J.B. 115,309,321,485
SKADHAUGE, E. 71
SMITH, G.P. 265
SOBOCIŃSKA, J. 97
SPAETH, H.J. 149
STEARDO, L. 367
SULLIVAN, M.J. 187
SZCZEPAŃSKA-SADOWSKA, E. 97
TARJAN, E. 1,321,473,485
THORNTON, S.N. 103
THRASHER, T.N. 83,301,327
TÜRKER, R.K. 205
UNGER, T. 123,199
VENTURI, F. 257
VIJANDE, M. 181,413
WAINER, I.W. 535
WEISINGER, R.S. 1,321,473,485
WEISS, M.L. 405
WRIGHT, J.W. 187,333
WURPEL, J.N.D. 149
YONG, Y. 135

SUBJECT INDEX

Acetyl salicylic acid
- and PGE drinking inhibition, 232
Acromegaly, 537–539
ADH see Arginine vasopressin
Adipsia
- following brain lesions in sheep, 321–323
Adrenalectomy
- , DOCA treatment and Na intake, 388–389
- , DOCA treatment and renin secretion, 388
- and Na appetite, 388, 447––449
- and Na appetite in Brattleboro rats, 447
- , Na appetite and captopril treatment, 388
- and Na appetite in Long Evans rats, 447
- and polydipsia in Brattleboro rats, 449
α_2-adrenoceptor antagonists
- and experimental drinking, 515
- and spontaneous drinking, 515
Aging
- and thirst in humans, 526
Aldosterone
- and Na appetite, 426, 427
Angiotensin I
- distribution in the brain, 127
- and drinking, 166
- and Na appetite, 443

Angiotensin I (continued)
- structure, 127
Angiotensin II
- and aldosterone release, 162, 205
antagonists of - and drinking, 172, 184, 205–211, 334––336
antagonists of - and Na appetite, 409
-/AVP comodulation, 43–54
- and AVP release, 162
- binding sites in gerbil brain, 143
- binding sites in monkey brain, 143
- binding sites in rabbit brain, 143
- binding sites in rat brain, 143, 155–160, 334–336
brain areas sensitive to -, 43, 49, 51, 135, 138
- converting enzyme, see Converting enzyme
- distribution in the brain, 127
effect of - on CSF pressure, 149
effect of - on Na intake, 155
effect of - on urine output, 155
effect of - on water intake, 155, 156
- immunoreactive fibers in the brain, 129
- induced drinking, 43, 44, 87, 88, 162, 164, 167, 187, 188, 193–198, 246, 258–262

551

Angiotensin II (continued)
- induced drinking in neo-
    natal rats, 13-14, 38-42
- induced drinking and the nu-
    cleus medianus, 44
- induced drinking and OVLT,
    43, 44
- induced drinking and the SFO,
    44, 310-317
- and lesions to zona incerta,
    358
- levels in experimental con-
    ditions, 167, 168, 193
- levels in pathological con-
    ditions, 167, 168
- levels in physiological con-
    ditions, 167, 168
- metabolism in the brain,
    144, 146
- and Na appetite, 384, 401,
    420, 421, 426, 427, 488
- and Na balance, 421-423
- and Na excretion, 155, 422
- pressor responses 162, 188,
    189, 193-198
- and prostaglandin release,
    207
- receptors in the brain, 128,
    155, 159
- repetitive injection and ta-
    chyphylaxis, 334, 337
- role in drinking behavior,
    163, 334
- role in human thirst, 523
- and SFO lesioned pigeons,
    252
- structure, 127
Angiotensin III
- binding sites in gerbil
    brain, 143
- binding sites in monkey
    brain, 143
- binding sites in rabbit
    brain, 143
- binding sites in rat brain,
    143, 334-336
effect of - on blood pressure,
    188, 189
- induced drinking, 187, 188

Angiotensin III (continued)
- induced drinking following
    repetitive injection,
    334, 337
- metabolism in the brain,
    144, 146
Angiotensinogen
- structure, 125
- synthesis in the brain, 125
- synthesis in the liver, 125
Anorexia
dehydration-induced -, 20
- induced by tachykinins in
    neonatal rats, 42
Anterior third periventricular
    area
- sensitivity to AVP, 45
Anterodorsal third ventricular
    region (AD3V)
- and drinking, 321-326
lesions of the - and Na appe-
    tite in sheep, 323-324
Anteroventral third ventricular
    region (AV3V)
- and ang II binding sites,
    143
- and ang III binding sites,
    143
- and drinking, 322, 323
lesions of the - and Na appe-
    tite in sheep, 323-324
- and SFO, 348
Antidipsogenic effect of
- AVP antagonists, 97-101
- bombesin, 217-219
-dermorphin, 219, 220, 258-
    -262
- eledoisin, 37, 215, 217
- eledoisin in neonatal rats,
    38, 42
- HYP^6-dermorphin, 219, 220
- kassinin, 245-249
- leu-enkephalin, 219, 220
- litorin, 215, 217
- met-enkephalin, 219, 220
- opioid peptides, 219, 220
- physalaemin, 37, 215, 217
- physalaemin in neonatal
    rats, 38, 42

Antidipsogenic effect of
    (continued)
 - prostaglandins E, 227-236
 - ranatensin, 215, 217
 - substance P, 215, 217
 - tachykinins, 37, 215-217
 - tachykinins in neonatal
    rats, 38, 42
Arachidonic acid
 - and drinking inhibition, 228
Area postrema
 lesions to - and drinking,
    279, 282-285
 neural projections of -, 350
 - and nucleus tractus solitari
    (NTS), 350
Area preoptica lateralis
 - and SFO, 348
Area preoptica medialis
 - sensitivity to ang II, 138
Area rostromedialis diencephali
 - and extracellular dehydra-
    tion, 43, 44
 - and intracellular dehydra-
    tion, 43, 44
Area septalis
 - sensitivity to ang II, 138
Area supraoptica
 - sensitivity to ang II, 138
Arginine vasopressin (AVP)
 -/ang II comodulation in dif-
    ferent brain areas, 51
 - antagonists and drinking,
    97-101
 brain areas, sensitivity to -,
    45, 51
 control of - secretion, 84,
    86, 91
 - and drinking, 97, 101
 - effect on CSF pressure, 149
 - levels following PEG, 119,
    120
 - levels in human thirst, 526
 - levels in hypovolemia, 119
 rapid satiety and - secretion,
    303
 - release, 103
 - secretion and lesions to
    septal area, 371

Arginine-vasotocin
 - in domestic fowls, 72
Arterial pressure
 - and volume regulation, 72
Atrial filling
 - and volume regulation, 72
Atrial receptors
 - and drinking, 6, 109
 - and Na intake, 109
 - and plasma renin activity,
    113
 stimulation of - and Na in-
    take, 459-463
Baroreceptors
 - in arteries, 162
 kidney - and extracellular-
    -thirst, 88
 kidney - and renal angiotensin
    system, 88
Benzodiazepines
 - effect on salt intake, 240
 - effect on water intake, 240
 - interactions with opiate an-
    tagonists, 241
Blood pressure
 effect of ang II on -, 188,
    189
 effect of ang III on -, 188,
    189
 effect of PEG on -, 201
 effect of PGE on -, 230
 increase in - and AVP secre-
    tion, 305
 increase in - and satiety, 305
 - and thirst, 193
Bombesin(s)
 - antidipsogenic effect, 217-
    219
 - dipsogenic effect, 220-222
 - dipsogenic effect in SFO le-
    sioned pigeons, 253
 - structure, 214
Brain lesions
 adipsia following - in sheep,
    321-323
 - , AD3V and drinking, 322
    323
 - , area postrema and drink-
    ing, 279, 282-285

Brain lesions (continued)
- , AV3V and drinking, 322,
    323
- and drinking, 279-285
- and Na appetite, 323, 324
- , OVLT and ang II-induced
    drinking, 328
- , OVLT and cell dehydration
    induced drinking, 328
- , OVLT and drinking follow-
    ing dehydration, 330
- , OVLT and drinking follow-
    ing hypovolemia, 329
- , posterior hypothalamus and
    drinking, 78
- , septal area and AVP secre-
    tion, 371
- , septal area and drinking,
    369, 370
- , SFO and drinking, 258-262,
    312, 313
- , solitary tract nucleus and
    drinking, 279, 282-285
- , zona incerta and ang II-
    induced drinking, 358
- , zona incerta and cell de-
    hydration thirst, 356
- , zona incerta and hypo-
    volemic thirst, 357
- , zona incerta and Na appe-
    tite, 358
- , zona incerta and extracel-
    lular dipsogens, 358
Brattleboro rats
- and Na intake, 119, 120

Captopril
- , adrenalectomy and Na appe-
    tite, 388-389
- and ang II-induced drinking,
    88, 173, 175, 183 195
- and brain ang II concentra-
    tion, 386
drinking responses following
    -, 384-388
- induced Na appetite, 443,
    444
- and Na appetite induced by
    ang II, 400-402

Captopril (continued)
- and Na appetite induced by
    DOCA, 408
- and Na appetite induced by
    Na depletion, 407, 408,
    431-438
- and PEG-induced drinking,
    199
- and pressor responses to
    ang II, 195, 196
- and pressure responses to
    PEG, 201
- and renin secretion, 384
-,ureteric ligation and drink-
    ing, 384, 386
- and urinary Na loss, 387
Carbachol
- dipsogenic effect, 43, 420
- and SFO, 314-317
- and Na appetite, 317-319,
    420, 421, 426, 428
- and Na balance, 421-423
- and Na excretion, 422
Cardiopulmonary receptors
low pressure - and drinking,
    87
Caval constriction
- and drinking, 87, 91, 167
Cellular dehydration
- and neuronal reactivity,
    43, 44
Cellular dehydration-induced
    drinking, 83-87, 97,
    114, 176, 247
- in neonatal rats, 38-42
ontogeny of -, 9, 10
phylogeny of -, 1-6
Central receptor systems
- and fluid intake control,
    309-319
- and Na appetite, 317-319
Cerebellum
ang II binding sites in -,
    143
ang III binding sites in -,
    143
Chorda tympani
stimulation of - and Na appe-
    tite, 481

Chorda tympani (continued)
- and taste reactivity, 25
Choroid plexus
- and converting enzyme, 126
Chronic renal failure, 537-539
Circumventricular organs, 346,
    353
- responses to hydromineral
    deficiencies, 43
Contraceptives
oral - and Na appetite, 444
Converting enzyme
- distribution in the brain,
    126
- inhibition and Na appetite,
    433
- inhibitors, 88, 173, 174,
    183, 195, 196, 199, 384-
    -392
Corticosterone
- and Na appetite, 426, 428
Crinia angiotensin II
- dipsogenic effect, 220-222
CSF
angiotensin II levels in -, 9
- [K] and Na appetite, 504
- [Na] and Na appetite, 473-
    -477, 504
- osmotic pressure, 3, 11
- pressure control by ang II,
    149-154
- pressure control by AVP, 149-
    -154
- [Na] and drinking, 2, 3-6, 8

Deoxycorticosterone acetate
    (DOCA)
- , ang II and Na appetite,
    396-399
- and Na appetite, 388, 389,
    396, 436, 438, 462
- and Na appetite in Brattle-
    boro rats, 450
- and Na appetite in Long
    Evans rats, 450
- and renin secretion, 388
Dermorphin
- antidipsogenic effect, 219,
    220, 258-263
- antidipsogenic effect in SFO
    lesioned rats, 259-260

Dermorphin (continued)
- structure, 214
Developing SHR
food intake in -, 541
water intake in -, 541
Diabetes insipidus
- and plasma ang II levels,
    167
Diagonal banda of Broca
- sensitivity to AVP, 45
Digitalis-like compounds, 538-
    -539
Dipsogenic effect of
- ang II, see Angiotensin II
- ang II in neonatal rats, 13,
    14
- ang III, 187, 188
- AVP, 97, 101
- bombesin(s), 217-219, 220-
    -222
- carbachol, 43
- crinia ang II, 220, 222
- eledoisin, 220-222
- insulin, 181-185
- isoproterenol, 88, 114, 164,
    165, 167, 176
- litorin, 252, 253
- physalaemin, 220-222
- ranatensin, 252, 253
- substance P, 220-222
- tachykinins, 220-222
Diuretics
- and Na appetite, 444
Dopamine
- D_1-antagonists and fluid
    intake, 453-457
- D_1-antagonists and salt pre-
    ference, 453-457
- D_2-antagonists and fluid in-
    take, 453-457
- D_2-antagonists and salt pre-
    ference, 453-457
Drinking
abdominal vagotomy and -, 269
- and α_2 adrenoreceptor anta-
    gonists, 515
- in amphibia, 7, 8
ang II-induced -, see Angio-
    tensin II
ang II repetitive injection,
    tachyphylaxis and -, 334-
    337

Drinking (continued)
  ang III-induced -, 187, 188,
      334
  - behavior and visceral neu-
      raxis, 352
  - and benzodiazepines, 240
  - in bony fishes, 7
  cervical vagotomy and -, 268
  - in chicken, 9
  coeliac vagotomy and -, 270-
      -273
  - in cows, 2
  - and CSF osmotic pressure, 3,
      11
  - in developing SHR, 541
  disturbances of - and brain
      lesions, 279-285
  disturbances of - and lesions
      to area postrema, 279,
      282-285
  disturbances of - and lesions
      to solitary tract nu-
      cleus, 279, 282, 285
  - in dogs, 2, 83, 89, 97-101,
      103, 301-307, 327-332
  - down regulation, 337
  - and endorphins, 240
  food associated -, 295-299
  gastric vagotomy and -, 270-
      -273
  - in goats, 2, 8, 84, 103
  hepatic vagotomy and -, 270-
      -273
  histamine, role in -, 295-298
  histamine, role in food-
      -associated -, 295-298
  - to hypertonic glucose, 3, 86
  - to hypertonic mannitol, 7,
      104
  - to hypertonic NaCl, 1, 79,
      83, 85, 86, 103
  - to hypertonic sucrose, 3, 7,
      8, 79, 86, 103, 104
  - to hypertonic urea, 3, 8, 83,
      86
  - in iguana, 8
  intravenous -, 266
  maintenance of -, peripheral
      mechanisms, 265-268
  - in man, 521-526

Drinking (continued)
  - in monkeys, 287-294
  - in neonatal rats, 13-24, 36-
      -42
  - and opiate receptor antago-
      nists, 240, 241
  - to PEG, 79, 111
  pharmacological vagotomy and -
      272
  - in pigeons, 104-108, 251-256
  - and posterior hypothalamus,
      80
  - in rabbits, 9, 84
  - in rats, 1, 6, 8, 9, 66-70,
      77-81, 84, 109-121, 181-
      -203, 215-220, 227-249,
      257-285, 295-299, 339-
      -374, 509-519, 527-534
  - in reptiles, 7
  rithmicity of -, 77
  - and septum, 367-374
  - in sheep, 3, 8, 104, 321-326
  - in SHR, 541-544
  termination of -, postabsorp-
      tive mechanisms, 267,
      268
  termination of -, postabsorp-
      tive sites, 267, 268
  termination of -, preabsorp-
      tive mechanisms, 266
  - and vanadate, 341, 343
  - and zona incerta, 356-358

Eating
  drinking elicited by -, 295-
      -299
Eledoisin
  - antidipsogenic effect, 37,
      215-217
  - antidipsogenic effect in
      neonatal rats, 38-42
  - dipsogenic effect, 220-222
  - dipsogenic effect in SFO
      lesioned pigeons, 254
  - structure, 214
Endorphins
  - effects on salt intake, 241
  - effects on water intake, 240
Essential hypertension, 537-538

Extracellular dehydration
   - and neuronal reactivity, 43,
      44
Extracellular dehydration-induced
      drinking, 87-90
   ontogeny of -, 9, 10
   phylogeny of -, 6, 8
Extracellular fluid
   - depletion, 120
   - osmolarity and drinking, 84

Food intake
   - in dehydrated rats, 61
   - in developing SHR, 541
   - in SHR, 541, 542
   - in thyroidectomized rats, 67
   - and water intake, 295-299

Globus pallidus
   - and converting enzyme, 126
Goldblatt hypertension
   - and Na appetite, 390-391,
      416, 417
   Na appetite and Na balance
      in -, 416
   Na appetite and polydipsia
      in -, 415
   Na appetite and renal function
      in -, 416
   Na appetite and renin in -, 417
Gustatory afferents
   - and Na appetite, 444, 479-
      -481

Haemorrhage
   - and drinking, 10, 89
Hippocampal pyramidal cells
   - sensitivity to ang II, 139
Histamine
   - role in drinking, 295-298
Human thirst
   aging and -, 526
   ang II, role in -, 523
   drinking termination in -, 521
   - and initiation of drinking,
      521
   - and ad libitum drinking, 523
   vasopressin concentration
      and -, 526
   water deprivation and -, 526

HYP[6]-dermorphin
   - antidipsogenic effect, 219,
      220
   - structure, 214
Hyperdipsia
   - and development of hyper-
      tension, 543
   food related -, 542
   mouth dryness and -, 543, 544
Hyperosmotic thirst, 83-87, 97,
      114, 176, 247
   - in neonatal rats, 38-42
   ontogeny of -, 9, 10
   - and OVLT, 328
   phylogeny of -, 1-6
   - and zona incerta, 356
Hypertension
   - and drinking, 176
   essential -, 537-538
   Goldblatt -, see Goldblatt
   hyperdipsia and development
      of -, 543
Hypertonic mannitol
   - and drinking, 473-478
   - and Na appetite in sheep,
      474
Hypertonic NaCl
   - and drinking, 473-478
   - and Na appetite in sheep,
      474
Hypotension
   - and drinking, 176, 193
Hypothalamus
   ang II binding sites in -,
      143
   ang II distribution in -, 127
   ang III binding sites in -,
      143
   posterior - and drinking, 80
   SFO projections to - later-
      alis, 348
Hypothyroidism
   - and Na appetite, 444
Hypovolemia
   - and AVP, 119
   - and drinking, 6, 8, 87-90,
      114, 176, 193, 202
   - and OVLT, 43
   - and SFO 43

Hypovolemia (continued)
- and suckling, 19
Hypovolemic thirst, 6, 8, 87–90,
      114, 176, 193, 202
- and OVLT, 329
- and zona incerta, 357

Indomethacin
- and PGE drinking inhibition,
      232–234
Ingestive behavior
effects of levorphanol on -,
      530
effects of methadone on -, 531
effects of morphine on -, 527
effects of opioid antagonists
      on -, 531, 532
effects of pentazocine on -,
      530
effects of pethidine on -, 531
effects of tilidine on -, 531
mesencephalic locomotor region
      and -, 362–365
- and neural correlates of
      water procurement, 362
neural substrates of -, 362
- and zona incerta, 358, 359
Insuline
- induced drinking, 181, 185
Isoprenaline see Isoproterenol
Isoproterenol
- dipsogenic effect, 88, 114,
      164, 165, 167, 176

K, see Potassium
Kassinin
- and ang II-induced drinking,
      246
- antidipsogenic effect, 245–
      -249
- and hyperosmotic thirst, 247,
      248
- structure, 245
- and water deprivation, 246,
      247

Lesions of the brain, see Brain
      lesions
Leu-enkephalin
- antidipsogenic effect, 219,
      220

Leu-enkephalin (continued)
- structure, 214
Limbic system
ang II in the -, 127
Litorin
- antidipsogenic effect, 217–
      -219
- dipsogenic effect, 252
- dipsogenic effect in SFO le-
      sioned pigeons, 253
- structure, 214
Locus coeruleus
ang II in the -, 127
Median eminence
- and ang II nerve terminals,
      127
- and converting enzyme, 126
Medulla oblongata
ang II in the -, 127
Met-enkephalin
- antidipsogenic effect, 219,
      220
- structure, 214
Methymazole, 61, 66
Milk
- intake, 13, 20, 22
- intake and tachykinins, 39,
      40
[Na] in -, 16
Mineralocorticoids
- and Na appetite, 388, 389,
      396, 426, 427, 436, 438,
      450
Motivation
- of Na intake, 471, 516
Mouth dryness
- and hyperdipsia, 543, 544
Nephrectomy
- and drinking, 20, 88
- and insulin-induced drinking,
      183
Nervus vagus
- and volume regulation, 72
Norepinephrine
- orexigenic effect, 14
Nucleus amygdaloideus centralis
ang II in the -, 127
Nucleus caudatus putamen
converting enzyme in the -,
      126

Nucleus dorsomedialis hypothalami
  ang II in the -, 127
Nucleus medialis corticohypo-
        thalami
  - and AVP/ang II comodulation,
    51
Nucleus medianus
  - and ang II-induced drinking,
    44
Nucleus paraventricularis hypo-
        thalami
  osmoreceptors in the -, 43, 45
  - and SFO, 347-348
Nucleus paraventricularis thalami
  - and AVP/ang II comodulation,
    51
Nucleus periventricularis hypo-
        thalami
  - and AVP/ang II comodulation,
    51
Nucleus preopticus medialis
  - and AVP/ang II comodulation,
    51
  - sensitivity to AVP, 45
Nucleus supraopticus
  osmoreceptors in the -, 43, 45
Nucleus tractus solitari
  - and area postrema, neural
      projections, 350
  lesions to - and drinking,
      279, 282-285
  - and taste reactivity, 25
Nucleus tractus spinalis n.
        trigemini
  ang II in the -, 127

Olfactory bulb
  ang II binding sites in the -,
      143
  ang III binding sites in the -,
      143
Ontogeny of
  - ang II-induced drinking, 13,
      14, 21
  - cellular dehydration-induced
      drinking, 9, 10, 20, 21
  - eledoisin antidipsogenic ef-
      fect, 38-42
  - extracellular dehydration-
      -induced drinking, 9,
      10, 20, 21

Ontogeny of (continued)
  - isoproterenol dipsogenic ef-
      fect, 21
  - Na appetite, 31
  - physalaemin antidipsogenic
      effect, 38-42
  - tachykinins antidipsogenic
      effect, 38-42
  - taste in rats, 25-28
  - taste in sheep, 25
  - thirst, 9, 10, 13-42
Opiate receptor agonists
  - effect on ingestive behav-
      ior, 527-531
Opiate receptor antagonists
  - effect on ingestive behav-
      ior, 531-532
  - effect on salt intake, 241
  - effect on water intake, 240,
      241
Opioid peptides
  - antidipsogenic effect, 219-
      -220
Orexia
  - and ang II in neonatal rats,
      42
Organum vasculosum laminae ter-
        minalis
  - and ang II-induced drinking,
      43, 44, 51, 52, 328
  - and AVP, 45, 51
  - and cell dehydration thirst,
      328
  - and hypovolemia, 43
  - and hypovolemic thirst, 87,
      329
  neural projections of the -,
      348
  - and water deprivation
      thirst, 330
Oropharyngeal receptors
  - and drinking, 1
Oropharyngeal stimuli
  - and rapid satiety, 302
Osmoreceptors
  - and drinking, 1, 3-5, 84,
      86, 103, 105, 107
  - in PVN, 43, 45, 51
  - in SON, 43, 45
Osmotic regulation
  - in birds, 72, 73

Ouabain, 537–538

Peptidase inhibitors
  - amastatin, 144
  - bacitracin, 144
  - bestatin, 144
  effect of - on ang II binding,
      144
  effect of - on ang II degra-
      dation, 144
  effect of - on ^{125}I-tyrosine
      uptake, 144
Phylogeny of
  - cellular dehydration-induced
      drinking, 1–6
  - extracellular dehydration-
      -induced drinking, 6, 8
Physalaemin
  - antidipsogenic effect, 37,
      215, 217
  - antidipsogenic effect in
      neonatal rats, 38, 42
  - dipsogenic effect, 220–222,
      254
  - dipsogenic effect in SFO-
      -lesioned pigeons, 254
  - structure, 214
Plasma
  - [Na] and drinking, 6, 8, 159
  - [Na] and volume regulation,
      72
  - osmolality and arginine
      vasotocin, 73, 74
  - osmolality and drinking, 84,
      86, 104
  - osmolality and volume regu-
      lation, 72
  - osmotic pressure and drink-
      ing, 10
  - prolactin concentration, 75
Plasma renin activity, 87, 92
  - and drinking, 165
  - in hypovolemic rats, 112
  - in normovolemic rats, 111
Polydipsia, 80
  Brattleboro rats, adrenalec-
      tomy and -, 449
  - in Goldblatt hypertension,
      415

Polyethylene glycol
  effect on blood pressure of -,
      201
  - and Na intake, 115, 118, 119
  - and plasma ang II levels,
      167
  - and water intake, 79, 111,
      115, 199
Potassium
  concentration of - in amniotic
      fluid, 16
  concentration of - in CSF and
      Na appetite, 504
  excretion of - and vanadate,
      340–342
Prostaglandins
  inhibition of - synthesis in
      the brain, 231
  inhibition of - synthesis and
      drinking, 232
  inhibition of - synthesis in
      the periphery, 231, 232
Prostaglandins E
  antidipsogenic effect of -,
      227–236
  pressor effect of -, 230
  thermogenic effect of -, 229

Ranatensin
  - antidipsogenic effect, 217–
      -219
  - dipsogenic effect, 252
  - dipsogenic effect in SFO-
      -lesioned pigeons, 253
  - structure, 214
Receptor complex
  - internalization, 337, 338
Renal failure
  chronic -, 537–539
Renin
  - induced drinking, 164
  - and Na appetite in Goldblatt
      hypertension, 417
  - plasma concentration, 165
  plasma - activity, see Plasma
  - release, 162, 388
  - structure, 123
  - synthesis in the brain, 124

Renin-angiotensin system
- blockade by ang II receptor
    antagonists, 171
- blockade by converting enzyme
    inhibitors, 173
- blockade by renin inhibitors,
    171
brain -, 123, 155
brain - and blood pressure
    control, 139
brain - and drinking, 8
- and hypovolemic thirst, 202
- in insulin-induced drinking,
    181-187
- and nephrectomy, 170
renal -, 161-177
role of the - in thirst, 164
- and Na appetite, 442
synthetic cascade of the -,
    169
- and volume regulation, 72

Salivectomy
- and Na appetite, 444
Salt intake, see Sodium appetite
Saralasin, 89, 172, 184
Satiety
blood pressure and -, 305
oropharyngeal stimuli and
    rapid -, 294
rapid - and AVP secretion, 303
Septum
ang II binding sites in -, 143
ang III binding sites in -,
    143
- and drinking, 367-374
Serum
- [K] and Na appetite, 505
- [Na] and Na appetite, 504
Serum osmolality
- , methymazole and dehydra-
    tion, 62
Sham drinking
duodenal -, 289
gastric -, 288
intestinal infusion during
    gastric -, 290
intravenous infusion during
    gastric -, 292
- in rhesus monkey, 287-294

Sodium appetite
- and adrenalectomy, 388, 442-
    -444, 447-449
- , adrenalectomy and capto-
    pril, 388-389
AD3V lesions and - in sheep,
    323-324
AD3V Na sensors and - in
    sheep, 475, 477
AD3V perfusion of hypertonic
    NaCl and - in sheep, 477
aldosterone and -, 426, 427
    and ang I, 443
- and angiotensin converting
    enzyme inhibition, 443
ang II, adrenalectomy and -,
    420-421
ang II antagonists and -, 409
ang II and -, 384, 420, 426
    427
- and atrial receptors, 109,
    459, 463
aversive taste and -, 465-470
AV3V lesions and - in sheep,
    323-324
AV3V Na sensors and - in sheep,
    475, 477
AV3V perfusion of hypertonic
    mannitol and - in sheep,
    474, 475
AV3V perfusion of hypertonic
    NaCl and - in sheep, 474
- and benzodiazepines, 240
- in Brattleboro rats, 119,
    120, 447-452
Brattleboro rats, adrenalec-
    tomy and -, 447
Brattleboro rats, deoxycorti-
    costerone and -, 450
captopril and -, 386, 400-402,
    405-410, 431-438, 443,
    444
carbachol, adrenalectomy
    and -, 420, 421
carbachol and -, 317-319, 420,
    426, 428
- in carnivores, 491
central receptor systems
    and -, 317-319

Sodium appetite (continued)

central representation of NaCl and -, 497, 499

cerebral Na sensors and - in sheep, 490

chorda tympani stimulation and -, 481

control mechanisms of -, 459--463, 494

corticosterone and -, 426, 428

CSF [Na] and - in sheep, 473--477

- in dams, 16

depletion-induced -, 433, 435, 460

- and diuretics, 444

DOCA and ang II, synergism in arousal of -, 396-399

DOCA and -, 396, 436, 438, 462

- in dogs, 495

- and endorphins, 241

evolution of -, 403

- and Goldblatt hypertension, 390, 391, 416, 417

- , Goldblatt hypertension and Na balance, 416

- , Goldblatt hypertension and renal function, 416

gustatory mechanisms and -, 444, 479-482

hedonic -, 486, 500

- in herbivores, 491

homeostasis regulation and -, 507

hormonal synergy and arousal of -, 395, 396, 429

- in humans, 491

- and hypothyroidism, 444

ICV infusions of hypertonic mannitol and - in sheep, 474

ICV infusion of hypertonic NaCl and - in sheep, 474, 486, 489

- induced by ang II, 401, 488

influence of brief exposure to NaCl on -, 498

intracarotid infusion of NaCl solutions and - in sheep, 486, 489

Sodium appetite (continued)

Long Evans rats, adrenalectomy and -, 447

Long Evans rats, DOCA and -, 450

mineralocorticoids-induced -, 450

- in monkeys, 492

motivation of -, 465, 471

Na depletion, CSF [K] and -, 504

Na depletion, CSF [Na] and -, 504

Na depletion, serum [K] and -, 505

Na depletion, serum [Na] and -, 504, 505

- in neonatal rats, 32

- in omnivores, 491

- and opiate receptor antagonists, 241

- and oral contraceptives, 444

peripheral mechanisms mediating changes in -, 479--481

- in pigeons, 492

- and polydipsia in Goldblatt hypertension, 415

- in rabbits, 491

rat behavior in -, 497, 498

- in rats, 383-446, 453-472, 479-483, 493-495, 497--508

- and renin-angiotensin-aldosterone system, 442

- and renin in Goldblatt hypertension, 417

- and salivectomy, 444

salty taste and -, 498, 499

- in sheep, 403, 473-478, 485--490, 491, 495

- and Na balance, 428

- and Na preference, 443, 444

Na receptors and -, 481

Na solutions, choise between, and -, 467-470

taste receptors and -, 481

- and thyroidectomy, 443

- and ureteric ligation, 386

Sodium appetite (continued)
- and vanadate, 340
- in wild rabbits, 403
zona incerta and -, 358
Sodium balance
adrenalectomy, ang II and -,
421-423
adrenalectomy, carbachol and -,
421-423
- and Na intake, 396, 399, 428
- and urinary Na output, 397
Sodium concentration
- in amniotic fluid, 16
- in CSF, 2, 3-6, 8
- in diet and plasma arginine
vasotocin, 73, 74
- in milk, 16
- in plasma, 6, 8
Sodium depletion
- and drinking, 6
- and Na appetite, 407, 408,
433, 435, 460
- and Na appetite in neonatal
rats, 31, 32
- and Na appetite in sheep, 486
Sodium excretion, 436
- and ang II, 422
- and carbachol, 422
- and vanadate, 340-342
Sodium homeostasis
- regulation, 423, 535
Na^+, K^+-ATPase inhibitors, 539
Sodium preference
- in adult rats, 115
dopamine receptor antagonists
and -, 453, 454, 457
- and Na appetite, 443, 444
- in neonatal rats, 31
SCH 23390 effects on -, 457,
458
sulpiride effects on -, 457
sultopride effects on -, 457
Sodium receptors
- and drinking, 2, 103
- and Na appetite, 481, 490
Spinal cord
ang II distribution in the -,
127
Spontaneously hypertensive rats
(SHR)
developing -, see Developing

Spontaneously hypertensive rats
(SHR) (continued)
food intake in -, 541, 542
ingestive behavior of -, 541,
542
meal-related drinking in -,
542
water intake in -, 540
SQ 20881, 173
Stretch receptors
- and drinking, 161
Striatum
ang II binding sites in -, 143
ang III binding sites in -,
143
Subfornical organ (SFO)
ang II binding sites in the -,
143
ang II immunoreactive fibers
in the -, 127, 129
- and ang II-induced drinking,
44, 310-317
ang III binding sites in
the -, 143
- and AVP/ang II comodulation,
51
- and AV3V, 348
- and cholinergic compounds,
314-317
- and converting enzyme, 126
- and dermorphin, 258-262
- and drinking, 251-256, 347
- and lateral hypothalamus,
348
- and lateral preoptic area,
348
lesions to - and ang II-induced drinking, 251-256,
312, 313
morphology of the cat -, 375-
-381
neural projections of the -,
347
- and nucleus paraventricularis, 347, 348
- responses to hydromineral
deficiencies, 43
- responses to hypovolemia, 43
- sensitivity to ang II, 135,
136

Substance P
  - antidipsogenic effect, 215-
      -217
  - dipsogenic effect, 220-222
  - structure, 214
Substantia gelatinosa
  ang II distribution in the -,
      127
Substantia nigra
  converting enzyme in the -,
      126
Suckling, 14, 26
  cellular dehydration and -, 19
  extracellular dehydration
      and -, 19
Superior colliculus
  ang II binding sites in the -,
      143
  ang III binding sites in
      the -, 143

Taste
  - afferents, 15
  - ontogeny in rats, 25-28
  - ontogeny in sheep, 25
  - reactivity and chorda tympa-
      ni, 25
  - reactivity to quinine, 25,
      27
  - reactivity to NaCl, 25
  - reactivity to NH₄Cl, 25
  - reactivity and nucleus
      tractus solitari, 25
  - receptors and Na appetite,
      481
  - for salt, 15
  salty - and Na appetite, 498,
      499
Tachykinins
  antidipsogenic effect of -, 37,
      215-217
  antidipsogenic effect of - in
      neonatal rats, 38-42
  dipsogenic effect of -, 220-222
Thirst
  - centers, 361
  development of -, nutritional
      factors, 15-17
  development of -, perinatal
      factors, 15-17

Thirst (continued)
  extracellular -, 87-93, 103
  human -, 521-526
  - and insulin, 181-187
  intracellular -, 83-87, 103
  primary -, 103
  secondary -, 103
Thyreotropic hormone
  - and dehydration, 62
Thyroid
  - activation following dehy-
      dration, 59-63
  - hormones and dehydration, 62
  - inhibition and drinking, 66
  - inhibition and feeding, 66
Thyroidectomy
  - and body weight, 61
  - and food intake, 61, 66, 67
  - and Na appetite, 443
  - and water intake, 66, 67
Thyroxine (T4)
  - and food intake, 67
  - and water intake, 67
Tractus medialis corticohypotha-
      lami
  - and AVP/ang II comodulation,
      51

Ureteric ligation
  - and drinking, 20, 78, 88
Urine excretion
  - and vanadate, 340, 342

Vagotomy
  abdominal - and drinking, 269
  cervical - and drinking, 268
  coeliac - and drinking, 270-
      -273
  gastric - and drinking, 270-
      -273
  hepatic - and drinking, 270-
      -273
  pharmacological - and drink-
      ing, 272
Vanadate
  - , Na depletion and drinking,
      341-343
  - , Na depletion and K excre-
      tion, 340-342
  - , Na depletion and Na excre-
      tion, 340-342

Vanadate (continued)
 - , Na depletion and Na intake, 340
 - , Na depletion and urine excretion, 340, 342
Venous pressure
 - and volume regulation, 72
Visceral neuraxis
 - and drinking behavior, 352
Volume regulation by
 - arterial pressure, 72
 - atrial filling, 72
 - nervus vagus, 72
 - plasma osmolality, 72
 - plasma [Na], 72
 - renin-angiotensin-aldosterone system, 72
 - venous pressure, 72

Water deprivation
 - and drinking, 168, 176
 - and human thirst, 526
 - and kassinin, 247
 - , methymazole and food intake, 61
 - , methymazole and water intake, 61
 - , thyroidectomy and food intake, 61
 - , thyroidectomy and water intake, 61
Water homeostasis regulation, 535
Water intake
 - and central clonidine, 511, 513, 515
 dopamine receptor antagonists and -, 446
 - and m-octopamine, 514, 515
 - and m-synephrine, 514, 515
 opioid agonists, effects on -, 516
 opioid antagonists, effects on -, 516
 - and peripheral clonidine, 511, 512, 515
 SCH 23390 and -, 457, 458
 sulpiride and -, 454
 sultopride and -, 457
 - and yohimbine, 513, 515

Water procurement
 neural correlates of -, 362
Zona incerta
 - anatomical connections, 360
 - and ingestive behavior, 358, 359
Zona reticulata
 converting enzyme in the -, 126